AF380105

GAP JUNCTIONS IN THE NERVOUS SYSTEM

David C. Spray

Albert Einstein College of Medicine
Bronx, New York, U.S.A.

Rolf Dermietzel

University of Regensburg
Regensburg, Germany

Springer-Verlag
Berlin Heidelberg GmbH

R.G. LANDES COMPANY
AUSTIN

Neuroscience Intelligence Unit

GAP JUNCTIONS IN THE NERVOUS SYSTEM

R.G. LANDES COMPANY
Austin, Texas, U.S.A.

ISBN 978-3-662-21937-9

Library of Congress Cataloging-in-Publication Data

Gap junctions in the nervous system / [edited by] David C. Spray, Rolf Dermeitzel.
      p. cm. — (Neuroscience intelligence unit)
      Includes bibliographical references and index.
      ISBN 978-3-662-21937-9   ISBN 978-3-662-21935-5 (eBook)
      DOI 10.1007/978-3-662-21935-5
      1. Molecular neurobiology. 2. Connexins. 3. Cell junctions (Cell
biology) I. Spray, David C. II. Dermeitzel, Rolf. III. Series.
      [DNLM: 1. Central Nervous System — physiology. 2. Gap Junctions
WL300 G211 1996
QP356.2.G37 1996
612.8—dc20
DNLM/DLC
for Library of Congress
                                      96-28050
                                        CIP

Publisher's Note

R.G. Landes Company publishes six book series: *Medical Intelligence Unit, Molecular Biology Intelligence Unit, Neuroscience Intelligence Unit, Tissue Engineering Intelligence Unit, Biotechnology Intelligence Unit and Environmental Intelligence Unit.* The authors of our books are acknowledged leaders in their fields and the topics are unique. Almost without exception, no other similar books exist on these topics.

Our goal is to publish books in important and rapidly changing areas of bioscience and environment for sophisticated researchers and clinicians. To achieve this goal, we have accelerated our publishing program to conform to the fast pace in which information grows in bioscience. Most of our books are published within 90 to 120 days of receipt of the manuscript. We would like to thank our readers for their continuing interest and welcome any comments or suggestions they may have for future books.

Shyamali Ghosh
Publications Director
R.G. Landes Company

CONTENTS

ABBREVIATIONS

Cx32	connexin32
Cx46	connexin46
Cx43	connexin43
F-BPE	forskolin combined with GGF from bovine pituitary extract
G_{ss}	steady state junctional conductance
GFAP	glial fibrillary acidic protein
GGF	glial growth factor
L-NGFR	low affinity nerve growth factor receptor
MAG	myelin-associated glycoprotein
MBP	myelin basic protein
N-CAM	neural cell adhesion molecules
NDFβ	neu differentiation factor beta
NGF	nerve growth factor
PCR	polymerase chain reaction
P_0	protein zero
PMP-22	peripheral myelin protein-22
TGFβ	transforming growth factor beta
TNFα	tumor necrosis factor alpha
V_j	transjunctional voltage

EDITORS

David C. Spray
Albert Einstein College of Medicine
Departments of Neuroscience and Medicine
Bronx, New York, U.S.A.
Chapter 1, 3, 14, 16

Rolf Dermietzel
Institute of Anatomy
University of Regensburg
Germany
Chapter 1, 2, 15

CONTRIBUTORS

Berj L. Bardakjian
Departments of Medicine (Neurology)
  and Physiology
Institute of Biomedical Engineering
University of Toronto
Toronto, Ontario, Canada
Chapter 18

John F. Becherger
Dept. of Anatomy and Cell Biology
The University of Western Ontario
London, Ontario, Canada
Chapter 11

Michael V. L. Bennett
Department of Neuroscience
Albert Einstein College of Medicine
Bronx, New York, U.S.A.
Chapter 4

Shari L. Bond
Department of Anatomy
  and Cell Biology
The University of Western Ontario
London, Ontario, Canada
Chapter 11

Linda Jo Bone
Department of Neurology
University of Pennsylvania
Philadelphia, Pennsylvania, U.S.A.
Chapter 13

Peter L. Carlen
Departments of Medicine (Neurology)
  and Physiology
Institute of Biomedical Engineering
University of Toronto
Toronto, Ontario, Canada
Chapter 18

Karen J. Chandross
Departments of Neuroscience
  and Neurology
Albert Einstein College of Medicine
Bronx, New York, U.S.A.
Chapter 14

Suzanne M. Deschênes
Department of Neurology
University of Pennsylvania
Philadelphia, Pennsylvania, U.S.A.
Chapter 13

Kenneth H. Fischbeck
Department of Neurology
University of Pennsylvania
Philadelphia, Pennsylvania, U.S.A.
Chapter 13

Christian Giaume
Collége de France
Paris, France
Chapter 8

Steven Goldman
Department of Neurology
  and Neuroscience
Cornell University Medical College
New York, New York, U.S.A.
Chapter 19

Shokrollah S. Jahromi
Departments of Medicine (Neurology)
  and Physiology
Institute of Biomedical Engineering
University of Toronto
Toronto, Ontario, Canada
Chapter 18

John A. Kessler
Departments of Neuroscience
  and Neurology
Albert Einstein College of Medicine
Bronx, New York, U.S.A.
Chapter 14

Gertrud Kurz-Isler
Institute of Pathology
University of Tübingen
Tübingen, Germany
Chapter 7

Fernando Miragall
Institut für Anatomie
Universität Regensburg
Regensburg, Germany
Chapter 15

Christian M. Müller
Max-Planck-Institute
  for Developmental Biology
Tübingen, Germany
Chapter 12

Christian C.G. Naus
Dept. of Anatomy and Cell Biology
The University of Western Ontario
London, Ontario, Canada
Chapter 11

Maiken Nedergaard
Departments of Cell Biology
  and Anatomy and Neurosurgery
New York Medical College
Valhalla, New York, U.S.A.
Chapter 19

Alejandro Peinado
Department of Neuroscience
Albert Einstein College of Medicine
Bronx, New York, U.S.A.
Chapter 17

Jose L. Perez-Velazquez
Departments of Medicine (Neurology)
    and Physiology
Institute of Biomedical Engineering
University of Toronto
Toronto, Ontario, Canada
Chapter 18

Bruce R. Ransom
University of Washington School
    of Medicine
Department of Neurology
Seattle, Washington, U.S.A.
Chapter 9

Astrid Rohlmann
Department of Molecular Genetics
University of Texas
Dallas, Texas, U.S.A.
Chapter 10

Renato Rozental
Deptartment of Neuroscience
Albert Einstein College of Medicine
Bronx, New York, U.S.A.
and
Institute of Biophysics "Carlos
    Chagas Filho"
Federal University of Rio de Janeiro
Rio de Janeiro, Brazil
and
Department of Internal Medicine
Federal University of Goiás
Goiána, 74.000, Brazil
Chapter 16

Steven S. Scherer
Department of Neurology
University of Pennsylvania
Philadelphia, Pennsylvania, U.S.A.
Chapter 13

Otto Traub
Institut fur Genetik
Universität Bonn
Bonn, Germany
Chapter 15

Taufik A. Valiante
Departments of Medicine (Neurology)
    and Physiology
Institute of Biomedical Engineering
University of Toronto
Toronto, Ontario, Canada
Chapter 18

David I. Vaney
Vision, Touch and Hearing
    Research Centre
Department of Physiology
    and Pharmacology
The University of Queensland
Brisbane, Queensland, Australia
Chapter 5

Laurent Venance
Collége de France
Paris, France
Chapter 8

Reto Weiler
University of Oldenburg
Oldenburg, Germany
Chapter 6

Hartwig Wolburg
Institute of Pathology
University of Tübingen
Tübingen, Germany
Chapter 7

J. R. Wolff
Georg-August-Universität Göttingen
Zentrum Anatomie-Fachbereich
 Medicizin
Göttingen, Germany
Chapter 10

PREFACE

Gap junctions are clusters of intercellular channels providing conduits for diffusional exchange of ions and small molecules directly from one cell to another. In the nervous system, gap junctions function as electrotonic synapses between neurons and as a pathway for the exchange of metabolites and second messenger molecules between glia. In recent years, the knowledge regarding nervous system gap junctions has grown extensively. Molecular genetic approaches have defined the connexin gene family of gap junction proteins and generated transgenic animals with ablated expression of individual connexins, physiological approaches have characterized differences in channel function in gap junctions formed of different connexins and have revealed roles of gap junction channels in tissue function, and anatomical/cell biological studies have detailed connexin distribution and changes under physiological and pathological conditions.

This monograph contains chapters by investigators who have led the way in studies of gap junction distribution and function in the nervous system. It focuses on different tissues of the brain as well as on changes in gap junction function and expression that occur in human genetic disease and in animal models of human pathology. The aim of this book is to provide reviews of recent advances in this field as well as an overview of the research efforts of many years that have led to these recent findings. These recent advances are due to the combined application of different methodologies, which are also emphasized in these chapters.

We are grateful to all the contributors for their manuscripts and especially to Ms. Fran Andrade for her editorial assistance and extraordinary perseverance in pulling this volume together. This book is dedicated to all of those with whom we have worked on these topics for the past two decades and all those from whom we have learned. Especially our children Amanda, Christopher, Katinka and Kaija.

David C. Spray and Rolf Dermietzel

GAP JUNCTIONS IN THE NERVOUS SYSTEM: AN INTRODUCTION

David C. Spray and Rolf Dermietzel

1. INTRODUCTION

A century ago, it was generally believed that consciousness and movement resulted from the flow of substances freely throughout the interconnected neural network. This Reticular Theory eventually gave way to the Neuronal Doctrine in which the nervous system was envisioned as a composite of discrete cells, where direct transfer of information among neurons is a rare event, occurring only in specialized nuclei and under specific circumstances. Nevertheless, as chapters in this volume testify, gap junctions, the structural elements responsible for direct intercellular communication, are increasingly detected between cells in both the central and peripheral nervous systems of mammals, including man.

Within the past decade, the genes encoding gap junctions have been identified, and at least a dozen mammalian members of this gene family are now known.[1] The gap junction gene products, the *connexins,* display overlapping cell type specificity and ontogenetic appearance within the same cells or cell types. Moreover, gap junction expression is now known to be regulated or influenced by a variety of stimuli, the gap junction proteins assembled into *hemichannels* or *connexons* have been found to interact with one another with varying affinities, and the channels that each of these connexin proteins forms have been shown to display physiological properties that are distinct from one another. The goal of this introductory chapter is to provide an overview of these areas of novel findings in the gap junction field in part to provide context for the chapters that follow and also to indicate to the reader where dogma is firm and where questions remain. Although this overview is necessarily painted

Gap Junctions in the Nervous System, edited by David C. Spray and Rolf Dermietzel.
© 1996 R.G. Landes Company.

with a broad brush, it is hoped that this summary might benefit those with only a passing interest in the field and that it may stimulate interest among those for whom gap junctions will offer intriguing possibilities for further study.

2. CONNEXIN BRAIN TOPOLOGY

It is now almost universally accepted that, in vertebrates, connexins (abbreviated Cx) are the molecules that gap junctions are made of. Although there have been proposals that other molecules may form the gap junction pore [including MIP, the most abundant membrane protein of lens fiber cells,[2] the 16 kDa protein isolated from lobster hepatopancreas and rat liver by Finbow and Pitts[3,4] that has turned out to be a vacuolar ATPase now termed *ductin*, and the 34 kDa protein isolated from rat brain that now appears to be a plasmalemmal VDAC protein and is termed *Br-VDAC*[5]], the exclusive role of connexins in the function of vertebrate intercellular communication is largely proven by the demonstrations that expression of connexins in exogenous systems results in the formation of gap junction channels with properties very similar to those of endogenous gap junctions.[6,7] There have also been recurrent suggestions that gap junction channels may not necessarily act alone, connexin proteins perhaps complexing with other molecules to thereby modify gap junction functions (e.g., calmodulin).[8] Although there remains no direct evidence for this latter hypothesis, it is difficult to disprove it entirely, and recent experiments on pH gating of Cx43 channels have suggested that channel closing may indeed involve the interaction between peptides that are part of the Cx43 molecule.[9] Nonetheless, apparently pure connexins do form channels in lipid bilayers,[10-14] supporting the concept that they may act alone to form the junctional channels of vertebrates. (Invertebrate, and in particular arthropod gap junction proteins are another story, that is considered below.)

Topological features of connexin molecules are almost as well understood as for any other membrane protein, with studies using region-specific antibodies and protease clipping of isolated junctional membranes having begun more than 8 years ago,[15] and X-ray diffraction of isolated junctional plaques having an even longer history.[16-18] From these studies, the connexin molecules are believed to be positioned with both carboxyl and amino termini on the cytoplasmic aspect of the cells (Fig. 1.1), with each molecule possessing four membrane-spanning α helical domains (designated M1-M4); segment M3 is most amphipathic and thus is most likely to line the pore. Segments connecting M1-M2 and M3-M4 (designated EL1 and EL2) are believed to form the extracellular loops that mate or dock connexons across the 10nm extracellular gap that defines the junctional structure in this section and is responsible for its name. The hinge or loop segment connecting M2 and M3 (designated CL) is believed to lie within the cytoplasm.

Regions of the connexin sequences that are most homologous throughout the gene family are the membrane spanning domains and the extracellular loops, the latter homologies probably explaining why hemichannels formed of so many types of connexins can pair with each other.[19-21] Regions of the sequences with most divergence are the CL region and the carboxyl terminal portion of the molecule. Unique amino acid sequences in the latter region are generally used to prepare connexin-specific antibody probes; differences in lengths of the CL region provide useful distinction between Group I(β) and Group II(α) connexins, as indicated below.

This dogma still allows room for improvement in assigning topology exactly. For example, G. Dahl and colleagues[22] have recently suggested on the basis of mapping connexin topology with glycosylation site mutations that a portion of the sequence generally assigned to M3 and EL2 may in fact dip into the membrane, forming a structure analogous with the so-called

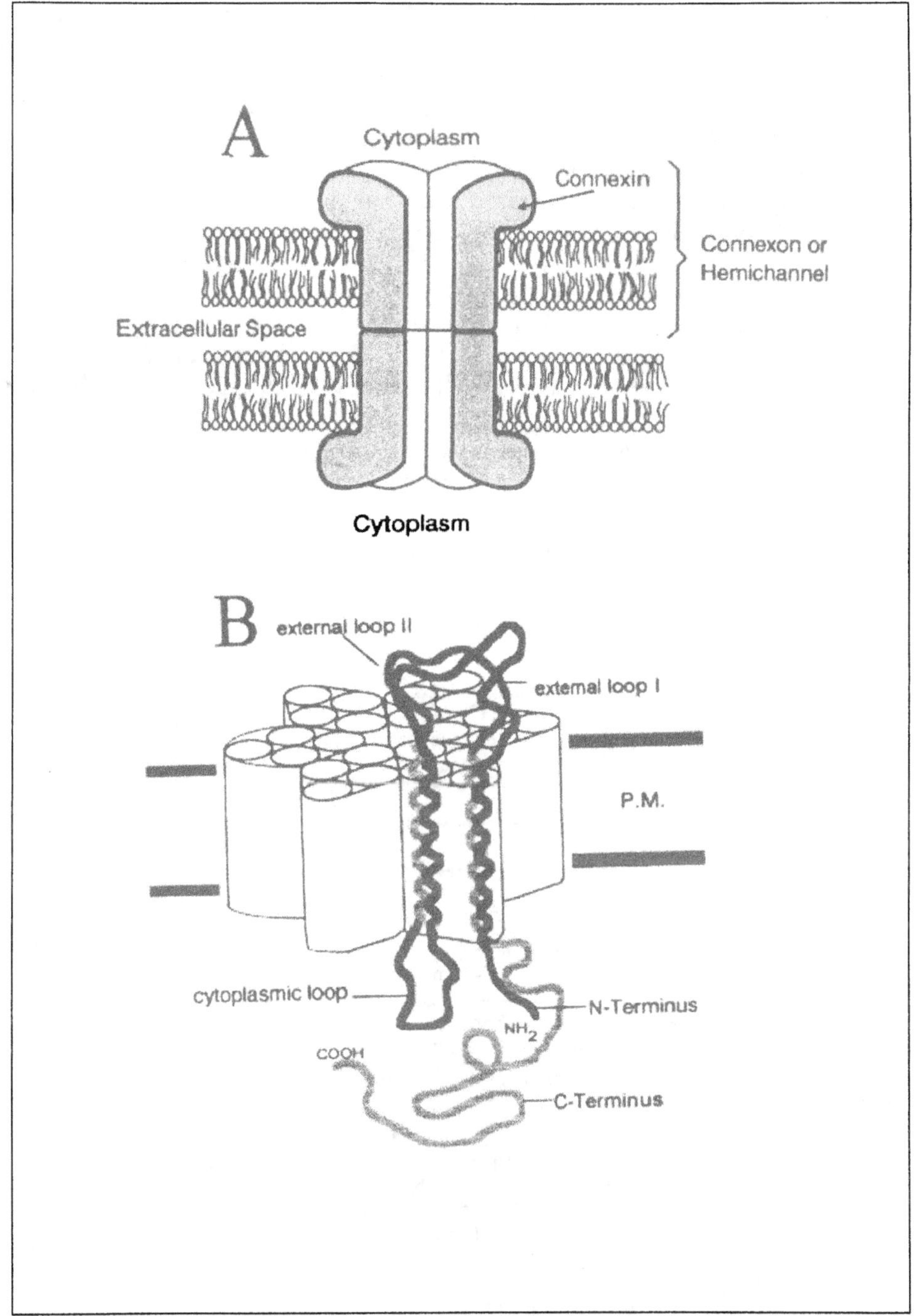

Fig. 1.1. Topology of gap junction channels. (A) Gap junction channels, extending from the cytoplasm of one cell to the cytoplasm of another, are formed by two connexons or hemichannels connected across extracellular space. (B) Each connexon is formed from six connexin subunits, each having four membrane-spanning domains and both amino and carboxyl termini within the cytoplasm. External loops (I and II) are believed to provide the high affinity interactions between the hemichannels.

P-region of voltage sensitive nonjunctional channels. And Delmar's group[9] has obtained evidence that intracellular acidification may result in a conformational change analogous to the ball and chain model of inactivation of voltage gated ionic channels, whereby the carboxyl terminal portion of connexin43 binds to CL, closing the channel.

Higher order structure of the channel is believed to consist of six connexins forming the hemichannel or connexon in a single cell, with connexons or hemichannels being connected across extracellular space at their EL regions with noncovalent, presumably hydrogen bonding. Three cysteine residues at conserved loci within each EL are characteristic features of the connexin molecules; these contributed disulfide bridges are presumably involved in intra-connexin and inter-EL loop tertiary structure. An old observation that should be repeated stoichiometrically with modern techniques is that gap junction channels can be split into connexons or hemichannels using hyperosmotic disaccharide solutions[23] again implying that linkage is not covalent.

3. EVOLUTIONARY ASPECTS OF THE CONNEXIN GENE FAMILY

There now are 13 rodent connexins in the gene family, with homologues found in vertebrates as phylogenetically far removed as fish, amphibia and birds[24] (Fig. 1.2). Connexins have not been found

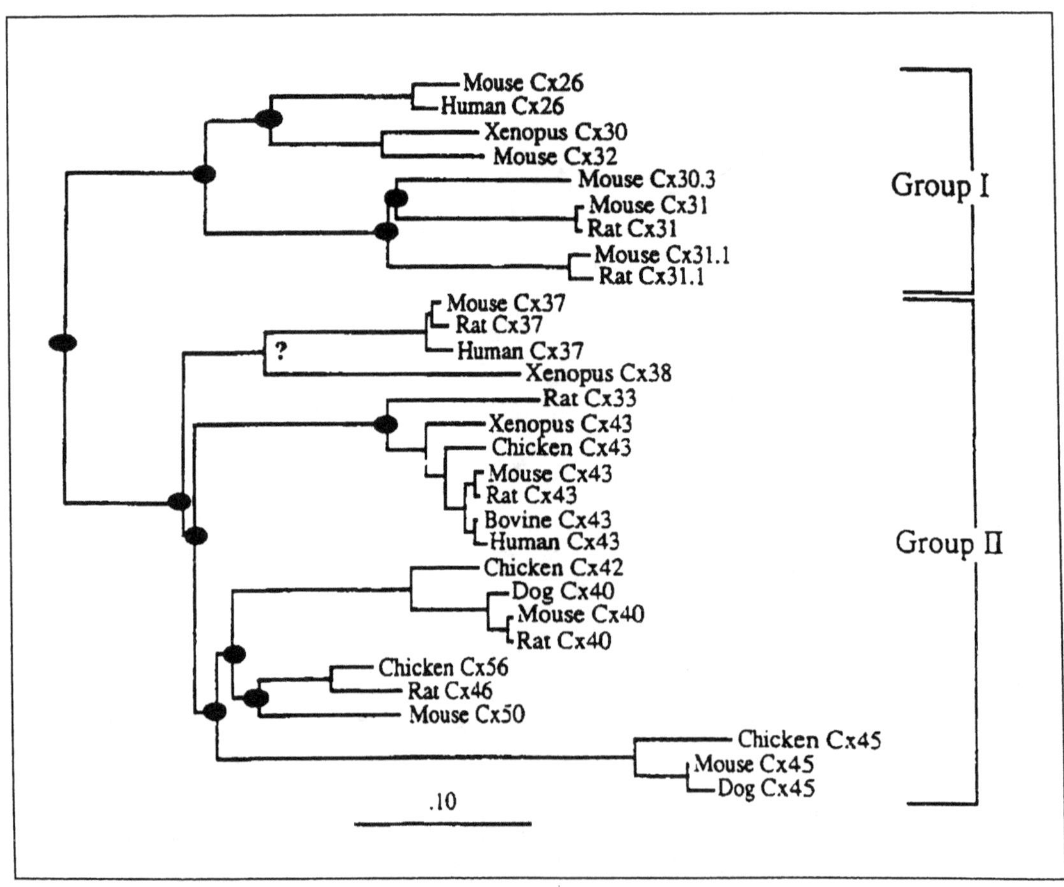

Fig. 1.2. Connexin family tree obtained by comparison of sequence differences (see ref. 24). Branch points with filled ovals represent gene duplications; nodes with open ovals represent speciation; one branch is ambiguous (indicated by "?"). Calibration bar represents 10% difference in sequence. Reprinted with permission from Bennett MVL et al, Soc Gen Physiol Series 1994; 49:223-233.

in invertebrates, and the grossly different morphology of the junctional plaques (with larger particles, cleaving to the E-face rather than to the P-face as in vertebrate junctions in freeze-fracture preparations, and with wider extracellular gaps visible in micrographs taken with conventional thin section electron microscopy) suggests that perhaps different proteins might fulfill this function in other phyla. Indeed it has recently been suggested that the *Drosophila* passover gene,[25] *Caenorhabditis* unc7[26] and wnt1[27] might comprise a novel gene family whose function is intercellular communication in invertebrates. The possibility that the intercellular connections of plant cells (*plasmodesmata*) might express related proteins and have similar structure appears less likely as the resolution of plasmodesmatal structure has reached higher definition.[28] Similarly, the dihydropyridine receptor and synaptophysin, both of which appear to mate with other proteins to form channels between intracellular compartments, have been suggested as gap junction homologues,[29,30] though there is no evidence that these proteins have structures similar to gap junction channels.

Phylogenetic history can be in theory recapitulated by comparing protein or DNA sequences. For gap junction cDNAs, such a comparison leads to the conclusion that there are two separate classes of connexins, that are more closely related to one another than to members of the other class.[31,32] Class I or the β family of connexins includes Cx26, Cx30.3, Cx31, Cx31.1 and Cx32. Class II or the α family includes Cx33, Cx37, Cx40, Cx43, Cx45, Cx46 and Cx50. *Xenopus* connexins that have thus far been cloned and sequenced include homologs of Cx32 and Cx43, as well as xCx38, which has no identified mammalian homologue. Many of these connexins are found in the nervous system, as summarized in Table 1.1.

4. COMMUNICATION COMPARTMENTS IN THE NERVOUS SYSTEM

The nervous system can be viewed as a series of communication compartments,[33] where there is intercellular communication within and between compartments that allows homo- and heterocellular communication, as well as autocellular communication between processes of the same cell (see chapter 7). Thus, astrocytes, oligodendrocytes, neurons, leptomeningeal cells, pinealocytes, ependymal cells and cells of the microvasculature all exhibit discrete patterns of connexin expression and most communicate between themselves and with cells of specific other compartments via gap junctions (Fig. 1.3). The strength of coupling within and between compartments differs greatly in different cell types.

Table 1.1. Expression of connexins in the nervous system

Connexin	Distribution in the nervous system
Cx26	Leptomeninges, ependyma, pinealocytes, mature neurons
Cx30	Present
Cx30.3	Unknown
Cx31	Unknown
Cx31.1	Unknown
Cx32	Oligodendrocytes, myelinating Schwann cells, some neurons
Cx33	Unknown; present in maturing neuroblasts
Cx37	Endothelium
Cx40	Astrocytes, neurons?
Cx43	Astrocytes, leptomeninges, neurons?, ependyma
Cx45	Astrocytes, early embryonic neuroblasts, oligodendrocytes
Cx46	Schwann cells
Cx50	?

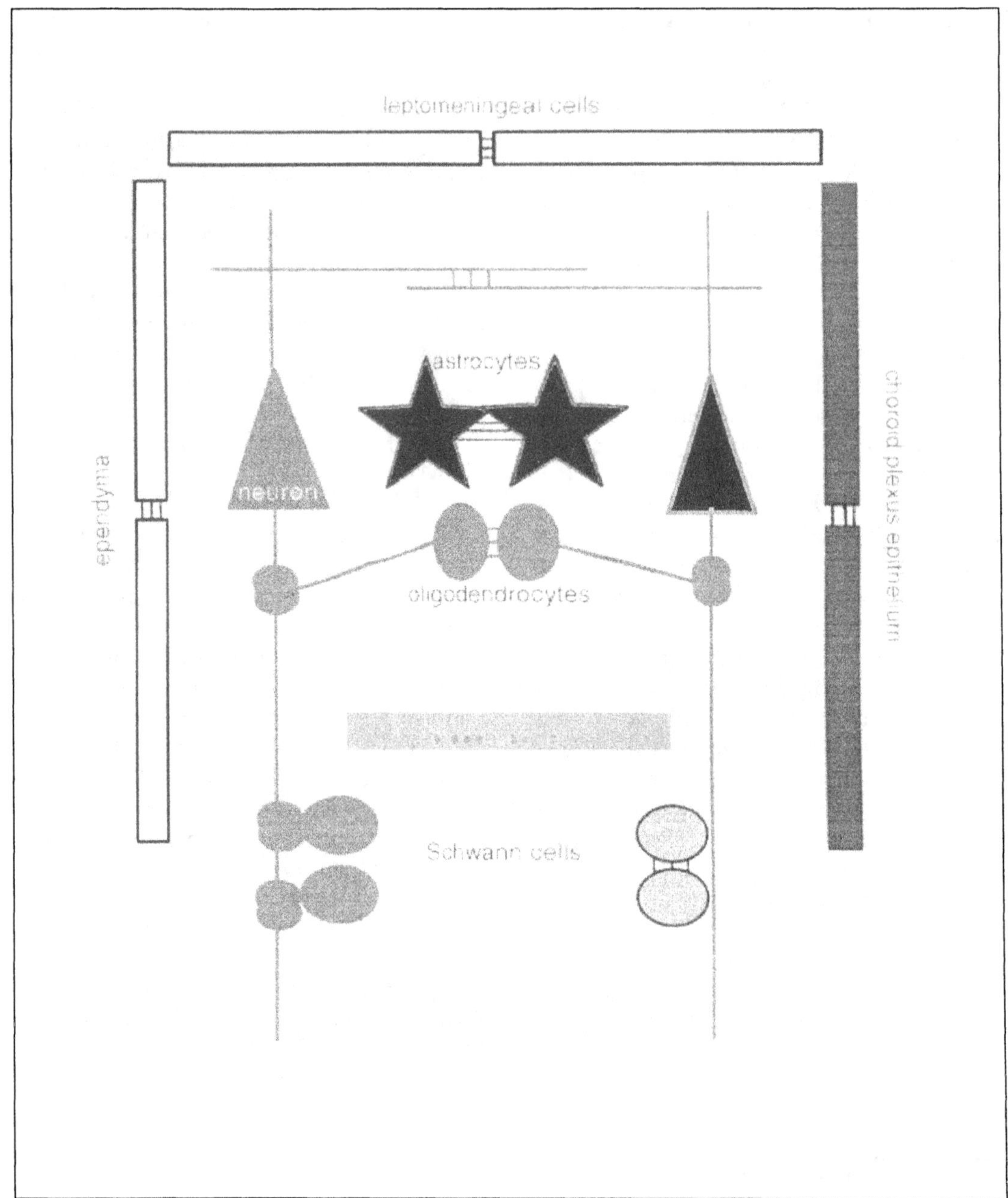

Fig. 1.3. Communication compartments in the nervous system. This diagram depicts coupling between astrocytes, between oligodendrocytes, between some neurons, and also between ependymal, leptomeningeal and vascular endothelial and choroid plexus cells. Gap junctions are also present between cytoplasmic processes of myelinating Schwann cells and between proliferating Schwann cells. Cell types are color-coded to correspond to major connexin types expressed in each population. Black represents Cx43; white with heavy outline, the coexpression of cx43 and Cx26; dark gray, Cx32; gray with heavy outline, Cx46; and light gray the coexpression of Cx37, Cx40 and Cx43.

Leptomeningeal cells and astrocytes are highly coupled, with the latter expressing as many as a million gap junctions/μm^3 (see chapter 7). Neurons and perhaps oligodendrocytes are weakly coupled, so that it is uncommon to see dye spread from one cell to another (even using positively charged and highly fluorescent tracer molecules), except under exceptional ionic conditions or during early developmental periods.

5. NOVEL APPROACHES TO STUDYING GAP JUNCTIONS

Work on gap junctions has accelerated in recent years, with the development of novel techniques and the availability of novel preparations in which to correlate and perhaps even definitively associate changes in gap junction expression and function with certain disease states. From the standpoint of techniques, the stable expression of wildtype and mutant gap junction channels in cells that normally lack them and the ability to insert or delete specific genes from specific cell types both are beginning to create new model systems in which to explore just what gap junction channels do in tissue function.

Combined with new experimental methods of single channel recording to determine channel gating properties and sensitive techniques such as Northern blotting, RT-PCR and connexin-specific antibodies with which to demonstrate which connexins are where and how their expression is regulated, these model systems should also reveal the extent to which regulated gap junction expression and function contribute to pathological conditions.

Numerous disease states are now associated with gap junction dysfunction.[34] These pathological conditions include various somatic disorders either resulting from or ultimately causing increases or decreases in functional gap junctions, and they also now include conditions where specific loss or compromise of function connexin mutations are directly linked to the cause of the disease (Table 1.2). Moreover, gene deletion by homologous recombination has now produced mouse models for further study for three of the connexins (Cx26, Cx32, Cx43). These studies have thus far been both revealing and surprising, and raise numerous questions for further study. For example, the Cx32 mutations

---

Table 1.2. Consequences of alterations in connexin expression

A. Pathological Conditions	Defect
X-linked Charcot-Marie-Tooth Disease	Coding region mutations in Cx32[35]
Chronic *T. cruzi* infection	Reduction in Cx43[40]
Acute cardiocyte *T. cruzi* infection	Rearrangement of Cx43[38]
Post infarct myocardium	Reduction and rearrangement of Cx43[41,42]
Stab wound in brain	Reduction in Cx43[43]
Facial nerve lesion	Increase in Cx43[44]
Epileptic cortex	Increase in Cx43[45]
Hypertensive rats	Increase in Cx40, decreased Cx43[46]
Hypertensive vessels	Abnormal endothelial gap junctions[47]
Heterovisceral atriotaxia	Cx43 phosphorylation site mutations[36]

B. Transgenic animals	
Cx43 knockout	Right ventricular hyperplasia; neonatal lethal[48]
Cx32 knockout	Decreased glucose mobilization in liver [49]
Cx46 knockout	Cataracts in lens[50]
Cx37 knockout	Impaired oogenesis[51]
Cx26 knockout	Early embryonic lethal[52]
Cx43 overexpression	Laterality defects[53]
Cx43 underexpression	Embryonic laterality defects[53]
Cx43 negative dominant	Embryonic lethal[54]

discovered in X-linked Charcot-Marie-Tooth disease[35] led to the demonstration of this connexin's expression in Schwann cells (although exactly how the mutations disrupt function remains unknown: see chapters 11 and 12), and the association of Cx43 phosphorylation site mutations with atriovisceral heterotaxia[36] has emphasized how important phosphorylation may be for normal function [although how the subtle biophysical changes that have demonstrated between phosphorylated and unphosphorylated Cx43[37] leads to such profound laterality defects remains to be determined, and whether abnormalities also exist in brain circuitry have not been explored at all]. The disappearance of Cx43 in cardiac myocytes, endothelial cells, astrocytes and leptomeningial cells after infection with *Trypanosoma cruzi* or *Toxoplasmosis gondii* in culture appears to involve a trafficking disorder, without change in connexin synthesis.[38] By contrast, both brain and cardiac infarct models (and, in the case of the heart, human disease as well) and at least certain types of tumors are characterized by decreased connexin expression, whereas in human epileptic cortical material and in rat facial nerve injury Cx43 expression is increased (Table 1.2 and chapter 7). And in one reported example (following glutamate toxicity), although connexin synthesis may remain constant or actually increase, recognition of Cx43 epitopes by certain sequence-specific antibodies virtually disappears, indicating topological reorganization of the protein.[39]

6. ORGANIZATION OF THIS VOLUME

The first overview chapter (chapter 2) provides a summary of which gap junction proteins are found where in the brain and illustrates the degree of plasticity of connexin expression; the second (chapter 3) provides information on gating behavior and single channel biophysics for gap junctions formed of each of connexins found in the nervous system as well as a summary of studies performed on brain cells that corroborate or call into question the

association of connexin type with functional properties. The next series of papers deals with the role of the gap functions in functional anatomy of the retina (chapters 4 through 6), emphasizing connectivity among cells within and between communication compartments, their functional properties and their plasticity. These papers emphasize that the retina may offer an advantageous experimental opportunity to better understand neuronal connectivity and its consequences.

The third series of chapters (chapters 7 through 11) describe gap junction organization and physiology and pharmacology among astrocytes, which comprise the largest communication compartment. The fourth group (chapters 13 and 14) describe the function of gap junctions in Schwann cells, the glial components of peripheral nerve. The last series of chapters (chapters 15 through 19) consider what is known of neuronal gap junctions, including developmental changes in gap junction organization and function between neurons, functional or anatomical changes in gap junctions that occur in the nervous system associated with neuronal hypo- or hyperexcitability, and the possibility of intercellular communication between neurons and glia.

Together, these chapters offer an overview of what is currently known about expression patterns and functional properties of gap junctions in the nervous system and also provide information regarding the role of gap junctions in normal and pathologic nervous system function. It is hoped that this collection of papers, as well as references to the background literature in each of the areas that is covered, will prove useful as the endeavor to understand the roles of gap junction-mediated intercellular communication in functions of the nervous system continues to expand.

REFERENCES

1. Kanno Y, Kataoko K, Shiba Y et al. Intercellular Communication through Gap Junctions. Prog Cell Res Vol. 4, Amsterdam: Elsevier, 1995.
2. Goodenough DA. The crystalline lens. A system networked by gap junctional inter-

cellular communication. Sem in Cell Biol 1992; 3:49-58.

3. Finbow ME, Harrison M, Jones P. Ductin—a proton pump component, a gap junction channel and a neurotransmitter release channel. Bioessays 1995; 17:247-255.

4. Finbow ME, Pitts JD. Is the gap junction channel-the connexon- made of connexin or ductin? J Cell Sci 1993; 106:463-471.

5. Dermietzel R, Hwang TK, Buettner R et al. Cloning and in situ localization of a brain-derived porin that constitutes a large conductance anion channel in astrocytic plasma membranes. Proc Nat Acad Sci USA 1994; 91:499-503.

6. Werner R, Miller T, Azarnia R et al. Translation and functional expression of cell-cell channel mRNA in Xenopus oocytes. J Membr Biol 1985; 87:253-268.

7. Eghbali B, Kessler JA, Spray DC. Expression of gap junction channels in a communication incompetent cell line after transfection with connexin32 cDNA. Proc Natl Acad Sci USA 1990; 87:1328-1331.

8. Peracchia C, Girsch SJ. Functional modulation of cell coupling: evidence for a calmodulin-driven channel gate. Am J Phys 1985; 248;H765-782.

9. Liu S, Taffet S, Stoner L et al. A structural basis for the unequal sensitivity of the major cardiac and liver gap junctions to intracellular acidification: the carboxyl tail length. Biophys J 1993; 64:1422-1433.

10. Spray DC, Saez JC, Brosius D et al. Isolated liver gap junctions: gating of transjunctional currents is similar to that in intact pairs of rat hepatocytes. Proc Natl Acad Sci USA 1986; 83:5494-5497.

11. Young JD, Cohn ZA, Gilula NB. Functional assemby of gap junction conductance in lipid bilayers: demonstration that the major 27 kd protein forms the junctional channel. Cell 1987; 48:733-743.

12. Campos de Carvalho AC, Hertzberg EL, Spray DC. Complex channel activity recorded from rat liver gap junctional membranes incorporated into lipid bilayers. Braz J Med & Biol Res 1991; 24:527-537.

13. Harris AL, Walter A, Paul D et al. Ion channels in single bilayers induced by rat connexin32. Bain Res Mol Brain Res 1992; 15:269-280.

14. Mazet JL, Jarry T, Gros D et al. Voltage dependence of liver gap-junction channels reconstituted into liposomes and incorporated into planar bilayers. Eur J Biochem 1992; 210:249-256.

15. Milks LC, Kumar NM, Houghten R et al. Topology of the 32-kd liver gap junction protein determined by site-directed antibody localizations. EMBO J 1988; 7:2967-1975.

16. Makowski L, Caspar DL, Phillips WC et al. Gap junction structures. VI. Variation and conservation in connexon conformation and packing. Biophys J 1984; 45:208-218.

17. Stauffer KA, Unwin N. Structure of gap junction channels. Sem in Cell Biol 1992; 3:17-20.

18. Yeager M, Gilula NB. Membrane topology and quaternary structure of cardiac gap junction ion channels. J Mol Biol 1992; 223:929-948.

19. Nicholson BJ, Suchyna T, Xu LX et al. Divergent properties of different connexins in *Xenopus* oocytes. In: Hall JE, Zampighi GA, Davis RM, eds. Progress in Cell Research, Vol. 3. Elsevier, 1993:3-13.

20. White TW, Bruzzone R, Wolfram S et al. Selective interactions among the multiple connexin proteins expressed in the vertebrate lens: the second extracellular domain is a determinant of compatibility between connexins. J Cell Biol 1994; 125:879-892.

21. Elfgang C, Eckert R, Lichtenberg-Frate H, Butterweck A et al. Specific permeability and selective formation of gap junction channels in connexin-transfected HeLa cells. J Cell Biol 1995; 129:805-817.

22. Dahl G, Nonner W, Werner R. Attempts to define functional domains of gap junction proteins with synthetic peptides. Biophys J 1994; 67:1816-1822.

23. Goodenough DA, Paul DL, Jesaitis L. Topological distribution of two connexin32 antigenic sites in intact and split rodent hepatocyte gap junctions. J Cell Biol 1988; 107:1817-1824.

24. Bennett MV, Zheng X, Sogin ML. The connexins and their family tree. (Review). Soc Gen Physiol Series 1994; 49:223-233.

25. Krishnan SN, Frei E, Swain GP et al. Passover: a gene required for synaptic connec-

tivity in the giant fiber system of Drosophilia. Cell 1993; 73:967-977.

26. Starich TA, Lee RYN, Panzarella C et al. *eat-5* and *unc-7* represent a multigene family in *Caenorhabditis elegans* involved in cell-cell coupling. J Cell Biol 1996; 134: 537-548.

27. Olson DJ, Christian JL, Moon RT. Effect of wnt-1 and related proteins on gap junctional communication in Xenopus embryos. Science 1991; 252:1173-1176.

28. Robards A, Lucas W, Pitts J et al. Plasmodesmata and Gap Junctions: Parallels in Evolution. Berlin, Heidelberg: Springer-Verlag, 1990.

29. Ma J, Fill M, Knudson CM et al. Ryanodine receptor of skeletal muscle is a gap junction type channel. Science 1988; 242:99-102.

30. Calakos N, Scheller RH. Vesicle-associated membrane protein and synaptophysin are associated on the synaptic vesicle. J Biol Chem 1994; 269:24534-24537.

31. Kumar NM, Gilula NB. Molecular biology and genetics of gap junction channels. Sem Cell Biol 1992; 3:3-16.

32. Bennett MVL, Barrio L, Barfiello TA et al. Gap junctions: New tools, new answers, new questions. Neuron 1991; 6:305-320.

33. Dermietzel R, Spray DC. Gap junctions in the brain: Where, what type, how many, and why? Trends in Neuroscience 1993; 16:185-192.

34. Spray DC, Dermietzel R. X-linked Charcot-Marie-Tooth Syndrome and other possible gap junction diseases of the nervous system. Trends in Neurosci 1995; 18:256-262.

35. Bergoffen J, Scherer SS, Wang S et al. Connexin mutations in X-linked Charcot-Marie-Tooth disease. Science 1993; 262: 2039-2042.

36. Britz-Cunningham SH, Shah MM, Zuppan CW, Fletcher WH. Mutations of the Connexin43 gap-junction gene in patients with heart malformations and defects of laterality. New Engl J Med 332:1323-1329.

37. Moreno AP, Saez JC, Fishman GI et al. Human connexin43 gap junction channels: Regulation of unitary conductances by phosphorylation. Circ Res 1994; 74:1050-1057.

38. Campos de Carvalho AC, Roy C, Dermietzel R et al. Gap junction distribution is altered between cardiac myocytes infected with Trypanosoma cruzi Circ Res 1991; 70: 733-742.

39. Vukelic JI, Yamamoto T, Hertzberg EL et al. Depletion of connexin43 immunoreactivity in astrocytes after kainic acid-induced lesions in rat brain. Neurosci Lett 1991; 130:120-124.

40. Saez J, Spray DC, Wittner M et al. Effect of verapramil on cardiac gap junctions in murine Chagas's disease: Memorias do Instituto Oswaldo Cruz 1993; 33:64.

41. Severs NJ. Pathophysiology of gap junctions in heart disease (Review). J Cardiov Electrophy 1994; 5(5):461-475.

42. Luke RA, Saffitz JE. Remodeling of ventricular conduction pathways in healed canine infarct border zones. J Clin Invest 1991; 87(5)1594-602.

43. Müller CM. Gap junction-mediated communication between astrocytes in mammalian cortical slices. In: Spray DC, Dermietzel R, eds. Gap Junctions in the Nervous System. Austin: R.G. Landes, 1996:203-212.

44. Rohlmann A, Laskawi R, Hofer A et al. Facial nerve lesions lead to increased immunostaining of the astrocytic gap junction protein (connexin43) in the corresponding facial nucleus of rats. Neurosci Lett 1993; 154:206-208.

45. Naus CC, Bechberger JF, Paul DL. Gap junction gene expression in human seizure disorder. Exp Neurol 1991; 111:198-203.

46. Bastide B, Neyses L, Ganten D et al. Gap junction protein connexin40 is preferentially expressed in vascular endothelium and conductive bundles of rat myocardium and is increased under hypertensive conditions. Circ Res 1993;(6)1138-1149.

47. Huttner I, Costabella PM, De Chastonay C et al. Volume, surface, and junctions of rat aortic endothelium during experimental hypertension: a morphometric and freeze fracture study. Lab Invest 1982; 46: 489-504.

48. Reaume AG, deSousa PA, Kulkarni S et al. Cardiac malformation in neonatal mice lacking connexin43. Science 1995 267: 1831-1834.

49. Nelles E, Bützler C, Jung D et al. Mice lacking connexin 32 gap junction show

normal nerve conduction but a defect in the propagation of sympathetic nerve signals in liver. Proc Nat Acad Sci USA 1996 (in press).

50. Kumar N, Gilula NB. Presentation at Keystone Conference, 1996.

51. Simon AM, Li En, Paul DL. Targeted inactivation of connexin37. Molecular Approaches to the Function of Intercellular Junctions. Keystone Conference Abstract), p. 323, 1996.

52. Willecke K, Winterhager E. Personal communication.

53. Lo CW, Ewart J, Sullivan R. Neural tube defects in transgenic mice with the gain or loss of connexin43 function: relevance to developmental pathologies. 1995 Gap Junction Conference.

54. Paul DL, Yu K, Bruzzone R et al. Expression of a dominant negative inhibitor of intercellular communication in the early Xenopus embryo causes delamination and extrusion of cells. Development 1995; 121:371-381.

CHAPTER 2

MOLECULAR DIVERSITY AND PLASTICITY OF GAP JUNCTIONS IN THE NERVOUS SYSTEM

Rolf Dermietzel

1. INTRODUCTION: AN EVOLUTIONARY PERSPECTIVE

Transfer of information between cells appears to be an essential step in the evolution of multicellular organisms. Prior to the evolution of communicative structures such as synapses and gap junctions, primitive eukaryotes secreted and recognized signal molecules and possessed the potential to associate via cell adhesion,[1] thereby assembling into social complexes of a higher order. In principle, cell adhesion and secretion of signal molecules is considered to be sufficient for coordinated events in such archaic cell complexes. However, the onset of differentiation necessitated direct communication between homotypic members and the creation of compartmental boundaries. Thus, the level and complexity of information transfer evolved in parallel with the differentiation of cell types within primitive organisms. The invention of direct pathways in form of communicative structures which allowed for the establishment of compartmental boundaries, e.g., the creation of diffusion gradients and directed signal exchange was undoubtedly a major step in the evolution of complex multicellular organisms. It is, therefore, not surprising that even archaic mesozoans exhibit gap junctions.[2]

During embryogenesis this phylogenetic trait is recapitulated. Direct transfer of molecules can be detected from the four cell stage embryo onwards, and gap junction forming proteins (connexins) are present as early as the four cell stage embryo in mouse (discussed in ref. 7), and in *Xenopus* oocytes. In the latter case connexins are presumably inherited from a maternal pool.[3] A further aspect of interest concerning the evolutionary history of junctional communication is its functional link to appropriate cell adhesion. Recent evidence from different sources show that

Gap Junctions in the Nervous System, edited by David C. Spray and Rolf Dermietzel.
© 1996 R.G. Landes Company.

effective adhesion is essential for the establishment of competent gap junctions.[4-6] Moreover, deficits in the organization of gap junctions also feed back on proper cell adhesion as has been shown by antibody injection in early mouse embryos[7] and transfection of *Xenopus* embryos with functional null mutations of a gap junction gene.[8]

Within the context of the evolutionary aspects of cellular communication, it is not surprising that signal transfer through gap junctions has been regarded as a basic but less advanced form of intercellular communication, especially in higher vertebrates. This prejudice has been exceptionally strong in relation to the nervous system and has found its way into the textbooks. Interneuronal signaling via chemical synaptic transmission has consistently been regarded as offering a superior evolutionary advantage, in contrast to electrotonic transmission via the electrical synapses formed by gap junctions. Thus, electrotonic signal transmission was formerly viewed as uniform, inflexible and mostly restricted to fast reflex propagation in phylogentically lower organisms and some synchronization events in vertebrate neural nuclei.[9] The multitude of ligand-receptor interactions inherent in chemical transmission seemed to suggest that chemical transmission is much better suited to fulfill the complex requirements of interneuronal communication in the evolved vertebrate brain.

Recent studies on the molecular diversity of the gap junction forming proteins (connexins) and their cell-specific expression patterns in the nervous system indicate that this dualistic view of a phylogenetically older electrical and a phylogenetically more modern chemical mode of impulse propagation in the brain is erroneous. Both apparently did not evolve in parallel forms, but rather evolved synergistically as interrelated mechanisms of intercellular signaling.[10]

This synergism becomes even more obvious if the brain is conceptualized as a complex of functionally interactive compartments (chapter 1). The degree and mode of intercellular communication within each compartment varies according to cell type, developmental stage and functional status. In the following I will elaborate on this concept of compartmentalized brain function and discuss the variety and the plasticity of gap junctions in the nervous system. In addition, I will address the role that gap junctional coupling may play in intra- and intercompartmental communication processes.

2. THE NEURAL COMPARTMENT

Conceptualizing the nervous system in terms of neural compartments oversimplifies the complexity of the adult vertebrate brain. However, for the sake of clarity and because of the lack of knowledge concerning expression of specific gap junction proteins in neuronal subpopulations, this simplification provides a useful conceptual framework. Evidence for the existence of gap junctions between mammalian CNS neurons has been obtained by several different strategies. Traditionally, direct visualization of gap junctions between neurons by means of conventional or freeze-fracture electron microscopy has been the method of choice for their evaluation (see Table 2.1 for a compiled list of references). These ultrastructural approaches allow conclusive identification of neuronal gap junctions as well as the types of neurons contributing to junction formation in cases where the junctional elements such as dendrites can be identified as belonging to a particular class of neurons.[11] Although the structural approach is descriptive and, thereby, does not allow one to draw functional conclusions, freeze-fracture data on neuronal gap junctions has provided data suggesting flexibility. Because each particle comprising the gap junction is believed to represent a single gap junction channel, the number of channels involved in interneuronal coupling (see chapter 3) can be estimated.

In contrast to highly ordered arrays of particles seen in glial gap junctions (as has been shown by freeze-fracture studies[12-17])

Table 2.1. Compiled list of neural nuclei where gap junctions have been described

	Electrophysiology/dye transfer	Morphology
Trigeminal mesencephalic nuc.[92,93]	+	
Cerebellar basket cells[94]		+
Lateral vestibular nuc.[95-97]	+	
Glomeruli/granular cells olfactory bulb[50,98-100]		+
Inferior olive[101-108]	+	+
Horizontal/amacrine cells retina[42,109-117]	+	+
Pyramidal cells/ interneurons cerebral cortex[1118-122]	+	+
Ventral cochlear nuc.[123,124]		+
Substantia nigra[125,126]	+	
Maganocelluar neurons, supraoptic/ paraventricular nuc.[38,127-136]	+	+
Pyramidal and granular cells hippocampus[21,22,39,137-149]	+	+
Motoneurons spinal cord/cranial nerve nuc.[150-156]	+	+

neuronal gap junctions consist preferentially of assemblies of particles or arrays of diffusely spread channels displaying variable degrees of order (see chapter 10). Although no systematic studies exist comparing the consistency and the degree of variability of gap junction assemblies among different neuronal subpopulations, the tentative conclusion is that neuronal gap junctions may represent a high degree of plasticity and/or functional flexibility. Extensive paracrystalline arrays which have been described among hepatocytes[18] have never been observed in mammalian neuronal membranes. Comparatively speaking, the neuronal gap junction assemblies are very much akin to the particle complexes described in dissociating and reassociating gap junction formations.[19,20] The generally small number of gap junction particles (channels) observed in neuronal junctions raises two important questions: (1) How many channels are necessary for effective coupling between neurons?

(2) What are the limitations for the detection of coupling by the standard dye injection methods?

With regard to the first question, we have recently pointed out[10] that the high resistance of neuronal cells provides enormous potential for relatively few gap junction channels to couple pre- and postsynaptic elements. The coupling coefficient (k) for two cells connected through a single channel was estimated to be 0.5 based on the unitary conductance of gap junction channels between 50 and 150 pS.[10] The threshold for detecting Lucifer Yellow dye-coupling between coupled cells, however, consistently appears to be about 2 nS, corresponding to 10-20 times as many gap junction channels. This may well be the reason why the number of coupled neurons traced by the Lucifer Yellow technique is relatively low and has been successful primarily in studying neuronal populations with high coupling efficiency, i.e., CA3 neurons of the hippocampus.[21-23]

Experiments with lower molecular weight tracer (Neurobiotin[R]) reveal that interneuronal coupling is much more extensive than formerly believed in the developing cortex and the neostriatum[24-26] as well as in the retina (see chapter 4).

2.1. GAP JUNCTIONS IN THE DEVELOPING BRAIN

Evidence for gap junctional-mediated interneuronal communication has been documented at various stages of CNS development. Sites where gap junctions have been described in the developing cortex are listed in Table 2.2. It has been suggested that nonsynaptic (nonchemical) communication during cortical development is responsible for the definition of discrete multicellular patterns that could define adult functional architecture of the brain.[24-26] Similar data have been described in the developing retina, demonstrating a tangential network of ganglion cells and amacrine cells which cooperate in the spontaneous generation of calcium oscillations.[28] These data suggest that gap junction communication represents a general functional mechanism that governs developmental and differentiation processes during neurogenesis (see chapter 16). Another possible function that gap junction may play in developing neuroblasts is the provision of a direct link for signaling cell death between neurons. In this context research on the embryonic anterior pagoda (AP) in the leech is of particular interest.[29] Dying cells in this model apparently initiate axonal sprouting only in neighboring cells which are coupled to the stressed neuron. The diffusion of the distress signal sent by the dying AP neuron seems to be mediated via gap junctions. Whether this pathway underlies a general avenue for death signals in the developing CNS and whether such signals are also utilized in the organized apoptotic program of neurogenesis in higher vertebrates remains further evaluation.

The functional implication of connexin expression during morphogensis of the brain makes the characterization of the connexin species a task of primary importance. Since emerging evidence indicates that different types of connexins are regulated in various fashions, both at the level of gating control (see chapter 3) and at the level of channel expression, information about the pattern of connexin expression during brain development will allow a better understanding of the molecular and functional background of junctional communication during corticogenesis. Our own data using immunofluorescence techniques on cryostat sections show that the amount and pattern of connexin expression during brain development changes.[30] A complementary up– and downregulation of connexin26 and connexin32 occurs during

---

Table 2.2. Compiled list of immature neurons where gap junctions have been described

	Electrophysiology	Dye transfer	Morphology
neuroblasts/ neonatal cortex[25,27,30,35,157,158]	+	+	+
Locus coeruleus[159]	+	+	
Striatum[26]	+	+	
Spinal cord[160-162]	+	+	
Sensory/parasympathetic ganglia[163-165]			+
Inferior olivary complex[166]	+		+
Retina[113,124,167]	+	+	+

embryonic and postnatal development of the brain, with connexin26 appearing present as early as E12, and connexin32 appearing postnatally at the time of oligodendrocytic maturation (around P7). Connexin43 is expressed at high levels throughout brain development with a peak during the first week postpartum (Fig. 2.1). A more refined evaluation by Fushiki and Kinoshita[31] demonstrates that the expression of the different connexins in the developing brain reveals regional specificity.

At least two other connexins must now be added to the list of putative gap junction proteins expressed in the embryonic brain. At the level of mRNA, connexin37 has been shown to have a biphasic course of expression, with levels four times higher in the immature brain as compared to the adult.[32] Immunostaining with an connexin37 specific antibody, however, has failed to detect this connexin in neuroblasts or glial precursor cells, but showed con-

finement of this connexin to brain endothelial cells (data not shown). Thus, the biphasic expression of this connexin during brain development is more likely to be related to the process of vasculogenesis of the brain rather than being a further developmentally regulated connexin of neuronal and/or glial cells. Recently, we also detected considerable levels of connexin45 in embryonic rat brains by immunocytochemistry (Fig. 2.2). The overall amount of this protein is lower than connexin43, although it does surpass that of connexin26, indicating that this connexin contributes substantially to the gap junction complement in immature brains (a list of the connexin forms expressed in the embryonic brain is given in Table 2.3). Connexin45 is mostly confined to the periventricular zone during E12 to E15 with a more distinct laminated expression as compared to connexin43 (Fig. 2.3). The biophysical properties of connexin45 are especially interesting with respect to its

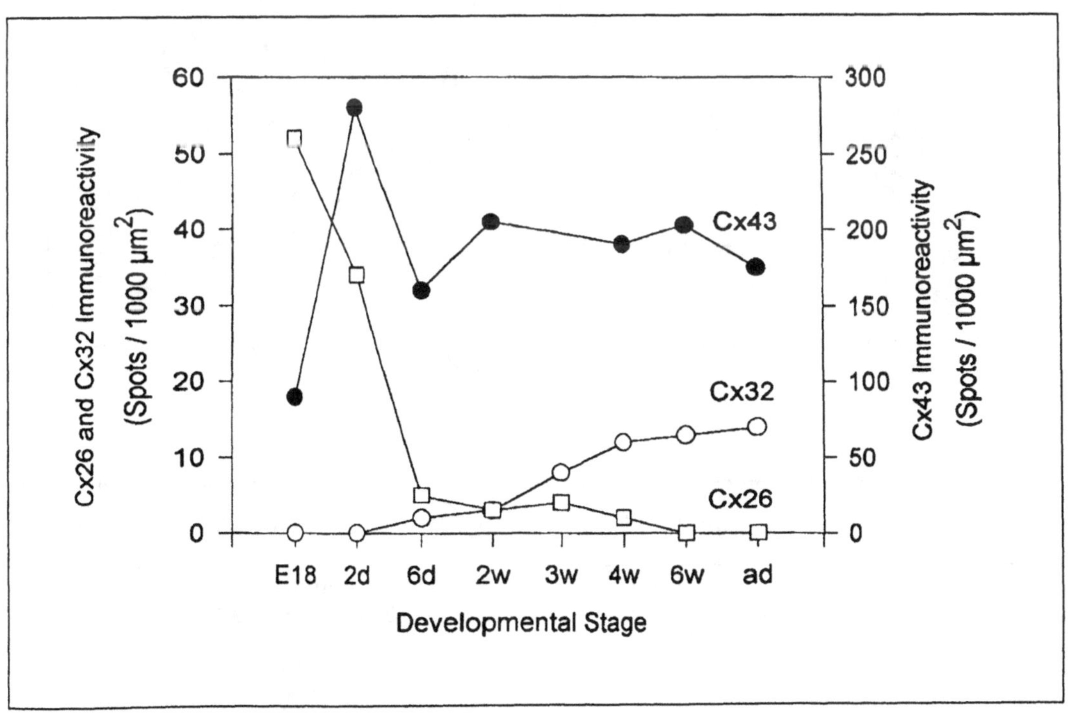

Fig. 2.1. Course of expression of three connexins (Cx32, Cx26, Cx43) in embryonic and postnatal brains. Expression was measured by counting immunofluorescence spots with a computer-aided digitizer over defined areas in the periventricular zones of embryonic brains and in the neostriatum of postnatal animals. Note the late onset of Cx32 expression which corresponds with the beginning of myelination.

expression in neuroepithelial cells. Studies of human connexin45 performed in a hepatoma cell line showed voltage sensitivity much steeper than that reported for any other mammalian connexin and unitary conductance in the range of 30pS.[33] Injection of Lucifer Yellow also revealed that connexin45 channels were virtually impermeable to anions.[33,34] One important implication of this finding is that gap junctions formed by different connexins may be capable of transmitting different types of signals. It is probable that for the developing brain the expression of particular sets of connexins are essential for the definition of cell compartment and communicative boundaries. The occurrence of clusters of coupled neuroblasts as reported by Lo Turco and Kriegstein[35] within the ventricular zone around E14 may represent such temporarily coupled compartments which serve as organizing units for later developmental processes.

In summary, growing evidence indicates that gap junctional coupling during neurogenesis of the brain is a spatially and temporally regulated event. Unveiling the molecular and biophysical properties of the connexin types and the exact expression pattern will certainly provide important

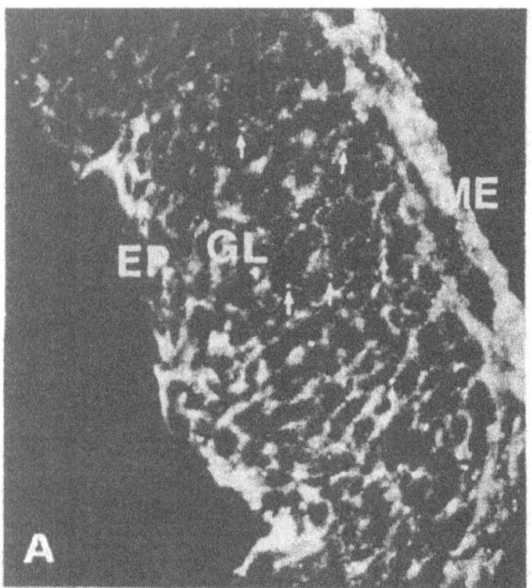

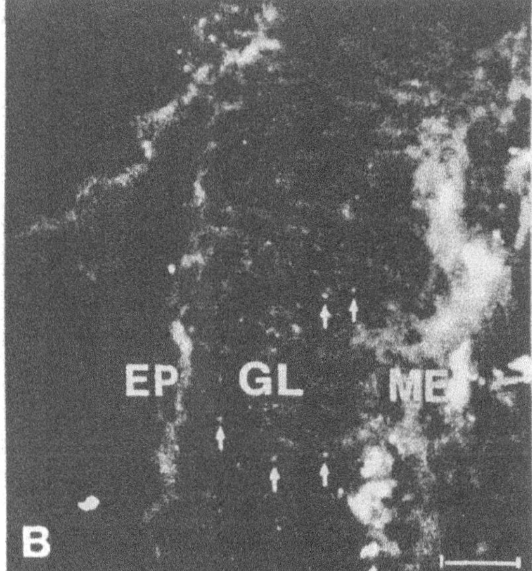

Fig. 2.2. Immunodetection of Cx43 (A) and Cx45 (B) in brain vesicles at stage E12. Arrows depict immunolabel which are considered to represent gap junction plaques. EP, ependyma, GL, germinal layer, ME, mesenchyme. Scale bar, 20 μm.

Table 2.3. Connexin forms expressed in the neuroepithelium and neonatal neurons

	Protein (immunocytochemistry)	mRNA (*in situ* hybridization)
cx43	+	+
cx45	+	n.d.
cx26	+	n.d
cx32	+/– *	–

Data compiled from[30,31,52,168]
* Inconsistency concerning the expression of cx32 in embryonic brains (see reference 30 and 31)

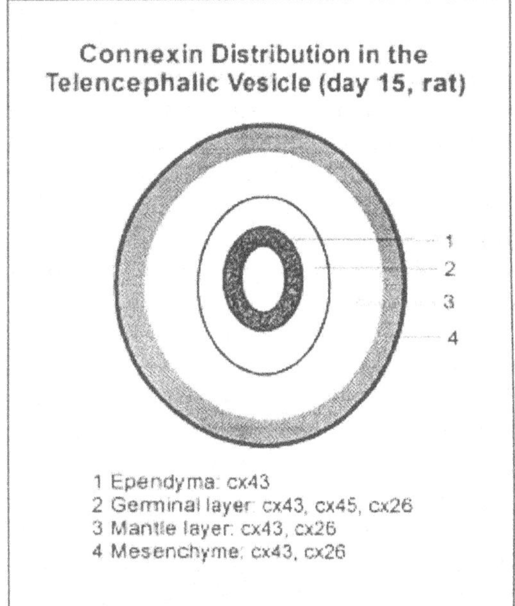

Fig. 2.3. Schematic drawing of connexin distribution in embryonic rats (E15).

clues regarding the functional significance of gap junction-mediated coupling for the morphogenesis of the nervous system.

2.2. GAP JUNCTIONS IN THE ADULT BRAIN

Conventional approaches (i.e., transmission and freeze-fracture electron microscopy, dye injections) do not readily permit quantitative characterization of direct interneuronal cell-to-cell communication in the adult brain. In addition, the possibility that gap junctions may be utilized in a use-dependent manner renders a general quantitative assessment even more difficult. As discussed above the structural organization of neuronal gap junctions in higher vertebrates implies a considerable degree of flexibility. Although the half lives of connexins in neurons have yet to be determined, evidence from in vitro measurements of astrocytic connexin43 suggests that the half life of this protein is short (between 1.5 and 2.5 hr; unpublished data). Similar short half lives have been reported for connexin32 and connexin26 in cultured hepatocytes,[36] although half times may be

somewhat longer for these connexins in the liver in vivo.[20,37] Thus, connexins can exhibit an exceptionally fast turnover in comparison to other integral membrane proteins.

Evidence for rapid functional recruitment of gap junctional coupling comes from electrophysiological and dye injection studies of in situ preparations of hippocampal slices. The hippocampal paradigm has been frequently used to evaluate the generation of synchronized population of spikes and its relevance for seizure discharge[38] (see chapters 17 and 18). Since synchronized action potentials can occur after chemical transmission has been blocked (in calcium-free solutions), electrical transmission is most likely to play a role in spike synchronization. In this model, field bursts and Lucifer Yellow coupling are rapidly initiated (within 1-2 min).[39] The strengthening of electrical coupling between hyperexcitable neurons can either be explained by gating mechanisms on the channel level or de novo assembly of functional channels from a pool of pre-existing connexons (either membranous or cytosolic). In situ hybridization at the single cell level indicates ubiquitous expression of neuronal connexin43 mRNA (see below).

In this context it is important to emphasize that the presence of mRNA does not necessarily imply a steady and high incidence of effective electrical and/or metabolic coupling between neurons. As has been shown in diverse other tissues, competent coupling is subject to control by a variety of posttranslational processes including phosphorylation, messenger stability, and appropriate expression of cell adhesion proteins.[40] There are also reports on exceptional discrepancies between connexin mRNA concentrations and protein levels. For instance, connexin37 and connexin40, both of which are abundantly expressed in the lung at the mRNA level, show remarkably low amounts of proteins in this organ.[32,41] Translational efficacy may, thus, be an important regulatory principle in connexin expression. However, from the point of metabolic effectiveness it seems likely that expression of high levels of mRNA is also followed by at least transient translations of the protein gene product.

Another problem that renders a reliable detection of electrotonic interaction between neurons difficult is its interaction with chemical transmission. For example, dye uncoupling between retinal horizontal cells can be inhibited by dopamine via activation of D1 receptors[42] (see chapters 4-6). Similar decoupling effects were demonstrated by norepinephrine application in cultured astrocytes.[43] On the other hand glutamate treatment seems to increase interglial coupling strength.[44] Effective electrophysiological measurements of electrical transmission and/or detection of junctional interneuronal coupling via dye injection has to take into account this interdependence of electrical-chemical transmission. The fact that consistent coupling has been encountered preferentially among certain subpopulations of neurons with degrees of synchronous activity (i.e., hippocampal CA3 neurons, neurons of the inferior accessory medial olive complex, etc.[23,45,46]) can therefore be explained by a more robust coupling efficiency of these cell populations due to either lower modification of their connexin complements by exogenous and/or endogenous factors or by expression of constitutively high numbers of communication-competent channels. The documentation of abundant connexin43

mRNA in diverse cortical, hippocampal and brain stem neurons makes coupling via gap junction channels a potent means of transient neuronal communication. Consequently, disruption of the metabolic balance required for adequate coupling may be a pathophysiological principle underlying diverse neuropathological mechanisms responsible for disorders of higher functions of the brain.[47]

Before I concentrate on cell-specific expression patterns in the nervous system, I will first give an account of the connexin candidates expressed in brain as evidenced from Northern blot analyses. By exploiting nine different connexin cDNAs to Northern blot analyses, connexin50, connexin33 and connexin46 proved negative in adult CNS tissues, while the other six yielded positive signals (Fig. 2.4, Table 2.4). The following data from the mRNA analyses seem notable with respect to a compartmentalized connectivity in brain tissues. First, the overall amount of the different connexin mRNAs is highly variable. Second, the expression pattern shows significant regional heterogeneity indicating local restriction or concentration of some of the connexins. For instance, connexin43 mRNA is abundant in all parts of the brain with little local variance, while connexin32 peaks in the spinal cord and the brain stem. On the other hand, connexin26 and connexin37 are expressed at low levels with constant concentrations.

Table 2.4. Distribution of connexin mRNAs in brain tissues

	Cortex	Cerebellum	Hippocampus	Brain Stem	Spinal Cord
Cx32	+	+	+	++	+++
Cx26	+	++	+	+	+
Cx43	++	++	++	++	++
Cx40	0/+	+	0/+	0/+	0/+
Cx45	+/0	++	+	+/0	+/0
Cx37	+	+	+	+	+
Cx33	0	0	0	0	0
Cx46 *	0	0	0	0	0
Cx50	0	0	0	0	0

Normalized to the following controls: For Cx32 and Cx26, liver RNA; for Cx43, Cx45, Cx40, heart RNA; for Cx37, lung RNA. *Positve signals were found in cultured astrocytes.

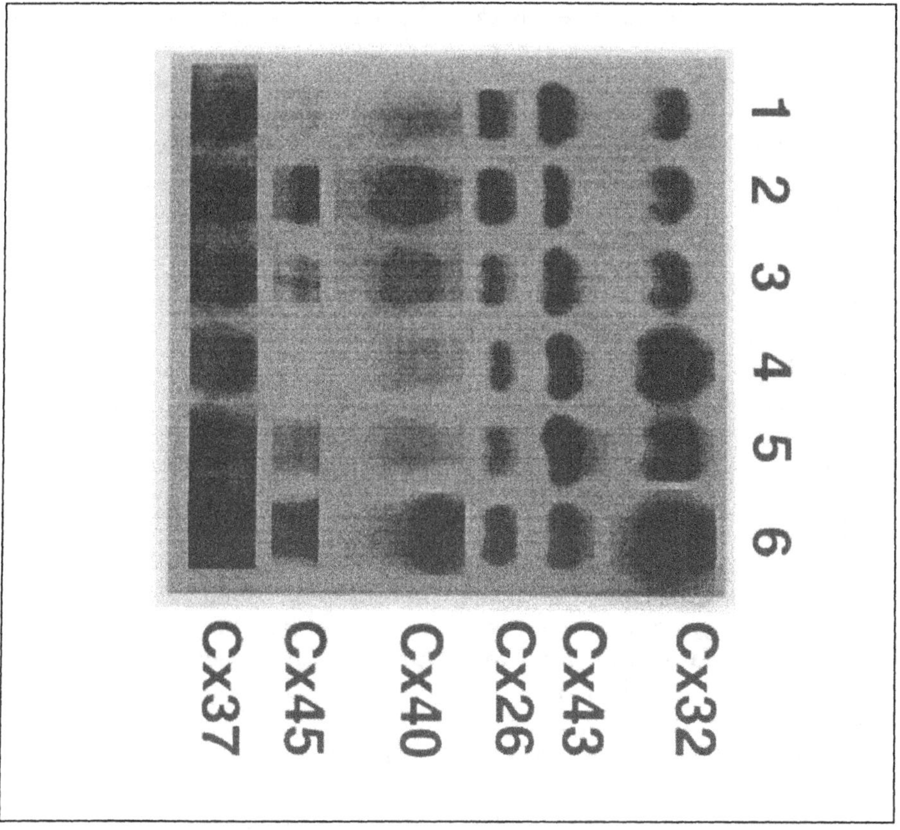

Fig. 2.4. Northern blot analyses of six different connexin mRNAs from different parts of the brain. Lanes are ordered in the following sequence: 1 total brains, 2 cerebellum, 3 hippocampus, 4 spinal cord, 5 hindbrain, 6 controls. The following controls were used: for Cx32 and Cx26, liver; Cx43, heart; Cx40 and Cx37, lung; Cx45, primary astrocytes.

These two connexins are most likely confined to nonneural/nonglial tissues in the adult brains, and the amount of mRNA can be taken as a means of leptomeningeal and vascular contaminants in the preparations. Connexin45 and connexin40 yielded maximal expressions in the cerebellum. Obviously, measurements of mRNA levels from bulk preparations give only a rough idea of the connexin species involved in gap junctional coupling in brain without providing any information regarding their cell-specific expression. The demonstrated regional concentrations of some of the connexins, however, strengthens the concept of a compartmentalized expression pattern, with likely implications for functional differences in intra- and intercompartmental connectivity.

2.3. WHICH TYPES OF CONNEXINS ARE EXPRESSED IN ADULT NEURONS?

Exactly which connexin species are expressed in adult neurons is still unresolved, although this remains a subject of extensive investigation. From what we know, we would expect expression of a heterogeneous complement of connexins in neurons. It would be not surprising if future studies reveal a highly diverse pattern of neuronal connexins with regional and functional specificity. Up to now, two connexin types have been definitely localized in neurons. By means of in situ immunocytochemistry, we have described that between 10% and 20% of neurons exhibit connexin32 immunoreactivity in various parts of the brain including the neostriatum, cortex and brain stem.[30] In addition, considerable amounts of connexin32 immunoreactivity have been described in the motoneurons of the spinal cord.[51] In an elaborate series of studies by Nagy's group the same connexin type was also identified as a neuronal connexin in gymnotid fish.[48] The occurrence of the class II (α) type connexin43 in these neurons could not be confirmed definitely by in situ immunocytochemistry. This was primarily due to the high amount of connexin43 immunoreactivity expressed in the surrounding astroglia.[30,49] Even at the electron microscopic level clear discrimination of neuronal versus astroglial expression sites proved difficult.[50] On the other hand, autoradiographic in situ hybridization initially performed by Miceviych and Abelson[51] with connexin specific riboprobes and more recently by Belliveau et al[52] and by our group (Simburger et al, in preparation) exploiting either autoradiographic or digoxigenin techniques, which allowed resolution at the cellular level, revealed ubiquituous presence of connexin43 mRNA in neurons of rodent cortex, hippocampus, brain stem and cerebellum (Fig. 2.5). As discussed above, the presence of the specific mRNA does not necessarily indicate constitutive translation of the protein. Regenerating hepatocytes, as has been shown

by Kren et al[53] which under normal conditions do not express connexin43, show transcriptional upregulation of this connexin under regenerating conditions without translation of detectable protein concentrations. This discrepancy makes the interpretation of the functional significance of the high level of connexin43 mRNA in neurons difficult. The rapid recruitment of interneuronal coupling under certain experimental conditions, however, lends support to the idea that a cytoplasmic pool of connexin43 mRNA exists which may be under control of translational regulatory mechanisms. The recent discovery that astroglial-neuronal coupling occurs, at least under in vitro conditions,[54] is indicative of the latent capability of neurons to establish functional gap junctions.

Tentative candidates for neuronal expression are connexin45 (Dermietzel, unpublished), connexin40 and a newly cloned connexin30 (K. Willecke, personal communication). The cell specific expression of the latter needs to be determined, while both class II (α) connexins, connexin40 and connexin45, are expressed in glial cells (see Table 2.5 for types of connexins expressed in adult neurons).

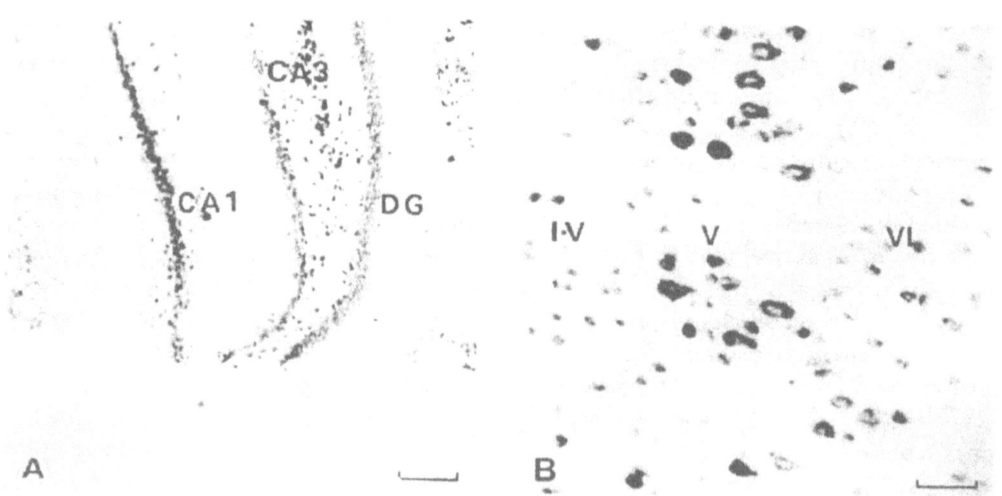

Fig. 2.5. In situ hybridization using a Cx43 specific riboprobe. Subcellular detection was achieved by exploiting digoxigenin immunocytochemistry. (A) shows a section through the hippocampal region. CA1, CA3 pyramidal layers of the hippocampus, DG dentate gyrus. Scale bar in (A) 40 μm, in (B) 20 μm.

Table 2.5. Connexin forms expressed in mature neurons

	Protein (immunocytochemistry)	mRNA (in situ hybridization)
cx43	?	+
cx32	+	+
cx26	o/+	n.d.
cx30	?	?

Data compiled from refs. 30,48,51,52.

How can the formation of functional channels be controlled to avoid promiscuous junctional interaction among neurons and glial cells? A powerful trait of neurons and/or glial cells to establish functionally correct gap junctions could occur via the interaction of homophilic cell adhesion molecules. One would expect that strict control mechanisms regulate the successful achievement of interneuronal and/or glial-neuronal gap junction formation in vivo. It was demonstrated that connexin43 is not phosphorylated to its higher isoforms in the communication deficient sarcoma cell line S180, resulting in a deficiency of plaque formation.[4] Similar results were obtained by us with a glial cell line transfected with the neuoncogene (c erbB/2).[55] In both instances the cells lack their constitutive cell adhesion molecules. The results suggest that the expression of an adequate complement of cell adhesion molecules may be essential for the formation of functional gap junction channels.[4]

3. ASTROCYTES EXPRESS A COMPLEMENT OF CLASS II (α) CONNEXINS

Astrocytes appear to form functionally coupled syncytia, which ionically and metabolically delimit and compartmentalize brain regions.[56-58] Gap junctions are considered to provide the molecular link for coordinated long-distance signaling among the individual members of these syncytia. Essential functions within the astrocytic compartment such as spatial buffering of potassium[59] and neurotransmitters, the generation of slowly changing electrical

fields,[60] and the propagation of calcium waves following stimulation by glutamate,[44,61-64] may depend on coupling mediated by gap junctions. Recent studies demonstrating both induced and spontaneous Ca^{2+} waves spreading throughout the astrocytic network have allowed speculations that glia and their gap-junction-mediated interactions may actually play an active role in neuromodulation.[65] Evidence is mounting that the coupling strength of astrocytes reveals regional heterogeneity[58] and is subject to metabolic regulation. Coupling strength can be modulated by neurotransmitter action, and the activation of protein kinase C.[66] Furthermore, a number of pathologically stressful situations, i.e., transient ischemia,[67] facial nerve ligation[68,69] and intraparenchymal injection of glutaminergic drugs[69] have been demonstrated to elicit rapid modulation of the expression of the major gap junction protein, connexin43, in astrocytes. These data indicate that gap junction channels responsible for the syncytial behavior of the astroglial network are regulated by various mechanisms maintaining a high degree of flexibility. The connectivity within the astrocytic compartments can, therefore, be considered to respond dynamically to local electrical and/or metabolic changes. A diagramatic scheme summarizing various factors which have been shown to influence junctional conductance in astrocytes is indicated in Figure 2.6.

A prerequisite for such a flexible response is the rapid recruitment of astrocytic gap junction channels. This can be achieved on different levels, as discussed

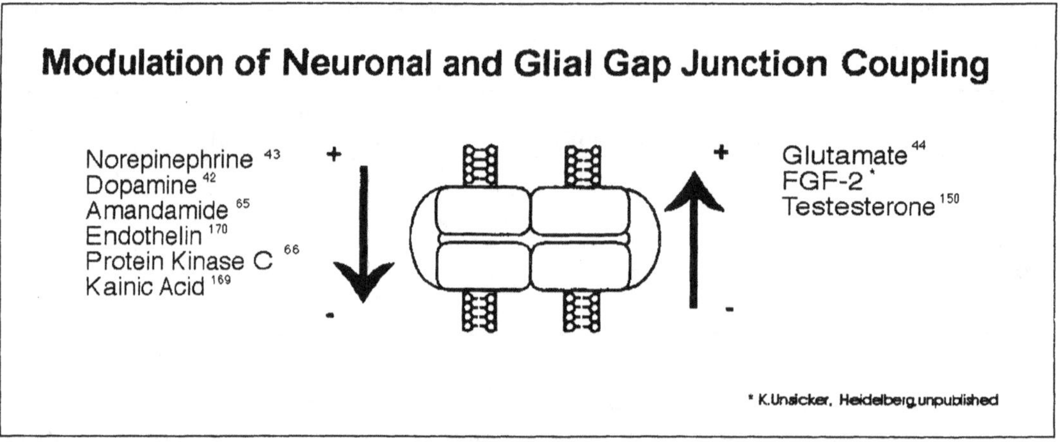

Fig. 2.6. Schematic drawing of different drugs and factors that have been shown to influence gap junction coupling in glial or neural cells. Left side indicates substances which exert a down regulation effect on coupling, while right side shows upregulating factors. Data are taken from the literature as indicated.

above. In the following I will give three examples for fast recruitment phenomena. Each of one may utilize a different level of regulation of connexin expression. For instance, in vivo data revealed evidence that the major astrocytic gap junction protein, connexin43, can be upregulated in response to ligation of the facial nerve. Within less than one hour the amount of immunodetectable connexin43 increases significantly in the facial nuclei[68,69] (see chapter 10). This is the earliest central response described thus far in this experimental paradigm. The dramatic increase of immunodetectable connexin43 speaks in favor of a transcriptional/translational control of this response mechanism.

A further clue to the understanding of fast recruitment phenomena of connexin43 in the astrocytic network is provided from immunocytochemical studies with antibodies directed to peptide sequences of this protein. High level of intracytoplasmic connexin43 immunoreactivity with considerable local variability have been described. The amount of cytoplasmic staining was found to be highest in protoplasmic astroglia as compared to fibrillar astrocytes.[70] If the cytoplasmic labeling is considered to represent a pool of hemichannels, this pool would provide a reliable source for flexible modeling and/or remodeling of junctional connections within the astrocytic network. Recruitment from such a pool would primarily embark on posttranslational mechanisms, i.e., trafficking and assembly mechanisms.

Finally, strength of interastrocytic coupling may be regulated by gating control mechanisms. Since the various substates of the connexin43 type channel correspond to the actual state of phosphorylation[71] of this protein, shifts in the degree of phosphorylation can directly influence the biophysical properties of the channels. A scenario seems feasible in which local activation of protein kinases through transduction proteins modulates the conductiveness of the astrocytic network. Potent physiological candidates affecting the strength of coupling in the glial network are neurotransmitters, growth factors and hormones. While the effects of neurotransmitters are discussed in chapter 8, I will focus primarily on the mechanisms by which growth factors and the activation of their receptor complement may influence intercellular coupling. Some evidence comes from studies on rat liver cells (T51B). When T51W cells are treated with EGF a transient disruption of gap junction communication is elicited.[72] This change apparently does not result from the gross disturbance of membrane gap junction plaques as measured by immunofluorescence microscopy, but instead correlates with markedly elevated phosphorylation of the connexin43 gap junction protein, a profound shift to predominantly phosphorylated forms of connexin43 and the appearance of a novel phosphorylated connexin43 protein. These changes in connexin43 phosphorylation involve only serine residues. Upon restoration of gap junctional communication, the alterations in connexin43 phosphorylation revert to the pre-EGF state.[72] Apparently, EGF via activation of kinase cascades is influential on the coupling strength. A further effect of tyrosine receptor activity has been shown by transfection with the v-src oncogene which results in inhibition of gap junction communication. In this case, however, changes correlate with phosphorylation of tyrosine residues.[73]

One may argue that these studies are irrelevant for a discussion of interglial coupling, since the described effects occur in highly neoplastic, nonglial cell lines. Evidence from two different sources, however, indicate that similar mechanisms may presumably apply to the astrocytic syncytium. The above mentioned transfection of an astrocytic cell line with the c-erbB/2 oncogene (which resulted in a constitutive overexpression of tyrosine kinase activity in this cell line) led to a significant decoupling effect, at least at the macroscopic level as imaged by dye transfer. In this case, no phosphorylation on tyrosine sites of connexin43 was detectable. Instead, a reduction of the higher phosphorylated isoforms of connexin43, similar to that in the S180 cell line[4] was found, indicating a tyrosine-kinase dependent activation of PKC events in the processing of connexin43.

The idea that activation of growth factor receptors are involved in the regulation of coupling strength concur with studies on MPTP (1-methyl-4-phenyl-1,2,3,6-tetrahydropyridine) treated mice (an established animal model of Parkinsonism). Significant increase of astrocytic connexin43 immunoreactivity was detectable after neostriatal application of bFGF (FGF-2) in MPTP mice indicating an influence of this growth factor on the recruitment of con-

nexin43 (K. Unsicker, in preparation). The regulation of gap junction communication seems to involve multisite phosphorylation orchestrated by a complex interplay of intracellular kinases, thereby working on distinct levels of the connexin pathway or gating control. Phosphorylation can, therefore, be considered to constitute a powerful tool for transient regulation of the strength of intercellular coupling within or among functional compartments of the brain. With respect to the astrocytic network this means that the efficacy and strength of coupling are subject to variable control mechanisms which involve transcellular as well as paracrine mechanisms of the growth factor-kinase cascade type.

Different connexins exhibit variable degrees of phosphorylation. Indeed some, e.g., connexin26, exhibit no phosphorylation.[36] When examining modulation of astrocytic coupling by phosphorylation it is important to reconsider the connexin complement expressed in astrocytes.

We as well as others have documented that the major gap junction protein in astrocytes is connexin43.[74,75] Recent data on transgenic "knock out" mice, which show a generally normal development of brain structures[76] (although degree of normality of higher order brain function has not been assessed because the animals die shortly after birth) have reinforced our search for additional connexins which might compensate the loss of connexin43, at least under developmental conditions. We found two other members of the class II (α) connexins expressed in vivo and in vitro in astrocytes, both at a much lower level then connexin43. These are connexin40 and connexin45. We do not know the functional implications of this diversity in the astrocytic syncytium, but preliminary data indicate that at least connexin43 and connexin45 show subcellular heterogeneity in their expression, in particular in the cerebellum (Fig. 2.7). Although only speculation at present, it seems feasible that local functional requirements necessitate site-restricted expression of connexins within the astrocytic network. Such site

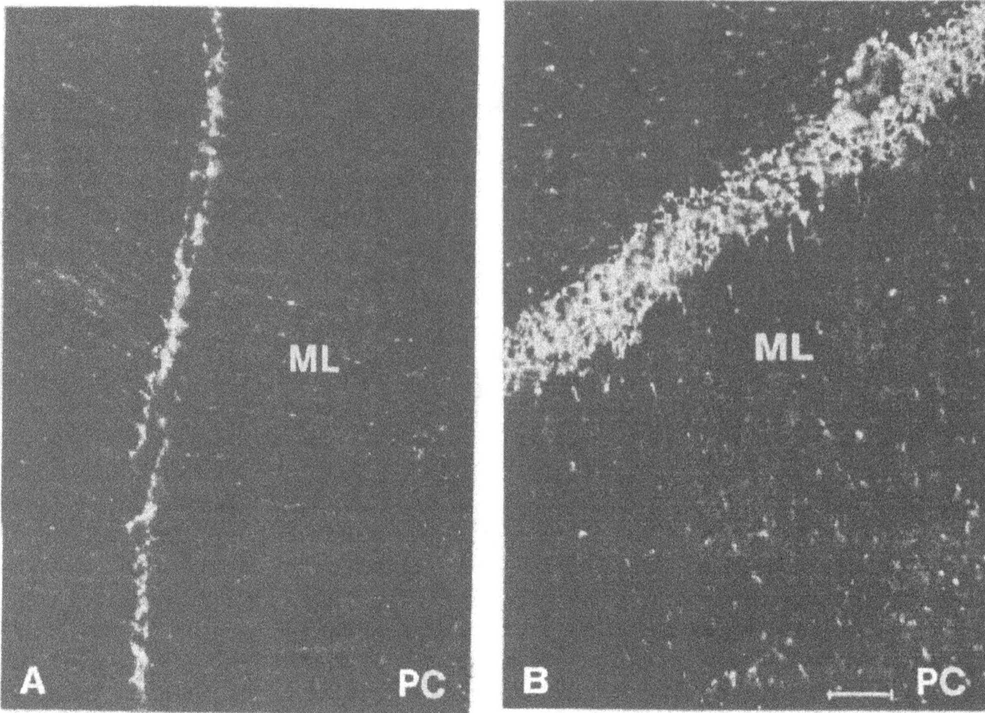

Fig. 2.7. Immunolocalization of Cx43 (A) and Cx45 (B) in the molecular layer of the cerebellum (ML). Note that both antibodies show considerable variation of immunolabel indicative for differences in subcellular localization of these connexin types. Scale bar 20 (m.

restriction, for example, could be responsible for the creation of functional boundaries if one of the connexins excludes heterotypic coupling, i.e., coupling of two hemichannels of different connexin types.

4. MYELINATING CELLS AND THE PRESENCE OF REFLEXIVE GAP JUNCTIONS

Within the central nervous system oligodendrocytes are responsible for myelin formation providing the biological machinery for maintenance and support of the mature myelin sheath.[77,78] Schwann cells represent the cellular analogs of oligodendrocytes in the peripheral nervous system. When considering the functional relevance of gap junction coupling between myelinating cells, it seems appropriate to recapitulate some essential differences in the cellular architecture of the central and peripheral myelin sheaths

which derive from differences in the mechanisms of the myelinating processes.

In the peripheral nervous system, a one to one ratio exists in the number of Schwann cells and myelinated internodes; the ratio of oligodendrocytes to the number of internodes is much smaller in the CNS. There are 20-40 times as many internodes as oligodendrocytes in the optic nerves.[79] This figure is even more variable in the spinal cord where the ratio varies between 18 and 60.[80] In addition, it is known from reconstruction of serial electron microscopical sections that more than one individual oligodendrocyte contributes to the formation of a single internode.[81] Thus, we must envision a complex arrangement of oligodendrocytes within a single internodal length of central myelin as compared to the stereotyped organization of Schwann cells in the peripheral nerves. One final aspect which should be addressed when considering differences in the organization of the central and peripheral myelin sheath is the presence of cytoplasmic relics in the mature sheath which originally derive from the myelinating cells during the process of compaction. In the peripheral sheath these processes are restricted to the inner cytoplasmic collar, the paranodal or lateral processes, and the Schmidt-Lantermann incisures. Central myelin sheath, however, does not exhibit Schmidt-Lantermann incisures; moreover, cytoplasmic relics are restricted to the inner, outer and lateral (paranodal) cytoplasmic loops. While in the peripheral myelin sheath the nucleus of the Schwann cells is always enclosed in the outer cytoplasmic collar, which surrounds the entire myelin sheath, the soma of the oligodendrocytes is distant from the myelin sheath connected via attenuated cytoplasmic processes with the outer cytoplasmic portions of the sheath.

With respect to the obligatory supply of the myelin sheath with its molecular constituents one can conceptually differentiate between long-living structural components, like the constitutive membrane proteins (proteo-lipid protein, myelin-basic protein, etc.) with half lives of more than 100 days, and short-lived metabolites including substances for obligatory energy supply. The internodal compartment of the myelin sheath is unique in its bipartite composition with an excessive part of bradytrophic material laid down in form of compact myelin and multiple cytoplasmic incisures which require a different schedule of energy and metabolite supply. In particular, the paranodal cytoplasmic loops are considered to be highly cooperative partners of the axon in propagating axon potentials,[82,83] and their molecular requirements are essentially different from the compact myelin sheath.

Channels of the gap junction type have been described exclusively between the various cytoplasmic parts of the myelin sheath in central and peripheral myelin.[84-86] However, the characterization of gap junctions in myelinated nerves has been neglected until recently, when the first gap

junction-related inherited human disease, the X-linked form of the Charcot-Marie-Tooth (CMTX1) syndrome was described.[87] Mutations in the connexin32 coding and noncoding region are responsible for the CMTX1 syndrome phenotype (see chapter 13). It was initially surprising that the phenotype resulting from apparent loss of function mutation of connexin32 is limited to peripheral nerves, because this connexin is a major component of gap junctions of diverse tissues including liver, exocrine pancreas and epithelium of the small intestine, etc.[88] Moreover, within the CNS connexin32 appears to be a major component of gap junctions in oligodendrocytes and some neurons,[30] yet no impairment of CNS function has been demonstrated in CMTX1 patients. The tissue-restricted phenotype of connexin32 mutations can be explained by either lack of a compensatory connexin in mature Schwann cells or the mutations of the connexin32 gene may result in specific loss of function in the peripheral myelin sheath because of this tissue's unique structural organization and/or metabolic requirements.

There are several functional roles that connexin32 might play in Schwann cell biology that could be affected by functional changes of the connexin32 gene resulting from the mutations. First, connexin32 hemichannels in paranodal membranes might operate per se, without assembly into complete double membrane gap junction structures. Such behavior has been observed for connexin46 channels expressed in oocytes,[89] for channels of an as yet unidentified connexin in retinal horizontal cells,[90] and for connexin43 hemichannels in macrophages stimulated with ATP.[91] Second, reflexive gap junction structures could provide necessary nutrient exchange between the Schwann perinuclear region and the distant cytoplasmic processes, which otherwise would be separated from the rest of the cytoplasm by a quite tortuous pathway. Third, if transnodal gap junctions occur (as observed in regenerating chick nerves by Tetzlaff[86]), they

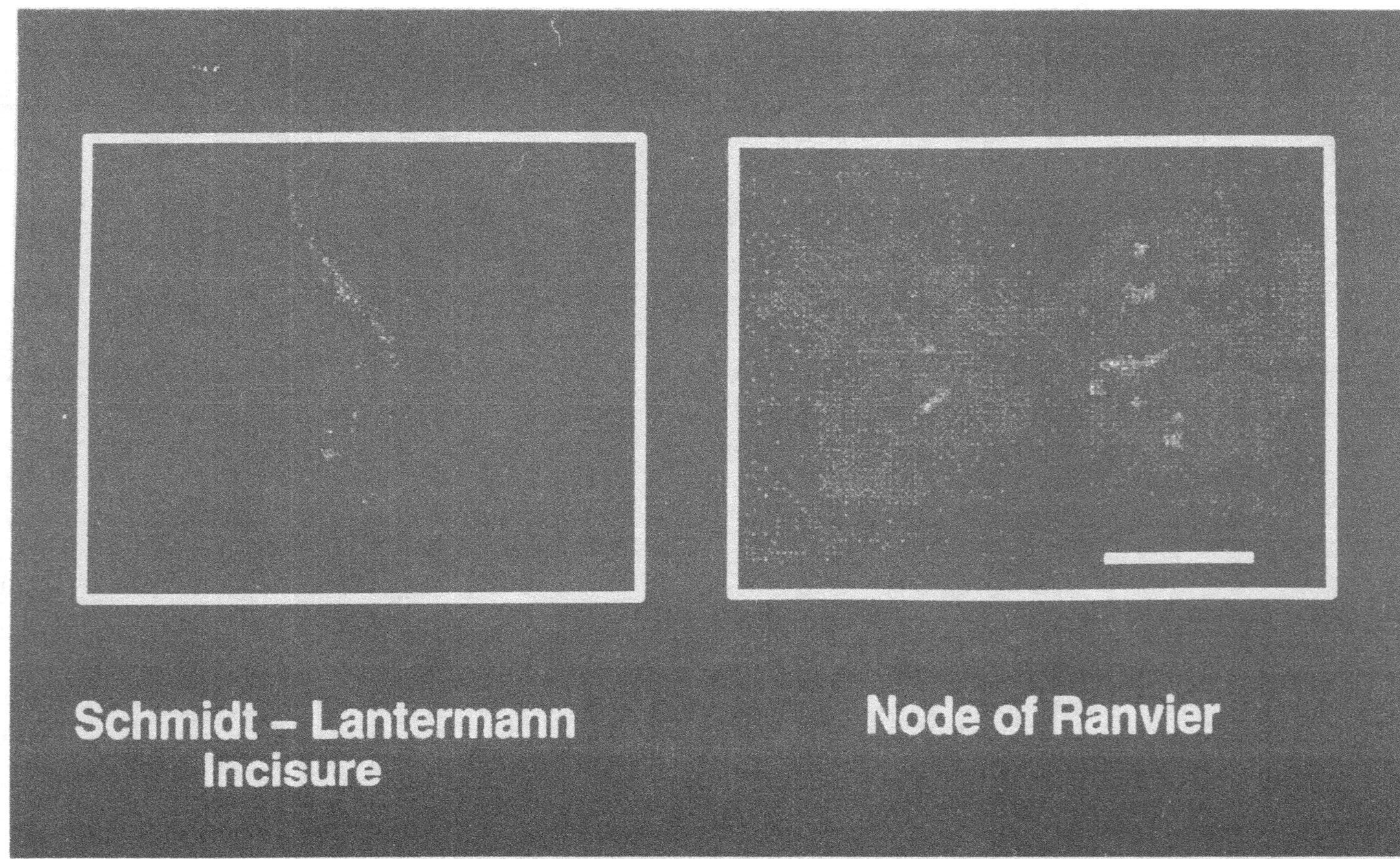

Schmidt – Lantermann Incisure
Node of Ranvier
A

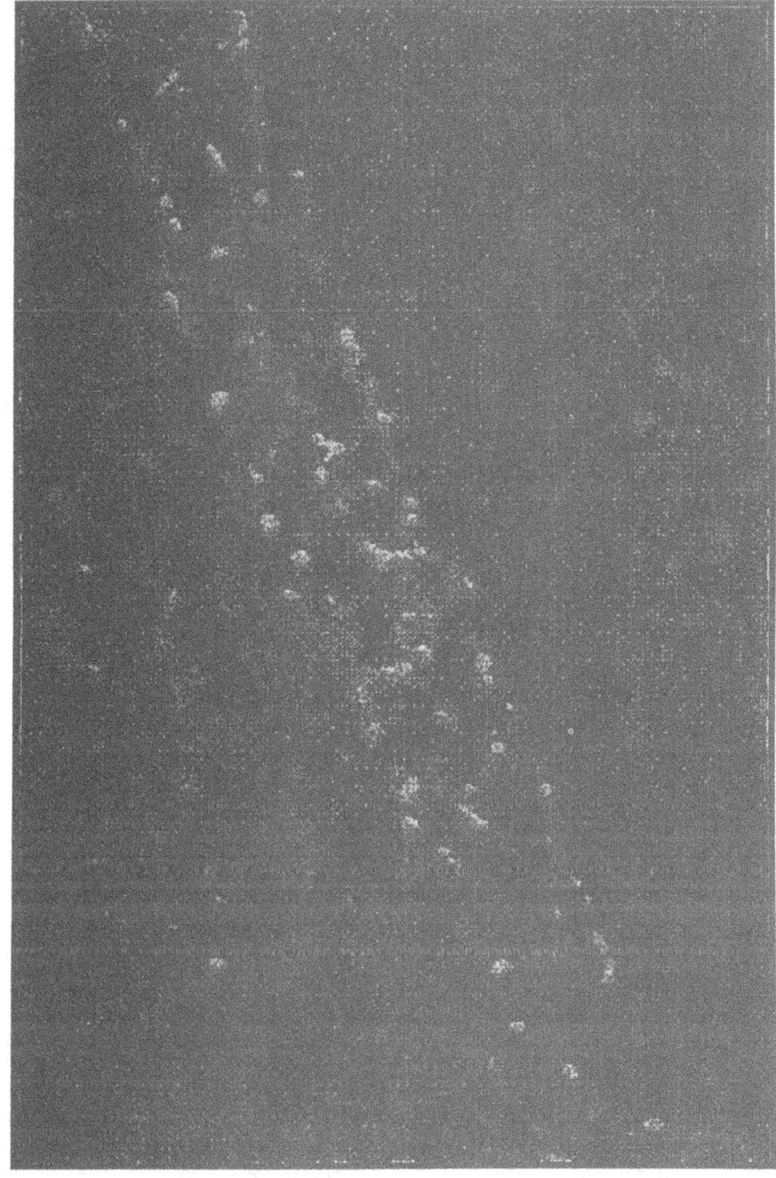

xFig. 2.8. Labeling of Cx32 immunoreactivity in peripheral nerve fibers (A, above) and central myelin (spinal cord). In peripheral nerves immunoreactivity is confined to Schmidt-Lantermann incisures (1) and nodes of Ranvier (2), while in control myelin (B, right) Cx32 immunolabeling occurs throughout the internode.

could provide a pathway for internodal signaling under either normal or injury-evoked conditions. Such a function might be important in response to injury including both de- and regenerating events.

Gap junctions in nonregenerating peripheral nerves, however, have been reported to be extremely rare, in contrast to the situation in central myelin where junctional assemblies have been reported between the outer cytoplasmic loops and the soma of quiescent oligodendrocytes[85] as well as between cytoplasmic portions between the paranodal processes.[84] Connexin32 immunoreactivity in the peripheral nerve was reported to be confined to Schmidt-Lanterman incisures and the paranodal portions of the node of Ranvier.[87] However, the low incidence of gap junction assemblies at these sites leaves open alternative interpretations concerning the structural identity of the immunolabel. Interestingly, immunolabeling with connexin32 antibodies in central myelin reveals an entirely different pattern as compared to peripheral myelin, with extensive immunoreactive material being present in the internodal region[47] (Fig. 2.8). In addition, its expression is highly variable with regional concentrations (Fig. 2.9). The highest amount of immunoreactive connexin32 is found in the white matter of the spinal cord with a decrease in order from brain stem>cerebellar white matter > corpus callosum. Studies on the time course of connexin32 expression during myelinogenesis in the CNS indicates that the onset of connexin32 expression coincides with the beginning of myelination, around P7 in rat spinal cord. A characteristic shift of connexin32 expression sites is prevalent showing a preferred cytoplasmic localization at early stages of myelination and a more peripheral location (in the internodal cytoplasmic processes) when myelination has succeeded. Apparently, the final assembly and organization of gap junctions is achieved in the mature myelin sheath.

Studies on peripheral nerves under de- and regenerating conditions (see chapters

13 and 14) also revealed a strong correlation of connexin32 expression with myelinating events. Initial downregulation of connexin32 expression is followed by upregulation during the remyelinating phase. Thus, the expression pattern of connexin32 in myelin-forming cells is apparently governed by the same genetic program as is in more classical myelin proteins.

As already discussed, the functional significance of gap junction channels within the myelin sheath remains to be established. The favored idea at present is that they in fact form reflexive junctions between parts of the same cell. They can thus be considered to be transcytoplasmic. The question that then arises is how the myelinating cells benefit from additional short cuts since large cytoplasmic continuities, albeit extended and tortuous, do already exist. One reasonable explanation is that gap junction channels provide regulated pathways between peripheral cytoplasmic portions of the myelin sheath which, because of their gating properties and selectivity, are responsive to changes in physiological conditions, i.e., transjunctional potentials, pH, etc. Furthermore, the extreme variability of connexin32 expression in central myelin indicates that local factors, either in form of the cytoarchitecture of the myelin sheath or of the microenvironment, require differences in the degree of "reflexiveness" (for a discussion of reflexive or autocellular gap junctions in astrocytes see chapter 10).

The interpretation that most of the gap junctions in central myelin sheath represent the reflexive type was corroborated by microinjection of Lucifer Yellow or Neurobiotin[R] in oligodendrocytes of spinal cord and corpus callosum slices. No dye transfer was achieved after injection of one of the tracers into electrophysiologically identified oligodendrocytes (H. Kettenmann, personal communication). The idea that the complement of myelinic gap junctions provide responsive short cuts for autocellular regulation is a compelling hypothesis

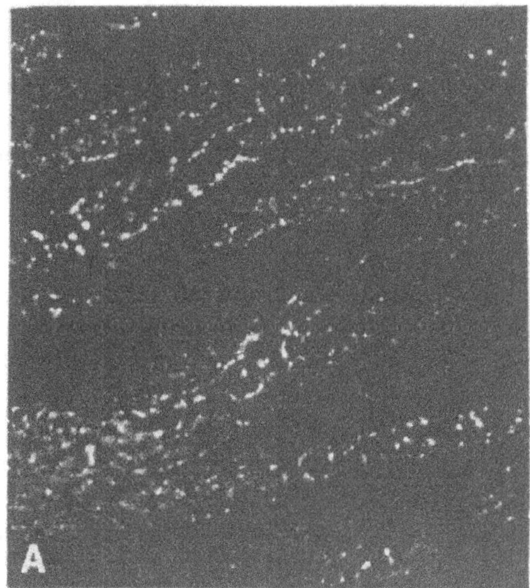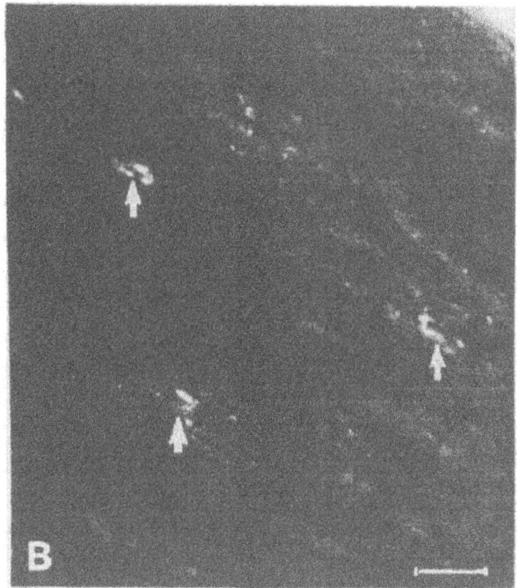

Fig. 2.9. Immunofluorescence micrographs depicting regional heterogeneity of Cx32 in white matter of the central nervous system. (A) indicates a sections through spinal cord white matter while (B) shows a micrograph of the corpus callosum. Arrows point to immunoreactive sites in the corpus callosum. Scale bar 20 (m.

which has to be proven by functional mutations that generate changes or interruptions of these pathways.

Acknowledgments

I gratefully acknowledge the help of my collaborators who contributed to this work: Dr. M. Davids, I. Hertting, Dr. A. Hofer, M. Kremer, Dr. F. Miragall, E. Simburger. Unpublished data were kindly provided by H. Kettenmann (Berlin), K. Unsicker (Heidelberg), K. Willecke (Bonn). The work was supported by a grant from the Deutsche Forschungsgemeinschaft (Schwerpunkt Glia) to R. D., and a NATO traveling grant which facilitated the extensive exchange between our laboratory and that of D.C. Spray.

References

1. Edelmann GM, Crossin KL. Cell adhesion molecules: Implications for a molecular histology. Ann Rev Biochem 1991; 60: 155-190.
2. Revel J. The oldest multicellular animal and its junctions. In: Hertzberg EL, Johnson RG eds. Gap Junctions. New York: Alan R. Liss, 1988:135-149.
3. Davids M, Hofer A, Heydrich U et al. Microinjection of antibody to connexin43 disrupts normal development of Xenopus embryos. (submitted)
4. Musil LS, Cunningham BA, Edelmann GM et al. Differential phosphorylation of the gap junction protein connexin43 in junctional communication-competent and -deficient cell lines. J Cell Biol 1990; 111:2077-2088.
5. Musil LS, Goodenough DA. Biochemical analysis of connexin43 intracellular transport, phosphorylation, and assembly into gap junctional plaques. J Cell Biol 1991; 115:1357- 1374.
6. Jongen WMF, Fitzgerald DJ, Asamoto M et al. Regulation of connexin43- mediated gap junctional intercellular communication by calcium in mouse epidermal cells is controlled by E-cadherin. J Cell Biol 1991; 114:545-556.
7. Becker DL, Leclerc D, Warner A. The relationship of gap junctions and compaction in the preimplantation mouse embryo. Develop 1992; Suppl.113-118.
8. Paul DL, Bruzzone K, Yu R et al. Expression of a dominant negativ inhibitor of intercellular communication in the early Xenopus embryo causes delamination and extrusion of cells. Development 1995;

121:371-381.

9. Bennett MVL, Goodenough DA. Gap junctions, electrotonic coupling, and intercellular communication. Neurosci Res Prog Bull 1978; 16:373-485.

10. Dermietzel R, Spray DC. Gap Junctions in the brain: where, what type, how many and why? Trend Neurosci 1993; 16:186-192.

11. Sotelo C, Korn H. Morphological correlates of electrical and other interactions through low-resistance pathways between neurons of the vertebrate central nervous system. Int Rev Cytol 1978; 55:67-107.

12. Dermietzel R. Junctions in the central nervous system of the cat. 3. Gap junctions and membrane-associated orthogonal particle complexes (MOPC) in astrocytic membranes. Cell Tissue Res.1974; 149:121-135.

13. Mugnaini E. Membrane spezialisations in neuroglial cells and at neuron-glia contacts. In: Sears TA, ed. Neuronal-glial Cell Interrelationships. Heidelberg: Springer, 1982: 39-56.

14. Massa PT, Mugnaini E. Cell junctions and intramembrane particles of astrocytes and oligodendrocytes: A freeze-fracture study. Neuroscience 1982; 7:523-538.

15. Massa PT, Mugnaini E. Cell-cell interactions and characteristic plasma membrane features of cultured rat glial cells. Neuroscience 1985; 14, 2:695-709.

16. Massa PT, Szuchet S, Mugnaini E. Cell-cell interactions of isolated and cultured oligodendrocytes: formation of linear occluding junctions and expression of peculiar intramembrane particles. J Neurosci 1984; 4:3128-3139.

17. Mugnaini, E. Cell junctions of astrocytes, ependyma and related cells in the mammalian central nervous system, with emphasis on the hypothesis of a generalized functional syncytium of supporting cell. Astroc Dev Morph Reg Spec 1986; 329-371.

18. Goodenough DA, Revel JP. A fine structural analysis of intercellular junctions in the mouse liver. J Cell Biol 1970; 45: 272-290.

19. Lane NJ, Swales LS. Dispersal of junctional particles, not internalization, during the in vivo disappearance of gap junctions. Cell 1980; 19:579-586.

20. Yancey SB, Nicholson BJ, Revel JP. The dynamic state of liver gap junctions. J Supramol Struct Cell Biochem 1981; 16:221-232.

21. MacVicar BA, Dudek EF. Electronic coupling between pyrymidal cells: A direct demonstration in rat hippocampal slices. Science 1981; 213:782-785.

22. MacVicar BA, Ropert N, Krnjevic K. Dye-coupling between pyramidal cells of rat hippocampus in vivo. Brain Res 1982; 238:239-244.

23. MacVicar BA, Jahnsen H. Uncoupling of CA3 pyramidal neurons by propionate. Brain Res 1985; 330:141-145.

24. Yuste R, Peinado A, Katz LC. Neuronal domains in developing neocortex. Science. 1992; 257:665-669.

25. Peinado A, Yuste R, Katz LC. Gap junctional communication and the development of local circuits in neocortex. Cereb Cortex 1993; 3:488-498.

26. Walsh JP, Cepeda C, Hull CD et al. Dye-coupling in the neostriatum of the rat: II. Decreased coupling between neurons during development. Synapse 1989; 4:238-247.

27. Yuste R, Nelson DA, Rubin WW et al. Neuronal domains in developing neocortex: Mechanisms of coactivation. Neuron 1995; 14:7-17.

28. Wong ROL, Chernjasky A, Smith SJ et al. Early functional neural networks in the developing retina. Nature 1995; 374: 716-718.

29. Wolszon LR, Rehder V, Kater SB et al. Calcium wave fronts that cross gap junctions may signal neuronal death during development. J Neurosci 1994; 14: 3437-3448.

30. Dermietzel R, Traub O, Hwang TK et al. Differential expression of three gap junction proteins in developing and mature brain tissue. Proc Natl Acad Sci USA 1989; 86:10148-10152.

31. Fushiki S, Kinoshita C. Spatial and temporal patterns of the distribution of the gap junction protein connexin32 and 43 during histogenesis of mouse cerebral cortex. In: Kanno Y, Kataoka K, Shiba Y et al. eds. Intercellular communication through gap junctions. Amsterdam: Elsevier, 1995:

239-243.

32. Willecke K, Heynkes R, Dahl E et al. Mouse connexin37: Cloning and functional expression of gap junction gene highly expressed in lung. J Cell Biol 1991; 114:1049- 1057.

33. Moreno AP, Laing JG, Beyer EC et al. Properties of gap junction channnels formed of connexin45 endogenously expressed in human hepatoma (SKHep1) cells. Am J Physiol 1995; 268:C356-365.

34. Steinberg TH, Civitelli R, Geist ST et al. Connexin43 and connexin45 form gap junctions with different molecular permeabilities in osteoblastic cells. EMBO 1994; 13: 744-750.

35. Lo Turco JJ, Kriegstein AR. Clusters of coupled neuroblasts in embryonic neocortex. Science 1991; 252:563-566.

36. Traub O, Look J, Dermietzel R et al. Comparative characterization of the 21 kDa and 26 kDa gap junction proteins in murine liver and cultured hepatocytes. J Cell Biol 1989; 108:1039-1051.

37. Fallon RF, Goodenough DA. Five-hour half-life of mouse liver gap junction protein. J Cell Biol 1981; 90:521-526.

38. Dudek FE, Andrew RD, MacVicar BA et al. Intracellular electrophysiology of mammalian peptidergic neurons in rat hypothalamic slices. Fed Proc 1982; 41:2953-2958.

39. Peres Velasquez JL, Valiante TA, Carlen PL. Modulation of gap junction mechanism in calcium-free induced field-burst acivity: A possible role for electrotonic coupling in epileptogenesis. J Neurosci 1994; 14: 4308-4317.

40. Spray DC, Saez JC, Hertzberg EL et al. Gap junctions in liver. In: Arias IM, Boyer JL, Fausto N ed al., eds. The Liver: Biology and Pathobiology. New York: Raven Press, 1994; 951-967.

41. Hennemann H, Suchyna T, Lichtenberg-Frate H et al. Molecular cloning and functional expression of mouse connexin40, a 2nd gap junction gene preferentially expressed in lung. J Cell Biol 1992; 117: 1299-1310.

42. Hampson ECGM, Vaney DI, Weiler R. Dopaminergic modulation of gap junction permeability between amacrine cells in mammalian retina. J Neurosci 1992;

43. Giaume C, Marin P, Cordier J et al. Adrenergic regulation of intercellular communication between cultured striatal astrocytes from the mouse. Proc Natl Acad Sci USA 1991; 88:5577-5581.

44. Enkvist MD, McCarthy KD. Astroglial gap junction communication is increased by treatment with either glutamat or high K⁺ concentration. J Neurochem 1994; 62: 489-495.

45. Dudek FE, Snow RW, Taylor CP. Role of electrical interactions in synchronization of epileptiform bursts. Adv Neurol 1986; 44:593-617.

46. Llinas RR. Electronic transmission in the mammalian central nervous system. In: Bennett MVL, Spray DC eds. Gap junctions. Cold Spring Harbor Laboratory, 1985; 337-353.

47. Spray DC, Dermietzel R. X-linked dominant Charcot-Marie-Tooth disease and other potential gap-junction diseases of the nervous system. Trend Neurosci 1995; 18: 256-262.

48. Yamamoto T, Maler L, Hertzberg EL et al. Gap junction protein in weakly electric fish (Gymnotide): Immunohistochemical localization with emphasis on structures of the electrosensory system. J Comp Neurol 1989; 289:509-536.

49. Yamamoto T, Ochalski A, Hertzberg EL et al. On the organization of astrocytic gap junctions in rat brain as suggested by LM and EM immunocytochemistry of connexin43 expression. J Comp Neurol 1990; 302:853-883.

50. Miragall F, Hwang TK, Traub O et al. Expression of connexins in the developing olfactory system of the mouse. J Comp Neurol 1992; 325:359-378.

51. Miceviych PE, Abelson L. Distribution of mRNA coding for liver and heart gap junction proteins in the rat central nervous system. J Comp Neurol 1991; 305:96-118.

52. Belliveau DJ, Naus CCG. Cellular localization of gap junction mRNAs in developing rat brain. Dev Neurosci 1995; 17:81-96.

53. Kren BT, Kumar NM, Wang S et al. Differential regulation of multiple gap junction transcripts and proteins during rat liver

regeneration. J Cell Biol 1993; 123: 707-718.

54. Nedergaard, M. Direct signaling from astrocytes to neurons in cultures of mammalian brain cells. Science 1994; 263: 1768-1771.

55. Hofer A, Seaz JC, Chang CC et al. cerB-2/neu transfection induces gap junctional communication-incompetence in glial cells. J Neurosci (in press).

56. Sontheimer H, Waxman SG, Ransom BR. Relationship between Na^+ current expression and cell-cell coupling in astrocytes cultured from rat hippocampus. J Neurophysiol 1991; 65:989-1002.

57. Binmöller F, Müller CM. Postnatal development of dye-coupling among astrocytes in rat visual cortex. Glia 1992; 6:127-137.

58. Batter DK, Corpina RA, Roy C et al. Heterogeneity in gap junction expression in astrocytes cultured from different brain regions. Glia 1992; 6:213-221.

59. Newman EA. High potassium conductance in astrocyte endfeet. Nature 1986; 233: 453-454.

60. Kettenmann H, Ransom BR. Electrical coupling between astrocytes and between oligodendrocytes studied in mammalian cell cultures. Glia 1988; 1:64-73.

61. Cornell-Bell AH, Finkbeiner SM, Cooper MS et al. Glutamate induces calcium waves in cultured astrocytes: Long-range glial signaling. Science 1990; 247:470-473.

62. Murphy TH, Blatter LA, Wier WG et al. Rapid communication between neurons and astrocytes in primary cortical cultures. J Neurosci 1993; 13:2672-2679.

63. Charles AC, Merrill JE, Dirksen ER et al. Intercellular signalling in glial cells: calcium waves and oscillations in response to mechanical stimulation and glutamate. Neuron 1991; 6:983-992.

64. Charles AC, Dirksen ER, Merrill JE et al. Mechanisms of intercellular calcium signaling in glial cells studied with dantrolene and thapsigargin. Glia 1993; 7:134-145.

65. Venance L, Piomelli D, Glowinski J et al. Inhibition by anandamide of gap junctions and intercellular calcium signalling in striatal astrocytes. Nature 1995; 376: 590-594.

66. Enkvist MOK, McCarthy KD. Activation of protein kinase-C blocks astroglial gap junction communication and inhibits the spread of calcium waves. J Neurochem 1992; 519-526.

67. Hossain MZ, Peeling J, Sutherland GR et al. Ischemia-induced cellular redistribution of the astrocytic gap junctional protein connexin43 in rat brain. Brain Research 1994; 652:311-322.

68. Rohlmann A, Lakawai R, Hofer A et al. Facial nerve lesions lead to increased immunostaining of the astrocytic gap junction protein (connexin43) in the corresponding facial nucleus of rats. Neurosci Lett 1993; 154:206-208.

69. Rohlmann A, Laskawi R, Hofer A et al. Astrocytes as rapid sensors of peripheral axotomy in the facial nucleus of rats. Neuroreport 1994; 5:409-412.

70. Nagy JL, Yamamoto T, Sawchuk MA et al. Quantitative immunohistochemical and biochemical correlates of connexin43 localization in rat brain. Glia 1992; 5:1-9.

71. Moreno AP, Fishman GI, Spray DC. Phosphorylation shifts unitary conductance and modifies voltage dependent kinetics of human connexin43 gap junction channels. Biophys J 1992; 62:51-53.

72. Lau AF, Kanemitsu MY, Kurata WE et al Epidermal growth factor disrupts gap-junctional communication and induces phosphorylation of connexin43 on serine. Mol Biol Cell. 1992; 3:865-874.

73. Swenson KL, Piwnica-Worms H, McNamee H et al. Tyrosine phosphorylation of the gap junction protein connexin43 is required for the pp60-v-src-induced inhibition of communication. Cell Regulation 1990; 1:989-1002.

74. Dermietzel R, Hertzberg EL, Kessler JA et al. Gap junctions between cultured astrocytes: Immunocytochemical, molecular, and electrophysiological analysis. J Neurosci 1991; 11:1421-1432.

75. Giaume C, Fromaget C, Aoumari AE et al. Gap junctions in cultured astrocytes: Single-channel currents and characterization of channel-forming protein. Neuron 1991; 6:133 -143.

76. Reaume AG, de Sousa PA, Kulkami S et al.

Cardiac malformation in neonatal mice lacking connexin43. Science 1995; 267: 1831-1834.

77. Bunge RP. Glial cells and the central myelin sheath. Physiol Rev 1968; 48:197-251.

78. Bunge, R. P. The development of myelin and myelin-related cells. Trends Neurosci 1981; 4:175-177.

79. Peters A, Vaughn JE. Morphology and the development of the myelin sheath. In: Davidson AN, Peters A, eds. Myelination. Springfield, Ill.: Charles C. Thomas, 1970:3-79.

80. Matthews MA, Duncan D. A quantitative study of the morphological changes accompanying the initiation of myelin production in the peripheral nervous system. J Comp Neurol 1971; 142:1-22.

81. Knobler RL, Stempak JG, Laurencin M. Nonuniformity of the oligodendriglial ensheathment of axons during myelination in the developing rat central nervous system. J Ultrastruct Res 1976; 55:417-432.

82. Neumcke B, Staempfli R. Sodium currents and sodium-current fluctuations in rat myelinated nerve fibres. J Physiol 1982; 329:163-184.

83. Waxmann SG, Ritchie JM. Molecular dissection of the myelin sheath. Ann Neurol 1993; 33:121-136.

84. Schnapp B, Mugnaini E. Membrane architecture of myelinated fibers as seen by freeze-fracture. In: Waxman SG, ed. Physiology and Pathobiology of axons. New York: Raven Press, 1978:83-123.

85. Dermietzel R, Schünke D, Leibstein A. The oligodendrocytic junctional complex. Cell Tissue Res 1978; 193:61-72.

86. Tetzlaff W. Tight junction contact events and temporary gap junctions in the sciatic nerve fibres of the chicken during Wallerian degeneration and subsequent regeneration. J Neurocytol 1982; 11:839-58.

87. Bergoffen J, Scherer SS, Wang S et al. Connexin mutations in X-linked Charcot-Marie-Tooth disease. Science 1993; 262: 2039-2042.

88. Dermietzel R, Leibstein A, Frixen U et al. Gap junctions in several tissues share antigenic determinants with liver gap junctions. EMBO 1984; 3:2261-2270.

89. Paul DL, Ebihara L, Takemoto LJ et al. Connexin46, a novel lens gap junction protein, induces voltage-gated currents in nonjunctional plasma membrane of Xenopus oocytes. J Cell Biol 1991; 115:1077-1089.

90. DeVries SH, Schwartz EA. Modulation of an electrical synapse betwen solitary pairs of catfish horizontal cells by dopamine and second messengers. J Physiol 1989; (Lond.) 414:351-375.

91. Beyer EC, Steinberg TH. Evidence that the gap junction protein connexin43 is the ATP-induced pore of mouse macrophages. J Biol Chem 1991; 266:7971-7974.

92. Baker R, Llinas R. Electrotonic coupling between neurons in the rat mesencephalic nucleus. J Physiol 1971; 212:45-63.

93. Hinrichsen CFL, Larramendi LMH. Synapses and cluster formation of the mesencephalic fifth nucleus. Brain Res 1968; 7:296-299.

94. Sotelo C, Llinas R. Specialized membrane junctions between neurons in the vertebrate cerebellar cortex. J Cell Biol 1972; 53:271-289.

95. Korn H, Sotelo C, Crepel F. Electrotonic coupling between neurons in the rat lateral vestibular nucleus. Exp Brain Res 1973; 16:255-275.

96. Sotelo C, Palay SL. The fine structure of the lateral vestibular nucleus in the rat. II. Synaptic organization. Brain Res 1970; 18:93-115.

97. Wylie RM. Evidence of electrotonic transmission in the vestibular nuclei of the rat. Brain Res 1973; 50:179-183.

98. Pinching AJ, Powell TPS. The neuropil of the glomeruli of the olfactory bulb. J Cell Sci 1971; 9:374-377.

99. Landis DMD, Reese TS, Raviola E. Differences in membrane structure between excitatory and inhibitory components of the reciprocal synapse in the olfactory bulb. J Comp Neurol 1974; 155:67-92.

100. Reyher CKH, Lübke J, Larsen WJ et al. Olfactory bulb granule cell aggregates: Morphological evidence for interperikaryal electronic coupling via gap jungtions. J Neurosci 1991; 11:1485-1495.

101. Llinas R, Baker R, Sotelo C. Electrotonic coupling between neurons in cat inferior olive. J Neurophysiol 1974; 37:560-571.

102. Sotelo C, Llinas R, Baker R. Structural study of inferior olivary nucleus of the cat: Morphological correlates of electrotonic coupling. J Neurophysiol 1974; 37:541-559.

103. King JS. The synaptic cluster (glomerulus) in the inferior olivary nucleus. J Comp Neurol 1976; 165:387-400.

104. Gwyn DG, Nicholson GP, Flumerfelt BA. The inferior olivary nucleus of the rat: A light and electron microscopic study. J Comp Neurol 1977; 174:489-520.

105. Rutherford JG, Gwyn DG. Gap junctions in the inferior olivary nucleus of the squirrel monkey, Saimiri sciureus. Brain Res 1977; 128:374-378.

106. Sotelo C, Gotow T, Wassef M. Localization of glutamic-acid decarboxylase immunoreactive axon terminals in the inferior olive of the rat, with emphasis on the anatomical relations between GABAergic synapses and dendrodendritic gap junctions. J Comp Neurol 1986; 252:32-50.

107. Benardo LS, Foster RE. Oscillatory behaviour in inferior olive neurons: Mechanism, modulation, cell aggregates. Brain Res Bull 1986; 17:773-784.

108. Zeeuw d CL, Holstege JC, Ruigrok TJ et al. Ultrastructural study of the GABAergic, cerebellar, and mesodiencephalic innervation of the cat medial accessory olive: anterograde tracing combined with immunocytochemistry. J Comp Neurol 1989; 284:12-35.

109. Raviola E, Gilula NB. Gap junctions between photoreceptor cells in the vertebrate retina. Proc Natl Acad Sci USA 1973; 70:1677-1681.

110. Raviola E, Gilula NB. Intramembrane organization of specialized contacts in the outer plexiform layer of the retina. A freeze-fracture study in monkeys and rabbits. J Cell Biol 1975; 65:192-222.

111. Kolb H. The inner plexiform layer in the retina of the cat: electron microscopic observations. J Neurocytol 1979; 8:295-329.

112. Nishimura Y, Smith RL, Shimai K. Junction-like structure appearing at apposing membranes in the double cone of chick retina. Cell Tissue Res 1981; 218:113-6.

113. Nishimura Y, Rakic P. Development of the rhesus monkey retina. I. Emergence of the inner plexiform layer and its synapses. J Comp Neurol 1985; 241:420-434.

114. Dacheux RF, Raviola E. The rod pathway in the rabbit retina: A depolarizing bipolar amacrine cell. J Neurosci 1986; 6:331-345.

115. Dowling JE. Retinal neuromodulation: The role of dapamine. Vis Neurosci 1991; 7:87-97.

116. Vaney DI. The coupling pattern of axon-bearing horizontal cells in the mammalian retina. Proc R Soc Lond B 1993; 252: 93-101.

117. Hampson EC, Weiler R, Vaney DI. pH-gated dopaminergic modulation of horizontal cell gap junctions in mammalian retina. Proc R Soc Lond B Biol Sci 1994; 255: 67-72.

118. Smith DE, Moskowitz N. Ultrastructure of the layer IV of the primary auditory cortex of the squirel monkey. Neuroscience 1979; 4:349-359.

119. Peters A. Morphological correlates of epilepsy. In: Glaser GH, Penry JK, Woodbury eds. Antiepileptic drugs: Mechanisms of action. New York: Raven Press, 1980:21-48.

120. Gutnick MJ, Prince DA. Dye coupling and possible electrotonic coupling in the guinea pig neocortical slice. Science 1981; 211: 67-70.

121. Sloper JJ. Gap junctions between dendrites in the primate neocortex. Brain Res 1972; 44:641-6.

122. Sloper JJ, Powell TP. Gap junctions between dendrites and somata of neurons in the primate sensori-motor cortex. Proc R Soc Lond Biol 1978; 203:39-47.

123. Sotelo C, Gentschev T, Zamora AJ. Gap junctions in ventral cochlear nucleus of the rat: A possible new example of electronic junctions in the mammalian CNS. Neuroscience 1976; 1:5-7.

124. Wouterlood FG, Mugnaini E, Osen KK et al. Stellate neurons in rat dorsal cochlear nucleus studied with combined Golgi impragnation and electron microscopy: Synaptic connections and mutual gap junctions. J Neurocytol 1984; 13:639-664.

125. Grace AA, Bunney BS. Intracellular and extracellular electrophysiology of nigral dopaminergic neurons. 3. Evidence for electrotonic coupling. Neuroscience 1983; 10:333-348.

126. Cepeda C, Walsh JP, Peacock W et al. Dye-

coupling in human neocortical tissue resected from children with intractable epilepsy. Cerebral Cortex 1993; 3:95-107.

127. Andrew RD, MacVicar BA, Dudek FE et al. Dye transfer through gap junctions between neuroendocrine cells of rat hypothalamus. Science 1981; 211:1187-1189.

128. Belin V, Moos F. Paired recordings form suparoptic and paraventricular oxytocin cells in suckled rats: recriutment and synchronization. J Physiol 1986; 377:369-390.

129. Cobbett P, Smithson KG, Hatton GI. Dye-coupled magnocellular peptidergic neurons of the rat paraventricular nucleus show homotypic immuoreactivity. Neuroscience 1985; 16:885-895.

130. Cobbett P, Yang QZ, Hatton GI. Incidence of dye coupling among magnocellular paraventricular nucleus neurons in male rats is testesterone dependent. Brain Res Bull 1987; 18:365-360.

131. Hatton GI. The hypothalamic slice approach to neuroendocrinology. Q J Exp Physiol 1983; 68:483-489.

132. Hatton GI. Some well-kept hypothalamic secrets disclosed. Fed Proc 1983; 42: 2869-2874.

133. Hatton GI, Yang QZ, Cobbett P. Dye coupling among immunocytochemically identified neurons in the supraoptic nucleus: increased incidence in lactating rats. Neuroscience 1987; 21:923-930.

134. Renaud LP. Magnocellular neuroendocrine neurons: Update on intrinsic properties, synaptic inputs and neuropharmacology. Trends Neurosci 1987; 10:498-502.

135. Theodosis DT, Poulain DA, Vincent JD. Possible morphological basis for synchronization of neuronal firing in the rat supraoptic nucleus during lactation. Neuroscience 1981; 6:919-929.

136. Theodosis DT, Poulain DA. Evidence for structural plasticity in the supraoptic nucleus of the rat hypothalamus in relation to gestation and lactation. Neuroscience 1984; 11:183-193.

137. Andrew RD, Taylor CP, Snow RW et al. Coupling in rat hippocampal slices: Dye tranfer between CA1 pyramidal cells. Brain Res Bull 1981; 8:211-222.

138. Jefferys JRG, Haas HL. Synchronized bursting of CA1 hippocampal pyramidal cells in the absence of synaptic transmission. Science 1982; 300:448-450.

139. Knowles WD, Funch PG, Schwartzkroin PA. Electrotonic and dye coupling in hippocampal CA1 pyramidal cells in vitro. Neuroscience 1982; 7:1713-1722.

140. Kosaka T. Gap junctions between nonpyramidal cell dendrites in the rat hippocampus (CA1 and CA3 regions). Brain Res 1983; 271:157-161.

141. Kosaka T. Neuronal gap junctions in the polymorph layer of the rat dentate gyrus. Brain Res 1983; 277:347-351.

142. Schmalbruch H, Jahnsen H. Gap junctions on CA3 pyramidal cells of guinea pig hippocampus shown by freeze-fracture. Brain Res 1981; 217:175-178.

143. Taylor CP, Dudek FE. A physiological test for electrotonic coupling between CA1 pyramidal cells in rat hippocampal slices. Brain Res 1982; 235:351-357.

144. Rao G, Barnes CA, McNaughton BL. Intracellular fluorescent staining with carboxyfluorescein: a rapid and reliable method for quantifying dye-coupling in mammlian central nervous system. J Neurosci 1986; 16:251-263.

145. Barnes CA, Rao G, McNaughton BL. Increased electrotonic coupling in aged rat hippocampus: a possible mechanism for cellular excitability changes. J Comp Neurol 1987; 259:549-558.

146. O'Beirne M, Bulloch AGM, MacVicar BA. Dye and electrotonic coupling between cultured hippocampal neurons. Neurosci Lett 1987; 78:265-270.

147. Katsumaru H, Kosaka T, Heizmann CW et al. Gap junctions on GABAergic neurons containing the calcium-binding protein parvalbumin in the rat hippocampus (CA1 region). Exp Brain Res 1988; 72:363-370.

148. Nunez A, Garcia-Austt E, Buno W. In vivo electrophysiological analysis of lucifer yellow-coupled hippocampal neurons. Exptl Neurol 1990; 108:76-82.

149. Baimbridge KG, Peet MJ, Mclennan H et al. Bursting responses to current-evoked depolarozation in rat CA1 pyramidal neurons is correlated with lucifer yellow dye coupling but not with the presence of

calbindin-28K. Synapses 1991; 7:269-277.

150. Nelson PG. Interaction between spinal motoneurons of the cat. J Neurophysiol 1966; 29:275-294.

151. Werman R, Carlen PL. Unusual behavior of the Ia EPSP in cat spinal motoneurons. Brain Res 1976; 112:395-401.

152. Rall W, Burke RE, Smith TG et al. Dendritic location of synapses and possible mechanisms for the monosynaptic EPSP in motoneurons. J Neurophysiol 1967; 30: 1169-1193.

153. Gogan P, Gueritaud JP, Horcholle-Bossavit G et al. Direct excitatory interaction between spinal motoneurons of the cat. J Physiol 1977; 272:755-767.

154. Matsumoto A, Arnold AP, Zampighi GA et al. Androgenic regulation of gap junctions between motoneurons in the rat spinal cord. J Neurosci 1988; 8:4177-4183.

155. Matsumoto A, Arnold AP Micevych PE. Gap junctions between lateral spinal motoneurons in the rat. Brain Res 1989; 495:362-366.

156. Ziegelgansberger W, Reiter C. Interneuronal movement of procion yellow in cat spinal cord. Exp. Brain Res 1974; 20:537-530.

157. Conners BW, Bernardo LS, Prince DA. Coupling between neurons of the developing rat neocortex. J Neurosci 1983; 3:773-782.

158. Bergmann M, Surchev L. Freeze-etching study of intercellular junctions in the rat developing neural tube. Acta Anat 1989; 136:12-15.

159. Christie MJ, Williams JT, North RA. Electrical coupling synchronizes subthreshold activity in locus coeruleus neurons in vitro from neonatal rats. J Neurosci 1989; 11:3584-3589.

160. Fulton BP, Miledi R, Takahashi T. Eletrical synapses between motoneurons in the spinal cord of the newborn rat. Proc R Soc Lond B 1980; 208:115-120.

161. Arasaki K, Kudo N, Nakanishi T. Firing of spinal motoneurons due to eletrical interaction in the rat. Exp Brain Res 1984; 54:437-445.

162. Walton KD, Navarette R. Postnatal changes in motoneuron electrotonic coupling studied in the in vitro lumbar spinal cord. J Physiol 1991; 433:283-305.

163. Pannese E, Luciano L, Iurato S et al. Intercellular junctions and other membrane specializations in developing spinal ganglia: A freeze-fracture study. J Ultrastruct Res 1977; 60:169-180.

164. Bonner PH. Gap junctions in culture between chick embryo neurons and skeletal muscle myoblasts. Brain Res 1988; 38: 233-244.

165. Fischbach GD. Synapse formation between dissociated nerve and muscle cells in low density cell culture. Dev Biol 1972; 38:407-429.

166. Gotow T, Sotelo C. Postnatal development of the inferior olivary complex in the rat: IV. Synaptogenesis of GABAergic afferents, analyzed by glutamic acid decarboxylase immunocytochemistry. J Comp Neurol 1987; 263:526-552.

167. Fisher SK, Linberg KA. Intercellular junctions in the human embryonic retina. J Ultrastruc Res 1975; 51:69-78.

168. Matsumoto A, Arai Y, Urano A et al. Cellular localization of the gap junction protein mRNA in the neonatal rat brain. Neurosci Lett 1991; 124:225-228.

169. Vulkelic JL, Yamamoto T, Hertzberg EL et al. Depletion of connexin43-immunoreactivity in astrocytes after kanaic acid-induced lesions in rat brain. Neurosci Lett 130: 120-124.

170. Giaume C, Cordier J, Glowinski J. Endothelins inhibit junctional permeability in cultured mouse astrocytes. Europ J Neurosci 1992; 4:877-881.

CHAPTER 3

PHYSIOLOGICAL PROPERTIES OF GAP JUNCTION CHANNELS IN THE NERVOUS SYSTEM

David C. Spray

1. INTRODUCTION

Before the patch clamp technique was developed, large cells were desirable for the multiple microelectrode impalements that physiological studies of gap junctions required. For that reason, as well as the inherent interest in understanding electrotonic synapses, much of the initial characterization of gap junction function, biophysical and pharmacological experiments were performed on the gigantic coupled neurons of molluscs, the fused axons and large cell bodies of arthropod neurons and the identifiable neurons and glia within the segmental ganglia of the leech. A few vertebrate preparations also fulfilled these requirements, such as the giant Mauthner cells[1] and electromotor nuclei of certain teleosts,[2] Rohon-Beard cells in the frog spinal cord[3] and neurons within the inferior olivary nucleus of mammals,[4] but the tedious dissection required for these vertebrate studies led most investigators to pursue the simpler nervous systems of invertebrates. Moreover, 20 years ago a focus of neurobiology was to determine neural circuits in organisms with a limited number of neurons, each of which was identifiable from one animal to the next.[5] If we understood the wiring diagram and activity patterns underlying the brains and behaviors of these animals, it was widely believed, we would progress in large measure toward understanding the human mind.

What was learned about the function and physiological properties of electrotonic synapses during this period continues to guide studies on mammalian neurons in tissue culture and in tissue slices, preparations in

Gap Junctions in the Nervous System, edited by David C. Spray and Rolf Dermietzel.
© 1996 R.G. Landes Company.

which small cells have been made accessible by a new generation of recording techniques. A major concept that was developed in these invertebrate and lower vertebrate studies is that electrotonic synapses synchronize outputs from coupled cells and provide extraordinarily rapid impulse conduction between pre- and postsynaptic elements, properties that have conveyed evolutionary advantage in certain behavioral repertoires. For example, the rapid impulse propagation between axonal segments in the crayfish and the electrotonic synapses onto motoneurons of both crayfish and hatchetfish allow quick prey evasion by a flip of the tail or pectoral fin;[6,7] coupled presynaptic motoneurons can enable a voracious carnivorous mollusc to synchronously expand its pharynx, ingesting slowly moving prey;[8] and in fish endowed with electric organs the synchrony through electrotonic coupling can provide coordinated output to stun nearby unwary victims.[2] The selective advantage of electrotonic synapses is obvious in these organisms; however, it may not be justifiable to extend these functions to gap junctions in the mammalian nervous system, where junctional expression between adult neurons appears in most cases to be low. The function served by weak or modest coupling is thus the issue in neurons of the mammalian nervous system, and it is therefore possible that the more common role of electrotonic synapses in mammalian brains is in biasing cells toward synchronous activity (as in reinforcing the convergence of ganglion cell inputs onto tectal targets),[9] or even in mediating second messenger or metabolite exchange (see below).

For glial cells of the nervous system, transmission of electrical signals is almost certainly not the primary function of gap junction channels. Rather, these cells pass second messengers such as IP_3 and Ca^{2+} in regenerating waves that may have consequences in changing local neuronal ionic environment or in signaling glucose mobilization, and because of the tremendous volume afforded by their coupled cytoplasms, may offer a virtually unlimited buffering capacity for K^+ ions, neurotransmitters and metabolites. Furthermore, gap junctions may even function effectively as water channels, allowing locally induced volume changes to be buffered by the entire syncytial population of cells.

The nervous system expresses most of the 13 connexins that have been cloned and sequenced from rodents (see chapter 1). In this tissue, these connexins are expressed in overlapping cell populations and at various times during development (see chapter 16). This chapter is intended to provide an overview of the properties of the channels formed by the connexins found in nervous tissue, as have been revealed through expression of these connexins in exogenous systems. In so far as possible, these properties are then compared with the characteristics of gap junctions directly determined from recordings from neurons and glia, topics dealt with in more detail in other chapters.

2. WHY ARE THERE SO MANY TYPES OF GAP JUNCTIONS?

Three reasons to explain the diversity of gap junction channels expressed in mammalian tissues have been proposed:

A. *That different connexins have different affinities for one another*, either allowing communication between diverse cell populations or segregating these cells into isolated communication compartments. Cell coupling is readily established between many different types of cells when they are cultured together.[10,11] Although this is accounted for in part by the same connexin being expressed by different cell types (Cx43 being the most widespread in most cell types under culture conditions), functional coupling between cells expressing different gap junction proteins has been confirmed by *Xenopus* oocyte expression of individual connexin cRNAs and in mammalian cells stably transfected with vectors containing individual connexin cDNAs.[12,13] From studies conducted on both of these exogenous expression systems, the most important message that emerges for consideration of the connexins expressed in brain is that Cx33 and Cx31.1 do not form functional channels either with themselves or with other connexins (and Cx33 may even block oligomers of other connexins in which it is incorporated), and that heterologous pairings of Cx32 with Cx43, Cx43 with Cx40 and Cx50 with Cx43 are not functional, whereas most of the other connexin combinations are. From the standpoint of the nervous system, compartmental interfaces where such heterologous pairings might have functional implications are between Cx32-expressing oligodendrocytes or certain neurons and Cx43-expressing astrocytes or certain neurons (chapters 1, 2; ref. 14). Although Cx40 and Cx46 are known to be expressed in brain, their precise anatomical location is not yet defined, and their functional significance as constituents of compartmental boundaries cannot yet be appreciated.

B. *That different connexins are differently affected by transcriptional and post-transcriptional control mechanisms* such that hormonal and other stimuli may have more profound effects on coupling in some tissues than in others. Moreover, different promoters may be generated by alternate splicing within the large intron separating the two connexin exons (as has recently been demonstrated for Cx32 in the brain[15]), providing a mechanism whereby the same connexin may be differentially regulated in different tissue targets. Numerous drugs, growth factors and hormones have been shown to affect connexin mRNA or protein levels and/or functional coupling over a time course of hours to days.[16] Both transcriptional and mRNA stabilizing signals differ among the few connexin proteins where gene regulatory sequences have been mapped or mRNA stability has been measured,[17,18] presumably allowing differential regulation of the same connexin in different cells and different connexins in the same cellular environment. This diversity is presumed to be critically important in the longterm plasticity of connexin expression in brain and is touched on in various chapters elsewhere in this volume.

C. *That channels formed by different connexins have different functional properties and are differently gated,* thereby providing functional advantage for the expression of certain connexins in specific tissues. Gap junction channels open and close in response to various stimuli, including transjunctional voltage, intracellular pH, phosphorylation of the connexin molecules and exposure to any of a wide variety of lipophilic (or amphipathic) molecules.[19] From the standpoint of pathophysiology, each of these gating stimuli may play significant roles during neural ischemia: appreciable transjunctional voltage gradients may develop, intracellular pH is lowered, availability of substrates for phosphorylation is reduced and phosphatases may be activated.[20] Moreover, in ischemic tissue, uncoupling lipophiles are produced through lipid peroxidation and phospholipase activity.[21] It is now quite clear that gap junction channels composed of different connexins are differentially sensitive to transjunctional voltage, to intracellular pH and to effects of phosphorylating agents. Furthermore, it has been recently suggested that Cx40 and Cx43 form channels that are variably sensitive to certain lipophiles.[22]

Additionally, the size of gap junction channels formed by different connexins differs, when measured either as the conductances of channels made of individual connexins or by the size limit or charge selectivities of the channels to permeant ions and molecules.[19] Because unitary conductance of a channel determines its current-carrying capacity, a higher or lower unitary conductance might be functionally facilitating or limiting in synchronizing neuronal inputs or rapidity of conduction along a multicellular strand (as in vertebrate giant axons).

4. BIOPHYSICAL PROPERTIES OF GAP JUNCTION CHANNELS

A. PERMEABILITY

Gap junction channels are permeable to ions and even small molecules, and their permeability to fluorescein and lissamine rhodamine derivatives and Lucifer Yellow[23,24] indicates that the minimal pore diameter exceeds 1 nm. Junctional channels would thus be expected to accommodate most molecules with molecular weights below about 1000 Da. This size range includes many physiological second messenger molecules (such as cyclic nucleotides, inositol trisphosphate, ATP, ADP and Ca^{2+}) as well as current-carrying ions (of which K^+ is most abundant and most mobile; in excitable systems where currently carrying capacity is the junctional membrane's functional role, the gap junction channel may operationally be regarded as a K^+ channel).[25] Essential tissue metabolites and enzyme substrates are also gap junction permeant, as was demonstrated two decades ago in recovery of function ("kiss of life") and in loss of function ("kiss of death") assays.[26,27] More recently, this transfer of toxic metabolites has been proposed to explain the "bystander effect", whereby gancyclivor-induced cell death extends to cells coupled to those in which *Herpes simplex* thymidine kinase is introduced.[28]

Ca^{2+} is also gap junction permeant, although the critical functions that could be fulfilled by intercellular Ca^{2+} diffusion were entirely overlooked until recently, due to the belief that Ca^{2+} closed junctional channels. Studies on *Obelia*, a hydrozoan with eyespot support cells containing an endogenous photoprotein (obelin) provided the first indication for junctional Ca^{2+} permeability, since these cells glowed in response to photic stimulation that elevated Ca^+ in the photoreceptors to which they were coupled.[29] Ca^{2+} waves flowing between mammalian cells were found soon thereafter in studies where intracellular Ca^{2+} and IP_3 injections revealed that both molecules could diffuse across the junctional membrane.[30] Gap junctions were strongly implicated in this cell to cell diffusion by the findings that the transfer was unaffected by reducing extracellular Ca^{2+} and was blocked by the uncoupling agent heptanol. Subsequent studies on astrocytes, tracheal epithelial cells, corpus cavernosum smooth muscle and aortic endothelial cells have now reported that evocation of intracellular Ca^{2+} elevations (by pharmacological and mechanical stimuli as well as by direct intracellular Ca^{2+} injection) can lead to propagated "Ca^{2+} waves," which are apparently due to regenerative Ca^{2+} or IP_3 induced calcium release triggered by transfer of these second messenger molecules through junctional channels.[31-34] In brain slices as in these other preparations, these waves cross long distances, with velocities on the order of 10-25 μm/sec. Interestingly, this velocity is similar to that of the spread of refractoriness through brain tissue ("spreading depression") resulting from strong mechanical or electrical stimulation (e.g., chapters 17, 19; ref. 35).

Until relatively recently, Lucifer Yellow was the dye of choice for demonstrating functional gap junction connections between cells. This molecule was synthesized by Walter Stewart for this purpose, and combined the desirable properties of high quantal yield (fluorescence output in response to UV irradiation), retention inside cells, and fixability with aldehydes.[24] The few drawbacks to its use (such as insolubility in high K^+ solutions, variable uptake into nuclei, toxicity under certain conditions) were either tolerated or ignored because there was no better tracer available. However, the recent applications of biotin and neurobiotin (see chapter 4) to studies of coupling in the brain and retina have strikingly demonstrated much more extensive intercellular connections than had been appreciated from studies using Lucifer Yellow in the same preparations. This more extensive spread of neurobiotin than Lucifer Yellow has generally been attributed to its smaller size, which would be expected to increase dye diffusion several fold compared to Lucifer Yellow. However, these new markers are positively charged, and thus may have greater diffusion through junctional channels favoring cations over anions (see discussion of ionic selectivity below).

One topic that recurs repeatedly in interpreting studies of junctional permeability involves the issue of whether diffusional exchange may rectify, so that molecules pass more readily in one direction than another.[36] Although rectification at equilibrium violates the Second Law of Thermodynamics,[37] geometric considerations can certainly result in asymmetry in dye coupling. Another issue is whether channel permeability may be reduced as cells uncouple, with only smaller molecules passing through partially closed channels. With regard to restricted permeability as channels close, recent evidence for conductance substates for most connexin channels might provide support for this possibility. However, substate and mainstate permeabilities of these channels may well be similar, as appears to be the case in Cx37 transfectants[38] and amphibian embryonic cells.[39]

B. Gap Junction Selectivity

Data on which to evaluate gap junction ionic selectivity are fewer than the extensive literature dealing with permeation by molecules of different sizes, being limited to comparisons of a few charged fluorescein derivatives and to only a few ionic substitutions within patch pi-pettes. With regard to large molecules, negatively charged dyes were initially observed to diffuse more rapidly than positively charged ones, and the diffusion rate was related to both the magnitude and polarity of the charge on the molecule.[11] More quantitative studies on channel permeability to fluorescein derivatives having different charges subsequently revealed that aminofluorescein was more restricted than were negatively charged compounds, leading to the hypothesis of a fixed negative charge within the gap junction channel.[40]

An alternate method for determining channel selectivity is to use electrophysiological methods, either to measure shifts in reversal potential under conditions of asymmetric internal solutions in the two cells or to measure unitary conductances under symmetrical conditions following ionic substitution. Although the range of ionic substitutions thus far employed in such analyses have been small, several studies have examined effects on currents through single gap junction channels. These studies have led to the remarkable discovery that the channels formed by many of the connexins favor the transfer of cations over anions. Thus, although Cx26, Cx32, Cx43 and Cx46 channels readily exchange both anions and cations, Cx37, Cx40, and Cx45 are much less anion-permeant.[41-43] This finding has important implications for the capacity of individual connexins for intercellular signaling. In the case of mice with ablated Cx43[44] and humans with modified Cx43,[45] the anion-impermeant junctional channels that continue to be expressed in heart apparently cannot functionally compensate, leading to profound developmental defects.

C. Unitary Conductances of Gap Junction Channels

Initial predictions considering the gap junction channel as a right cylindrical pore filled with physiological salt solution were that unitary conductances should be on the order of 100 pS[46] (paradoxically, however, the volume of such a cylinder is on the order of 10^{-25} to 10^{-26} l, so that the probability of a single ion occupying the channel at any time is less than 1). Studies of unitary conductances of gap junction channels in pairs of voltage clamped cells bore out the expectation of 100 pS channels, although the large range of unitary conductances for different connexins has been somewhat surprising. The single channel conductance (γ_j) values obtained for gap junctions in various cell types and after transfection with wild type and mutant connexins are summarized in Table 3.1. These values span the range from 30 pS (Cx45 and substate of Cx43 channels) to more than 300 pS (Cx37 and channels between blastomeres in early sea urchin embryos). One conclusion from this table is that each connexin type has a characteristic set of unitary conductance values and large conductance channels such as Cx37 may be less permeant to Lucifer Yellow than a small channel such as Cx46. Unitary conductance is thus a connexin-specific property, even though a given connexin may exhibit multiple unitary conductance values. In principle, this specificity may allow identification of the connexin types forming active channels between cells purely from electrophysiological recordings, in much the same way as connexin-specific antibody probes identify connexin proteins in cell membranes or Northern blots with connexin-specific oligonucleotide sequences are used to determine connexin mRNAs. In practice, however, such identification is often ambiguous, due to the coexpression of more than one connexin type in most cells (and the possibility of heterotypic as well as heteromeric oligomerization), the variety of conductance states for some of the connexin channels, and the similarities of unitary conductances for some of the channels.

The significance for tissue function of the distinctive unitary conductances (γ_j) of the different connexin channels is not clear. However, for the role of gap junction channels in mediating current flow, γ_j will govern how many channels are required, when matched with input resistance of the cells. Thus, in cells such as neurons and other small mammalian cells where input resistance may be exceptionally high (even >10 GOhms), a single open gap junction channel with $\gamma_j = 100$ pS may provide coupling coefficients of 0.5 or higher, and thus at least 50% of the steady state current generated in one cell will be transferred across the junctional membrane (transient events are more complicated, due to the time constant of the junctional membrane[47]).

4. GATING PROPERTIES OF GAP JUNCTION CHANNELS

A. COMMENTS ON UNCOUPLING

Electrophysiological evaluation of coupling strength has often relied on measurements of coupling coefficient (k), the ratio of voltage (V) recorded in the postsynaptic to that in the presynaptic cell ($k = V_2/V_1$). Because these voltages depend on both junctional and nonjunctional conductances (in nonisopotential cells, the latter gener-

Table 3.1. Unitary conductances and voltage sensitivities of mammalian gap junction channels

| | γ_j (substate) | P_a/P_c* | Boltzmann parameters of voltage sensitivity | | |
			A(n)	V_0	g_{min}/g_{max}
Cx32	125 pS	0.5	0.07 (1.7)	27 mV	0.1
Cx37	300 (70) pS	2	0.09 (2.1)	30 mV	0.25
Cx40	180 (50) pS	2	0.16 (4)	40 mV	0.2
Cx43	60,90 (30) pS	1	0.08 (1.9)	55 mV	0.4
Cx45	30 (10) pS	0.3	0.12 (2.9)	14 mV	0.06
Cx46	40 pS	?	0.10 (2.4)	17 mV	0

* Relative permeability to cations compared to ions

ally including shunts between cell body and junctional membrane), cells may be quite well coupled by only a few junctional channels or may be totally uncoupled by nonjunctional effects.

One example of a behavioral repertoire where a neural circuit makes use of coupling strength changes due to nonjunctional effects is in the feeding behavior of the opisthobranch mollusc *Navanax* (Fig. 3.1). This animal follows spoors of other molluscan prey and upon contact explosively expands its muscular pharyngeal cavity, sucking prey inside. The same muscular mass that expanded synchronously to engulf the prey subsequently expands and contracts regionally, peristaltically moving the prey into the digestive portion of the gut. This entire behavioral sequence appears to be mediated by a coupled pool of motoneurons and a synaptically connected pool of pharyngeal stretch receptors with somata located in the buccal ganglia.[48-50] Sensory inputs from the mouth excite the coupled motoneurons, causing the radial muscles that they innervate to contract, enlarging the buccal cavity. Subsequently, inhibitory sensory input onto the same motoneurons from pharyngeal proprioceptors shunts current flow so that the cells can fire asynchronously, participating with circumferential motoneurons in generating peristalsis, sweeping prey into the gut.[51] An analogous circuit exists in mammalian inferior olive, presumably allowing modulation of neuronal recruitment and frequency of the rhythmic discharge that characterizes activity patterns in the olivary nucleus.[4]

In addition to effects on nonjunctional membranes, coupling strength can also be modulated by stimuli that act specifically on junctional conductance, as has been directly demonstrated by recordings between voltage clamped cell pairs, where high resolution measurements demonstrated that gap junction channels continuously open and close.[46] Total junctional conductance is the product of a number of channels, unitary conductance, and fraction of time that each channel is open. Although it would be of immense interest to quantify effects on

these parameters for each of the types of stimuli that have been shown to affect junctional membranes, thus far only a few studies have succeeded in dissociating the mechanisms of uncoupling at the single channel level. The technical problem to be overcome in such studies is the large number of junctional channels typically connecting cell pairs. Ways around this difficulty have been to attempt to seal onto junctional regions in freshly dissociated cells, thereby isolating only a few channels or hemichannels in the patch,[53,54] or to choose poorly coupled cells, either freshly paired,[55] or cell types in which total expression is either normally low[56] or experimentally manipulated.[57]

B. pH AND CA

In terms of number of studies, and the generation of controversy, the agents whose effects on junctional conductance have been studied the most are the ions H^+ and Ca^{2+}. Numerous publications by Loewenstein's group, primarily in insect salivary gland cells (although also performed in amphibian embryonic and mammalian cells in tissue culture) indicated that lowering extracellular Ca^{2+} or raising intracellular Ca^{2+} led to uncoupling, with most estimates of effective intracellular concentrations being in the range above $10\,\mu M$.[58] Studies on cardiac myocytes, molluscan neurons and crayfish septate axon have been interpreted as supporting the generality of the Ca^{2+} effect,[59-61] and an hypothesis of calmodulin involvement has been advanced as an explanation.[62] Despite this abundant literature, however, the ready passage of Ca^{2+} waves throughout coupled networks provides clear evidence that levels of Ca^{2+} sufficient for intercellular signaling do not block coupling completely, as does the maintenance of conduction through contracting cardiac and smooth muscle tissue, where Ca^{2+} concentrations are physiologically elevated to high levels.

The data sets may be reconciled by the hypothesis that it is not Ca^{2+} that is causing the uncoupling that results from Ca^{2+} injection, but secondary effects on other processes to which the junctional membrane is sensitive. For example, in the initial studies on insect cells, Ca^{2+} injection and manipulations of extracellular Ca^{2+} may have depolarized the cells and thus reduced junctional conductance due to the sensitivity of arthropod gap junction channels to depolarization, a phenomenon discussed in more detail below. Such an explanation would be consistent with the restoration of coupling by hyperpolarization,[63] the relative insensitivity of mammalian cells to comparable treatments, and might also explain why such high levels of Ca^{2+} are necessary for the reported effects in the salivary gland cells. In the case of vertebrate cells, where junctions are in most cases relatively insensitive to cell resting potential, Ca^{2+} may exert its effect through intercellular acidification[64,65] or lipophile generation through phospholipase effects (see below). In some cell types, the simultaneous measurement of intracellular pH (pH_i) and junctional conductance (g_j) has revealed that the pH_i-g_j relation followed a simple titration curve,[65-67] and it has been suggested that the parameters defining pH sensitivity may be connexin-specific.[66,68]

Gating by H^+ (and Ca^{2+}) apparently involves transitions between only the fully conducting and fully closed states of the channel, as is also the case for uncoupling by lipophiles (see below). In pairs of cardiac myocytes uncoupled by CO_2 exposure, open times of channels became very short, while channel size was unchanged, which was interpreted as reflecting a change in channel open probability.[46] Regardless of whether the substantial changes in intracellular concentration of these ions necessary to uncouple are achieved physiologically, unraveling the mechanism of action will likely extend our understanding of how these channels are regulated at the level of channel structure. For example, recent site-directed mutagenesis studies suggest that the hinge (or cytoplasmic loop) regions of Cx32 and Cx43 may interact with the cytoplasmic tail in a ball and chain mechanism.[67,69]

C. LIPOPHILIC MOLECULES

Lipophilic molecules provide another set of compounds that affect whether gap junction channels are fully conducting or completely closed, and it is conceivable that therapeutically useful lipophiles may eventually be found. The generation of oleic acid and arachidonic acids, acylcarnitines and other lipophilic compounds under ischemic conditions is a possible example of naturally occurring, albeit pathological, action. This group of agents also includes the general anesthetics halothane and enfluorane, certain alcohols (heptanol and octanol are most potent, hexanol is not) and can be used for control for some nonspecific alcohol effects,[70] and the doxyl stearate spin resonance probes.[71] Although arachidonic acid acts in part through generation of lipoxygenase products and in astrocytes through amantadine metabolites,[72] arachidonic acid may also exert an uncoupling effect by itself.[73]

Several modes of action of the lipophiles have been proposed. Based on studies with heptanol, uncoupling has been correlated with fluidity changes in cholesterol-rich membrane domains.[75,76] Such studies reinforce the notion that the lipid environment of the gap junction may be unusual, consistent with older studies indicating that junctional membranes were particularly rich in cholesterol (as judged by filipin incorporation into junctional complexes,[77] and by studies quantifying lipid composition in isolated junctional structures.[78,79] Alternatively, lipophiles may act by intercalating into hydrophobic domains of connexin molecules, or at the lipid-protein interface surrounding the channel.[68] Because connexins are fatty acylated,[80] it is even possible that the lipophilic uncoupling agents destabilize the anchoring function of the myristolate or palmitate linked to the connexin molecule.

A few studies have begun to address the action of lipophiles at the single channel level. Heptanol, octanol, halothane, arachidonic acid, oleic acid, doxyl stearic acid all apparently act on a variety of cell types without inducing substate activity.

One reason that these agents are so commonly used to reduce junctional conductance in order to record activity of single channels is that the channel open times in partially uncoupled cells appear to be longer than those obtained with CO_2 uncoupling.[81] Thus, γ_j can be measured easily at even the modest time resolution of strip chart recordings. Although this channel behavior in the presence of lipophiles has been commented upon only phenomenologically,[76] it suggests the possibility that channel closed time is more effectively reduced than open time.

D. VOLTAGE SENSITIVITY OF GAP JUNCTION CHANNELS

Voltage sensitivity of gap junction channels was initially viewed as a curiosity confined to certain "rectifying" gap junctions in the nervous system that are involved particularly in escape behaviors of vertebrates and invertebrates.[6,82] However, the discovery of voltage dependence in blastomeres from early amphibian embryos[83] and subsequent analysis of the steady state and kinetic properties[84,85] made it clear that this gating property of junctional channels was akin to voltage

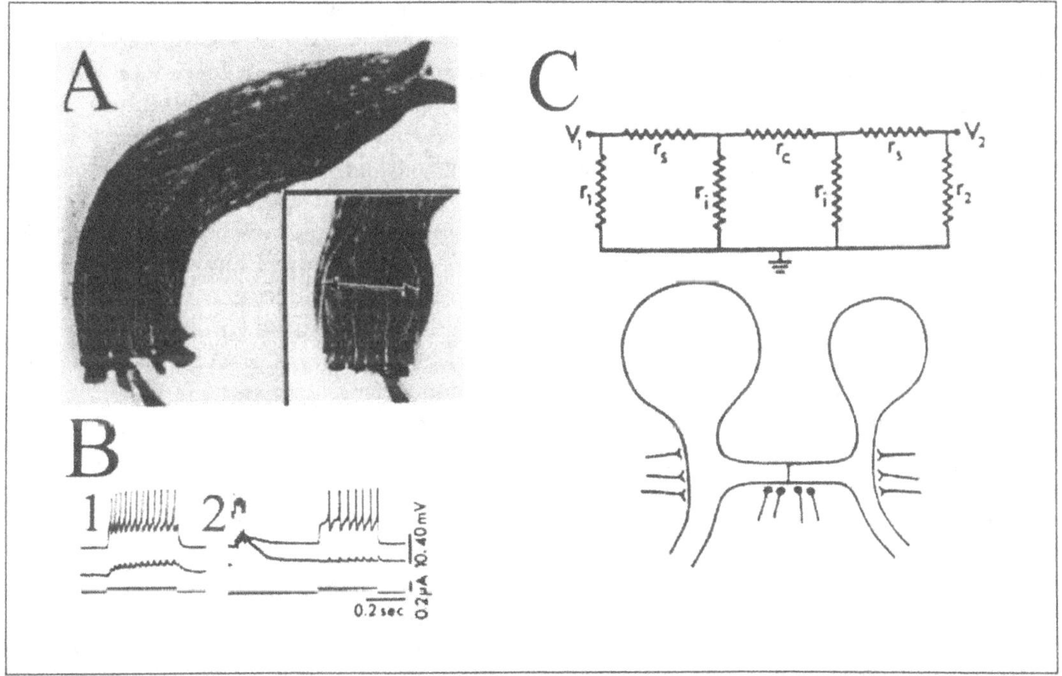

Fig. 3.1. Neural control of feeding in the opisthobranch mollusc Navanax. After contact of prey with receptors in the lips, the pharynx rapidly expands (inset, arrows indicating enlargement of the pharyngeal cavity), sucking prey inside. (B) The motoneurons responsible for pharyngeal expansion are located in the buccal ganglion and are electrically coupled so that depolarization spreads from one cell to another (1). In response to pharyngeal expansion (simulated by electrical stimulation of the pharyngeal nerve in 2), the cells uncouple, attenuating the electrotonic spread of currents. This uncoupling allows expansion motoneurons to fire out of phase with one another, enabling peristalsis of prey from pharynx to esophagus. (C) Equivalent circuit and schematic of coupling and uncoupling circuit for two expression motoneurons. Cells are uncoupled by coupling resistance (r_s) located at a distance from the soma (represented in the circuit by the access resistances r_s), and the coupling is shunted by inhibitory synaptic inputs (represented by variable resistances and by closed synaptic endings near the site of coupling in the lower schematic). Cells are excited by depolarizating electrical current delivered at the somas (V_1, V_2) or by excitatory chemical inputs from oral receptors (Y-shaped inputs near the somas in the lower diagram).

sensitivity of other membrane channels. Subsequent studies on a variety of cell types and on connexins exogenously expressed in *Xenopus* oocytes and in mammalian cells have clearly established the point that all mammalian gap junction channels are voltage sensitive, with the degree of voltage sensitivity a feature that distinguishes one connexin from another.

Because each gap junction channel spans two membranes, with hemichannels docking across extracellular space, it presents two distinct regions for voltage drops to be sensed by the channel proteins. One of these is *transjunctional*, such that the potential drops from inside one cell to inside the other cell, along the axis of the channel without influence of the extracellular field. Such a drop will only occur when a potential difference is present between the two cells (when one cell is depolarized relative to the other). The other possible potential that can be sensed is from the inside to the outside of the cell, and was initially referred to as *inside-outside* voltage dependence (V_{i-o})[86] to distinguish it from transjunctional (V_j) voltage sensitivity. Although V_{i-o} sensitivity is prominent in arthropod junctions,[86-90] where depolarization readily uncouples cell pairs, V_{i-o} sensitivity is only weakly present in certain mammalian connexins expressed in oocytes,[91] and has not been shown to exert a major influence on junctional conductance in mammalian cells.

The initial demonstration of gating by V_j in amphibian embryos described several features which are now known to be generalizable to most mammalian connexins. These studies introduced the dual voltage clamp technique,[83] where each cell is voltage clamped to the same holding potential. Then, as one cell is stepped to a command potential, the clamp on the other cell injects current (I_2) to hold its potential construct. This current is equivalent to current flowing through the junctional membrane (I_j), but is of opposite sign $(I_j = -I_2)$; when I_j is divided by V_j, junctional conductance is measured directly. For small and brief V_j pulses, most gap junctions exhibit linear I_j-V_j relations (the clearest exceptions being saturation behavior of rodent Cx37, ref. 38) and "fast" voltage sensitivity of Cx32-Cx26 pairings.[92] However, when increasing V_j of either polarity and long duration (>10 sec, typically) is applied to either cell of the pair, I_j relaxes during the pulse, reaching steady state values at the end of the pulse that depend on the magnitude of the V_j pulse. For the amphibian blastomeres (and for most of the connexins expressed either endogenously or exogenously in mammalian cells), the relaxations during the V_j pulses were found to be described well by single exponential declines, indicating that (for a single polarity of voltage) a two state kinetic process was involved in voltage dependent gating.[85] For amphibian blastomeres and most other cell types, when steady-state g_j is plotted as a function of V_j of both polarities, the decrease is symmetrical about the 0 mV axis (see Fig. 3.3). Even at high voltages, however, a detectable residual conductance remains, which was originally termed minimum conductance, g_{min}. This residual voltage-insensitive conductance component is apparently a connexin-specific gating characteristic, and the ratio of minimum conductance, g_{min}, to maximal conductance at the start of the pulse, g_{max}, varies from a high of about 40% in the case of cardiac myocytes and for Cx43 expressed in numerous other cell types or in transfected cells to a low of 0 in cultured rat Schwann cells,[93] which express Cx46.[94] After subtraction of the residual conductance, g_{min}, the data for V_j of each polarity can be fit by a form of the Boltzmann equation, where assumptions include a population of channels with a single open and a single closed state, where open probability is governed by the applied voltage.[83,84] Such fits determine V_0, the voltage at which the voltage sensitive component of g_j declines to half of its maximal value, and n, the equivalent number of gating charges; moreover, the amount of work on the channel necessary for the open-closed transition

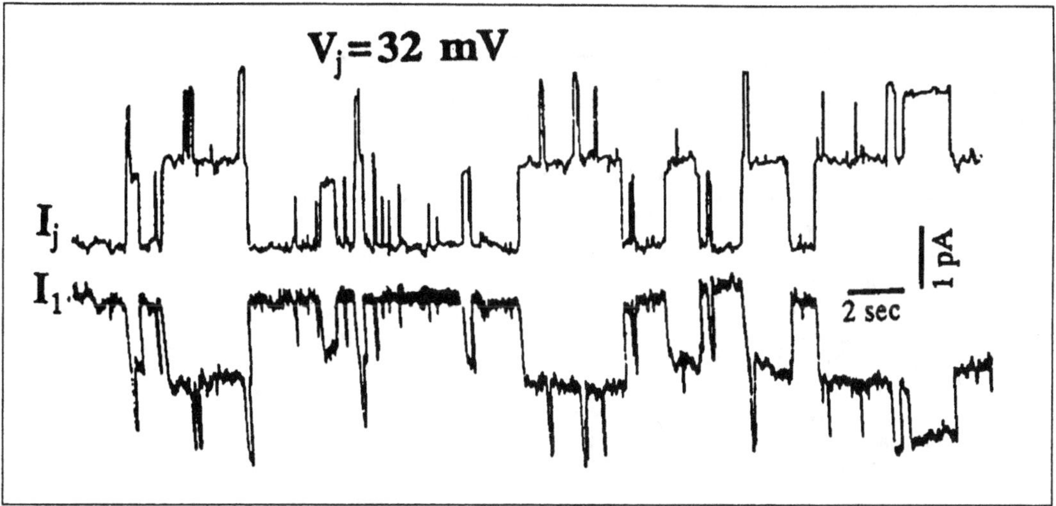

Fig. 3.2. Single channel recording from a pair of neonatal mouse astrocytes where junctional conductance had been reduced by applying 3 mM halothane. Lower trace current recording (I_1) is from cell in which a steady holding potential of -32 mV was applied. Note discrete equal sized but opposite polarity events, which correspond to openings and closures of single junctional channels. At the beginning of the recording, all channels are closed; divergence of tracings indicates channel opening.

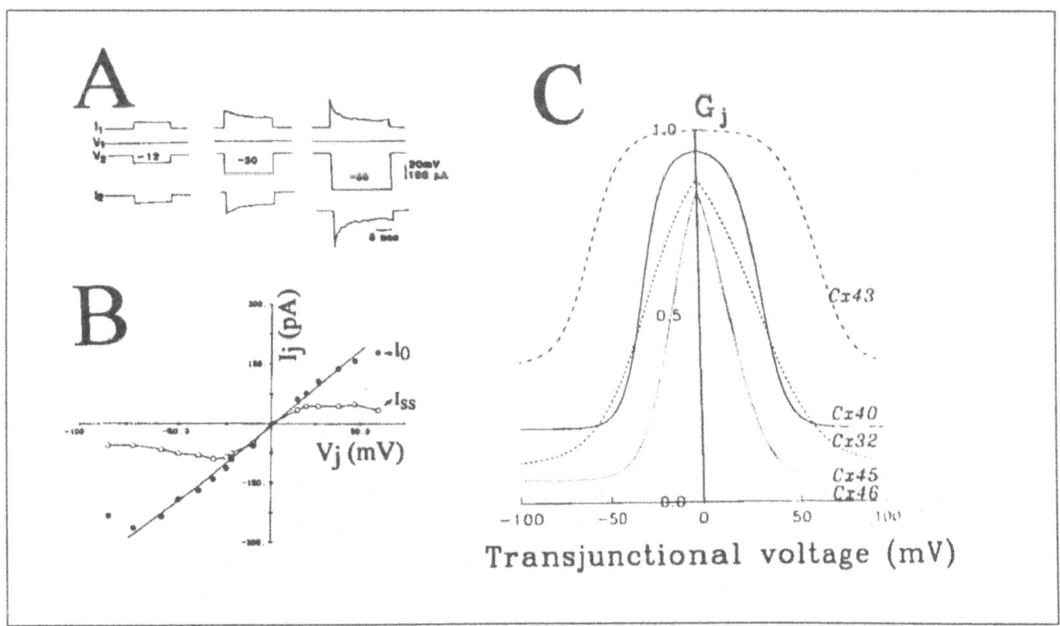

Fig. 3.3. Voltage sensitivity of junctional conductance for mammalian connexins. (A) In response to increasing amplitude to 20 sec voltage clamp pulses applied to the lower trace cell (I_2), current in the other cell's voltage clamp recording (I_1) relaxes to steady state levels that are lower for larger voltages. (B) For most connexins, including Cx32 in transfectants, initial current (I_o) is linear with voltage over most of the transjunctional voltage (V_j) range. Steady state current (I_{ss}) relaxation is seen as rectification that is symmetric about the 0 V_j axis. Normalized junctional conductance (G_j)V_j relations for each connexin shows a different voltage sensitivity, with Cx45 and Cx46 being most voltage sensitive and Cx43 the least.

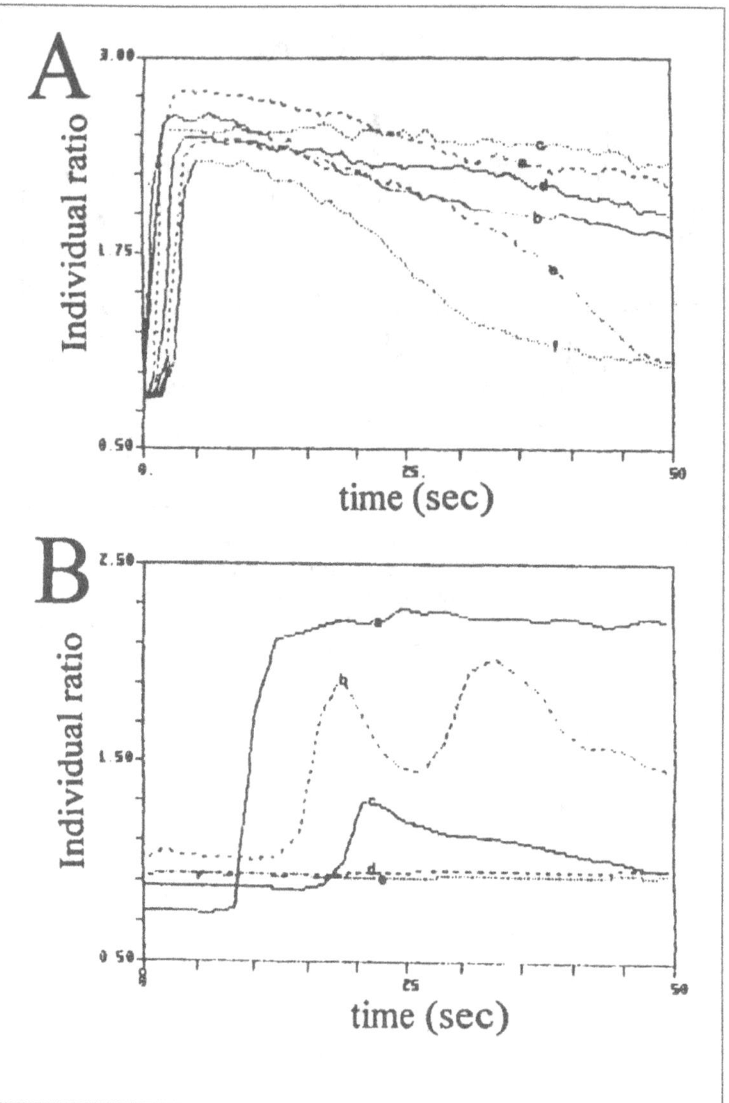

Fig. 3.4. Ca²⁺ wave transfer in confluent dishes of wildtype and Cx43 KO astrocytes. Recordings were obtained from Indo1AM-loaded cells using a Nikon RCM 8000 scanning confocal microscope. Traces in top panel are from six wildtype astrocytes chosen to fall at progressively larger distances from the stimulated cell and were acquired at 1 sec intervals. In response to a brief mechanical stimulus, Ca²⁺ signals spread from one cell to the next with a velocity of about 1 cell diameter/sec. Traces in the bottom panel were obtained from the field in which we observed the strongest coupling of all Cx43 KO astrocyte cultures examined (four litters). Images here showed less rapid spread, with many cells totally refractory to the Ca²⁺ wave entry. Moreover, the plot shows that the delay between Ca²⁺ wave spread in contiguous cells was on the order of 5 sec, much longer than the latency between invasion of neighboring wildtype astrocytes. Note also the presence of secondary Ca²⁺ spread into one of the regions, resembling the phenomenon of electrical re-entry seen in cardiac tissues under pathological conditions.

can be estimated from the product of these parameters.

Quantitative studies of voltage dependence have now been performed on gap junctions between numerous types of cells expressing most of the known connexins. In these studies, the decays in I_j with increasing V_j of one polarity have generally been found to be described well by single exponential kinetics. V_0 values have ranged from 15 mV or less to more than 60 mV and n has ranged from about 2 to 6 gating charges (Table 3.1).

Studies first performed in amphibian blastomeres (and subsequently observed in other systems) indicated that V_j dependent gating involved two gates in series.[84,85] When g_j was first reduced by a strong V_j pulse of one polarity and then V_j was reversed, g_j increased slowly and then decreased to the low conductance state. The kinetics of this process were consistent with those predicted for a three-state model in which a closed channel must first open before it can sense the potential of the other sign and close again. The contingent model also fits data well for Cx32[95] and Cx45 junctional channels,[96] but cardiac junctional channels are deviant,[97] perhaps due to the substantial contribution of the voltage-insensitive component of junctional conductance contributed by Cx43 in these cells.

Recent studies have clarified several issues regarding the polarity and single channel components of voltage sensitivity of g_j. First, acid injection into amphibian embryonic cells led to an asymmetric voltage sensitivity, revealing that the voltage sensor on each hemichannel responded to relative positivity of each cell.[98] Asymmetric pairings of Cx32 and Cx26 expressed in oocytes led to steady state g_j-V_j relations which were initially reported not to be the geometric sum of each hemichannel's properties as determined in the symmetric case;[99] however, it is now realized that the properties do match those predicted if gates are assumed to possess opposite polarities of voltage sensitivity.[100] Most recently, charge substitutions in the first extracellular loop and in the amino-terminal domain of Cx32 and Cx26 have reversed field sensitivity, suggesting a role of these domains in the voltage sensor.[101] In addition, studies at the single channel level have now revealed that g_{min} arises from a voltage-induced substate of the channel and that the voltage sensitivity described by the Boltzmann fits to macroscopic data is due to decreased open probability of the fully open state at higher voltages.[93,102] Gating by voltage is thus distinct from gating by low pH or by lipophiles, in that transjunctional voltage induces a partially conducting state, rather than completely closing the channel.

E. CONNEXIN PHOSPHORYLATION

Cx32 was one of the first membrane channel-forming proteins to be demonstrated to be a phosphoprotein[103] and the major site of phosphorylation in response to either cAMP-dependent kinase (PKA) or protein kinase C (PKC) was mapped to SER232 using synthetic peptides and matching these peptides to protease finger-printing of phosphorylated Cx32.[104] Stoichiometry of ^{32}P incorporation into Cx32 in response to cAMP is rather low, about 10%, although the stoichiometry of incorporation into the corresponding peptides is almost 100% with PKC. Whether the low stoichiometry in vivo represents phosphorylation of only a limited number of connexin molecules within the connexon remains to be determined. The time course of phosphorylation matches that of increased junctional conductance in these cells in response to cAMP-elevating treatments; comparable physiological effect of PKC activation, though predicted from the phosphorylation remain to be demonstrated, as do whether Cx32 phosphorylation affects single channel properties or steady state or kinetic parameters of voltage sensitivity.

Cx26 is apparently not a phosphoprotein in vivo and is not phosphorylated by any of the attempted kinases.

Phosphorylation of Cx43 has received the greatest attention, primarily because of the finding that phosphorylation results in altered gel mobility that can be monitored on Western blots without the necessity of autoradiograms.[105] Of the numerous potential phosphorylation sites within the murine Cx43 sequence, there is moderately strong evidence that four are utilized: SER262, TYR286, SER364 and SER368. With regard to the tyrosine residue, David Paul's group demonstrated TYR phosphorylation in oocytes injected with the tyrosine kinase Src, which correlated with uncoupling.[106] Evidence that TYR286 was the residue involved was provided by mutation of this site to PHE, after which the protein was no longer phosphorylated and coupling was no longer responsive to the tyrosine kinase.

Phosphorylation of SER364 and SER368 on connexin43 also results in physiological effects on the junctional channels. In cardiac cells, junctional conductance increased in response to PKA or PKC activation and is inhibited by inhibitors of these kinases.[107,108] In studies using synthetic peptides, these serine residues were found to be phosphorylated by PKC, with stoichiometry of 1 mole phosphate per mole peptide per site.[109] Studies on human Cx43 transfectants indicate that phosphorylation state, presumably involving these same two SER residues, affects single channel conductance.[110] Treatment with okadaic acid, an inhibitor of phosphatase, results in phosphorylation of the protein and over a similar time course okadaic acid and activators of PKA and PKC lead to smaller unitary conductance values (60-70 vs 90-110 pS events measured in CsCl internal solution at moderate V_j's). Dephosphorylation by intracellularly applied phosphatase or the protein kinase inhibitor staurosporine increased unitary conductance. Interestingly, phosphatase treatment resulted in much faster decline in junctional current during voltage steps, although the steady state Boltzmann distribution was not strongly altered,[111] indicating that both opening and closing rate constants were affected to the same extent by dephosphorylation. Mutational analysis of the human Cx43 sequence supports the assignment of phosphorylation to these SER residues: substitution by ALA or truncation of the molecule to remove these phosphorylatable residues renders the channels insensitive to phosphorylating treatments.[112]

Evidence that SER 262 is phosphorylated in the rat sequence comes from studies in which cGMP stimulation increases phosphate incorporation into Cx43 from myocytes or from rat Cx43 transfectants, but does not affect incorporation into the human sequence, which does not contain this putative cGMP phosphorylation site.[108]

It remains to be determined whether gap junction proteins formed of other connexins will show the remarkable range of plasticity in response to phosphorylation that is exhibited by Cx43. However, comparison of c-terminal sequences within the α of Group II connexin family does indicate that the SER-rich carboxyl terminus is highly conserved suggesting that PKA and PKC phosphorylation sites physiological effects will be demonstrable.

5. GATING OF GAP JUNCTION CHANNELS IN NERVOUS TISSUE

A. ASTROCYTES FROM WILDTYPE AND Cx43 KO ANIMALS

Astrocytes make Cx43, and, as demonstrated in several chapters in this volume, they do so abundantly. It is therefore not surprising that two studies describing properties of gap junction channels in astrocytes reported unitary conductances in the same range as those now reported for Cx43 in transfectants, and that these channels displayed only modest sensitivity to transjunctional voltage.[112,113] Nevertheless, more recent studies with sensitive RT-PCR techniques have demonstrated that astrocytes also express Cx40, Cx45 and Cx46 (see chapter 2). The relatively low expression of these connexins presumably accounts for their distinctive unitary conductances having not been detected in previous studies.

However, the availability of animals in which Cx43 is deleted through homologous recombination[44] has provided a preparation in which the abundant background expression of this connexin is totally eliminated. As expected, junctional channels with unitary conductances and voltage sensitivity characteristic of Cx43 are absent from astrocytes prepared from these animals. The channels that are present display quite different properties,[114] being very voltage sensitive (V_0 about 15 mV) and exhibiting low unitary conductances (about 30 and 40 pS). These unitary conductances are consistent with presence of Cx45 and Cx46, as detected using RT-PCR.

Consistent with the substantially reduced macroscopic junctional conductance recorded in Cx43 KO astrocytes compared to wildtype siblings, Lucifer Yellow transfer between Cx43 KO astrocytes is virtually never observed.[114,115] Although Ca^{2+} waves are sometimes seen between these cells (and might be attributable to extracellular release of ATP or other active substance: see chapter 11), velocity of propagation is severely attenuated compared to that between astrocytes of wildtype littermates. Curiously, large unitary conductances characteristic of Cx40 junctional channels have not been recorded in astrocytes from Cx43 knockout mice, indicating that the detected Cx40 mRNA may not be processed into functional channels in this tissue or that its expression is simply too low to play a significant role in astrocyte function.

B. OLIGODENDROCYTES

In vivo, oligodendrocytes express Cx32;[116] In vitro, oligodendrocytes form elaborate networks that are dye coupled, although such coupling is less extensive than for astrocytes.[117-119] Electrophysiological studies on pairs of cultured oligodendrocytes reveal that the properties of the junctional channels are not entirely as expected for Cx32 channels (R. Dermietzel and D.C. Spray, unpublished).[118] Coupling is totally absent in many cell pairs; when junctional channels are present, they are generally of lower conductance and are less voltage sensitive than those described in mammalian cells transfected with Cx32. It remains to be clarified whether the different electrophysiological properties of oligodendrocyte junctional channels are simply due to attenuation of the currents, arising from the nonisopotentiality due to long processes that generally separate these cells or whether the channels recorded are from an additional connexin type that is expressed in these cells but has not yet recognized using biochemical and molecular techniques (see chapter 2).

C. NEURONS, NEURON-LIKE CELLS, AND NEURONAL PRECURSOR CELLS

As is discussed in chapter 16, the electrophysiological identification of neuronal connexins under voltage clamp conditions has been hampered by the lack of coupling observed between neurons in tissue culture. Although certain types of neurons can be manipulated to increase their incidence of coupling by exposure to specific sets of hormones and growth factors,[14,119] such studies have not yet definitively revealed the expected functional or biochemical expression of specific connexins in specific types of neurons. Prior to differentiation, hippocampal neuroblasts express Cx43 and, as summarized in chapter 16, Cx43 expression decreases during differentiation. Interestingly, in partially differentiated neural progenitor cells, large junctional channels have been detected with properties similar to those expected for Cx40, although this expression has not been confirmed by use of antibody or oligonucleotide probes. Thus, the critical issues of which connexin connects neurons and how this expression may change during development remain totally unresolved.

D. SCHWANN CELLS

Schwann cells in vivo express Cx32,[120] and as summarized in Chapter 12 mutations in the coding region of Cx32 are associated with most families of patients suffering from the X-linked form of Charcot-Marie-Tooth disease. The expression pattern within the myelinating Schwann cell, however, appears to reflect the function of providing communication between the paranodal loops of a single Schwann cell, rather than between Schwann cells, which come into close apposition at nodes of Ranvier.[119,120] In response to nerve injury, however, and in tissue culture, previously meylinating Schwann cells dedifferentiate and express a novel connexin, which as summarized in chapter 14 appears to be Cx46.

Studies in tissue culture have revealed that Cx46 in Schwann cells forms small channels (about 40 pS), which exhibit the steepest voltage dependence of any connexin channel yet studied.[93] These channels are permeable to Lucifer Yellow and the diffusion of second messengers between these cells is hypothesized to coordinate Schwann cell proliferation and migration in response to axonal regeneration (see chapter 14).

6. CONCLUSIONS

In conclusion, the functional importance of gap junctions in the nervous system lies in their behavior as channels. This chapter summarizes what we now know about the properties of gap junction channels formed of the individual connexin proteins from the perspectives of what goes through the channel (and consequent relevance for nervous system function), how the channels open and close, and what physiological and pharmacological stimuli modify whether these channels are in the conducting or nonconducting states. These studies reveal that gap junction family of channel forming proteins is endowed with a remarkable range of functional plasticity.

Preparations and techniques for evaluating properties and functions of gap junctions in neural tissue have now evolved to the point where in vivo phenotypes characteristic of most of the cell types can be attained in tissue culture, where ontogeny of defined cell types is accessible to patch electrodes, where sophisticated imaging and electrophysiological techniques can be applied to both brain slices and in vivo brain, and where molecular genetics is providing organisms where gap junction properties may be inferred from the animal's behavior in their absence. The stage is now set for studies that will determine how the properties of the individual connexins are matched with their expression patterns in brain, how plasticity in expression and function govern developmental and adult patterns of neural activity, and exactly what functions are served by these channels in the brain and peripheral nervous system and how these functions are altered under pathological conditions.

ACKNOWLEDGMENTS

Supported in part by NIH grants NS07512 and NS34931 and by a grant from the Muscular Dystrophy Association. A NATO traveling grant facilitated the exchange of personnel between our laboratory and that of R. Dermietzel.

REFERENCES

1. Pereda AE, Nairn AC, Wolszon LR et al. Postsynaptic modulation of synaptic efficacy at mixed synapses on the Mauthner Cell. J Neurosci 1994; 14:3704-3712.
2. Bennett MV, Pappas GD, Aljure E et al. Physiology and ultrastructure of electrotonic junctions. II. Spinal and medullary electromotor nuclei in mormyrid fish. J Neurophys 1967; 30:180-208.
3. Spitzer NC. Voltage-and stage-dependent uncoupling of Rohon-Beard neurones during embryonic development of Xenopus tadpoles. J Physiol 1982; 330:145-162.
4. Llinas R, Baker R, Sotelo C. Electrotonic coupling between neurons in cat inferior olive. J Neurophys 1974; 37:560-571.
5. Kennedy D. Comparative strategies in the investigation of neural networks. J Exp Zool 1975; 194:35-49.
6. Auerbach AA, Bennett MVL. A rectifying electrotonic synapse in the central nervous system of a vertebrate. J Gen Physiol 1969; 53:211-237.
7. Zucker RS, Kennedy D, Selverston AI. Neuronal circuit mediating escape responses in crayfish. Science 1971; 173:645-650.
8. Susswein AJ, Achituv Y, Cappell MS et al. Pharyngeal movements during feeding sequences of Navanax inermis (Gastropoda: Opisthobranchia) in successive stages of dissection. J Exp Biol 1987; 128:323-333.
9. Penn AA, Wong RO, Shatz CJ. Neuronal coupling in the developing mammalian retina. J Neurosci 1994; 14:3805-3815.
10. Michalke W, Loewenstein WR. Communication between cells of different type. Nature 1971; 232:121-122.
11. Flagg-Newton J, Simpson I, Loewenstein WR. Permeability of the cell-to-cell membrane channels in mammalian cell junction. Science 1979; 205:404-407.
12. Nicholson BJ, Suchyna T, Xu LX et al. Divergent properties of different connexins in *Xenopus* oocytes. In: Hall JE, Zampighi GA, Davis RM, eds. Progress in Cell Research, Vol. 3. Elsevier, 1993:3-13.
13. Elfgang C, Eckert R, Lichtemberg-Frate H et al. Specific permeability and selective formation of gap junction channels in

connexin-transfected HeLa cells. J Cell Biol 1995; 129:805-817.

14. Dermietzel R, Spray DC. Gap junctions in the brain: Where, what type, how many, and why? Trends in Neuroscience 1993; 16:185-192.

15. Neuhaus IM, Dahl G, Werner R. Use of alternate promoters for tissue-specific expression of the gene coding for connexin32. Gene 1995; 158:257-262.

16. Spray DC, Saez JC. Agents that regulate gap junctional conductance: Sites of action and specificities. In: Milman H, Elmore E, eds. Biochemical Mechanisms and Regulation of Intercellular Communication. Princeton, N.J.: Princeton Scientific Publishing Co, 1987:1-20.

17. Bai S, Spray DC, Burk R. Characterization of rat connexin32 gene regulatory elements. In: Hall JE, Zampighi GA, Davis RM, eds. Gap Junctions (Progress in Cell Research, Vol. 3. Amsterdam: Elsevier, 1993:291-297.

18. DeLeon JR, Buttrick PM, Fishman GI. Functional analysis of the connexin43 gene promoter in vivo and in vitro. J Mol & Cell Card 1994; 26:379-389.

19. Spray DC. Physiological and pharmacological regulation of gap junction channels. In: Citi S, ed. Molecular Mechanisms of Epithelial Cell Junctions: From Development to Disease. Austin, TX: R.G. Landes Co., 1994:195-215.

20. Siesjo BK, Katsura K. Ischemic brain damage: focus on lipids and lipid mediators. (Review). Adv Exp Med & Biol 1992; 318:41-56.

21. Yavin E, Kunievsky B, Bazan NG et al. Regulation of arachidonic acid metabolism in the perinatal brain during development and under ischemic stress (Review). Adv in Exp Med & Biol 1992; 318:315-323.

22. Hirschi KK, Minnich BN, Moore LK et al. Oleic acid differentially affects gap junction-mediated communication in heart and vascular smooth muscle cells. Am J Physiol 1993; 265:C1517-1526.

23. Schwarzmann G, Wiegandt H, Rose B et al. Diameter of the cell-to-cell junctional membrane channels as probed with neural molecules. Science 1981; 213:551-553.

24. Stewart WW. Functional connections between cells as revealed by dye-coupling with a highly fluorescent naphthalimide tracer. Cell 1978; 14:741-759.

25. Spray DC, Vink MJ. Cardiac gap junctions as K^+ (and Ca^{2+}) channels. In: Vereecke J, Verdonck F, van Bogaert P-P. Potassium Channels in Normal and Pathological Conditions. Leuven University Press, 1996 (in press).

26. Smith TA, Box KM, Hooper ML. Clones defective in metabolic cooperation selected from a pluripotent feeder-dependent mouse embryonal carcinoma cell line. Exptl Cell Res 1986; 167:106-118.

27. Martin W, Zempel G, Hulser D et al. Growth inhibition of oncogene-transformed rat fibroblasts by cocultured normal cells: relevance of metabolic cooperation mediated by gap junctions. Cancer Res 1991; 51:5348-5351.

28. Elshami AA, Saavedra A, Zhang H et al. Gap junctions play a role in the "bystander effect" of the herpes simplex virus thymidine kinase/gancylovir system in vitro. Gene Therapy 1996; 3:85-92.

29. Dunlap K, Takeda K, Brehm P. Activation of a calcium-dependent phosphoprotein by chemical signalling through gap junctions. Nature 1987; 325:60-62.

30. Saez JC, Connor JA, Spray DC et al. Hepatocyte gap junctions are permeable to the second messenger, inositol 1,4,5-triphosphate, and to calcium ions. Proc Natl Acad Sci USA 1981; 86:2708-2712.

31. Charles DC, Merrill JE, Dirksen ER et al. Intercellular signalling in glial cells: Calcium waves and oscillations in response to mechanical stimulation and glutamate. Neuron 1991; 6:983-992.

32. Sanderson MJ, Charles AC, Dirksen ER. Mechanical stimulation and intercellular communication increases intracellular Ca^{2+} in epithelial cells. Cell Regulation 1990; 1:585-596.

33. Nathanson M, Padfield P, O'Sullivan A et al. Mechanism of Ca^{2+} wave propagation in pancreatic acinar cells. J Biol Chem 1992; 267:18118-18121.

34. Cornell-Bell A, Finkbeiner S, Cooper M. Glutamate induces calcium waves in cultured astrocytes: Long-range glial signalling. Science 1990; 247:470-473.

35. Martins-Ferreira H, Ribeiro LJC. Biphasic effects of gap junctional uncoupling agents on the propagation of retinal spreading depression. Brazailian J Biol Med 1996, in press.

36. Robinson SR, Hampson EC, Munro MN et al. Unidirectional coupling of gap junctions between neuroglia. Science 1994; 262: 1072-1074.

37. Finkelstein A. Gap junctions and intercellular communications. Science 1994; 265:1017-1018.

38. Waltzmann M, Bai S, Spray DC. Stable transfection of a gap junction protein, connexin37 (Cx37) in a communication deficient cell line. Biophys J 1994; 66:A260.

39. Verselis V, White RL, Spray DC et al. Gap junctional conductance and permeability are linearly related. Science 1986; 234:461-464.

40. Brink PR, Dewey MM. Nexal membrane permeability to anions. J Gen Physiol 1978; 72:69-78.

41. Veenstra RD, Wang HZ, Beyer EC et al. Connexin37 forms high conductance gap junction channels with subconductance state activity and selective dye and ionic permeabilities. Biophys J 1994; 66: 1915-1928.

42. Veenstra RD, Wang HZ, Beyer EC. Selective dye and ionic permeability of gap junction channels formed by connexin45. Circ Res 1994; 75:483-490.

43. Veenstra RD, Wang HZ, Beblo DA et al. Selectivity of connexin-specific gap junctions does not correlate with channel conductance. Circ Research 1995; 77:1156-1165.

44. Reaume AG, de Sousa PA, Kulkarni S et al. Cardiac malformations in neonatal mice lacking connexin43. Science 1995; 267: 1831-1834.

45. Britz-Cunningham SJ, Shah MM, Zuppan CW et al. Mutations of the connexin43 gap junction gene in patients with heart malformations and defects of laterality. N Eng J Med 1995; 332:1323-1329.

46. Burt JM, Spray DC. Single channel events and gating behavior of the cardiac gap junction channel. Proc Natl Acad Sci USA 1988; 85:3431-3434.

47. Bennett MVL. Physiology of electrotonic junctions. Ann NY Acad Sci 1996; 137: 509-539.

48. Spira ME, Spray DC, Bennett MV. Synaptic organization of expansion motoneurons of Navanax inermis. Brain Res 1980; 195:241-269.

49. Spray DC, Spira ME, Bennett MV. Peripheral fields and branching patterns of buccal mechanosensory neurons in the opisthobranch mollusc, Navanax inermis. Brain Res 1980; 182:253-270.

50. Spray DC, Spira ME, Bennett MV. Synaptic connections of buccal mechanosensory neurons in the opisthobranch mollusc, Navanax inermis. Brain Res 1980; 182: 271-286.

51. Spira ME, Spray DC, Bennett MV. Electrotonic coupling: effective sign reversal by inhibitory neurons. Science 1976; 194: 1065-1067.

52. Neyton NJ, Trautmann A. Single-channel currents of an intercellular junction. Nature 1985; 317:331-335

53. Manivannan K, Ramanan SV, Mathias RT. Multichannel recordings from membranes which contain gap junctions. Biophys J 1992; 61:216-227.

54. Dahl G, Nonner W, Werner R. Attempts to define function of gap junction proteins with synthetic peptides. Biophys J 1994; 67:1816-1822.

55. Bukauskas FF, Weingart R. Multiple conductance states of newly formed single gap junction channels between insect cells. Pflugers Archiv Eur J Physiol 1993; 423:152-154.

56. Chanson M, Chandross K, Rook MB et al. Gating characteristics of a steeply voltage dependent gap junction channel in rat Schwann cells. J Gen Physiol 1993; 102:925-946.

57. Fishman GI, Gao Y, Hertzberg EL et al. Reversible intercellular coupling by regulated expression of a gap junction channel gene. Cell Adhesion and Communication 1995; 3:353-365.

58. Loewenstein WR. Junctional intercellular communication: the cell-to-cell membrane channel. Physiol Rev 1995; 61:829-913.

59. Noma A, Tsuboi N. Dependence of junctional conductance on proton, calcium and magnesium ions in cardiac paired cells of

guinea pig. J Physiol 1987; 382:193-211.

60. Arellano RO, Rivera A, Ramon F. Protein phosphorylation and hydrogen ions modulate calcium-induced closure of gap junction channels. Biophys J 1990; 57:363-367.

61. Baux G, Simonneau M, Tauc L et al. Uncoupling of electrotonic synapses by calcium. Proc Natl Acad Sci USA 1978; 75: 4577-4581.

62. Peracchia C. The calmodulin hypothesis- six years later. In: Hertzberg EL, Johnson RG, eds. Gap Junctions. New York: Alan R. Liss, 1988:267-282.

63. Socolar SJ, Politoff AL. Uncoupling cell junctions in glandular epithelium by depolarizing current. Science 1971; 172:492-494.

64. Turin L, Warner AE. Intracellular pH in early Xenopus embryos: its effect on current flow between blastomeres. J Physiol (Lond) 1980; 300:489-504.

65. Spray DC, Harris AL, Bennett MVL. Gap junctional conductance is a simple and sensitive function of intracellular pH. Science 1981; 211:712-715.

66. Campos de Carvalho A, Spray DC, Bennett, MVL. pH dependence of transmission at electrotonic synapses of the crayfish septate axon. Brain Research 1984; 321: 276-286.

67. Liu S, Taffet S, Stoner L et al. A structural basis for the unequal sensitivity of the major cardiac and liver gap junctions to intracellular acidification. The carboxyl tail length. Biophys J 1993; 64:1422-1433.

68. Spray DC, Burt JM. Structure-activity relations of the cardiac gap junction channel. Amer J Physiol 1990; 258:C195-C207.

69. Ek JF, Delmar M, Perzova R, Taffet SM. Role of histidine 95 on pH gating of the cardiac gap junction protein connexin43. Circ Res 1994; 74:1058-1064.

70. Johnston MF, Simon SA, Ramon F. Interaction of anaesthtics with electrical synapses. Nature 1980; 286:498-500.

71. Burt JM. Uncoupling of cardiac cells by doxyl stearic acids: specificity and mechanism of action. Am J Physiol 1989; 256:C913-C924.

72. Venance L, Piomelli D, Glowinski J et al. Inhibition by anandamide of gap junctions and intercellular calcium signalling in striatal astrocytes. Nature 1995; 376: 590-594.

73. Massey KD, Minnich BN, Burt JM. Arachidonic acid and lipoxygenase metabolites uncouple neonatal rat cardiac myocyte pairs. Am J Physiol 1992; 263:C494-C501.

74. Omitted in proofs.

75. Bastiaanse EM, Jongsma HJ, van der Laarse A et al. Heptanol-induced decrease in cardiac gap junctional conductance is mediated by a decrease in the fluidity of membranous cholesterol-rich domains. J Memb Biol 1993; 136:135-145.

76. Takens-Kwak BR, Jongsma HJ, Rook MB et al. Mechanism of heptanol-induced uncoupling of cardiac gap junctions; a perforated-patch clamp study. Am J Physiol 1992; 262,1531-1538.

77. Mazet F. Filipin and digitonin studies of membrane cholesterol in frog atrial fibers with unusual gap junction configurations. J Mol & Cell Card 1987; 19:1121-1128.

78. Spray DC, Saez JC, Brosius D et al. Isolated liver gap junctions: Gating of transjunctional currents is similar to that in intact pairs of hepatocytes. Proc Natl Acad Sci USA 1986; 83:5494-5497.

79. Evans WH. A biochemical dissection of the functional polarity of the plasma membrane of th hepatocyte. Biochim Biophys Acta 1980, 604:27-64.

80. Willecke K, Traub O, Look J et al. Different protein components contribute to the structure and function of hepatic gap junctions. Modern Cell Biol 1988; 7:41-52.

81. Burt JM, Spray DC. Volatile anesthetics reversibly reduce gap junctional conductance between cardiac myocytes. Circ Research 1989; 65:829-837.

82. Furshpan EJ, DD Potter. Transmission at the giant motor synapses of the crayfish. J Physiol 1959; 145:289-325.

83. Spray DC, Harris AL, Bennett MVL. Voltage dependence of junctional conductance in early amphibian embryos. Science 1979; 204:432-434.

84. Spray DC, Harris AL, Bennett MVL. Equilibrium properties of a voltage dependent junctional conductance. J Gen Physiol 1981; 77:75-94.

85. Harris AL, Spray DC, Bennett MVL. Kinetic properties of a voltage dependent junc-

tional conductance. J Gen Physiol 1981; 77:95-120.

86. Spray DC, White RL, Campos de Carvalho A et al. Gating of gap junctional conductance. Biophys J 1984; 45:219-230.

87. Chanson M, Roy C, Spray DC. Voltage-dependent gap junctional conductance in hepatopancreatic cells of *Procambarus clarkii*. Amer J Physiol 1994; 35:C569-C577.

88. Verselis VK, Bennett MVL, Bargiello TA. A voltage-dependent gap junction channel in Drosophilia melanogaster. Biophys J 1992; 59:114-122.

89. Obaid AL, Socolar SJ, Rose B. Cell-to-cell channels with two independently regulated gates in series: Analysis of junctional conductance modulation by membrane potential, calcium and pH. J Membr Biol 1983; 73,69-89.

90. Churchill D, Caveney S. Double whole-cell patch-clamp of gap junctions in insect epidermal cell pairs: single channel conductance, voltage dependence, and spontaneous uncoupling. Progress in Cell Research 1993; 3,239-245.

91. Bennett MVL, Rubin JB, Bargiello TA et al. Structure-function studies of voltage sensitivity of connexins, the family of gap junction forming proteins. Jap J Physiol 1993; 1:S301-310.

92. Bukauskas FF, Elfang C, Willecke K et al. Heterotypic gap junction channels (connexin26-connexin32) violate the paradigm of unitary conductance. Pflugers Archiv Eur J Phys 1995; 429:870-872.

93. Chanson M, Chandross K, Rook MB et al. Gating characteristics of a steeply voltage dependent gap junction channel in rat Schwann cells. J Gen Physiol 1993; 102:925-946.

94. Chandross KJ, Spray DC, Cohen RI et al. Cytokine regulation of Schwann cell phenotype, proliferation and gap junctional communication. J Cell Mol Neurosci 1996, in press..

95. Moreno AP, Eghbali B, Spray DC. Connexin32 gap junction channels in stably transfected cells. Equilibrium and kinetic properties. Biophys J 1991; 60: 1267-1277

96. Moreno AP, Laing JG, Beyer EC et al. Properties of gap junction channels formed of connexin45 endogenously expressed in human hepatoma (SKHep1) cells. Amer J Physiol 1995; 268:C356-C365.

97. Wang H-Z, Li J, Lemanski LF et al. Gating of mammalian cardiac gap junction channels by transjunctional voltage. Biophysical J 1992; 63:139-151.

98. Bennett MVL, Verselis V, White RL et al. Gap junctional conductance: Gating. In: Hertzberg EL, Johnson RG, eds. Gap Junctions. A.R.Liss, 1988:287-304.

99. Barrio LC, Suchyna T, Bargiello T et al. Gap junctions formed by connexins26 and 32 alone and in combination are differently affected by applied voltage. Proc Natl Acad Sci USA 1991; 88:8410-8414.

100. Rubin JB, Verselis VK, Bennett MV and Bargiello TA: Molecular analysis of voltage dependence of heterotypic gap junctions formed by connexins 26 and 32. Biophys J 1992; 62:183-93

101. Verselis VK, Ginter CS, Bargiello TA. Opposite voltage gating polarities of two closely related connexins. Nature 1994; 368:348-351.

102. Moreno AP, Rook MB, Fishman GI et al. Gap junction channels: Distinct voltage-sensitive and insensitive conductance states. Biophys J 1994; 67:113-119.

103. Saez JC, Spray DC, Mairn A et al. cAMP increases junctional conductance and stimulates phosphorylation of the 27 kDa principal gap junction polypeptide. Proc Natl Acad Sci USA 1986; 83:2473-2477.

104. Saez JC, Nairn AC, Spray DC. The main hepatocyte gap junction protein (MP27) is phosphorylated by cAMP and Ca^{2+} dependent kinases. Eur J Biochem 1990; 192: 263-273.

105. Musil LS, Beyer EC, Goodenough DA: Expression of the gap junction protein connexin43 in embryonic chick lens: molecular cloning, ultrastructural localization, and posttranslational phosphorylation. J Membr Biol 1990; 116:163-175

106. Swenson KI, Piwnica-Worms H, McNamee H et al. Tyrosine phosphorylation of the gap-junction protein connexin43 is required for the pp 60 v-*src*-induced inhibition of communication. Cell Regulation 1990; 1, 989-1002.

107. Burt JM, Spray DC. Inotropic agents modulate gap junctional conductance between cardiac myocytes. Amer J Physiol 1988; 254: H1206-H1210.
108. Kwak BR, Sáez JC, Wilders R et al. Effects of cGMP-dependent phosphorylation on rat and human connexin43 gap junction channels. Eur J Physiol 1995; 430:770-778.
109. Saez JC, Nairn A, Spray DC, Hertzberg EL. Rat connexin43: Regulation by phosphorylation in heart. In: Hall JE, Zampighi GA, Davis RM, eds. Gap Junctions. Progress in Cell Research, Vol. 3. Amsterdam: Elsevier, 1993:211-217.
110. Moreno AP, Saez JC, Fishman GI et al. Human connexin43 gap junction channels: Regulation of unitary conductances by phosphorylation. Circ Research 1994; 4: 1050-1057.
111. Moreno AP, Fishman GI, Spray DC. Phosphorylation shifts unitary conductance and modifies voltage dependent kinetics of human connexin43 gap junction channels. Biophys J 1992; 62:51-53.
112. Moore LK, Moreno AP, Fishman GI et al. Human connexin43 (HCx43) phosphorylation site mutants: Unitary conductance and voltage sensitivity of channels in stable transfectants. Biophys J 66:A260.
112a. Dermietzel R, Kessler JA, Hertzberg EL et al. Gap junctions between cultured astrocytes: Immunocytochemical, molecular and electrophysiological analysis. J Neuroscience 1991; 11:1421-1432.
113. Giaume C, Fromaget C, Aoumari AE et al. Gap junctions in cultured astrocytes, single-channel currents and characterization of channel-forming protein. Neuron 1991; 6,133-143.
114. Spray DC, Vieria D, El-Sabban M et al. Gap junction properties in astrocytes from connexin43 (Cx43) knockout (KO) mice. Soc Neurosci Abstr 1995; 21:563.
115. Bechberger JF, Naus CCG, Giaume C et al. Functional characterization of astrocytes deficient in connexin43. Soc Neurosci Abstr 1995; 21:563.
116. Dermietzel R, Traub O, Hwang TK et al. Differential expression of three gap junction proteins in developing and mature brain tissues. Proc Natl Acad Sci USA 1989; 86:10148-10152.
117. Kettenmann H, Ransom BR. Electrical coupling between astrocytes and between oligodendrocytes studied in mammalian cell cultures. Glia 1988; 64-73.
118. Ransom BR, Kettenmann H. Electrical coupling, without dye coupling, between mammalian astrocytes and oligodendrocytes in cell culture. Glia 1990; 3:258-266.
119. Spray DC, Dermietzel R. X-Linked Charcot-Marie-Tooth Syndrome and other possible gap junction diseases of the nervous system. Trends in Neuroscience 1995; 18:256-262.
120. Bergoffen J, Scherer SS, Wang S et al. Connexin mutations in X-linked Charcot-Marie-Tooth disease. Science 1993; 252: 2039-2042.

CHAPTER 4

GAP JUNCTIONS AS ELECTRICAL SYNAPSES

Michael V. L. Bennett

1. INTRODUCTION

A synapse can be defined as a specialized site of functional interaction between neurons. By this definition gap junctions form one class of electrical synapse.[1] There is another kind of electrical synapse that mediates short latency inhibition of the Mauthner cell of teleost fishes; this form of electrical transmission is not mediated by gap junctions, but involves a different kind of junctional specialization.[2] In addition, there probably are electrical effects that occur between closely apposed cells without obvious gap junctions or specialization other than the absence of interposed glia.[2,3] Whether these sites are to be considered synapses or incidental or accidental sites of interaction is a matter of opinion. I have no difficulty in using the term electrical synapse only for gap junctions between neurons and not for gap junctions between nonneuronal cells. Admittedly, this terminology leads to different names for gap junctions depending on where they are located, even when they are comprised of the same protein. However, I find less attractive the alternatives of calling the neuronal interactions mediated by gap junctions nonsynaptic or of using synapses to denote the junctions between such cells as hepatocytes. As will be seen below, transmission mediated at electrical synapses of the gap junction type can be very synaptic in its properties.

One reason for stating at the outset that gap junctions between neurons are synapses is to avoid some of the contradictory material in the literature. For example, J.G.R. Jeffereys in his recent *Physiological Review*[3] considers "four classes of nonsynaptic interaction, mainly in the mammalian brain" of which the first is "Electrotonic (and chemical) coupling through gap junctions". Yet he also writes of "'gap junctions,' which

---

Gap Junctions in the Nervous System, edited by David C. Spray and Rolf Dermietzel.
© 1996 R.G. Landes Company.

commonly serve as electrical synapses in invertebrates but appear to be used less often for electrical signaling in vertebrates." The relative incidence of chemically transmitting and gap junction synapses in mammalian brains, as well as submammalian and invertebrate brains is not clearly resolved, as discussed below. It is certain that you, dear Reader, have gap junctions between neurons in your brain; it is likely that they are important to you; and it is likely that they were important to you while you were wiring up your brain.

Although most of the data in this chapter antedate the cloning of connexins, the discussion will be simpler if some of the newer basic facts are accepted.[4,5] Gap junctions in vertebrates are made of proteins termed connexins that are encoded by a gene family. A junctional channel consists of two hemichannels (or connexons) in series, one provided by each of the apposed cells. Hemichannels can be homomeric, i.e., comprised of a single type of connexin. It is likely that some hemichannels are heteromeric.[5A] Junctions can be homotypic, i.e., formed by two hemichannels of the same kind, and they can be heterotypic, i.e., formed by hemichannels of different kinds or in earlier usage formed by different cell types, which we now know may express the same or different connexins. A given cell type can express more than one connexin. A given connexin can be expressed by more than one cell type. Connexins expressed by neurons and forming electrical synapses can also be expressed by other cells. Thus, gap junctions between inexcitable cells can serve as models for electrical synapses.

2. SOME HISTORICAL OBSERVATIONS

A prominent controversy in the history of Neuroscience was whether synaptic transmission was chemical or electrical, a dichotomy sometimes affectionately characterized as soup versus sparks. After a long period of debate between pharmacologists and electrophysiologists, each group having not quite convincing arguments (ap-

plied chemicals mimic the effect of nerve stimulation, nerve stimulation releases acetylcholine; chemicals would be to slow too mediate transmission in the CNS, the action potential is propagated electrically and should be able to cross the synapse in the same way, so who needs chemicals), it became clear that CNS inhibition and neuromuscular transmission were chemically mediated. The pendulum swung, dragging with it the generalization to excitatory transmission between neurons, although here the initial data were much less convincing. Nevertheless, Paul Fatt (and no doubt Bernard Katz) were aware that synapses between large fiber systems might well be electrical (on the unreliable assumption that large size means large conductance, see Fatt's *Physiological Review* in 1954[6]), and in a paper that almost everyone knows about (but few have read) Ed Furshpan and David Potter reported from Katz's laboratory that fast excitatory transmission in the crayfish grant fiber system was electrically mediated, although there were also chemically mediated inhibitory inputs.[7] Independently and more or less simultaneously on the other side of the world Akira Watanabe reported in a paper that almost no one knows about that neurons in the cardiac ganglion of the mantid shrimp, *Squilla*, were electrically coupled.[8] He suggested that the coupling was responsible for synchronous firing (certainly right) and was mediated by cytoplasmic continuity of axonal processes (probably wrong). In the crayfish there is fast transmission for escape responses and rectification in the junctional membrane, so that impulses cross the giant to motor synapses in only one direction. In the shrimp there is slow, possibly bidirectional transmission leading to synchronization of motoneuron activity and cardiac contraction. Thus, electrical transmission is mediating two quite different functions.

At the same time, Stanley Crain, Harry Grundfest and I were working on the pufferfish, a teleost with large neurons that sit on top of the spinal cord just behind the cerebellum (Fig. 4.1).[9] Back before

Fig. 4.1. Anterior spinal cord of the puffer viewed from the dorsal side. The posterior limit of the cerebellum is to the left. The supramedullary neurons are the large round cells, about 250 μm in diameter, that are located on the surface of the cord. Several of the cells that had been penetrated for intracellular recording are dark because of increased toluidine blue staining.

patch clamping, large accessible neurons were particularly attractive. These so called supramedullary neurons are monopolar and have extensively branching axons that run out through the dorsal roots to the skin. They are effector cells, but their action is still unknown. What is relevant to this chapter is: (1) they are electrically coupled via gap junctions[10] (the ultrastructure was done by Yasuko Nakajima and George Pappas); (2) they fire synchronously;[9] but (3) the degree of synchronization is not very precise, a point to which I will return (Fig. 4.2). It is a somewhat self serving aside to point out recent evidence that arthropod and vertebrate gap junctions are made of completely unrelated proteins, although there are a number of convergent features (cf. Barnes, 1994). Thus, we were the first to identify vertebrate, and NIH fundable, electrical synapses.

In collaboration with Harry Grundfest a number of us were working on mecha-nisms by which electric organs generate electricity,[12] and I and several others went on to look at neural control of the dis-charges.[13] The problem was particularly interesting, because in some of these fishes the frequency of discharge is very high, greater that 1 kHz, and the requirements for synchronization of activity can be very stringent. As an example, consider the electric catfish. Its electric organ, which is found in the skin, is controlled by only two neurons, one on either side of the me-dulla.[14] Graded stimulation of afferents produces smoothly graded EPSPs in them, but always excites them together; no stimulus can be found to excite one at a time (Fig. 4.3A). The reason is that if one cell fires, its action potential propagates through electrical synapses to excite the other (Fig. 4.3B, spread of hyperpolariza-tion between cells is shown in Fig. 4.3C). Here and in many other electric fishes, electrical transmission between neurons is required because of its speed; the electric organ discharges are so brief that chemical transmission with a 1/2 msec delay between neurons would not give adequate precision of firing. Reciprocal excitation is necessary for all the controlling cells to fire together; to get really precise synchronization the excitation must be electrical. In this situation the speed and reciprocity of electrical synapses are both required. Several other points are illustrated here, given the additional evidence that transmission to the electromotor neurons is via gap junctions and the cells are coupled to each other not

directly but rather by way of presynaptic fibers that end on both of them (Fig. 4.4, lower diagram, other electromotor neurons are coupled via dendrodendritic gap junctions as in the upper diagram). An impulse in one or a few afferent fibers will not excite either cell. As stimulus strength is increased, both cells are depolarized; when one of them reaches threshold, the other cell is already near threshold and the impulse rapidly propagates between them (Fig. 4.3A). An action potential evoked by directly stimulating one cell takes a longer time to excite the other cell

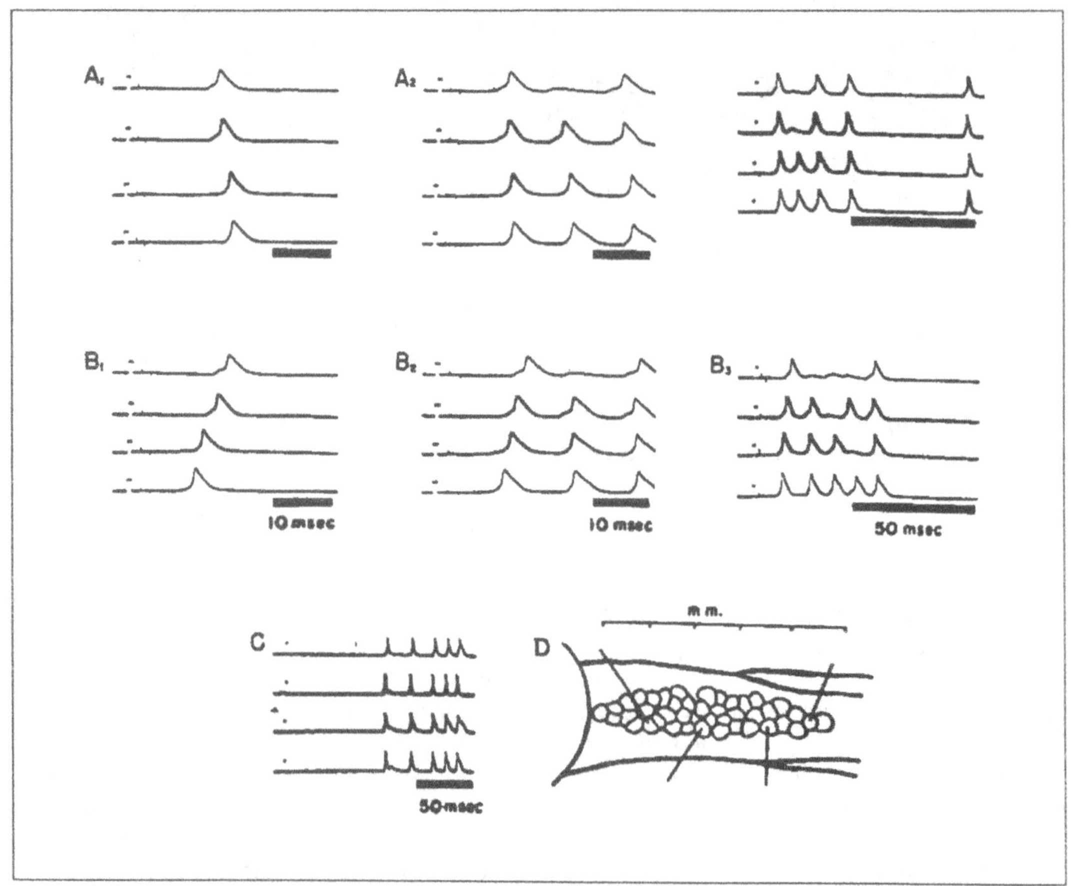

Fig. 4.2. Synchronous activity in four supramedullary neurons diagrammed in D. The most rostral cell is shown on the uppermost trace. Electrical stimuli were given to the cranial nerves (A₁-A₃) and cauda equina (B₁-B₃); the skin was stimulated tactilely (C). Weak, near threshold stimuli in A₁ and B₁; strong stimuli at the same sweep speed in A₂ and B₂ and at a slower sweep speed in A₃ and B₃. Calibration pulses (50 mV, 1 msec) occur at the beginning of each of the four traces followed after 1 msec by the electrical stimulus artifact. All cells generate the same number of impulses, although the action potential fails to invade the somata in some cases. Impulses tend to arise earlier rostrally for the cranial nerve stimulus and earlier caudally for the cauda equina stimulus. Reprinted from Bennett MVL et al, J Gen Physiol 1959; 43:221-250 with copyright permission of The Rockefeller University Press.

(Fig. 4.3B). Thus, coupling that is weak when measured between a single afferent and the postsynaptic cells can be very effective when enough afferent fibers are active.

Once a central "command" nucleus has initiated a synchronous signal for the electric organ to discharge, other adaptations are involved in getting the activation to arrive at different regions of the electric organ simultaneously.[13] There may be relay nuclei intercalated between command nucleus and electromotor neurons; the "relay neurons" commonly are electrically coupled which would tend to synchronize their firing and correct for any loss of synchronization that arose in conduction of the signal from the command nucleus. From a teleological perspective, one might argue that a single cell command nucleus would not require electrical synapses; that is true and the electric catfish command nucleus is down to only two cells. However, electrical synapses allow multiple cells to act with nearly the precision of a single cell, and having multiple cells in parallel reduces the number of efferent synapses that a single cell must support.

With the insight that a fast synchronous response required electrical transmission, it was reasonable to extend the study to motor responses. One motor system that we investigated was the sonic motor system of the toadfish, *Opsanus tau*.[15] As in number of independently evolved species, this fish generates sound by rapid synchronous contractions of its swim bladder musculature. In the breeding season, males generate boat whistle calls in which the fundamental frequency generated by repetitive contractions of the swim bladder muscle is about 200 Hz. These motoneurons proved to be electrically coupled by gap junctions.

Extension of this idea to mammals was only moderately successful. Vestibular

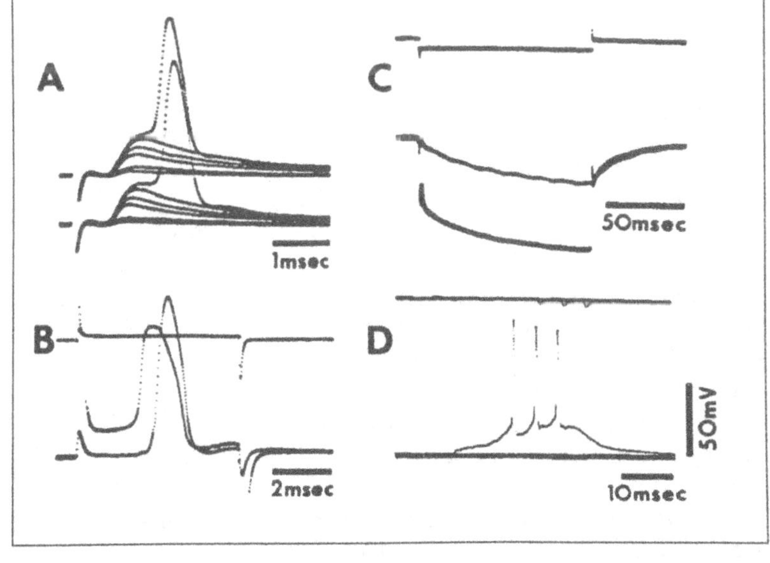

Fig. 4.3. Properties of the giant electromotor neurons of the electric catfish. (A) Upper and lower traces, recordings from right and left cells, respectively. Brief stimuli of gradually increasing strength are applied to the nearby medulla (several superimposed sweeps; the stimulus artifact occurs near the beginning of the sweep). Depolarizations of successively increasing amplitude are evoked until in one sweep both cells generate spikes. (B) Two electrodes in the right cell, one for passing current (shown on the upper trace) and one for recording; one recording electrode in the left cell. The lower traces, which are from the recording electrodes, start from the same base line. When an impulse is evoked in the right cell by a depolarizing pulse, the left cell also generates a spike after a short delay. (C) When a hyperpolarizing current is passed in the right cell, the left cell also becomes hyperpolarized, but more slowly and to a lesser degree (display as in B). (D) When organ discharge is evoked by irritating the skin, a depolarization gradually rises up to the threshold of the giant cell and initiates a burst of three spikes (lower traces, base line indicated by superimposed sweeps). Each spike produces a response in the organ (upper trace, recorded at high gain and greatly reduced in amplitude because curare, used to prevent movement, also blocked transmission from nerve to electrocyte). Reprinted with permission from Bennett MVL et al, J Neurophysiol 1967; 30:209-235.

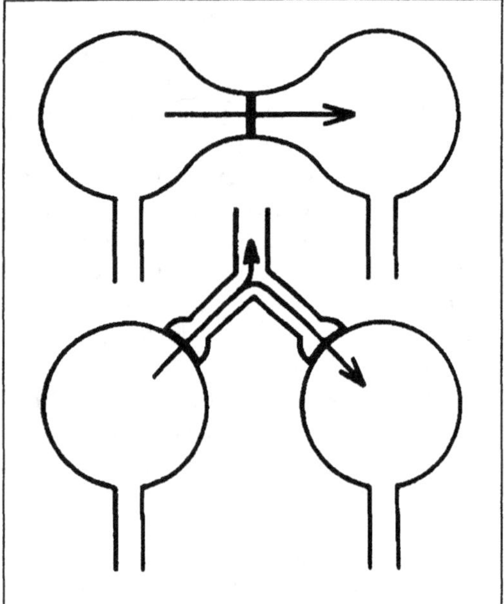

Fig. 4.4. Pathways of electrotonic coupling of cell bodies. Upper diagram: dendrodendritic pathway. Lower diagram: coupling by way of presynaptic fibers coupled to both cells. If the arrows indicate the direction of current flow, the cell at the arrow head is excited and the cell at the tail is inhibited; thus, the synapses are both excitatory and inhibitory.

inputs to the Mauthner cell have an electrical component mediated by gap junctions.[2] The teleological explanation is that the latency of the escape reflex of tail flip is decreased. Henri Korn and colleagues did find electrical inputs to vestibular neurons in the rat,[16] but apparently they do not occur in larger mammals. One possible reason is that in a large animal nerve conduction distances and the inertia are so great that a response cannot be rapid enough to require electrical transmission, particularly since chemical transmission is faster in warm blooded animals. Another site is the mesencephalic nucleus of the fifth cranial nerve,[17] where gap junctions were first identified morphologically.[18] In rodents mastication is fast and synchronous, but it would be difficult to conclude that electrical transmission was required for its speed. Although it is true in principal that "Direct electrical interactions can alter neuronal activity on a shorter time scale then

chemical synaptic transmission" it is not true that "there is *much* (italics mine) less synaptic delay associated with electrical interactions",[69] and the delay of electrical interactions can be considerably longer than that of chemical interactions, even not considering conduction delays which are electrical.

Firing of the supramedullary neurons of the pufferfish is not highly synchronous (Fig. 4.2), and reciprocal excitation mediated by chemical synapses would probably be able to provide an adequate degree of synchronization; gap junctions may do it better because subthreshold potentials are transmitted, and all neighboring cells tend to reach threshold at the same time. If one really pushes functional arguments, both chemical and electrical transmission could be made to work at most synapses.[1] Where electrical synapses are clearly better is in speed and in reciprocity. Where chemical synapses are clearly better is in temporal changes as a result of prior activity and in inhibition.

Another point is that gap junctions can transmit small molecules in addition to ions.[4] In this manner gap junctions can mediate chemical communication between cells. The early hypothesis that gap junctions mediated developmental gradients has not yet been borne out. It appears that most signaling factors are peptides that act extracellularly. Gap junctions are implicated in the transmission of Ca^{2+} waves that propagate between glia and may involve neurons.[19]

3. HOW MANY ELECTRICAL SYNAPSES ARE THERE?

At this stage in the development of neuroscience, it was clear that there were electrical as well as chemical synapses. Eccles[20] did suggest that they would not be found in mammals, which is not the case. But it is difficult to determine how many electrical synapses of the gap junction type there are. Although a very short synaptic delay indicates electrical transmission, the delay at electrical synapses because of time to charge the postsynaptic membrane capacity (and propagation time in the presynaptic fibers) can exceed the delay at chemical synapses, particularly at mammalian body temperature.[1] Direct measurement of electrotonic coupling by simultaneous recording from pairs of neurons remains difficult, except where cells are large. Dye coupling can be convincing, but a negative result is not, and some positive results may be artifactual. Conversely, reversal by polarization is diagnostic of a chemically mediated PSP, and usually is easily done for inhibition. It may be hard to reverse an EPSP, particularly if the synapses are located on the dendrites distant from the recording site. Now that most central excitatory synapses are known to be glutamatergic, pharmacological approaches become possible, but failure of glutamate antagonists to block transmission would not be considered very strong evidence of electrical transmission. Anatomical methods including electron microscopy, immunostaining and in situ hybridization all have their problems. Electron microscopy requires good fixation, not always easy in CNS tissues, and gap junctions that are small compared to section thickness, perhaps five junctional particles across, are difficult to identify. Freeze fracture permits identification of small junctions, but in neuropil it may be difficult to determine the identity of the structures forming them. Presynaptic vesicles, which define an active zone, are also found at axosomatic and axodendritic synapses where electrophysiological findings indicate that transmission is purely electrical.[14-16] In fact, all axodendritic and axosomatic synapses with gap junctions also have active zones, i.e., they are morphologically mixed synapses. Conversely, correlation of the number of gap junction channels at club endings on the Mauthner cell with junctional conductance suggests that most of the channels are closed.[21] Antibodies to many connexin proteins are not yet available and presence of mRNA shown by in situ hybridization does not guarantee protein synthesis, nor does protein synthesis guarantee junction formation. My summary conclusion based on currently available data is that gap junctions are a small minority of synapses in the brain, but that there are probably quite a few more still to be found. I believe this characterization applies to mammals and lower vertebrates and also to animals of the arthropod line, which apparently have gap junction forming proteins unrelated to connexins.

The widely held opinion that electrical transmission is characteristic of lower forms probably derives from the large cell systems that were studied in the initial period of intracellular recording, which hardly constitute a reasonable sample. There may be a kernel of truth in the idea, since synaptic delay is shorter at mammalian body temperature and the advantage of electrical transmission is less. In many of the sites of electrical transmission in mammals, such as sensory motor cortex,[22] olfactory bulb[23] and cerebellum,[24] the functional advantages are more likely to be in reciprocity and transmission of subthreshold potentials. Actually, one can argue that unicellular organisms have evolved the basic machinery of chemical transmission for release of and response to chemicals, but have no functional equivalent of electrical synapses. Thus, gap junctions are the more advanced form of transmission. As more genomes of lower life forms are sequenced, it may become clearer how gap junction forming proteins evolved. Many of the proteins recently implicated in neurotransmitter release have homologs involved in secretion in yeast.

4. FLEXIBILITY IN ELECTROTONICALLY COUPLED SYSTEMS

Although one could argue about the relative merits, functionally, of chemical and electrical synapses, my main point was that one needed to examine the mode of transmission carefully at new sites rather than assume that it was chemical. There was also self interest in that I did not want to be working on a second class or more primitive synapse. The real stretch in arguing for basic equality of electrical transmission was in synaptic plasticity. Gap junctions exhibited very little modifiability or dependence on prior activity, in part perhaps because they were being studied in systems where synchrony and constancy were important. One could demonstrate temporal summation of electrical PSPs, which depends on charging the postsynaptic membrane capacity, and there were several examples of temporal changes that were unexpectedly long lasting. For example, in our initial studies of the puffer, stimulation of one supramedullary neuron would usually not excite the others; however, paired stimulation sometimes would; the effect of the first stimulus could last for several hundred milliseconds, a duration that we thought indicated chemical mediation.[9] We later showed that the facilitation was due to a long lasting depolarizing afterpotential that facilitated propagation across the electrical synapses.[25] In the stargazer, another electric fish, antidromic stimulation evokes electrical PSPs in the electromotor neurons (which are derived from the oculomotor nucleus). These PSPs show pronounced summation and facilitation with repetitive stimulation; the facilitation is due to more extensive invasion of the antidromic impulses closer to the sites of electrical coupling in the dendrites.[26] These examples indicated that electrical synapses can exhibit activity dependent changes in effectiveness of unexpectedly long duration, although there was no indication that the phenomena were functional at these sites or at all common in nervous systems in general.

We then found that transmission at electrical synapses can be under the control of chemical synapses on the same cells. We first observed this phenomenon in the mollusc, *Navanax*.[27] This animal can expand its pharynx very rapidly as an ingestive response that sucks in prey. Under resting conditions, the pharyngeal motoneurons are electrically coupled and their firing tends to be synchronous, which accounts for the rapid pharyngeal expansion. However, stimulation of the pharyngeal nerve or stimulation of the pharyngeal wall reduces the coupling between them. The loss of coupling appears to be due to inhibitory synapses along the pathway connecting the neurons; increase in conductance simply short circuits the electrotonic spread. Under these circumstances, the neurons can be activated asynchronously by other excitatory inputs. Our functional explanation of these data was that all the expansion motoneurons should be active for prey ingestion but that during peristaltic swallowing expansion should involve subsets of the expansion musculature. A similar situation of inhibitory control of coupling has been suggested for the mammalian inferior olive.[28] In this system dendrodendritic gap junctions synchronize firing, but inhibitory inputs localized to the dendrites are in a strategic position to shortcircuit the coupling and permit asynchronous firing to somatic inputs.

A further variant was provided by goldfish oculomotor neurons; these cells are coupled in the cell body region, and synapses depolarize this region to mediate the relatively synchronous action of eye withdrawal and perhaps saccadic eye movements.[29] Vestibular inputs end out on the dendrites where the cells presumably are not coupled, and these inputs may mediate the slow phase of nystagmus (Fig. 4.5). In this case localization of inputs to regions that were and were not coupled appeared to determine the degree of functional coupling and whether firing was synchronous or asynchronous. Weak coupling of the somata would permit asynchronous firing of impulses arising in the dendrites, but this same coupling would synchronize firing if the somata were depolarized by other inputs.

A dramatic and distinctly different form of synaptic control of coupling was observed in the turtle and teleost retina.[30,31] In the unstimulated preparation dye coupling between horizontal cells is extensive, and dye injected into one cell spreads to many neighbors. Application of dopamine to the retina greatly restricts dye coupling. Subsequently, dopamine and cAMP increasing agents were shown to decrease junctional conductance and dopaminergic interplexiform cells are thought to be the

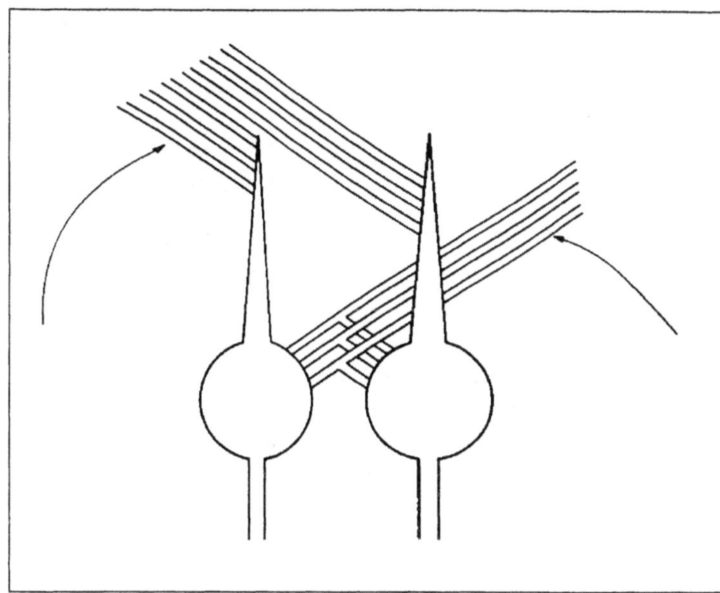

Fig. 4.5. Diagram of somatic and dendritic inputs to medial rectus motoneurons in a teleost. Dendritic inputs (left arrow) are activated by stimulation of the ipsilateral eighth nerve. There is no coupling, and movements are smoothly graded in amplitude. The somatic inputs (right arrow) are activated by stimulation of the ophthalmic nerve or contralateral eighth nerve. There is weak coupling between the cell bodies by way of the presynaptic fibers and some increase in synchronization of firing results.

endogenous source.[32] These cells are presumed to be active during light adaptation, when they mediate the observed reduction in receptive field size. This pronounced modulation of junctional conductance may be due to phosphorylation of the channel protein, itself. Sequences of the retinal gap junctions in these animals have not yet been reported. Rodent Cx32 has a phosphorylation site for cAMP dependent kinase, but the effect of phosphorylation is to increase rather than decrease coupling.[33] Similar phenomena may occur in the mammalian retina,[34] which has many electrical synapses.[35]

Extensive coupling between cortical neurons in the neonatal mouse has been observed using a gap junctional permeable tracer, neurobiotin.[36] This communication largely disappears during development, and it is hypothesized to be involved in synapse formation. In the young animal application of dopamine to brain slices reduces the tracer coupling, but the relation, if any, between this process and the loss of coupling with maturation is unclear.[37]

More recently Faber and colleagues[38] showed LTP of electrical PSPs at club endings on the Mauthner cell. The pharma-

cology suggests involvement of NMDA receptors as in some forms of mammalian LTP. Increase in junctional conductance may involve hemichannels in either pre- or postsynaptic elements considering that most of the gap junction channels appear to be closed under normal conditions.[38]

The incidence of gap junctions certainly has all the controls of other cell proteins. In different experimental systems controls have been demonstrated at transcriptional, translational and posttranslational levels.[4] Posttranslational modifications include phosphorylation and myristoylation, and how connexin assembly, membrane insertion and junction formation occur are subjects of active inquiry. In some culture systems at least, turnover is rapid and control of junctional conductance by removal is a possible if slow form of modulation.[4] Coupling may change as a function of hormonal state.[39]

5. CA²⁺, H⁺ AND OTHER EXTRINSIC MODULATORS OF JUNCTIONAL CONDUCTANCE

Experimental modulation of junctional conductance was an early goal. Treatments that increased intracellular Ca^{2+} decrease junctional conductance in insect salivary gland cells, but Ca^{2+} levels were not accurately measured.[40] Moreover, rise in cytoplasmic Ca^{2+} may cause rise in cytoplasmic H^+ because of shared buffers. Given the probable lack of homology between connexins and gap junctions in the arthropod line,[11] the old data are of limited predictive value for connexins in any case. Although it is likely that rises in Ca^{2+} do decrease conductance of connexin channels, there is still no good dose response curve or evidence of direct versus indirect action. The new methods of single channel recording will allow, hopefully, resolution of this question. What is (and was) clear is that low levels of Ca^{2+} will permeate gap junctions, and waves of increase in intracellular Ca^{2+} have been observed to propagate between many kinds of coupled cells.[41] The low levels of intracellular Ca^{2+} that healthy cells permit may never reach the levels required to close gap junctions.

A point of purely historical interest: Y. Asada working with George Pappas and me observed that replacing extracellular Cl^- with propionate decreased junctional conductance in the septate axon of the crayfish.[42] We ascribed the effect to removal of Cl^- and did not at the time appreciate that propionate is a weak acid which permeates the surface membrane and acidifies the cell interior. (George did show that the gap junctions disappeared when the junctional conductance was lost and that recovery was associated with their reappearance.[43] This early result indicates that although acidification acts rapidly, there can be subsequent slow changes that involve junction removal, and during recovery, reformation.)

After Turin and Warner[44] showed uncoupling in amphibian blastulae by treatment with weak acids, we confirmed that cytoplasmic acidification decreases junctional conductance in a variety of cell types and using intracellular electrodes were able to obtain well behaved titration curves.[45] Application of CO_2 or other weak acid was for some time the best way to decrease junctional conductance between cells. As

one would expect from knowledge of the connexin gene family, not all junctions are equally sensitive, but none appear to be completely insensitive. (Now Delmar and colleagues[47] are defining sites in the Cx43 molecule responsible for the H^+ effect on conductance, which does appear to be a form of gating distinct from voltage gating, although as noted below H^+ affects voltage sensitivity in amphibian blastomeres, which express Cx38.) pH uncoupling is obviously not a normal way of controlling junctional conductance, but junctional coupling may be affected by rises in acidity during pathological conditions such as anoxia, ischemia and seizures.

After pH uncoupling came uncoupling with long chain alcohols, n-heptanol and n-octanol,[48] and still more recently halothane.[49] These agents also appear quite nonspecific in the junctions on which they act and are potent in both arthropod and vertebrate junctions. Again, the convergence is remarkable. These agents have provided the best approach to reducing junctional conductance, either to determine that coupling is gap junction mediated or what the effect of loss of coupling is. The major problem with these agents, as with pH, is action on other cell properties; treatment tends to block excitability, for example, which greatly limits their usefulness with respect to excitable cells such as neurons. Newer approaches dependent on molecular biology offer more specificity. Connexins can be knocked out by homologous recombination,[50] connexin expression can be reduced by treatment with antisense oligonucleotides[51] and antibodies to extracellular loops can block junction formation.[52]

6. VOLTAGE DEPENDENCE OF JUNCTIONAL CONDUCTANCE

In 1979 another form of modulation of vertebrate gap junctions was discovered. Dave Spray, Andy Harris and I observed that the junctional conductance between amphibian blastomeres was strongly dependent on transjunctional voltage.[53] Again, Warner's laboratory[46] had made an initial suggestive observation. Previously, certain electrical synapses in the hatchetfish had been found to rectify, i.e., their conductance was increased by one sign of transjunctional voltage and decreased by the opposite sign;[54,55] the rectification tended to make the transmission unidirectional as in the rectifying synapses of crayfish.[7] When we voltage clamped pairs of blastomeres we found that transjunctional voltage, V_j, of either sign decreased junctional conductance, g_j (Fig. 4.6). (g_j was little affected by the voltage between the interior of the cells and the outside, V_{i-o}, as is commonly true of connexins, but untrue of some invertebrate junctions.[56]) The symmetry of g_j/V_j relation is consistent with the symmetry of the coupled cells. The steady state conductance for one polarity of transjunctional voltage was well fit by a Boltzmann relation (Fig. 4.6C), which suggests a voltage gate that can be in either of two states with the energy difference between them linearly dependent on voltage (but Boltzmann relations can give a reasonable fit to almost any sigmoid and the predictive value is not great). We hypothesized that there were two gates, one in each hemichannel, one closed by one polarity of transjunctional voltage, the other closed by the opposite polarity (Fig. 4.7). Because of the independence of V_{i-o}, we placed the gates at the cytoplasmic ends of the channel. Now we know that residues at the end of the cytoplasmic N-terminal of the connexin are part of the gating charge, but that residues at the beginning of the first extracellular loop also contribute.[57] Actually, we did not know which gate was closed by which polarity of V_j until Vytas Verselis injected acid into one of a pair of coupled cells;[58] this procedure modified gating properties for one polarity of voltage, and we assumed that the effect was on the gate on the acidified side. Our heuristic diagram turned out to be right for these junctions; transjunctional voltage closes the gate on the positive side. We were, and are, uncertain why amphibian blastomeres have this voltage sensitivity, which we observed in a urodele, several ranids and *Xenopus*. (It also occurs in ascidian blastomeres, but not in teleost blastomeres.[59]) The degree of sensitivity is comparable to that of Na channels, although the transitions are much slower. We did show that if there were modest resting potential differences between cells, the voltage dependence could lead to bistability in which the cells existed in stable coupled or uncoupled conditions depending on prior history (Fig. 4.6D).[60] We thought that this property might be involved in setting up embryonic compartments.

The study of voltage dependence has come a long way from our initial work. The cloned connexins can be expressed in *Xenopus* oocytes or communication deficient cell lines for determination of macroscopic and single channel properties, respectively. All vertebrate connexins tested have some degree of voltage dependence (except Cx33 which apparently does not form functional junctions and interferes with formation of junctions by Cx37 when the two are expressed together[61]). For some, e.g., Cx43, the dependence is weak, and in heart muscle, constancy of transmission in the face of depolarization is likely to be functional. Conversely Cx37 is quite voltage sensitive[62] and is found in neural progenitor cells (R. Rozental, personal communication) as well as skin.[63] This sensitivity might give rise to functional changes in synaptic transmission. Hemichannels formed of different connexins can close in response to opposite polarities of transjunctional voltage.[57] When channels of opposite polarity form junctions, their channels respond to the same polarity of V_j, giving rise to strongly rectifying g_j/V_j relations.[57,64] These observations suggest that rectifying synapses arise because the pre- and postsynaptic cells express connexins with opposite polarities of voltage dependence. Structure function studies are revealing components of the voltage gating mechanism.[57] Remarkably, a charge at the N-terminus of the connexin molecule is part of the gating charge and changing the sign of this charge by site directed mutagenesis reverses the polarity of gating of Cx32. Another charge change at the border between the first membrane spanning and first extra-cellular loop domains can also reverse po-larity of gating or alternatively increase its voltage sensitivity. The changes are con-sistent with movement of both of these charges towards the cytoplasm. Thus, the voltage sensor appears to extend quite far within the molecule and the reptation model of gating[4] in which a transmem-brane domain slides along the axis of the

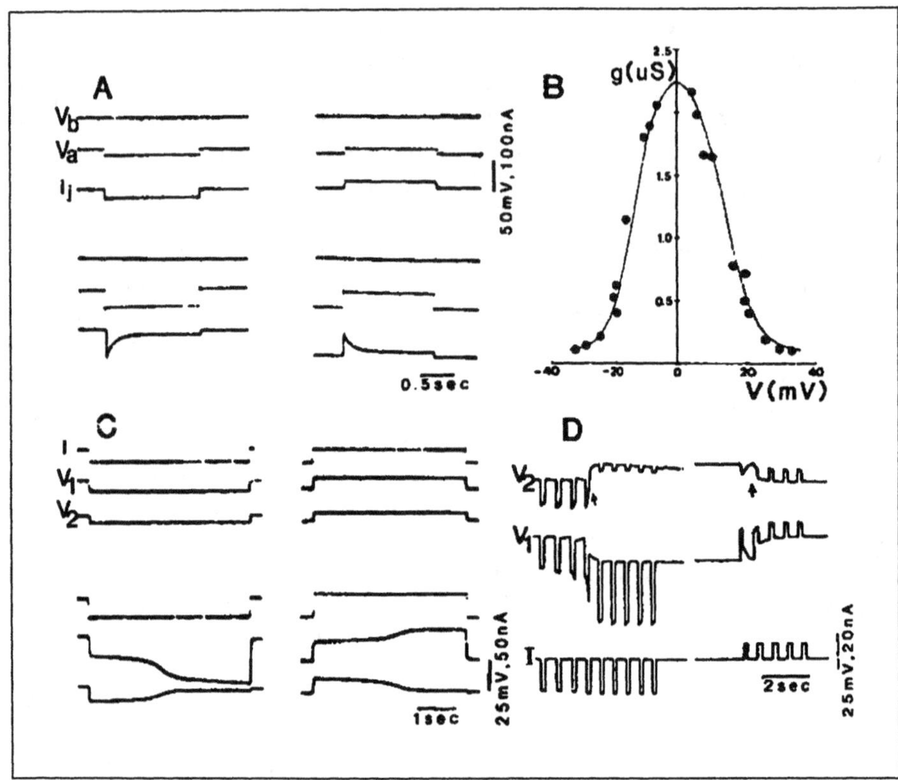

Fig. 4.6. Voltage dependence of junctional conductance in amphibian blastomeres. (A) Records from a dual voltage clamp experiment on a coupled cell pair. One cell was stepped to a new voltage, while the second cell was held constant. The current applied in the second cell represents the transjunctional current flowing as a result of the voltage step. For small voltages of either sign, the current decreases only slightly during the step. For larger voltages of either sign, the current relaxes exponentially to a lower value. (B) Steady-state junctional conduc-tance, g_j, as a function of transjunctional voltage, V_j. g_j decreases steeply for V_j of either sign. The curve is a Boltzmann relation as would be observed if the energy difference between open and closed states were a linear function of transjunctional voltage. The parameters of the particular curve are for the equivalent of about 6 electron charges moving through the entire transjunctional voltage. (C) Effects of current pulses. During large enough current pulses of either sign, the voltage in the directly polarized cell increases in a regenerative manner while the voltage in the other cell declines. The cells change from a well-coupled to a poorly-coupled state as a result of voltage dependence of the junctional conductance. (D) Bistability of coupling resulting from voltage dependence and difference in resting potentials between two cells of a pair. A train of pulses in one cell (V_2) leads to progressive decrease in junctional conductance until the cells uncouple, the pulsed cell having the more negative resting potential. The break in the record represents 100 sec. Near the end of the record an oppositely direct pulse was applied that reduced the transjunctional voltage and allowed the cells to recouple. Reprinted with permission from Harris AL et al, J Gen Physiol 1981; 77:95-117 by copyright permission of The Rockefeller University Press and Harris AL et al, J Neurosci 1983; 3:79-100.

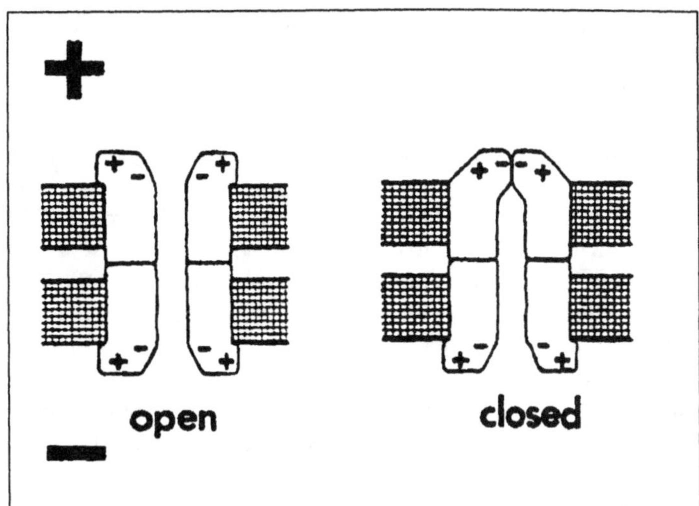

Fig. 4.7. Diagram of a gap junction channel with voltage gates at either end. V$_j$s close the hemichannel on the relatively positive side.

channel gains some support. The third membrane spanning domain is amphiphilic, which led to the suggestion that it is the pore lining domain. Rotation of this domain could introduce tyrosines into the channel lumen and thereby close the channel. Massaging these possibly divergent structure function data into a coherent view of connexin gating is a major task confronting the field.

At least seven connexins are expressed in rodent nervous system, Cx26, Cx32, Cx37, Cx40, Cx43, Cx45 and Cx46; how cells expressing one or more of these connexins interact with other cells expressing the same or different connexins is an area of active exploration. Some connexins are compatible in that they form functional heterotypic junctions, which may be quite symmetrical or very rectifying. Other connexins are incompatible and will not form functional junctions and may even interfere with junction formation by otherwise compatible connexins.[61,71] Junctional properties are quite variable and are usually, but not always, predictable from the properties of their constitutive connexins in other combinations. Single channel recording is illuminating aspects of the gating mechanism; many but not all of these aspects would have been predicted from macroscopic properties of the junctions.

Several surprises have appeared. Gap junctions may be more permselective than previously believed, e.g., Cx46 is quite cation selective and can function as an isolated hemichannel without an apposed hemichannel.[65] The voltage sensitivity of opening of the isolated hemichannel is within the physiological range of membrane potential changes in neurons. The residual conductance of gap junctions in the face of strong transjunctional voltage is, in Cx43 and Cx46 at least, due to a conductance substate induced at these voltages.[65-67] When channels first form and open, the initial opening transition is slow compared to gating transitions of formed junctions to a substate.[68] Similar slow transitions are seen in Cx46 hemichannels and can be explained as unresolved transitions between multiple substates associated with opening of the extracellular loops, which requires another hemichannel in most connexins but occurs without an apposed hemichannel in Cx46.[65] In situ hybridization and sequence specific antibodies will reveal which connexins are expressed in the nervous system and where. Application of patch clamping (and probably optical methods) to dissociated cells and brain slices in culture will provide evidence as to the actual function of the gap junctions at these synapses. Knock-out by homologous recombination, antisense techniques to reduce connexin expression and specific antibodies to block formation should provide important adjuncts to the direct measurement of electrical coupling.

7. SUMMARY

Gap junctions are the morphological substrate of one class of electrical synapse. The history of the debate on electrical vs. chemical transmission is entertaining. One lesson is that Occam's razor sometimes cuts too deep; the nervous system does its operations in a number of different ways and a unitarian approach can lead one astray.[70] Electrical synapses can do many things that chemical synapses can do, and do them just as slowly. More intriguing are the modulatory actions that chemical synapses can have on electrical synapses. Voltage dependence provides an important window on structure function relations of the connexins, even where the dependence may have no physiological role. The new molecular approaches will greatly advance our knowledge of where gap junctions occur and permit experimental manipulation with high specificity.

ACKNOWLEDGMENTS

This work depended on numerous colleagues most of whose names can be found in the reference list; I am deeply indebted to them. Funding came from many NIH and NSF grants over the years, initially to Harry Grundfest. A major support for 26 years has been NS-07512. I am the Sylvia and Robert S. Olnick Professor of Neuroscience.

REFERENCES

1. Bennett MVL. Electrical transmission: a functional analysis and comparison with chemical transmission. In: Kandel ER, ed. Cellular Biology of Neurons Vol. I, Sec. I, Handbook of Physiology. The Nervous System. Baltimore: Williams and Wilkins, 1977:357-416.

2. Faber DS, Korn H. Electrical field effects: their relevance in central neural networks. Physiol Rev 1989; 69:821-63.

3. Jefferys JG. Nonsynaptic modulation of neuronal activity in the brain: electric currents and extracellular ions. Physiol Rev 1995; 75:689-723.

4. Bennett MVL, Barrio L, Bargiello TA, Spray DC et al. Gap junctions: new tools, new answers, new questions. Neuron 1991; 6:305-320.

5. White TW, Bruzzone R, Paul DL. The connexin family of intercellular channel forming proteins. Kidney Int 1995; 48:1148-57.

5A. Stauffer KA. The gap junction proteins beta 1-connexin (connexin32) and beta 2-connexin (connexin26) can form heteromeric hemichannels. J Biol Chem 1995; 270: 6768-6772.

6. Fatt P. Biophysics of junctional transmission. Physiol Rev 1954; 34:674-710.

7. Furshpan EJ, Potter DD. Transmission at the giant motor synapses of the crayfish. J Physiol London 1959; 145:289-325.

8. Watanabe A. The interaction of elecrical activity among neurons of lobster cardiac ganglion. Japan J Physiol 1958; 8:305-318.

9. Bennett MVL, Crain SM, Grundfest H. Electrophysiology of supramedullary neurons in *Spheroides maculatus*. III. Organization of the supramedullary neurons. J Gen Physiol 1959; 43:221-250.

10. Bennett MVL, Nakajima Y, Pappas GD. Physiology and ultrastructure of electrotonic junctions. I. Supramedullary neurons. J Neurophysiol 1967; 30:161-179.

11. Barnes TM. OPUS: a growing family of gap junction proteins? Trends Genet 1994; 10:303-5.

12. Bennett MVL. Electric organs. In: Hoar WS, D.J. Randall DJ, eds. Fish Physiology, Vol. 5. New York: Academic Press, 1971: 347-491.

13. Bennett MVL. Neural control of electric organs. In: Ingle D, ed. The Central Nervous System and Fish Behavior. Chicago: Univ. of Chicago Press, 1968:147-169.

14. Bennett MVL, Nakajima Y, Pappas GD. Physiology and ultrastructure of electrotonic junctions. III. Giant electromotor neurons of *Malapterurus electricus*. J Neurophysiol 1967; 30:209-235.

15. Pappas GD, Bennett MVL. Specialized junctions in electrical transmission between

neurons. Ann NY Acad Sci 1966; 37: 495-508.

16. Korn H, Sotelo C, Crepel F. Electronic coupling between neurons in the rat lateral vestibular nucleus. Exp Brain Res 1973; 16:255-75.

17. Baker R, Llinas R. Electrotonic coupling between neurones in the rat mesencephalic nucleus. J Physiol London 1971; 212:45-63.

18. Hinrichsen CFL, Larramendi LMH. Synapses and cluster formation of the mouse mesencephalic fifth nucleus. Brain Res 1968; 7:296-299.

19. Dani JW, Smith SJ. The triggering of astrocytic calcium waves by NMDA-induced neuronal activation. Ciba Found Symp 1995; 188:195-205; discussion 205-9.

20. Eccles JC. The Physiology of Synapses. Berlin: Springer Verlag, 1964.

21. Tuttle R, Masuko S, Nakajima Y. Small vesicle bouton synapses on the distal half of the lateral dendrite of the goldfish Mauthner cell: freeze-fracture and thin section study. J Comp Neurol 1987; 265:254-74.

22. Sloper JJ. Gap junctions between dendrites in the primate neocortex. Brain Res 1972; 44:641-646.

23. Pinching AJ, Powell TPS. The neuropil of the glomeruli of the olfactory bulb. J Cell Sci 1971, 9.347-377.

24. Sotelo C, Llinas R. Specialized membrane junctions between neurons in vertebrate cerebellar cortex. J Cell Biol 1972; 53:271-289.

25. Bennett MVL. Physiology of electrotonic junctions. Ann NY Acad Sci 1966; 37: 509-539.

26. Bennett MVL, Pappas GD. The electromotor system of the stargazer: a model for integrative actions at electrotonic synapses. J Neurosci 1983; 3:748-761.

27. Spira ME, Bennett MVL. Synaptic control of electrotonic coupling between neurons. Brain Res 1972; 37:294-300.

28. Llinas R, Baker R, Sotelo C. Electrotonic coupling between neurons in cat inferior olive. J Neurophysiol 1971; 37:560-571.

29. Korn H, Bennett MVL. Vestibular nystagmus and teleost oculomotor neurons: functions of electrotonic coupling and dendritic impulse initiation. J Neurophysiol 1975;

38:430-451.

30. Piccolino M, Neyton J, Gerschenfeld HM. Decrease of gap junction permeability induced by dopamine and cyclic adenosine 3',5'-monophosphate in horizontal cells of turtle retina. J Neurosci 1984; 4:2477-88.

31. Teranishi T, Negishi K, Kato S. Dopamine modulates S-potential amplitude and dye-coupling between external horizontal cells in carp retina. Nature 1983; 301:243-2466.

32. Dowling JE. Retinal neuromodulation: the role of dopamine. Vis Neurosci 1991; 7:87-97.

33. Saez JC, Nairn AC, Czernik AJ et al. Phosphorylation of connexin32, the hepatocyte gap junction protein, by cAMP-dependent protein kinase, protein kinase C and Ca^{2+}/calmodulin-dependent protein kinase II. Eur J Biochem 1990; 192:263-273.

34. Hampson EC, Weiler R, Vaney DI. pH-gated dopaminergic modulation of horizontal cell gap junctions in mammalian retina. Proc R Soc Lond B Biol Sci 1994; 255: 67-72.

35. Vaney DI. Many diverse types of retinal neurons show tracer coupling when injected with biocytin or neurobiotin. Neurosci Lett 1991; 125:187-90.

36. Peinado A, Yuste R, Katz LC. Extensive dye coupling between rat neocortical neurons during the period of circuit formation. Neuron 1993; 10:103-14.

37. Roerig B, Klausa G, Sutor B. Dye coupling between pyramidal neurons in developing rat prefrontal and frontal cortex is reduced by protein kinase A activation and dopamine. J Neurosci 1995; 15:7386-400.

38. Pereda AE, Faber DS. Activity-dependent short-term enhancement of intercellular coupling. J Neurosci 1996; 16:983-92.

39. Hatton GI, Yang QZ. Incidence of neuronal coupling in supraoptic nuclei of virgin and lactating rats: estimation by neurobiotin and lucifer yellow. Brain Res 1994; 650:63-9.

40. Oliveira-Castro GM, Loewenstein WR. Junctional membrane permeability. Effects of divalent cations. J Membrane Biol 1971; 5:51-77.

41. Sanderson MJ. Intercellular calcium waves mediated by inositol trisphosphate. Ciba

Found Symp 1995; 188:175-89, discussion 189-94.

42. Asada Y, Bennett MVL. Experimental alteration of coupling resistance at an electrotonic synapse. J Cell Biol 1971; 49:159-172.

43. Pappas GD, Asada Y, Bennett MVL. Morphological correlates of increased coupling resistance at an electrotonic synapse. J Cell Biol 1971; 49:173-188.

44. Turin L, Warner A. Carbon dioxide reversibly abolishes ionic communication between cells of early amphibian embryo. Nature 1977; 270:56-7.

45. Spray DC, Harris AL, Bennett MVL. Gap junctional conductance is a simple and sensitive function of intracellular pH. Science 1981; 211:712-715.

46. Blackshaw SE, Warner AE. Alterations in resting membrane properties during neural plate stages of development of the nervous system. J Physiol London 1976; 255:231-47.

47. Ek JF, Delmar M, Perzova R et al. Role of histidine 95 on pH gating of the cardiac gap junction protein connexin43. Circ Res 1994; 74:1058-64.

48. Johnston MF, Simon SA, Ramon F. Interaction of anaesthetics with electrical synapses. Nature 1980; 286:498-500.

49. Burt JM, Spray DC. Volatile anesthetics block intercellular communication between neonatal rat myocardial cells. Circ Res 1989; 65:829-37.

50. Reaume AG, de Sousa PA, Kulkarni S et al. Cardiac malformation in neonatal mice lacking connexin43. Science 1995; 267: 1831-4.

51. Moore LK, Burt JM. Selective block of gap junction channel expression with connexin-specific antisense oligodeoxynucleotides. Am J Physiol 1994; 267:C1371-80.

52. Meyer RA, Laird DW, Revel JP et al. Inhibition of gap junction and adherens junction assembly by connexin and A-CAM antibodies. J Cell Biol 1992; 119:179-89.

53. Spray DC, Harris AL, Bennett MVL. Voltage dependence of junctional conductance in early amphibian embryos. Science 1979; 204:432-434.

54. Auerbach AA, Bennett MVL. A rectifying synapse in the central nervous system of a vertebrate. J Gen Physiol 1969; 53:211-237.

55. Hall DH, Gilat E, Bennett MVL. Ultrastructure of the rectifying (electrical) synapses between giant fibers and pectoral fin adductor motoneurons in the hatchetfish. J Neurocytol 1985; 14:825-834.

56. Verselis VK, Bennett MVL, Bargiello TA. A voltage dependent gap junction in *Drosophila melanogaster*. Biophys J 1991; 59: 114-126.

57. Verselis VK, Ginter CS, Bargiello TA. Opposite voltage gating polarities of two closely related connexins. Nature 1994; 368:348-51.

58. Bennett MVL, Verselis VK, White RL et al. Gap junctional conductance: gating. In: Hertzberg EL, Johnson R eds. Gap Junctions. New York: Alan Liss, Inc., 1988: 287-304.

59. Knier J, Verselis VK, Spray DC et al. Gap junctions between tunicate blastomeres: gating similarities and differences compared to amphibia. Biophys J 1986; 49:203a.

60. Harris AL, Spray DC, Bennett MVL. Control of intercellular communication by voltage dependence of gap junctional conductance. J Neurosci 1983; 3:79-100.

61. Chang M, Dahl C, Werner R. Is the role of connexin33 an inhibitory one? Biophys J 1994; 66:A20.

62. Reed KE, Westphale EM, Larson DM et al. Molecular cloning and functional expression of human connexin37, an endothelial cell gap junction protein. J Clin Invest 1993; 91:997-1004.

63. Goliger JA, Paul DL. Expression of gap junction proteins Cx26, Cx31.1, Cx37, and Cx43 in developing and mature rat epidermis. Dev Dyn 1994; 200:1-13.

64. Barrio LC, Suchyna T, Bargiello TA et al. Gap junctions formed by connexin26 and 32 alone and in combination are differently affected by applied voltage. Proc Natl Acad Sci USA 1991; 88:8410-8414.

65. Trexler EB, Bennett MVL, Bargiello TA et al. Voltage gating and permeation in a gap junction hemichannel. Proc Natl Acad Sci USA 1996; 93: in press.

66. Perez-Armendariz EM, Romano M, Luna J et al. Characterization of gap junctions between pairs of Leydig cells from mouse testis. Amer J Physiol 1994; 267:C570-C580.

67. Moreno AP, Rook MB, Fishman GI et al. Gap junction channels: distinct voltage-sensitive and -insensitive conductance states. Biophys J 1994; 67:113-119.
68. Bukauskas FF, Elfgang C, Willecke K et al. Heterotypic gap junction channels (connexin26-connexin32) violate the paradigm of unitary conductance. Pflugers Arch 1995; 429:870-2.
69. Valiante TA, Perez Velazquez JL, Jahromi SS et al. Coupling potentials in CA1 neurons during calcium-free-induced field burst activity. J Neurosci 1995; 15:6946-56.
70. Bennett MVL. Nicked by Occam's razor: unitarianism in the investigation of synaptic transmission. Biol Bull 1985; 168 (Suppl.):159-167.
71. White TW, Paul DL, Goodenough DA et al. Functional analysis of selective interactions among rodent connexins. Mol Biol Cell 1995; 6:459-70.

CHAPTER 5

CELL COUPLING IN THE RETINA

David I. Vaney

1. INTRODUCTION: A MODEL SYSTEM FOR STUDYING CELL COUPLING IN THE CENTRAL NERVOUS SYSTEM

The vertebrate retina has many features that make it an outstanding model for studying both neuronal coupling and glial coupling in the central nervous system (CNS).[1] The neural retina forms a transparent sheet, about 100 μm thick, which lines the back of the vitreous chamber, and the retina's only link with the brain is provided by the optic nerve, which carries the axons of the retinal ganglion cells. Consequently, the intact retina can be readily isolated both from the rest of the CNS (eyecup preparations) and, if need be, from the underlying pigment epithelium and sclera (wholemount preparations). The retina has thus been described as "nature's brain slice" and the ability to preserve the structural integrity of the retina in vitro is a critical feature of this model system, particularly given the mounting evidence that tissue sectioning reduces neuronal coupling in both the brain[2] and the retina.[3] The isolated retina thus retains many of the important advantages of in vivo preparations (intact neuronal circuitry; responsiveness to natural stimuli), while gaining the significant benefits of in vitro preparations (recording stability; microscopic identification of cells; controlled application of drugs).

The structure and function of the retina have been elucidated in remarkable detail and, thus, the retina is a particularly familiar part of the CNS.[4,5] Consequently, new observations can be related readily to countless previous studies and their significance assessed informatively. This is facilitated by the highly ordered architecture of the retina, both in the plane of the retina (horizontal) and through its depth (vertical). Identification of the major classes of retinal neurons is guided by the retina's laminar structure, which comprises three somatic layers separated by two synaptic layers (Fig. 5.1). The photoreceptors are located in the outer nuclear layer, adjacent to the retinal pigment epithelium, whereas most of the retinal interneurons are located medially in the inner nuclear layer;

Gap Junctions in the Nervous System, edited by David C. Spray and Rolf Dermietzel.
© 1996 R.G. Landes Company.

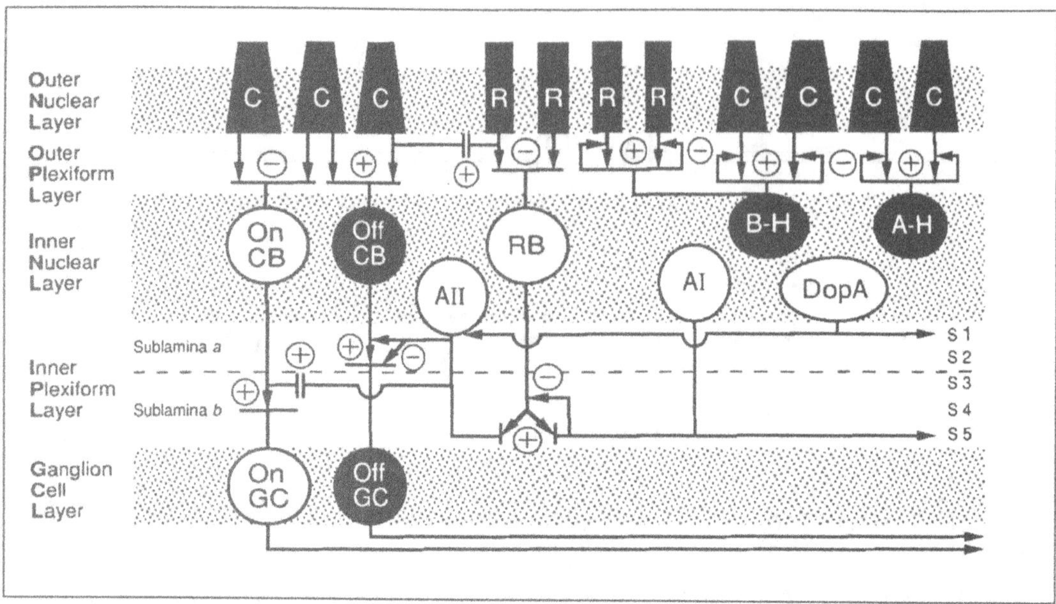

Fig. 5.1. Schematic circuit of the mammalian retina showing the direct neuronal pathways from the cone (C) and rod (R) photoreceptors to the ganglion cells (GC), together with some of the interneurons that modulate the signal flow. Excitatory (+) and inhibitory (–) chemical synapses are marked with arrows; electrical synapses between different types of neurons are marked with parallel lines, representing gap junctions. The white neurons depolarize at light On, while the black neurons depolarize at light Off. Other abbreviations: cone bipolar cell (CB), rod bipolar cell (RB), A-type horizontal cell (A-H), B-type horizontal cell (B-H), rod amacrine cell (AII), reciprocal rod amacrine cell (AI), dopaminergic amacrine cell (DopA).

the ganglion cell layer contains the projection neurons of the retina as well as some interneurons. A direct "vertical" pathway connects photoreceptors to bipolar cells to ganglion cells, with the axons of photoreceptors synapsing on the dendrites of bipolar cells in the outer plexiform layer, and the axons of bipolar cells synapsing on the dendrites of ganglion cells in the inner plexiform layer. Two classes of interneurons provide "lateral" connections in the retina: the horizontal cells are located at the scleral margin of the inner nuclear layer and branch in the outer plexiform layer, whereas most of the amacrine cells are located at the vitread margin and branch in the inner plexiform layer. In general terms, the visual information transduced by the photoreceptors undergoes simple spatial processing in the outer retina and then more complex temporal processing in the inner retina.

The five major classes of retinal neurons together comprise as many as one hundred "natural types", each of which is defined by the regular association of particular morphological, neurochemical and physiological properties.[6-9] Serial reconstructions from electron micrographs indicated that different types of retinal neurons make characteristic patterns of synaptic connections, whether mediated by chemical synapses or gap junctions, and the selective nature of neuronal coupling in the retina has been graphically confirmed by tracer-coupling experiments.[10-12] Although there are important examples in the retina of coupling between different types of neurons which overlap in retinal space, such serial "heterologous" coupling is much less common than parallel "homologous" coupling between neighboring neurons of the same type. Consequently, neuronal coupling in the retina can usually be considered as an isotropic phenomenon occurring in only two dimensions,[13,14] which greatly facilitates both the documentation and the analysis of the coupling, whether considered from morphological or physiological perspectives.

Because the cell coupling is isotropic, retinal neurons injected with junction-permeant dyes or tracers show symmetrical patterns of coupling to surrounding neurons, thus enabling artifactual coupling or "inappropriate" coupling[11] to be reliably detected in retinal wholemounts. Artifactual coupling results from the fusion of cell membranes by the micropipette, producing strong labeling of one or two isolated cells in addition to the injected cell.[15] In the retina, artifactual coupling commonly labels radial glial cells adjacent to the injected cell body, and this occurs more commonly with the depolarizing currents used to eject biotinylated tracers than with the hyperpolarizing currents used to eject fluorescent dyes. In brain-slice preparations, by contrast, injected neurons usually show dye coupling to only one or two other neurons, and it is difficult to exclude the possibility that these cells have been labeled by artifactual coupling.

Some of the features that have made the retina a model system for studying neuronal coupling are now proving advantageous for studying glial coupling. The retina contains specialized radial glia, the Müller cells, which show strong coupling in some lower vertebrates[16] but insignificant coupling in mammals.[17] Müller cell coupling would facilitate the spatial buffering of extracellular potassium by providing a pathway for lateral buffering,[16] prior to the radial "siphoning" of potassium through the Müller cell endfeet to the vitreous.[18] In addition, there is a striking two-dimensional network of astrocytes in the nerve fiber layer of vascularized retina;[19,20] the retinal astrocytes are coupled by prominent gap junctions[21,22] and show extensive dye coupling when injected with Lucifer Yellow.[17]

In most species, the ganglion cell axons remain unmyelinated within the eyecup, thereby preserving the retina's transparency. Most retina are thus devoid of oligodendrocytes, but an important exception is provided by the rabbit retina, which con-

tains a horizontal band of myelinated fibers on each side of the optic nerve head. The diversity of astrocytes and oligodendrocytes in the myelinated band[17,23-25] is similar to that in the optic nerve,[26,27] and thus the band provides an elegant model of CNS white matter that is both accessible and intact. Optical recordings from the glial cells in the myelinated band have recently revealed the presence of slow calcium waves,[28] comparable to those imaged in cultured astrocytes[29-31] and brain slices,[32,33] but it remains to be demonstrated that the calcium signaling in the myelinated band is mediated by the gap junctions between glial cells.

The coupling patterns shown by the major classes of retinal neurons have been systematically described and illustrated in a recent review.[11] This chapter addresses several areas of controversy, where there are outstanding questions concerning the nature and function of cell coupling in the retina. In chapter 6, Reto Weiler reviews the extensive literature on the modulation of gap-junction permeability between retinal neurons.

2. HORIZONTAL CELLS PROVE THE RULES OF NEURONAL COUPLING

The largest gap junctions in the retina occur between horizontal cells and the resultant electrical coupling endows these outer retinal neurons with extensive receptive fields, thus enabling photoreceptors that are widely separated to provide negative feedback to each other. Two years before Revel and Karnovsky[34] coined the term "gap junction", Yamada and Ishikawa[35] demonstrated that horizontal cells of the same morphological type are joined by "fused-membrane structures" in teleost and shark retina. They noted that, in other tissues, such junctions were a device for electrical transmission of stimuli,[36] indicating that the horizontal cells in fish retina are "morphologically separate but functionally a single unit". This insightful conclusion was supported by intracellular recordings from fish horizontal cells, which showed that their hyperpolarization to visual stimuli could be elicited over a very large area.[13,37,38] Dual-recording experiments by Kaneko[39] confirmed that the horizontal cells are electrically coupled and provided the first images of dye coupling between retinal neurons. Parallel experiments on mammalian retina indicated that their horizontal cells are coupled in a similar manner.[40,41]

In a wide range of vertebrate retina, it now appears that each functional type of horizontal cell is homologously coupled, to greater or lesser extent, although for some types the evidence is incomplete or even contradictory.[11] As such, the horizontal cells provide the most suitable case for defining the criteria that should be met to establish the presence of neuronal coupling, both in the retina and elsewhere in the CNS. This will be illustrated here with reference to the horizontal cells in rabbit retina (Fig. 5.2), which show the typical mammalian pattern, although examples from other vertebrate retina will be drawn on to underline the conclusions.

There are two well-defined types of horizontal cells in the rabbit retina: the A-type horizontal cell is axonless whereas the B-type horizontal cell has a short intraretinal axon, 300-600 μm long, which gives rise to a profuse terminal arborization; the dendritic tree of both cell types contacts the synaptic pedicles of cones whereas the axon terminal of the B-type cells contacts the synaptic spherules of rods.[42,43] The A-type cells are one of the few types of horizontal cells in vertebrate retina that have been demonstrated to: (1) make gap junctions; (2) show dye coupling and (3) have extended receptive fields indicative of electrical coupling. Other types of horizontal cells that meet all three of these established criteria include the external horizontal cell in shark retina, the cone horizontal cells in teleost retina, and the luminosity horizontal cell in turtle retina. The receptive field of A-type cells, as mapped with relatively dim stimuli under dark-adapted conditions, is about 20 times wider than the dendritic field,[44,45]

reflecting the relatively large gap junctions between the overlapping dendrites of neighboring A-type cells.[43,46,47] Correspondingly, the A-type cells are the only neurons in the mammalian retina that have been reported to show dye coupling[45] when injected with Lucifer Yellow.[48]

The evidence that the B-type horizontal cells are coupled homologously was less compelling than for the A-type horizontal cells. Somatic recordings from the B-type cells in rabbit retina showed that the receptive field is 5-10 times wider than the dendritic field and this indicated that these cells are electrically coupled, since neither light scattering nor electrical coupling of the presynaptic photoreceptors could account for the mismatch in field size.[44,45] However, the B-type horizontal cells did not show dye coupling when injected with Lucifer Yellow and electron microscopy of ultrathin sections did not reveal gap junctions between the dendrites of B-type cells in cat retina.[47] The latter negative evidence must be viewed with caution because electron microscopy of freeze-fractured primate retina[46] revealed that horizontal cell processes make minute gap junctions (arrays of connexon particles) that were not apparent in thin-sectioned material. Direct evidence that the B-type horizontal cells are coupled was recently obtained by injecting biotinylated tracers into the soma, thus revealing strong dendro-dendritic tracer coupling between neighboring B-type cells.[10,49,50]

Rather surprisingly, the axon terminal of each injected cell also showed tracer coupling to the overlapping terminals of other B-type cells.[49,51] This finding had not been predicted by previous studies on these cells but it is quite consistent with the findings on nonmammalian horizontal cells, whose axon terminals are electrically coupled by extensive gap junctions, which are either permeable (carp, turtle) or impermeable (*Xenopus*) to Lucifer Yellow. In the rabbit retina, the receptive field of the axon terminal is only twice as wide as the arborization and, therefore, it usually does not overlap the spatially offset dendritic

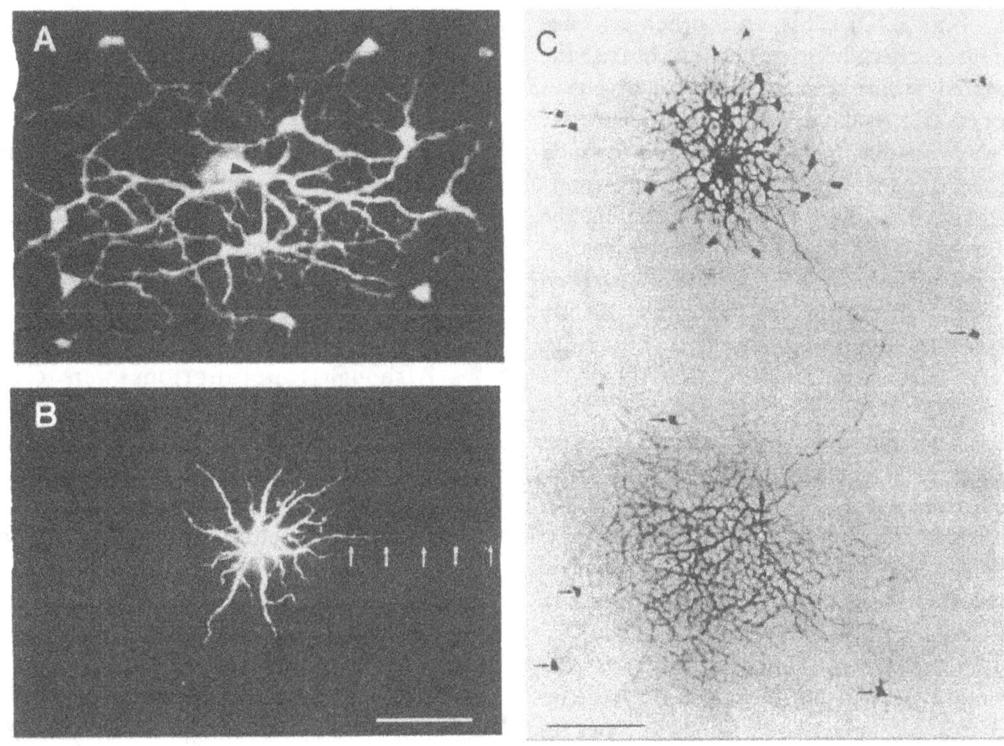

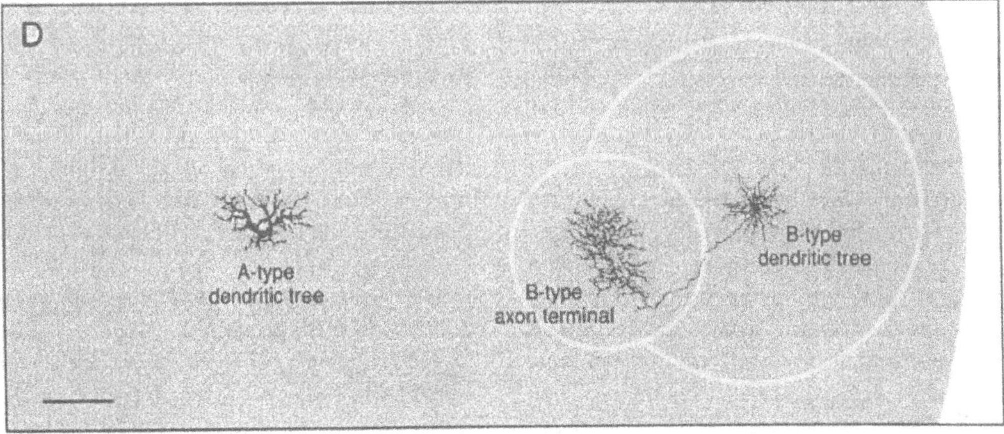

Fig. 5.2. Horizontal cell coupling in the rabbit retina. (A) An axonless A-type horizontal cell injected with Lucifer yellow (arrow head) shows strong dye coupling to surrounding A-type cells; (B) a B-type horizontal cell, identified by its short axon (arrows), does not show dye coupling. (C) The dendritic tree of a B-type horizontal cell injected with Neurobiotin (asterisk) shows tracer coupling to surrounding B-type cells, whose somata form a regular array. The large arrow marks the transition between the thin axon of the injected cell and the thick branches of the terminal arborization, which shows tracer coupling to the overlapping terminals of other B-type cells; the small arrows mark B-type somata that were retrogradely labeled from the coupled terminals. (D) Diagram showing the relative sizes of the receptive fields (grey circles) for the A-type and B-type dendritic trees and the B-type axon terminal in the visual streak of the rabbit retina, based on the physiological data of Bloomfield at al.[51]; strong electrical coupling between the A-type horizontal cells gives rise to a large receptive field (only part of which is shown) that is about is about 20 times wider than the dendritic field. Scale bars = 100 μm. Reprinted with permission from: (A) Vaney DI, Progress Retinal Eye Res 1994; 13:301-55; (B) Vaney DI, Neurosci Lett 1991; 125:187-90; (C) Vaney DI, Proc R Soc Lond B 1993; 252:93-101.

tree (Fig. 5.2D). This and other evidence indicates that, in mammalian retina, the terminal arborization of horizontal cells is electrically isolated from the dendritic tree.[44,52] In mammalian and other tetrapod retina, the horizontal cell dendrites intermingle with the axon terminals in the outer plexiform layer, but the tracer-coupling studies on the rabbit retina showed that the different processes are not directly coupled,[49] supporting the earlier findings of dye-coupling studies on the turtle retina.[48,53] Moreover, although the tracer-coupling technique is exquisitely sensitive when used in combination with photochromic intensification,[54] there was no indication that the well-labeled axon of an injected horizontal cell was directly tracer coupled to either the dendrites, axons or terminals of other B-type cells. The distinctive pattern of coupling suggests that the axon-bearing horizontal cells in tetrapod retina may produce two types of connexin proteins that are segregated in different cellular compartments (dendrites and terminals); alternatively, the synchronous activation of contiguous processes that are connected to the same type of photoreceptors (rods or cones) may stimulate the formation of functional connexons.[49]

The situation may be simpler in the fish retina, where the dendrites and the axon terminals of horizontal cells are spatially segregated. In carp retina, the axons of the luminosity horizontal cells are directly coupled to the axons and terminals of other luminosity cells through gap junctions that are permeable to Lucifer Yellow.[55] These connections provide a "short-circuit" from the dendritic syncitium in the outer plexiform layer to the terminal syncytium in the inner nuclear layer, thus facilitating the spread of the light-evoked responses to the axon terminals, which do not contact photoreceptors directly.[56,57]

Thus, some types of horizontal cells provide clear examples showing that retinal neurons that do not show dye coupling when injected with Lucifer Yellow may, nevertheless, be extensively coupled

through gap junctions. In addition to the B-type horizontal cells in mammalian retina, other examples are provided by both the luminosity and chromatic horizontal cells in amphibian retina.[58-61] Moreover, the weak dye coupling of the rod horizontal cells in some teleost retina (cyprinids, perch) is not commensurate with the strong electrical coupling that is required to produce the extensive receptive fields of these cells.[62-64]

Although gap junctions were commonly thought to be permeable to fluorescent dyes such as Lucifer Yellow,[48] procion Yellow[65] and fluorescein and its derivatives,[66,67] it is now recognized that these dyes have only limited utility for studying cell coupling in the nervous system.[1,11] Studies on a variety of neuronal and nonneuronal tissues have shown that cells that are electrically coupled may not show dye coupling with Lucifer Yellow.[68-72] This dye is sub-optimal for probing the permeability of gap junctions in three respects: (1) The molecule's abaxial width of 1.3 nm[73] is close to the 1.6 nm channel diameter determined empirically for mammalian connexons;[74] consequently, the permeability of the channel to Lucifer Yellow is at least two orders of magnitude less than to current-carrying potassium ions,[75] which have an hydrated diameter of 0.46 nm.[76] (2) Lucifer Yellow's negative charge probably reduces its channel permeancy compared to electroneutral or positively charged probes of similar size.[67,77] (3) Lucifer Yellow binds irreversibly to the cytoplasm and, thus, the dye may be freely diffusable for only a few minutes after injection;[78] consequently, insufficient dye may pass through the gap junctions to be detected in the coupled cells, regardless of the incubation time.

These diverse factors make it difficult to determine whether the absence of dye coupling between neurons that are known to be electrically coupled or connected by gap junctions actually reflects weak coupling, or whether the cells are coupled by connexons that are impermeable to classical fluorescent dyes. Thus it cannot be excluded that the A-type coupling and the B-type coupling are both mediated by only one type of connexon that is marginally permeable to Lucifer Yellow. However, biophysical measurements on pairs of coupled blastomeres demonstrated a linear relationship between the junctional permeability to a tracer molecule and the junctional conductance to all ions.[76] Given that the B-type somata appear to be well coupled electrically but show no evidence of dye coupling, this strongly suggests that the connexons between B-type dendrites differ in either connexin type or subconductance state[77] from the Lucifer-permeable connexons of the A-type cells.[49] Amongst other things, this would (1) allow for differential metabolic coupling of second-messenger molecules, depending on their size and charge, and (2) enable the degree of coupling to be differentially modulated for each cell type. For example, the uncoupling of horizontal cells by cAMP[63,79,111] which is negatively charged, may be confined locally for the B-type cells but spread laterally for the A-type cells. In summary, the retinal horizontal cells provide a well-defined case for examining the expression, formation and regulation of connexon heterogeneity in the CNS.

3. PROMISES AND PITFALLS OF THE TRACER-COUPLING TECHNIQUE

The use of small biotinylated tracers to reveal patterns of cellular coupling[10] has overcome many of the problems associated with the fluorescent molecules used in dye-coupling studies. Both biocytin[80] (373 D) and Neurobiotin™[81] (286 D) are smaller than Lucifer Yellow (443 D) and they are positively charged at physiological pH. The biotin moiety enables the tracers to be secondarily labeled with avidin conjugates of relatively large reporter molecules, such as fluorescein isothiocyanate (FITC) or horseradish peroxidase (HRP). Thus the biotinylated tracers can be visualized by the avidin-biotin-peroxidase complex (ABC)

procedure commonly used for immunohistochemistry,[82] resulting in an opaque reaction product that is stable and very sensitive, particularly when intensified.[54,83]

Many of the neuronal types in the retina show homologous tracer coupling when injected with biocytin or Neurobiotin, thus enabling more populations of neurons to be selectively labeled by the tracer-coupling technique than by immunocytochemistry with presently available antibodies. Moreover, tracer injection often reveals the complete morphology of the injected cell together with the local array of coupled somata; in immunolabeled retina, by contrast, single cells are usually stained incompletely or obscured by the surrounding cells. Unlike Lucifer Yellow,[84,85] biotinylated tracers do not mask the antigenicity of other molecules,[86-88] thus allowing for quantitative immunocytochemical analysis of the neuroactive substances in the tracer-coupled cells; the injected cell should be excluded from analysis because its contents may washout if the cell is damaged by the micropipette. The scope for such double-labeling studies has been enhanced by the recent production of sensitive antisera against formaldehyde conjugates of amino acid transmitters, which are effective on lightly fixed retinal wholemounts and brain slices.[89] Thus, for many types of retinal neurons, the tracer-coupling technique enables the morphology, distribution and transmitter content to be characterized in a single preparation (Fig. 5.3).

Tracer injection revealed that neuronal coupling in the retina is more common, more diverse and more complex than indicated by previous studies,[10,11] raising the question of whether the tracer coupling is actually mediated by gap junctions. It may be difficult to provide a definitive answer if tracer coupling is actually more sensitive than other techniques for demonstrating cellular coupling in intact tissue. Certainly, retinal neurons that appeared to be coupled on the basis of ultrastructural or physiological evidence, now show appropriate patterns of tracer coupling when injected with biocytin or Neurobiotin. The clearest example is provided by the AII (rod) amacrine cell, which was the first specific type of neuron shown to make gap junctions in the inner retina.[90] The AII amacrine cell is the third-order neuron in the rod-signal pathway of the mammalian retina (Fig. 5.1), and its dendrites make both homologous gap junctions with neighboring AII cells and heterologous gap junctions with several types of cone bipolar cells that terminate in sublamina _b_ of the inner plexiform layer.[7,90-92] The AII amacrine cells do not show dye coupling,[93,94] but show a complex pattern of tracer coupling that accurately reflects their gap-junction connectivity[10,95,96] (Fig. 5.4).

By analogy, tracer coupling between other types of neurons can be parsimoniously attributed to tracer diffusion through gap junctions. Moreover, this conclusion can be reconciled with the apparent absence of gap junctions on some neurons that show tracer coupling, such as retinal ganglion cells. Freeze-fracture studies[97] on monkey and rabbit retina revealed that many of the gap junctions in the inner plexiform layer are either quite small or comprise loosely arranged connexons enclosing islands of particle-free membrane, and thus they might not be recognized in ultrathin sections. In freeze fractures, it is often difficult to identify the class of neuron, let alone the specific type, that gives rise to a particular gap junction. Moreover, a junction of minimal size could, in principle, consist of only a single connexon or a couple of connexons associated with a narrowing of the intercellular cleft.[98,99]

Nevertheless, it cannot be excluded that tracer coupling occurs by two mechanisms, one mediated by gap junctions and the other by a form of intercellular communication that has yet to be characterized. However, given (1) that each neuronal type shows a specific pattern of tracer coupling following intracellular injection, and (2) that such patterns are not seen following

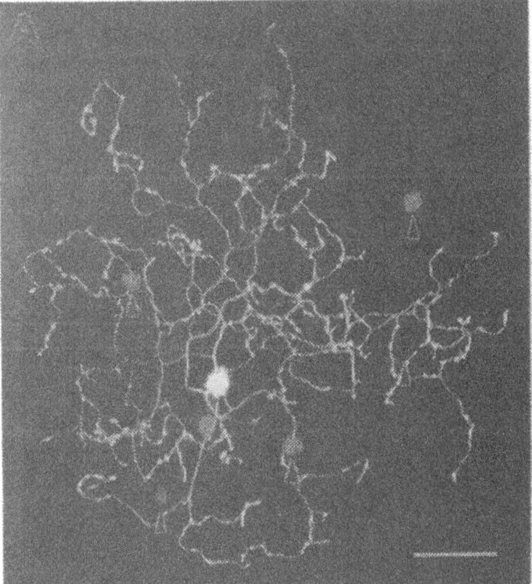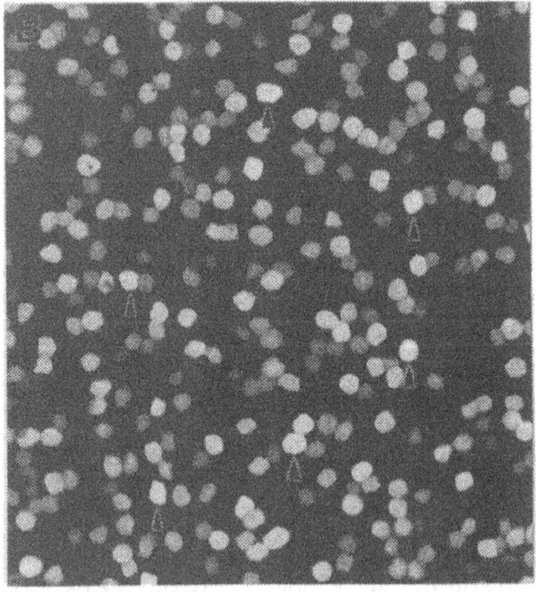

Fig. 5.3. Confocal fluorescence micrographs of a rabbit retinal wholemount that was double-labeled to reveal a Neurobiotin-injected DAPI-3 amacrine cell[10,116] (A) and the underlying mosaic of glycinergic amacrine somata (B); the Neurobiotin was visualized with streptavidin-Texas Red whereas the glycine antibody[89] was visualized with an FITC-tagged secondary antibody. The injected cell, which was reconstructed from a Z-series of 15 images, shows homologous tracer coupling to six neighboring DAPI-3 cells (arrow heads), each of which shows strong glycine immunoreactivity. Scale bar = 20 μm. Provided by LL Wright.[150]

extracellular application of the tracer, then an alternative mechanism would also require a "direct line" between the tracer-coupled neurons, such as might be provided by juxtaposed transporter proteins. This in turn would probably require the close apposition or fusion of adjacent membranes, similar in appearance to gap junctions, and thus an absence of structures resembling gap junctions does not provide positive support for alternative mechanisms of cellular coupling.

With some types of retinal neurons, the tracer coupling is confined to cells that are in direct contact with the injected neuron and, even then, the labeling may be apparent only in the somata (biotinylated tracers appear to bind preferentially to the

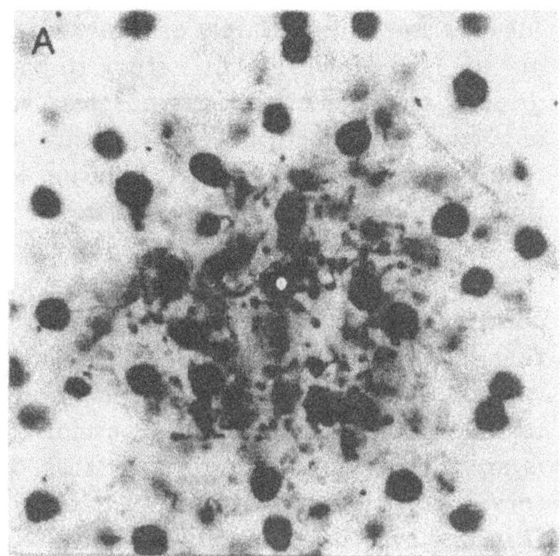
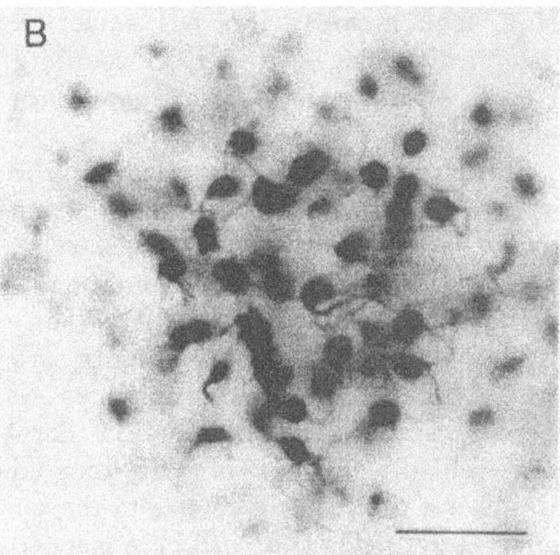

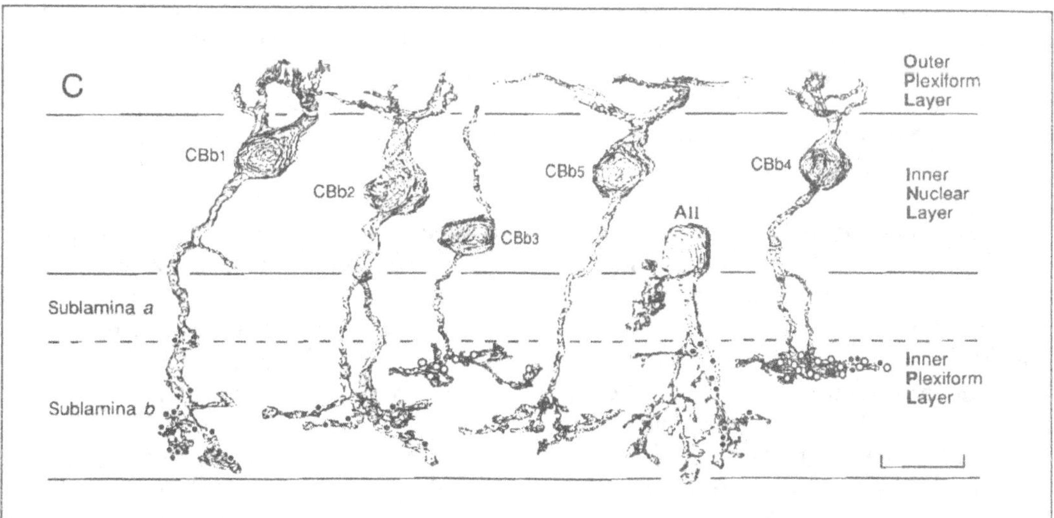

Fig. 5.4. Coupling pattern of AII (rod) amacrine cells in the cat retina. An AII amacrine cell injected with Neurobiotin (white dot) shows strong tracer coupling to surrounding AII cells (A; focus at the IPL-INL border) and to several types of cone bipolar cells whose somata are located deeper in the INL (B; focus at the INL-OPL border). Scale bar = 20 μm. (C) Electron-micrograph reconstructions show that the AII amacrine cells make gap junctions (black dots) with three types of cone bipolar cells in sublamina b of the IPL, and that the b4 bipolar cells are both homologously coupled to each other and heterologously coupled to the b3 bipolar cells (white dots); the dendrites of AII amacrine cells also make homologous gap junctions but these are not shown. Scale bar = 10 μm. (C refigured from Cohen and Sterling[91] and Sterling et al[149]).

nucleus). The functional significance of this limited tracer coupling is unclear, although it now appears that the coupling is not induced by the iontophoretic current used to inject the tracer.[100,101] By contrast, neurons that are known to be electrically coupled or joined by prominent gap junctions, show more extensive tracer coupling, requiring tracer movement through multiple sets of gap junctions. Well-characterized examples are provided by fish bipolar cells[102] and mammalian AII amacrine cells,[10] which may show tracer coupling through at least six sets of gap junctions.[95,103]

Similarly, Vaney[10] reported that the A-type horizontal cells in rabbit retina showed tracer coupling to about 900 cells, covering an area of 2.5 mm diameter; under the same conditions, the B-type horizontal cells showed tracer coupling to about 100 cells, covering an area of 0.6 mm diameter. For each type of horizontal cell, the network of tracer-coupled neurons was comparable in size to the visual receptive fields that had been mapped previously,[44,45] suggesting that tracer coupling may provide a useful index of the strength of electrical coupling.[11] Combined physiological and morphological measurements on rabbit horizontal cells[51] have since indicated that the extent of tracer coupling is linearly related to the receptive-field diameter over a 10-fold range. This was somewhat surprising because the space constant of a horizontal cell network is determined by both the internal (junctional) resistance of the neural sheet (Rs) and the leakage resistance across the surface membrane (Rm).[13,14] Bloomfield et al[51] accounted for the linearity by suggesting that the movement of Neurobiotin is largely driven by the iontophoretic current, and thus its spread would be prescribed by the same factors (Rs and Rm) that limit the light-evoked currents underlying the receptive-field responses. Although Kamermans[104] made a similar argument regarding the spread of Lucifer Yellow through horizontal cells, direct microscopic observation indicates that this

fluorescent dye continues to diffuse through the network of A-type cells for a few minutes after iontophoresis has stopped.[79]

Bloomfield et al[51] argued that passive diffusion of Neurobiotin cannot account for the spatio-temporal characteristics of tracer spread through the A-type horizontal cells in rabbit retina: the maximal spread (2.0-2.5 mm diameter) was reached after 8-10 min of iontophoresis (+0.4 nA), and did not vary significantly whether the tissue was fixed immediately or incubated for up to 3 hr. However, this study did not directly test whether the tracer spread varied with the iontophoretic current, nor whether there was significant diffusion following shorter injection times, as suggested by the results of earlier studies.[10,95] Thus, the results of Bloomfield et al[51] are also compatible with an alternative hypothesis that Neurobiotin diffuses passively for a limited time, after which it becomes bound to the cytoplasm, with a time constant that is an order of magnitude greater than observed for Lucifer Yellow. Neither this "binding hypothesis" nor Bloomfield's "current hypothesis" exclude the possibility that biotinylated tracers may be further spread by active transport processes, thus accounting for the extensive movement of tracer along the axons of B-type horizontal cells, wide-field amacrine cells and retinal ganglion cells.

Studies on the retina have usually assumed that changes in dye coupling or tracer coupling reflect comparable changes in electrical coupling.[63,95,105] However, if the spread of biotinylated tracers through a network is driven by the iontophoretic current to a significant extent, then changes in tracer coupling under different conditions could also reflect changes in the membrane resistance.[51,106] Unfortunately, the present evidence is both incomplete and contradictory, perhaps because different injection paradigms favor different mechanisms of tracer movement. This clouds the interpretation of some problematic examples of tracer coupling in the inner retina that do not appear to be associated with an enlarged receptive field.

The most publicized case is provided by the On-Off direction-selective (DS) ganglion cells in rabbit retina, which comprise four discrete subtypes that orthogonally code the direction of image motion.[107] The first tracer-coupling experiments on the retina[10,108] revealed that a minority of the presumptive DS cells show homologous coupling, which is sometimes strong enough to label both the neighboring cells and an outer ring of cells that do not contact the injected neuron directly (Fig. 5.7C). Tracer injection into two or three closely spaced DS cells with overlapping dendritic trees, which presumably comprised different physiological subtypes, indicated that only one subtype of DS cell shows tracer coupling, but its preferred direction has yet to be determined.[109] By contrast, electrophysiological recordings[110] in isolated rabbit retina showed that all On-Off DS cells have a receptive field that is comparable in size to the dendritic field, thus providing no evidence that electrical coupling enlarges the receptive field of any of the four subtypes; however, Yang and Masland[110] did not establish whether any of the recorded DS cells showed tracer coupling under their experimental conditions. This leaves open the possibility that the strength of neuronal coupling may depend on the level of background illumination,[105,111] as recently reported for the horizontal cells and the AII amacrine cells in rabbit retina;[112] at one extreme, the tracer coupling of the DS cells might have been triggered by the intense epi-illumination used to visualize the fluorescently labeled neurons.[109]

These qualifications do not apply to comparable experiments by Bloomfield and Xin[103] on amacrine cells and ganglion cells recorded in a rabbit eyecup preparation, because they measured the receptive field both before and after Neurobiotin injection. Rather surprisingly, most types of tracer-coupled neurons in the inner retina had a receptive-field center that approximately matched the size of the dendritic field, confirming the results of an earlier HRP-injection study.[113] If this tracer movement is mediated by gap junctions, it is unclear why the tracer coupling is not associated with concomitant electrical coupling, which would be reflected in an enlarged receptive field. While these paradoxical findings could reflect transport mechanisms that do not involve gap junctions,[11,110] studies on the α ganglion cells in cat retina provide direct evidence that such tracer coupling is actually associated with weak electrical coupling, as described in the following section. It is possible that the visually evoked currents are shunted by tonically activated chloride channels located in close proximity to the gap junctions, whereas the passive diffusion of membrane-impermeant tracers would be unaffected by an inhibitory shunt. Such a mechanism would thus limit electrical coupling but not metabolic coupling.

4. OCCURRENCE AND FUNCTION OF HETEROLOGOUS COUPLING IN THE RETINA

Tracer injection revealed numerous examples of heterologous coupling in the retina, not only between different neuronal types of the same class, but also between different classes of neurons, including bipolar-amacrine coupling and amacrine-ganglion coupling. Heterologous coupling is of particular interest because its serial nature contributes to the temporal processing of visual information, whereas the parallel nature of homologous coupling is more suited to spatial processing. Moreover, heterologous coupling has greater potential than homologous coupling for the differential expression or regulation of connexins in the coupled cells, which may be a necessary prerequisite for the unidirectional coupling of ions or metabolites.

The pronounced gap junctions between AII amacrine dendrites and cone bipolar terminals provide the definitive example of heterologous coupling in the retina[7,90-92] (Fig. 5.4C). In mammals, the rod bipolar cells do not contact the ganglion cells directly but synapse on the AII amacrine cells. These narrow-field neurons respond to illumination with a transient-sustained depolarisation,[113-115] and this On response is then fed into the cone-signal pathways through either sign-inverting glycinergic synapses with Off-center cone bipolars or sign-conserving electrical synapses with On-center cone bipolars (Fig. 5.1). Gap junctions thus enable neurons that contain an inhibitory transmitter to make excitatory synapses, and the prevalence of neuronal coupling in the retina may be linked to the finding that most types of amacrine cells and horizontal cells contain either GABA or glycine.[116]

Intracellular recordings[117] in the cat retina showed that the response characteristics of the On-center cone bipolar cell resemble those of the AII amacrine cell rather than the cone photoreceptor, thus providing indirect evidence that the AII cells are electrically coupled to the bipolar cells, under illumination conditions where both the rods and cones are responsive. However, such coupling may not operate in the reverse direction, at least in the light-adapted retina. In both cat and rabbit retina, the responses of the AII cells are virtually abolished as the background illumination approaches rod saturation,[114,115] whereas the On cone bipolars show a strong sustained depolarization. If these bipolar cells are symmetrically coupled to the AII cells, then the graded responses of the bipolar cells to photopic stimuli of increasing intensity should be mirrored in the AII cells, but this is not the case in rabbit retina.[115] Intracellular recordings thus suggest that the heterologous gap junctions between AII cells and cone bipolar cells may be rectifying synapses, as has been occasionally demonstrated for heterologously coupled neurons in other systems.[118-121]

Additional support for this hypothesis is provided by preliminary evidence that cone bipolar cells injected with Neurobiotin do not show tracer coupling to AII amacrine cells.[11] If confirmed, the asymmetric tracer permeability of the heterologous gap junctions should be interpreted with caution because such "chemical rectification" may not be synonymous with electrical rectification, which is dependent on a potential difference across the gap junction.[122-124] The heterologous gap junctions have an asymmetric morphology, with "fluffy material" lining the amacrine side,[92,125] and this may represent a structural correlate of signal rectification, as shown for the electrical synapses between giant fibers and pectoral-fin adductor motor neurons in the hatchetfish.[126] In the latter system, however, the electron-dense material is associated with the postsynaptic side of the heterologous gap junction whereas, in the retina, it is associated with the amacrine side, which is presumed to be presynaptic.[92,125] A similar effect to unidirectional coupling could also be achieved by simply uncoupling the On cone bipolar cells from the AII amacrine cells under light-adapted conditions. Hampson et al[95] showed that dopamine greatly reduces the tracer coupling between AII amacrine cells but, for methodological reasons, it was not possible to determine whether the heterologous coupling is modulated in a similar fashion.

When retinal ganglion cells are injected with Neurobiotin, many types show heterologous tracer coupling to distinct populations of amacrine cells and, in some cases, the injected cell also shows homologous tracer coupling to neighboring ganglion cells.[10,11] While it cannot be excluded that the ganglion cell labeling is mediated indirectly by the coupled amacrine cells,[10,12] heterologous coupling can produce strong labeling of the amacrine cells' dendrites, without giving rise to secondary labeling of overlying ganglion cells.[11] Moreover, there appears to be little evidence that Neurobiotin-injected amacrine cells show tracer coupling to ganglion cells, suggesting that the coupling between ganglion cells and amacrine cells may be unidirectional. Thus, the heterologous coupling would enable these classic projection neurons to provide feedback to the inner plexiform layer, while preserving the conventional postsynaptic and presynaptic functions of the dendrites and axons, respectively.

The tracer-coupling findings were unexpected because the hundreds of papers on retinal ultrastructure appear to contain no reports that ganglion cell dendrites make gap junctions, with the notable exception of a recent abstract.[127] However, a decade before tracer coupling sparked the present interest in neuronal coupling in the retina, a series of elegant electrophysiological experiments by Mastronarde[128] indicated that the α ganglion cells in cat retina are electrically coupled (Fig. 5.5A). Antidromic activation of single α cells, which are the largest ganglion cells in the retina, increased the firing rate of neighboring α cells of the same subtype (On-center or Off-center) with a 1-4% success rate. When almost all α cells were activated by stimulating the contralateral optic tract, this increased the firing rate of isolated, ipsilaterally projecting α cells with a 25% success rate. The timing of the interactions indicated that they did not involve chemical synapses and did not occur along axons. Mastronarde[128,129] thus proposed that the interactions were mediated by electrical synapses, either directly between the ganglion cells or between each ganglion cell and dedicated interneurons. The effect was sufficiently weak that it would not be reflected in an enlarged receptive field, although it is possible that the strength of coupling varies with the level of illumination, thus accounting for the 2-fold increase in the receptive-field diameter of α cells, going from light-adapted to dark-adapted conditions.[130]

Direct support for Mastronarde's hypothesis is now provided by the tracer-coupling pattern of the α ganglion cells, which show both homologous coupling to 5-7 surrounding α cells of the same subtype and heterologous coupling to several types of stratified amacrine cells[10,11] (Fig. 5.5B). The parasol ganglion cells in macaque retina[12] and the α ganglion cells in ferret retina[131] show similar patterns of tracer coupling. Ultrastructural studies indicate that the neuronal connections mediated by gap junctions are as stereotyped as those mediated by chemical synapses,[91] and thus

tracer coupling provides readily obtained "signatures" of synaptic connectivity, without resorting to the laborious reconstruction of serial electron micrographs. The relatively complex signatures associated with heterologous coupling may be used both to distinguish different types of neurons in the same retina and to identify equivalent types in diverse retina, thus complementing other morphological, neurochemical and physiological criteria.[11]

Ultrastructural studies[132] on CNS white matter demonstrated that astrocytes and oligodendrocytes are connected by gap junctions and, correspondingly, recordings[71] from explant cultures of spinal cord indicated that the two classes of macroglia are electrically coupled, albeit weakly. By contrast, macroglial cells injected with Lucifer Yellow in either the optic nerve[133] or cell cultures[71] showed only homologous dye coupling, and did not show any heterologous dye coupling between astrocytes and oligodendrocytes. However, recent studies on the myelinated band of rabbit retina have revealed complex patterns of coupling between retinal macroglia.[17,25] Periaxonal astrocytes injected with Lucifer Yellow showed prominent dye coupling to both astrocytes and oligodendrocytes, whereas Lucifer-filled oligodendrocytes rarely showed either homologous or heterologous dye coupling (Fig.5.6); when oligodendrocytes were injected with biocytin, about one-third showed weak tracer coupling to both classes of macroglia. Taken together, these results[17] indicate that the heterologous coupling has a unidirectional bias, with larger metabolites moving more freely from the astrocytes to the oligodendrocytes than vice-versa. This would reduce the potential for reciprocal signaling, thus imposing a hierarchical organization on the "metabolic syncytium"[71] of CNS macroglia.

5. PATTERNS AND DEVELOPMENT OF COUPLING IN TERRITORIAL NETWORKS

The somatic distribution and dendritic spread of each neuronal type are generally organized so as to provide uniform cover-

Fig. 5.5. Neuronal coupling of α ganglion cells in the cat retina. (A) Correlations in the unstimulated firing of a pair (P and Q) of overlapping Off-center α cells; the schematic circuit shows the connections that may underlie each part of the correlogram: fast excitatory effects of P on Q and of Q on P give rise to peaks 1 and 2, respectively, whereas shared excitatory input gives rise to the central peak 3. (B) The Neurobiotin-injected α cell in the center of this low-power micrograph shows homologous tracer coupling to five neighboring α cells and heterologous coupling to about 20 underlying amacrine cells. Scale bar = 100 μm. Reprinted with permission from: (A) Mastronarde DN, Trends Neurosci 1989; 12:75-80; (B) Vaney DI, Neurosci Lett 1991; 125:187-90.

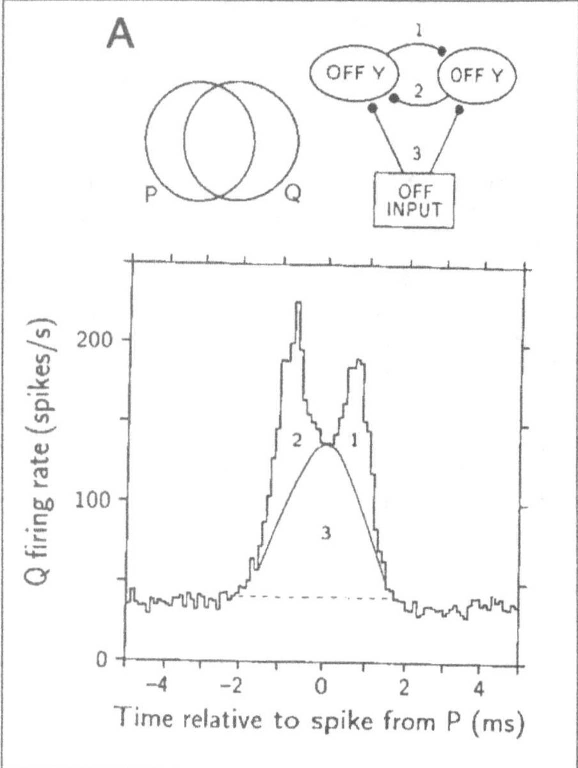

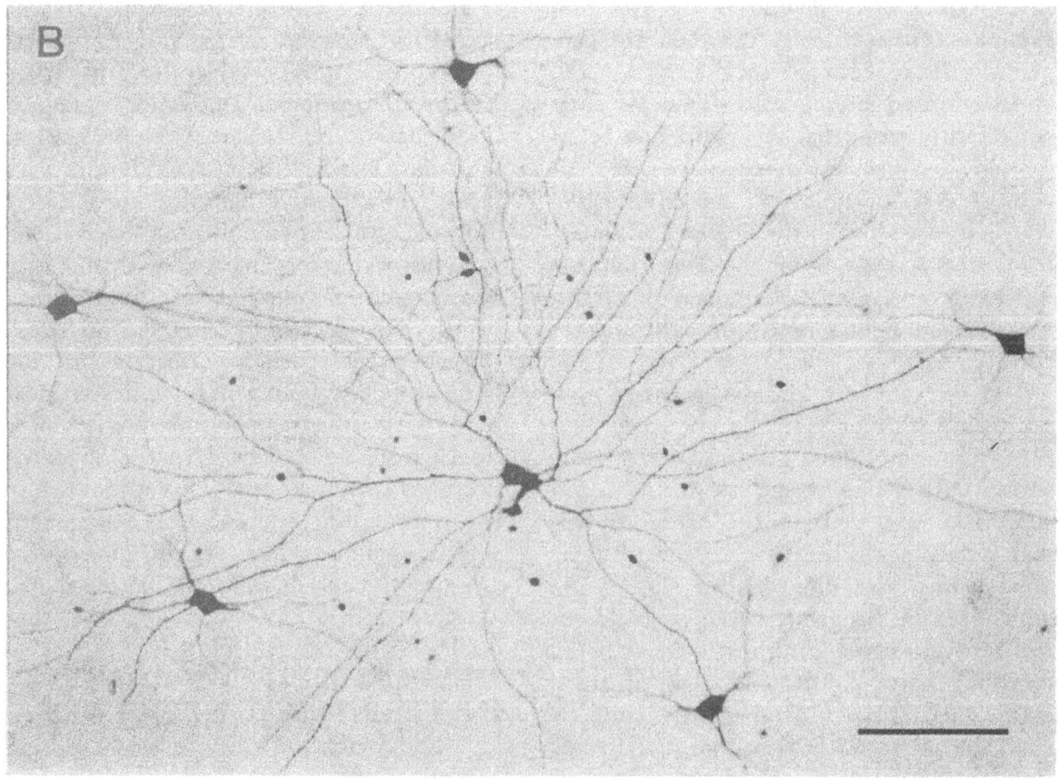

age of the retina.[134] In many cases, this is achieved by dendritic-field overlap, which may range in extent from the 2-fold coverage of the AII amacrine cells[93,94] to the almost 1000-fold coverage of the AI amacrine cells.[135,136] Both types of amacrine cells, which form the postsynaptic elements at the ribbon synapses of rod bipolar cells,[137] show strong homologous tracer coupling, even though they have markedly different dendritic architectures.[10,136] By contrast, other types of retinal neurons with either limited coverage (e.g., β ganglion cells) or extensive coverage (e.g., cholinergic amacrine cells) show no tracer coupling, either in the mature retina or prior to eye opening.[10,11,131,138] Taken together, these findings indicate that neuronal coupling is unlikely to be essential for the development or maintenance of the spatial organization of neuronal arrays in general, although such a role cannot be discounted

where coupling is present at an appropriate stage of development.[139,140]

Important exceptions may be provided by those retinal neurons whose spatial extent appears to be physically bounded by dendritic contact with neighboring cells: this territorial organization achieves seamless coverage of the retina with minimal overlap of the dendritic trees. The best characterized example is provided by the On-Off transient amacrine cells in fish retina, which are electrically coupled by large gap junctions that are permeable to Lucifer Yellow.[141,142] The coupled amacrine cells make dendritic contacts either from tip-to-shaft or from tip-to-tip, forming a remarkable pattern of "closed loops" in which it is difficult to tell where the coupled cells begin and end.[87,143,144] Such "exclusive territoriality" is exhibited by diverse types of bipolar cells, amacrine cells and ganglion cells, most of which show strong homologous tracer coupling (Fig. 5.7). The gap junctions on the distal

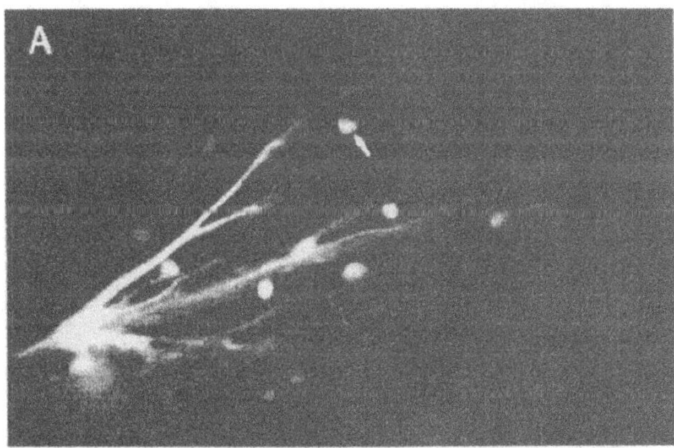
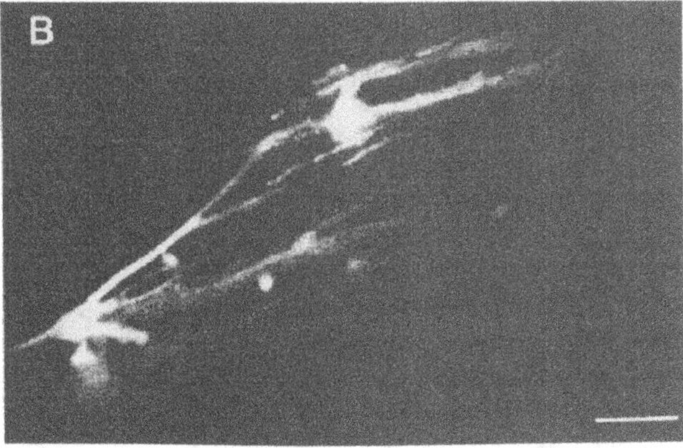

Fig. 5.6. *Unidirectional dye coupling between retinal macroglia. (A) A periaxonal astrocyte in the myelinated band of rabbit retina was injected with Lucifer yellow and it shows dye coupling to surrounding astrocytes and oligodendrocytes; (B) one of the dye-coupled oligodendrocytes was subsequently injected with Lucifer yellow, but this did not result in further dye coupling. Scale bar = 50 μm. Reprinted with permission from Robinson SR et al, Science 1993; 262:1072-4.*

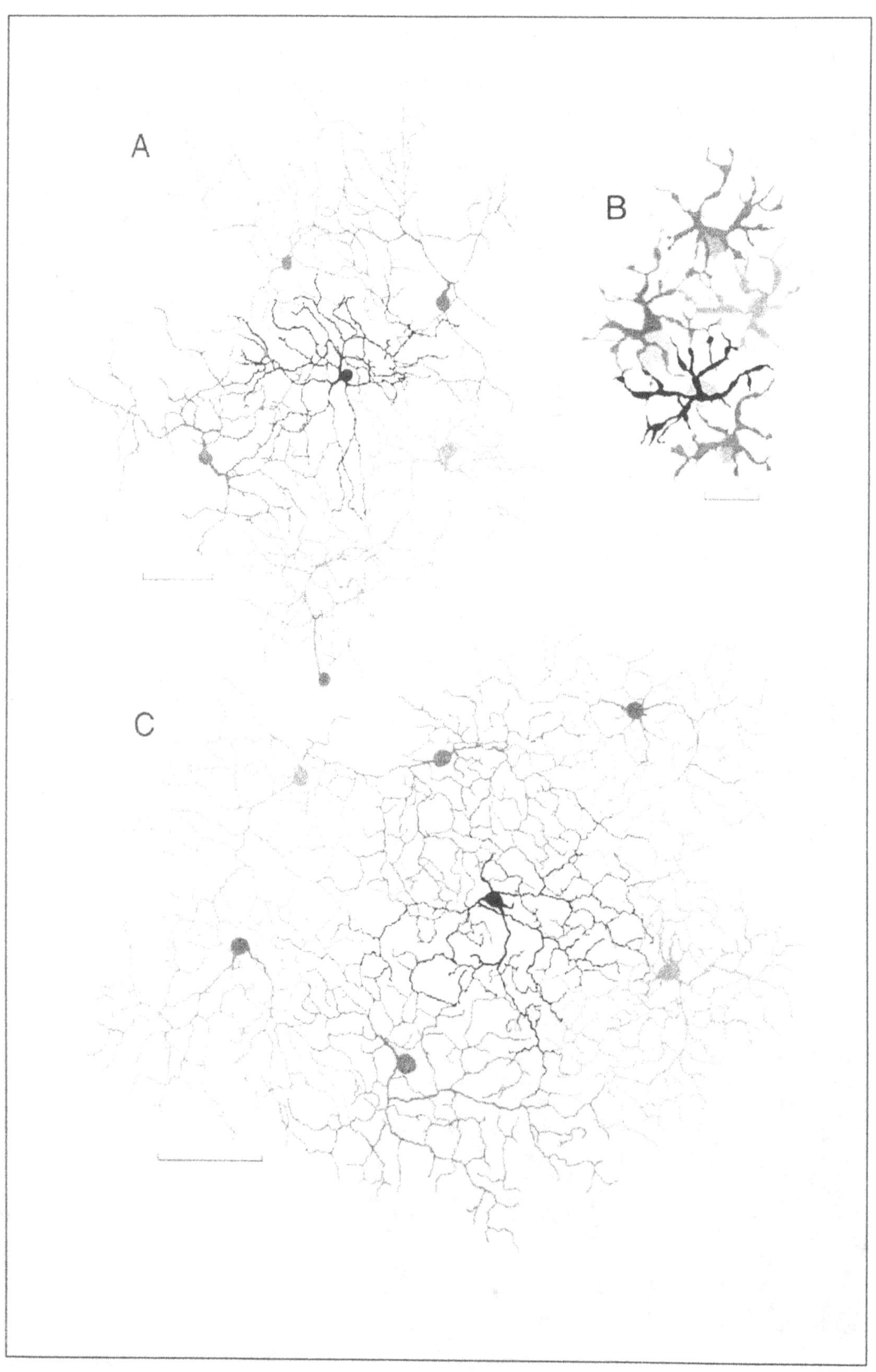
A
B
C

dendrites that mediate this coupling are thus appropriately placed to transmit intercellular signals that would inhibit further dendritic growth, as proposed for the anterior pagoda neurons in the leech embryo.[145]

The exclusive territoriality of the dendritic trees of the On-Off direction-selective (DS) ganglion cells (Fig. 5.7C) resembles, on a larger scale, the territorial organization of the dendritic branching systems of individual cells.[109,146] The terminal dendrites of a DS ganglion cell are distributed throughout the dendritic field, creating a regular space-filling lattice in which some dendrites appear to form closed loops, which may be an internal expression of the tip-to-shaft contacts that are formed between the dendrites of neighboring DS cells of the same subtype. This suggests that both the branching pattern and the spatial extent of these neurons are shaped by common mechanisms, which may be mediated by gap junctions. However, most of the terminal dendrites located inside the dendritic field do not form closed loops, perhaps because self-recognition constrains the formation and/or retention of reflexive coupling.[147]

Although each of the four subtypes of On-Off DS cells exhibit exclusive territoriality,[148] it appears that only one subtype shows tracer coupling in the adult retina.[109] If the territorial organization is shaped by neuronal coupling, then all of the subtypes should show tracer coupling in the developing retina. DeBoer and Vaney[138] recently reported that the DS ganglion cells in rabbit retina have an adult-like organization by postnatal-day 10 (PD 10), at the time of eye opening. At PD 5 (and PD 3), however, most of the DS ganglion cells injected

with Neurobiotin appeared to be coupled not only to a regular array of surrounding DS cells, but also to one (or more) arrays of overlying DS cells (Fig. 5.8). Consequently, tracer-coupled somata were located inside the dendritic field of the injected cell, unlike the adult pattern, resulting in multiple overlap of their dendritic fields. Thus, there is neuronal coupling both within and between subtypes of developing DS cells, with subsequent uncoupling (either partial or complete) to produce the adult pattern.[138] The immature coupling pattern appears to be more complex than would be required to simply shape the territorial organization of the DS ganglion cells; therefore, this coupling may serve additional processes requiring coordinated communication between the different subtypes of DS cells, such as the appropriate selection of asymmetric inputs to generate the four preferred directions.

These and other findings now indicate that the patterns of neuronal coupling in the developing retina are unexpectedly intricate. However, this would be in keeping with the numerous previous studies on retinal gap junctions, which have revealed both remarkable complexity and intriguing subtlety in their structure and function. Retinal studies have contributed greatly to the emerging perception that gap junctions are dynamic sophisticated components of neuronal circuits and they promise to play an equally important role in the future.

ACKNOWLEDGMENT
D.I. Vaney is an NHMRC Principal Research Fellow.

Fig. 5.7. Reconstructions of territorial networks labeled by homologous tracer coupling; the cell injected with the biotinylated tracer is drawn in black while the coupled cells are drawn in grayscale. (A) Bistratified GABAergic amacrine cells[150] in the rabbit retina, showing the dendritic arborization in sublamina b of the inner plexiform layer. Scale bar = 50 μm. (B) Off-center bipolar cells in the black bass retina, showing the dendritic trees in the outer plexiform layer. Scale bar = 50 μm. (C) Subtype of On-Off direction-selective ganglion cells in the rabbit retina, showing the dendritic arborization in sublamina b of the inner plexiform layer. Scale bar = 100 μm. Reprinted with permission from: (A) provided by LL Wright; (B) refigured from Umino et al.[102]

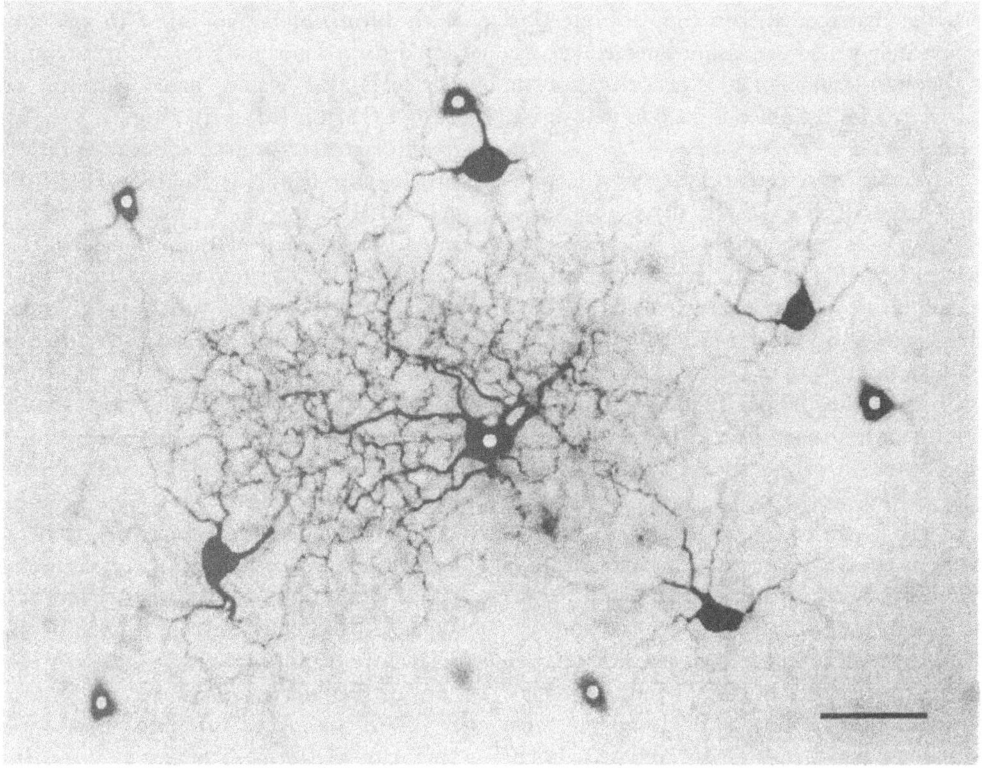

Fig. 5.8. Tracer-coupling pattern of an On-Off direction-selective (DS) ganglion cell in the retina of a five-day old rabbit. The Neurobiotin-injected DS cell in the center of the field shows weak coupling to five surrounding DS cells of the same subtype (white dots) and stronger coupling to four overlapping DS cells of a different subtype. Scale bar = 50 μm. Provided by DJ DeBoer.

REFERENCES

1. Dermietzel R, Spray DC. Gap junctions in the brain: where, what type, how many and why? Trends Neurosci 1993; 16:186-92.
2. Gutnick MJ, Lobel-Yaakov R, Rimon G. Incidence of neurons dye-coupling in neocortical slices depends on the plane of section. Neurosci 1985; 15:659-66.
3. Boos R, Schneider H, Wässle H. Voltage- and transmitter-gated currents of AII-amacrine cells in a slice preparation of the rat retina. J Neurosci 1993; 13:2874-88.
4. Dowling JE. The Retina: An Approachable Part of the Brain. Cambridge, Mass: Belknap Press, 1987:282.
5. Wässle H, Boycott BB. Functional architecture of the mammalian retina. Physiol Rev 1991; 71:447-80.
6. Rodieck RW, Brening RK. Retinal ganglion cells: properties, types, genera, pathways and trans-species comparisons. Brain Behav Evol 1983; 23:121-64.
7. Sterling P. Microcircuitry of the cat retina. Ann Rev Neurosci 1983; 6:149-85.
8. Vaney DI, Hughes AA. Is there more than meets the eye? In: Blakemore C, ed. Vision: Coding and Efficiency. Cambridge: Cambridge University Press, 1990:74-83.
9. Kolb H. The architecture of functional neural circuits in the vertebrate retina. Invest Ophthalmol Vis Sci 1994; 35:2385-404.
10. Vaney DI. Many diverse types of retinal neurons show tracer coupling when injected with biocytin or Neurobiotin. Neurosci Lett 1991; 125:187-90.
11. Vaney DI. Patterns of neuronal coupling in the retina. Progress Retinal Eye Res 1994; 13:301-55.
12. Dacey DM, Brace S. A coupled network for parasol but not midget ganglion cells in

the primate retina. Vis Neurosci 1992; 9:279-90.

13. Naka K-I, Rushton WAH. The generation and spread of S-potentials in fish (Cyprinidae). J Physiol 1967; 192:437-61.

14. Lamb TD. Spatial properties of horizontal cell responses in the turtle retina. J Physiol 1976; 263:239-55.

15. Kaneko A, Nishimura Y, Tauchi M et al. Morphological observation of retinal cells presumably made syncitial by an electrode penetration. J Neurosci Methods 1981; 4:299-303.

16. Mobbs P, Brew H, Attwell D. A quantitative analysis of glial cell coupling in the retina of the axolotl (Ambystoma mexicanum). Brain Res 1988; 460:235-45.

17. Robinson SR, Hampson ECGM, Munro MN et al. Unidirectional coupling of gap junctions between neuroglia. Science 1993; 262:1072-4.

18. Newman EA, Frambach DA, Odette LL. Control of extracellular potassium levels by retinal glial cell K+ siphoning. Science 1984; 225:1174-5.

19. Schnitzer J. Astrocytes in mammalian retina. Prog Retinal Res 1988; 7:209-31.

20. Chan-Ling T. Glial, neuronal and vascular interactions in the mammalian retina. Progress Retinal Eye Res 1994; 13:357-89.

21. Büssow H. The astrocytes in the retina and optic nerve head of mammals: a special glia for the ganglion cell axons. Cell Tissue Res 1980; 206:367-78.

22. Holländer H, Makarov F, Dreher Z et al. Structure of the macroglia of the retina: sharing and division of labour between astrocytes and Müller cells. J Comp Neurol 1991; 313:587-603.

23. Schnitzer J, Karschin A. The shape and distribution of astrocytes in the retina of the adult rabbit. Cell Tissue Res 1986; 246:91-102.

24. Robinson SR, Dreher Z. Evidence for three morphological classes of astrocyte in the adult rabbit retina: functional and developmental implications. Neurosci Lett 1989; 106:261-8.

25. Hampson ECGM, Robinson SR. Heterogenous morphology and tracer-coupling patterns of retinal oligodendrocytes. Phil Trans R Soc Lond B 1995; 349:353-364.

26. Raff MC. Glial cell diversification in the rat optic nerve. Science 1989; 243:1450-55.

27. Butt AM, Colquhoun K, Tutton M et al. Three-dimensional morphology of astrocytes and oligodendrocytes in the intact mouse optic nerve. J Neurocytol 1994; 23:469-485.

28. Hampson ECGM. Calcium imaging reveals signalling between glial cells in mammalian retina. Proc Aust Neurosci Soc 1995; 6:54.

29. Cornell-Bell AH, Finkbeiner SM, Cooper MS et al. Glutamate induces calcium waves in cultured astrocytes: long-range glial signalling. Science 1990; 247:470-3.

30. Finkbeiner SM. Calcium waves in astrocytes: filling in the gaps. Neuron 1992; 8:1101-8.

31. Nedergaard M. Direct signalling from astrocytes to neurons in cultures of mammalian brain cells. Science 1994; 263:1768-71.

32. Dani JW, Chernjavsky A, Smith SJ. Neuronal activity triggers calcium waves in hippocampal astrocyte networks. Neuron 1992; 8:429-40.

33. van den Pol A, Finkbeiner SM, Cornell-Bell AH. Calcium excitability and oscillations in suprachiasmatic nucleus neurons and glia in vitro. J Neurosci 1992; 12:2648-64.

34. Revel JP, Karnovsky MJ. Hexagonal array of subunits in intercellular junctions of the mouse heart and liver. J Cell Biol 1969; 33: C7-C10.

35. Yamada E, Ishikawa T. The fine structure of the horizontal cells in some vertebrate retinas. Cold Spring Harb Symp Quant Biol 1965; 30:383-92.

36. Bennett MVL, Aljure E, Nakajima Y et al. Electrotonic junctions between teleost spinal neurons: electrophysiology and ultrastructure. Science 1963; 141:262-4.

37. Tomita T, Tosaka T, Watanabe K et al. The fish EIRG in response to different types of illumination. Jap J Physiol 1958; 8:41-50.

38. Norton AL, Spekreijse H, Wolbarsht ML et al. Receptive field organization of the S-potential. Science 1968; 160:1021-2.

39. Kaneko A. Electrical connexions between horizontal cells in the dogfish retina. J Physiol 1971; 213:95-105.

40. Raviola G, Raviola E. Light and electron

microscopic observations on the inner plexiform layer of the rabbit retina. Am J Anat 1967; 120:403-26.

41. Steinberg RH. Rod and cone contributions to S-potentials from the cat retina. Vision Res 1969; 9:1319-29.

42. Dowling JE, Brown JE, Major D. Synapses of horizontal cells in rabbit and cat retinas. Science 1966; 153:1639-41.

43. Fisher SK, Boycott BB. Synaptic connexions made by horizontal cells within the outer plexiform layer of the retina of the cat and the rabbit. Proc R Soc Lond B 1974; 186:317-31.

44. Bloomfield SA, Miller RF. A physiological and morphological study of the horizontal cell types of the rabbit retina. J Comp Neurol 1982; 208:288-303.

45. Dacheux RF, Raviola E. Horizontal cells in the retina of the rabbit. J Neurosci 1982; 2:1486-93.

46. Raviola E, Gilula NB. Intramembrane organization of specialized contacts in the outer plexiform layer of the retina: a freeze-fracture study in monkey and rabbit. J Cell Biol 1975; 65:192-222.

47. Kolb H. The organization of the outer plexiform layer in the retina of the cat: electron microscopic observations. J Neurocytol 1977; 6:131-53.

48. Stewart WW. Functional connections between cells as revealed by dye-coupling with a highly fluorescent naphthalimide tracer. Cell 1978; 14:741-59.

49. Vaney DI. The coupling pattern of axon-bearing horizontal cells in the mammalian retina. Proc R Soc Londs B 1993; 252: 93-101.

50. Mills SL, Massey SC. Distribution and coverage of A- and B-type horizontal cells stained with Neurobiotin in the rabbit retina. Vis Neurosci 1994; 11:549-560.

51. Bloomfield SA, Xin D, Persky SE. A comparison of receptive field and tracer coupling size of horizontal cells in the rabbit retina. Vis Neurosci 1995; 12:985-999.

52. Nelson R, Lützow AV, Kolb H et al. Horizontal cells in cat retina with independent dendritic systems. Science 1975; 189:137-9.

53. Piccolino M, Neyton J, Witkovsky P et al. g-Aminobutyric acid antagonists decrease junctional communication between L-horizontal cells of the retina. Proc Natl Acad Sci USA 1982; 79:3671-5.

54. Vaney DI. Photochromic intensification of diaminobenzidine reaction product in the presence of tetrazolium salts: applications for intracellular labelling and immunohistochemistry. J Neurosci Methods 1992; 44:217-23.

55. Kouyama N, Watanabe K. Gap-junctional contacts of luminosity-type horizontal cells in the carp retina: a novel pathway of signal conduction from the cell body to the axon terminal. J Comp Neurol 1986; 249:404-10.

56. Stell WK. Horizontal cell axons and axon terminals in goldfish retina. J Comp Neurol 1975; 159:503-20.

57. Yagi T. Interaction between the soma and the axon terminal of retinal horizontal cells in Cyprinus carpio. J Physiol 1986; 375:121-35.

58. Hassin G, Witkovsky P. Intracellular recording from identified photoreceptors and horizontal cells of the Xenopus retina. Vision Res 1983; 23:921-31.

59. Stone S, Witkovsky P. Center-surround organization of Xenopus horizontal cells and its modification by g-aminobutyric acid and strontium. Exp Biol 1987; 47:1-12.

60. Stone S, Witkovsky P, Schütte M. A chromatic horizontal cell in the Xenopus retina: intracellular staining and synaptic pharmacology. J Neurophysiol 1990; 64:1683-94.

61. Kim HG, Miller RF. Physiological and morphological correlations of horizontal cells in the mudpuppy retina. J Neurophysiol 1992; 67:829-40.

62. Kaneko A, Stuart AE. Coupling between horizontal cells in the carp retina revealed by diffusion of Lucifer Yellow. Neurosci Lett 1984; 47:1-7.

63. Teranishi T, Negishi K, Kato S. Regulatory effect of dopamine on spatial properties of horizontal cells in carp retina. J Neurosci 1984; 4:1271-80.

64. Tornqvist K, Yang X-L, Dowling JE. Modulation of cone horizontal cell activity in the telesot fish retina: III. effects of prolonged darkness and dopamine on electrical coupling between horizontal cells. J Neurosci 1988; 8:2279-88.

65. Stretton AOW, Kravitz EA. Neuronal geometry: determination with a technique of intracellular dye injection. Science 1968; 162:132-4.

66. Loewenstein WR, Kanno Y. Studies on an epithelial (gland) cell junction: I. modifications of surface membrane permeability. J Cell Biol 1964; 22:565-86.

67. Flagg-Newton JL, Loewenstein WR. Experimental depression of junctional membrane permeability in mammalian cell culture: a study with tracer molecules in the 300 to 800 Dalton range. J Membr Biol 1979; 50:65-100.

68. Lo CW, Gilula NB. Gap junctional communication in the postimplantation mouse embryo. Cell 1979; 18:411-22.

69. Audesirk G, Audesirk T, Bowsher PJ. Variability and frequent failure of Lucifer Yellow to pass betwen two electrically coupled neurons in Lymnaea stagnalis. J Neurobiol 1982; 13:369-75.

70. Warner AE, Lawrence PA. Permeability of gap junctions at the segmental border in insect epidermis. Cell 1982; 28:243-52.

71. Ransom BR, Kettenmann H. Electrical coupling, without dye coupling, between mammalian astrocytes and oligodendrocytes in cell culture. Glia 1990; 3:258-66.

72. Pérez-Armendariz M, Roy C, Spray DC et al. Biophysical properties of gap junctions between freshly dispersed pairs of mouse pancreatic beta cells. Biophys J 1991; 59:76-92.

73. Bennett MVL, Spira ME, Spray DC. Permeability of gap junctions between embryonic cells of Fundulus: a reevaluation. Dev Biol 1978; 65:114-25.

74. Schwarzmann G, Wiegandt H, Rose B et al. Diameter of the cell-to-cell junctional membrane channels as probed with neutral molecules. Science 1981; 213:551-3.

75. Spray DC, Saez JC. Agents that regulate gap junctional conductance: sites of action and specificities. Adv Mod Environ Toxicol 1987; 14:1-26.

76. Verselis V, White RL, Spray DC et al. Gap junctional conductance and permeability are linearly related. Science 1986; 234:461-4.

77. Veenstra RD, Wang H-Z, Beyer EC et al. Connexin37 forms high conductance gap junction channels with subconductance state activity and selective dye and ionic permeabilities. Biophys J 1994; 66:1915-28.

78. Brink PR, Ramanan SV. A model for the diffusion of fluorescent probes in the septate giant axon of earthworm: axoplasmic diffusion and junctional membrane permeability. Biophys J 1985; 48:299-309.

79. Hampson ECGM, Weiler R, Vaney DI. pH-gated dopaminergic modulation of horizontal cell gap junctions in mammalian retina. Proc R Soc Lond B 1994; 255:67-72.

80. Horikawa K, Armstrong WE. A versatile means of intracellular labeling: injection of biocytin and its detection with avidin conjugates. J Neurosci Methods 1988; 25:1-11.

81. Kita H, Armstrong W. A biotin-containing compound N-(2-aminoethyl)biotinamide for intracellular labeling and neuronal tracing studies: comparison with biocytin. J Neurosci Methods 1991; 37:141-50.

82. Hsu S-M, Raine L, Fanger H. Use of avidin-biotin-peroxidase complex (ABC) in immunoperoxidase techniques: a comparison between ABC and unlabeled antibody (PAP) procedures. J Histochem Cytochem 1981; 33:857-66.

83. Adams JC. Heavy metal intensification of DAB-based HRP reaction product. J Histochem Cytochem 1981; 29:775.

84. Voigt T. Cholinergic amacrine cells in the rat retina. J Comp Neurol 1986; 248:19-35.

85. Scharfman HE, Kunkel DD, Schwartzkroin PA. Intracellular dyes mask immunoreactivity of hippocampal interneurons. Neurosci Lett 1989; 96:23-28.

86. Rønnekleiv OK, Loose MD, Erickson KR et al. A method for immunocytochemical identification of biocytin-labeled neurons following intracellular recording. BioTechniques 1990; 9:432-8.

87. Goddard JC, Behrens UD, Wagner H-J et al. Biocytin: intracellular staining, dye-coupling and imunocytochemistry in carp retina. NeuroReport 1991; 2:755-8.

88. Tasker JG, Hoffman NW, Dudek FE. Comparison of three intracellular markers for combined electrophysiological, morphological and immunohistochemical analyses. J Neurosci Methods 1991; 38:129-43.

89. Pow DV, Wright LL, Vaney DI. The im-

munocytochemical detection of amino-acid neurotransmitters in paraformaldehyde-fixed tissues. J Neurosci Methods 1995; 56: 115-23.

90. Famiglietti EV, Kolb H. A bistratified amacrine cell and synaptic circuitry in the inner plexiform layer of the retina. Brain Res 1975; 84:293-300.

91. Cohen E, Sterling P. Demonstration of cell types among cone bipolar neurons of cat retina. Phil Trans R Soc Lond B 1990; 330:305-21.

92. Strettoi E, Raviola E, Dacheux RF. Synaptic connections of the narrow-field, bistratified rod amacrine cell (AII) in the rabbit retina. J Comp Neurol 1992; 325:152-68.

93. Vaney DI. The morphology and topographic distribution of AII amacrine cells in the cat retina. Proc R Soc Lond B 1985; 224: 475-88.

94. Vaney DI, Gynther IC, Young HM. Rod-signal interneurons in the rabbit retina: 2. AII amacrine cells. J Comp Neurol 1991; 310:154-69.

95. Hampson ECGM, Vaney DI, Weiler R. Dopaminergic modulation of gap junction permeability between amacrine cells in mammalian retina. J Neurosci 1992; 12:4911-22.

96. Mills SL, Massey SC. Dye coupling between AII amacrine cells and bipolar cells in the rabbit retina. Invest Ophthalmol Vis Sci 1994; 35:1822.

97. Raviola E, Raviola G. Structure of the synaptic membranes in the inner plexiform layer of the retina: a freeze-fracture study in monkeys and rabbits. J Comp Neurol 1982; 209:233-48.

98. Williams EH, DeHann RL. Electrical coupling among heart cells in the absence of ultrastructurally defined gap junctions. J Membr Biol 1981; 60:237-48.

99. Ginzberg RD, Morales EA, Spray DC et al. Cell junctions in early embryos of squid (Loligo pealei). Cell Tissue Res 1985; 239:477-84.

100. Picanço-Dinez CW, Silveira LCL, Yamada ES et al. Biocytin as a retrograde tracer in the mammalian visual system. Braz J Med Biol Res 1992; 25:57-62.

101. Saltarelli DP, Ball AK. Passive diffusion of Neurobiotin between ganglion cells and amacrine cells in the goldfish retina. Invest Ophthalmol Vis Sci 1995; 36:S602.

102. Umino O, Maehara M, Hidaka S et al. The network properties of bipolar-bipolar cell coupling in the retina of telesot fishes. Vis Neurosci 1994; 11:533-48.

103. Bloomfield SA, Xin D. Relationship between receptive field and tracer-coupling size of amacrine and ganglion cells in the rabbit retina. Invest Ophthalmol Vis Sci 1994; 35:1822.

104. Kamermans M. The Functional Organization of the Horizontal Cell Layers in Carp Retina (PhD thesis). Amsterdam: University of Amsterdam, 1989:159.

105. Baldridge WH, Ball AK. Background illumination reduces horizontal cell receptive-field size in both normal and 6-hydroxy-dopamine-lesioned goldfish retinas. Vis Neurosci 1991; 7:441-50.

106. Jamieson MS, Baldridge WH, Ball AK. Modulation of horizontal cell receptive field size in light-adapted goldfish retinas. Invest Ophthalmol Vis Sci 1994; 35:1821.

107. Oyster CW, Barlow HB. Direction-selective units in rabbit retina: distribution of preferred directions. Science 1967; 155: 841-2.

108. Vaney DI. Biocytin dye-coupling reveals subpopulations of direction-selective ganglion cells in rabbit retina. Proc Aust Neurosci Soc 1990; 1:136.

109. Vaney DI. Territorial organization of direction-selective ganglion cells in rabbit retina. J Neurosci 1994; 14:6301-16.

110. Yang G, Masland RH. Receptive fields and dendritic structure of directionally selective retinal ganglion cells. J Neurosci 1994; 14:5267-80.

111. Mangel SC, Dowling JE. Responsiveness and receptive field size of carp horizontal cells are reduced by prolonged darkness and dopamine. Science 1985; 229:1107-09.

112. Xin D, Bloomfield SA, Persky SE. Effect of background illumination on receptive field and tracer-coupling size of horizontal cells and AII amacrine cells in the rabbit retina. Invest Ophthalmol Vis Sci 1994; 35:1363.

113. Bloomfield SA. Relationship between receptive and dendritic field size of amacrine cells in the rabbit retina. J Neurophysiol 1992;

68:711-25.

114. Nelson R. AII amacrine cells quicken time course of rod signals in the cat retina. J Neurophysiol 1982; 47:928-47.

115. Dacheux RF, Raviola E. The rod pathway in the rabbit retina: a depolarizing bipolar and amacrine cell. J Neurosci 1986; 6:331-45.

116. Vaney DI. The mosaic of amacrine cells in the mammalian retina. Prog Retinal Res 1990; 9:49-100.

117. Kolb H, Nelson R. Rod pathways in the retina of the cat. Vision Res 1983; 23:301-12.

118. Furshpan EJ, Potter DD. Transmission at the giant motor synapses of crayfish. J Physiol 1959; 145:289-325.

119. Auerbach AA, Bennett MVL. A rectifying electrotonic synapse in the central nervous system of a vertebrate. J Gen Physiol 1969; 53:211-37.

120. Nicholls JG, Purves D. Monosynaptic chemical and electrical connexions between sensory and motor cells in the central nervous system of the leech. J Physiol 1970; 209:647-67.

121. Ringham GL. Localization and electrical characteristics of a giant synapse in the spinal cord of the lamprey. J Physiol 1975; 251:395-407.

122. Flagg-Newton JL, Loewenstein WR. Asymmetrically permeable membrane channels in cell junction. Science 1980; 207:771-3.

123. Loewenstein WR. Junctional intercellular communication: the cell-to-cell membrane channel. Physiol Rev 1981; 61:829-913.

124. Giaume C, Korn H. Voltage-dependent dye coupling at a rectifying electrotonic synapse of the crayfish. J Physiol 1984; 356:151-67.

125. Kolb H. The inner plexiform layer in the retina of the cat: electron microscopic observations. J Neurocytol 1979; 8:295-329.

126. Hall DH, Gilat E, Bennett MVL. Ultrastructure of the rectifying electrotonic synapses between giant fibres and pectoral fin adductor motor neurons in the hatchetfish. J Neurocytol 1985; 14:825-34.

127. Marshak DW, Jacoby RA, Stafford DK et al. Synaptic inputs to parasol ganglion cells in primate retina. Invest Ophthalmol Vis Sci 1995; 36:S602.

128. Mastronarde DN. Interactions between ganglion cells in cat retina. J Neurophysiol 1983; 49:350-65.

129. Mastronarde DN. Correlated firing of retinal ganglion cells. Trends Neurosci 1989; 12:75-80.

130. Peichl L, Wässle H. The structural correlate of the receptive field centre of a ganglion cells in the cat retina. J Physiol 1983; 341:309-24.

131. Penn AA, Wong ROL, Shatz CJ. Neuronal coupling in the developing mammalian retina. J. Neurosci 1994; 14:3805-15.

132. Massa PT, Mugnaini E. Cell junctions and intramembrane particles of astrocytes and oligodendrocytes: a freeze-fracture study. Neurosci 1982; 7:523-38.

133. Butt AM, Ransom BR. Visualization of oligodendrocytes and astrocytes in the intact rat optic nerve by intracellular injection of Lucifer Yellow and horseradish peroxidase. Glia 1989; 2:470-5.

134. Wässle H, Peichl L, Boycott BB. Dendritic territories of cat retinal ganglion cells. Nature 1981; 292:344-5.

135. Vaney DI. Morphological identification of serotonin-accumulating neurons in the living retina. Science 1986; 233:444-6.

136. Vaney DI. The AI (A17) amacrine cells of the cat retina: a tracer-coupling study. Proc Aust Neurosci Soc 1995; 6:77.

137. Kolb H, Famiglietti EV. Rod and cone pathways in the inner plexiform layer of cat retina. Science 1974; 186:47-9.

138. DeBoer DJ, Vaney DI. Development of neuronal coupling between direction-selective ganglion cells in rabbit retina: a Neurobiotin tracer-coupling study. Proc Aust Neurosci Soc 1995; 6:78.

139. Hitchcock PF. Neurobiotin coupling betwen developing ganglion cells in the retina of the goldfish. Invest Ophthalmol Vis Sci 1993; 34:878.

140. Cook JE, Becker DL. Gap junctions in the vertebrate retina. Microsc Res Tech 1995; 31:408-419.

141. Naka KI, Christensen BN. Direct electrical connections between transient amacrine cells in the catfish retina. Science 1981; 214:462-4.

142. Teranishi T, Negishi K, Kato S. Dye coupling between amacrine cells in carp retina. Neurosci Lett 1984; 51:73-8.

143. Negishi K, Teranishi T. Close tip-to-tip contacts between dendrites of transient amacrine cells in carp retina. Neurosci Lett 1990; 115:1-6.

144. Teranishi T, Negishi K. Dendritic morphology of a class of interstitial and normally placed amacrine cells revealed by intracellular Lucifer Yellow injection in carp retina. Vision Res 1991; 31:463-75.

145. Wolszon LR, Gao W-Q, Passani MB et al. Growth cone "collapse" in vivo: are inhibitory interactions mediated by gap junctions? J Neurosci 1994; 14:999-1010.

146. Oyster CW, Amthor FR, Takahashi ES. Dendritic architecture of ON-OFF direction-selective ganglion cells in the rabbit retina. Vision Res 1993; 33:579-608.

147. Guthrie PB, Lee RE, Rehder V et al. Self-recognition: a constraint on the formation of electrical coupling in neurons. J Neurosci 1994; 14:1477-85.

148. Amthor FR, Oyster CW. Relation between preferred direction and dendritic organization of identified, contacting On-Off DS ganglion cells and Ach amacrine cells in rabbit retina. Soc Neurosci Abstr 1993; 19:1258.

149. Sterling P, Freed MA, Smith RG. Architecture of rod and cone circuits to the *On*-beta ganglion cell. J Neurosci 1988; 8:623-42.

150. Wright LL. A new type of GABAergic amacrine cell with a bistratified morphology. Proc Aust Neurosci Soc 1995; 6:198.

THE MODULATION OF GAP JUNCTION PERMEABILITY IN THE RETINA

Reto Weiler

1. INTRODUCTION

Electrical coupling through gap junctions is very widespread in the retina (see chapter 5) and is found in all cell classes that form its neuronal network. There is not only homologous coupling but also heterologous coupling and the implementation of neurobiotin as a tracer molecule in the retina[1] has unraveled an unexpected incidence of coupled cells and diversity of coupling patterns so far unmatched by other parts of the brain. But it is not only this abundance and variety of coupled networks in the retina that has attracted much interest. Indeed, neuronal coupling in the retina has come into the limelight due to the discovery that coupling can be modulated through neurotransmitters. Neuronal control of coupling resistances might explain why such extensive coupling that appears to be a counter-intuitive arrangement for a sensory organ responsible for high spatial resolution is copious in the retina. The vertebrate retina is therefore a unique preparation to study the functional role of coupling modulation in sensory processing. Furthermore, it offers the chances to analyze the neurotransmitter pathways involved in such modulation and the intracellular signal cascades. Last, but not least, the probable co-occurrence of multiple types of abundant connexin proteins in the retina should facilitate the search for their identity and modulatory properties at the molecular level. Knowledge we obtain from studying the modulation of retinal gap junctions will therefore provide not only useful information for vision researchers but for the neuroscientist who is interested in the plasticity of the brain.

Gap Junctions in the Nervous System, edited by David C. Spray and Rolf Dermietzel.
© 1996 R.G. Landes Company.

2. HORIZONTAL CELL COUPLING

2.1. MODULATION BY DOPAMINE

In 1978 Hedden and Dowling[2] first applied dopamine as an aerosol to the isolated goldfish retina. Among other observations they also noted that dopamine reduced the light response of horizontal cells. Around the same time, Negishi and Drujan[3-6] reported that dopamine altered the receptive field properties of horizontal cells. As a result of an extensive coupling through gap junctions, horizontal cells exhibit a receptive field that is much larger than their anatomical receptive field. In their study they used a centered spot and an annulus as stimuli. In such an experimental design, the spot of light produces a center response in a horizontal cell that predominantly reflects the direct cone input to the impaled cell, while the annulus evokes a surround response that predominantly reflects the input from neighboring horizontal cells mediated through the gap junctions. This protocol therefore allows separation of the contribution of the electrical signal spread in the lateral sheet formed by the coupled horizontal cells and the local chemically transmitted signals. Negishi and Drujan[3-6] found that dopamine produced reciprocal changes in the center and surround responses of cone-driven horizontal cells: It attenuated the annulus response but increased the spot response. This indicated that dopamine was affecting the lateral spread of the electrical signal through the gap junctions, and was the first hint of neuromodulatory control of gap junctional conductance in the central nervous system.

The idea of modulation of the conductance of the gap junctions by dopamine was further pushed by experiments of Laufer and Salas[7] who recorded an increase in the coupling and membrane resistance following application of dopamine. But it was the elegant combination of electrical recording, tracer coupling, and pharmacological knock-out experiments done by Teranishi, Negishi and Kato[8] that unequivocally demonstrated a direct effect of dopamine on the coupling between a set of horizontal cells in the carp retina known as external or H1 cells. They used the specific neurotoxic effect of 6-OHDA on dopaminergic neurons to produce retinas depleted of such neurons. The autofluorescence of dopaminergic neurons after formaldehyde fixation facilitated in this case the control of successful depletions. There was a clear difference in the size of the receptive field between horizontal cells recorded from control retinas and from depleted retinas (Fig. 6.1A). In the depleted retinas, the field size was larger than in control retinas. Superfusion of control retinas with dopamine decreased the size even more. The change of the overall size of the receptive field was paralleled by an increase of the light response to stimulation of its center with a small spot. The conclusion that the electrophysiologically recorded changes of the receptive field size was not the sole result of modulation of membrane resistance but also involved a modulation of the coupling resistance was confirmed by the dye coupling experiments. Dopamine-depletion resulted in dye coupling of about a dozen H1 cells, whereas dopamine application restricted the dye to a single cell. Results similar to those obtained from dopamine-depleted retinas were also obtained when haloperidol, a well-known dopamine receptor antagonist, was applied. By contrast, the effects of dopamine could be mimicked through the application of dibutyryl-cyclic-adenosine-monophosphase, a cell-permeable analog of cAMP.

The landmark study of Teranishi and coworkers and their follow-up paper[9] certainly provoked a large body of work concentrating on the role of dopamine in the retina and the molecular mechanisms of neuronal gap junction regulation. Their key finding was subsequently corroborated in the turtle retina by Piccolino et al[10] and in the mudpuppy retina,[11] but it took another ten years until it was possible to demonstrate dopaminergic modulation of gap junction conductivity between horizontal cells in the mammalian retina.[12]

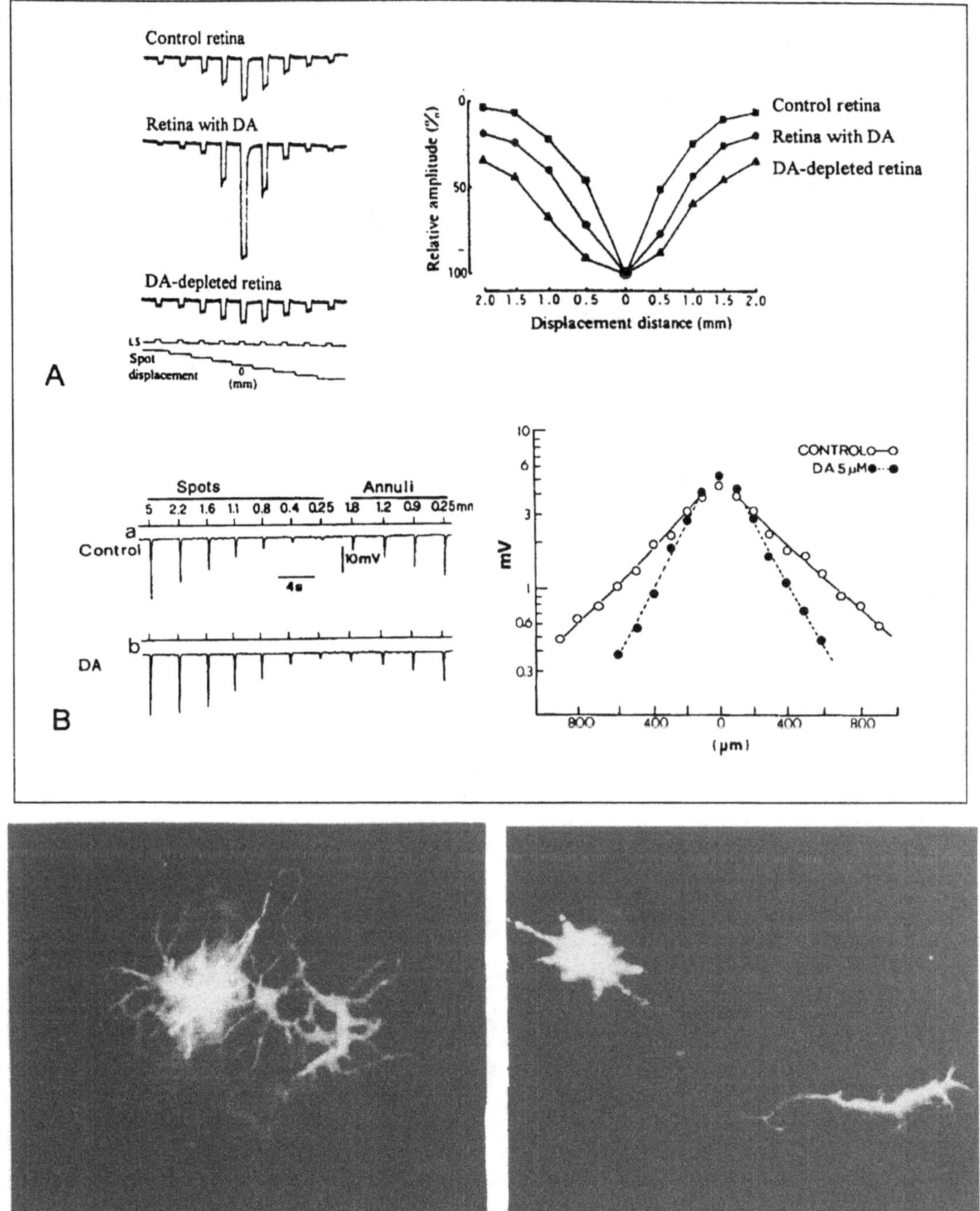

Fig. 6.1. The effect of dopamine (DA) on the receptive field size of horizontal cells from the carp (A) and turtle (B) retina. Receptive fields were determined by intracellularly recording the light responses to a moving spot (in A) or to a moving slit (in B). Dopamine enhanced the responses to the stimulation of the center and attenuated the responses to stimulation of the surround. (Modified from refs. 8,10). (C) Dye-coupling of horizontal cells in the turtle retina. In a control retina (left), numerous horizontal cell somata and axon terminals are labeled. Lucifer Yellow, injected into one soma, passed through gap junctions into neighboring somata and through the short axon into the terminal from where it passed through gap junctions into neighboring axon terminals. Dopaminergic activity (right) restricted the dye to the injected cell. Only a few axon terminals are very faintly labeled. (Courtesy J. Ammermüller). Portions of the figure are reprinted with permission from Teranishi et al, Nature 1983; 301:243-246, copyright 1983 Macmillan Magazines Limited, and Piccolino et al, J Neuroscience 1984: 10:2477-2488, copyright 1984 Oxford University Press.

In the turtle retina, Piccolino and co-workers did a very thorough pharmacological study to identify the source of retinal dopamine and the type and location of dopamine receptors involved in the modulation of the gap junctional resistance between luminosity-type horizontal cell axon terminals.[10,13] Dopamine narrowed the receptive field profile (Fig. 6.1B) and the space constant of the network was reduced by about 50%. Again, dye coupling was almost completely abolished when the retina is superfused with dopamine (Fig. 6.1C). As in the fish retina, dopamine was shown to act through a D1 receptor located on the horizontal cell axon terminal itself.

However, while interplexiform cells that synapse onto horizontal cells are the source of dopamine in the fish retina,[14] such cells do not exist in the turtle retina. There, dopamine is located in amacrine cells[15] that do not contact horizontal cells. This implied an active role of dopamine through a diffusional process, a process that has gained momentum over the past years as volume transmission and that has been treated more analytically in the retina in a recent study.[16]

Not only do the luminosity-type horizontal cells form coupled networks in the turtle retina but also the chromaticity-type horizontal cells, a class of horizontal cells that respond to light of different wavelengths with different polarities.[17] Dopamine was confirmed to modulate the receptive field size also of this class of horizontal cells.[18] Direct injection of cAMP into chromaticity-type horizontal cells also decreased the coupling conductance and blocked dye coupling[19] demonstrating that again the dopaminergic action is most likely linked to the cAMP cascade. There is evidence that also in the fish retina, dopamine does act through volume transmission onto the rod-driven horizontal cells. In contrast to the cone-driven horizontal cells, rod-driven horizontal cells appear not to be contacted by the dopamine-containing interplexiform cells,[20] but dopamine does decrease their receptive field size.[21] On the other hand, coupled rod horizontal cells in the skate

retina did not respond to dopamine or internal increase of cAMP.[22] An effect of dopamine on the receptive field properties has also been shown for the mudpuppy retina. Like in the fish and in the turtle retina, dopamine increased the response to a stimulation of the center, but decreased the response to a stimulation of the surround recorded from horizontal cells of this amphibian.[11] This effect of dopamine was also blocked by D1-receptor antagonists. The uncoupling effect of dopamine was much greater in dark-adapted than in light-adapted animals.

Horizontal cells in the mammalian retina are also extensively coupled but many attempts to demonstrate modulation through dopamine initially failed. The reason for this failure was recently discovered: The dopaminergic modulation of horizontal cell coupling in the mammalian retina is pH-gated.[12] Dopamine does uncouple A-type horizontal cells, but only at a pH of 7.2. At pH 7.4 it does not have any effect (Fig. 6.2). The pH effect is a gating effect, since a pH of 7.2 alone does not uncouple the horizontal cells. It is very important to recall that the pH within the outer retina varies depending upon the adaptational level from about 7.3 to about 7.1,[23] mainly due to the metabolic activity of the photoreceptors. The gating effect of the pH thus reflects an activity-dependent gating of the dopaminergic modulation. As in the nonmammalian retina, dopamine exerts its effect on gap junction permeability through a D1 receptor and a decrease of the coupling is also obtained through intracellular increase of cAMP. Since this effect is not pH-dependent, the pH gating must take place somewhere within the receptor-G-protein complex.

2.2. CELLULAR AND MOLECULAR EVENTS UNDERLYING THE DOPAMINERGIC MODULATION OF GAP JUNCTION PERMEABILITY

In all cases so far studied the uncoupling effect of dopamine on the horizontal cell network could be mimicked by either a membrane-permeable analog of cAMP or

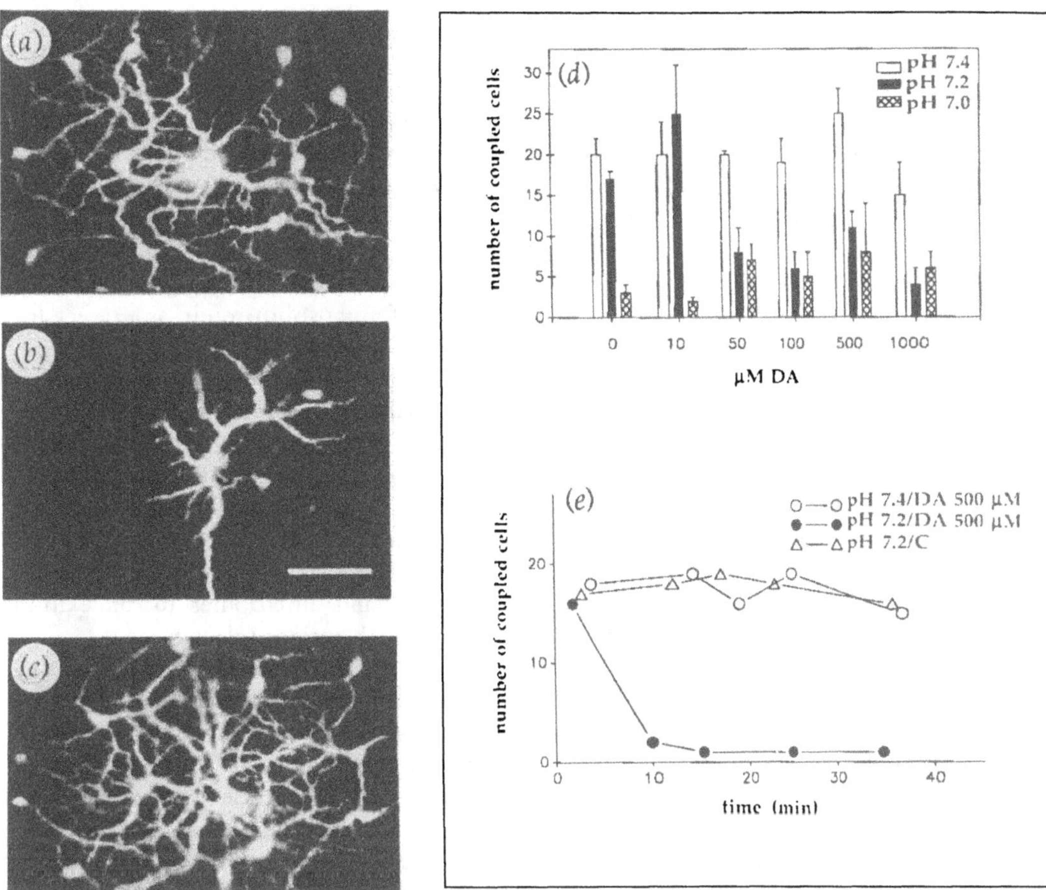

Fig. 6.2. Dopaminergic modulation of gap junctions between horizontal cells in the rabbit retina. Isolated retinas were superfused with control Ringer at pH 7.2 (a), with dopamine containing Ringer at pH 7.2 (b) and at pH 7.4 (c). Scale bar = 50 µm. Histograms (d) of the number of dye-coupled cells for increasing dopamine concentrations at three different pH levels (open bars pH 7.4, filled bars pH 7.2, hatched bars pH 7.0). Timecourse (e) for the action of dopamine at pH 7.2 (filled circles) and at pH 7.4 (open circles). Under control conditions at pH 7.2 (open triangles), dye-coupling did not change. Reprinted with permission from Hampson ECGM et al, Proc R Soc Lond [Biol] 1994; 255:67-72.

an increase of cAMP through activating adenylate cyclase and/or inhibiting the cAMP degrading enzyme phosphodiesterase. Together with an early report that dopamine receptors in the retina are linked to adenylate cyclase[24] this suggested the following scenario for the uncoupling effect of dopamine. Dopamine released either from amacrine cells in the inner retina and diffused to the outer retina or released from interplexiform cells synapsing onto horizontal cells binds to D1 receptors on horizon-

tal cells. As a result, adenylate cyclase is internally and locally activated leading to an increase of cAMP. The next important question was now whether cAMP was activating protein kinase A which in turn might phosphorylate a protein leading ultimately to a reduction of the conductance of the gap junction channels.

An important step towards answering these questions was made through the introduction of an isolated cell preparation of horizontal cells. For this purpose a retina is excised from the eye and placed for a short period in Ringer solution containing papaine or a similar enzyme. After the digestion process has been stopped the tissue is chopped and pieces are gently triturated through a pasteur pipette. This leads to a cell suspension in which horizontal cells are easily recognized due to their unique morphology. Very often coupled horizontal cell pairs persist and are an ideal object to study the intercellular communication through gap junctions. In this case both cells are penetrated with an electrode or patch-clamped. Each cell can then be voltage-clamped and the coupling resistance can be measured. This is achieved by stepping the membrane potential of the driver cell and measuring the current flow across the junction into the follower cell. At the same time various drugs can be injected into the cells or added to the bathing medium.

Using this approach, the role of dopamine and cAMP has been confirmed. Bath application of dopamine and direct injection of cAMP in one cell decreased the coupling conductance, g_j (Fig. 6.3A).[25,26] Injection of protein kinase A at nanomolar levels rapidly and reversibly uncoupled the horizontal cell pairs.[27] The uncoupling effect was recordable about 30 sec after PKA had been pressure injected into one cell, and complete uncoupling was reached after about 5 min. Bath-applied dopamine also uncoupled the pairs within this time frame. This effect was blockable through the prior injection of an inhibitor of PKA (Walsh inhibitor). In another preparation, isolated pairs of horizontal cells from the channel catfish were used. Dopamine uncoupled the pairs in a dose-dependent fashion and a concentration of 10 nM dopamine reduced coupling by 50%. The injection of cAMP also reduced the coupling, an effect that was quickly reversible indicating a high phosphodiesterase activity in the cytoplasm.[28]

These results strongly supported the notion that g_j was regulated through a phosphorylation process. The identity of the corresponding phosphoproteins, however, have not yet been unraveled. Biochemical attempts using in vitro phosphorylation protocols with isolated and enriched horizontal cell fractions from the carp retina have indicated that proteins of 47 kDa and 26-35 kDa are targets for cAMP- and dopamine-dependent phosphorylation (Fig. 6.4A).[29,30] A similar protocol using the white perch retina also revealed a 43/44 kDa phosphoprotein as a target for cAMP-dependent phosphorylation.[31]

The identity of these phosphoproteins still remains to be elucidated. Their molecular range, of course, gives rise to speculations that they are connexins. Indeed, antibodies raised to connexin32 strongly label horizontal cells and their axon terminals in the fish retina (Fig. 6.4B).[32] On the other hand, antibodies to connexin43 did not label horizontal cells.

Besides the biochemical steps of the signal transduction process, the functional properties of the junctional channel have attracted much interest. The electrophysiological analysis of the conductance and gating properties of single channels and their modulation through dopamine is partially impaired by the high densities of channels within the junctional areas and the relative low resistance of the nonjunctional membrane. Despite these limitations it was possible to quantify some unitary junctional events using variance analysis of the junctional current noise and direct observations of unitary junctional gating events in poorly coupled cell pairs.[33] Channels between isolated horizontal cell pairs of the white perch have a unitary conductance of about 50 pS (Fig. 6.3B). This conductance was not significantly affected by dopamine. The open probability of the channels, however, decreased from 0.75 to 0.14. These results suggest that dopamine reduces the open probability of junctional channels by decreasing their open duration.

Similar results were obtained in a study using isolated pairs of horizontal cells from the zebrafish.[34] These neurons do not have the limitations mentioned above. They are very poorly coupled and have a high nonjunctional membrane resistance allowing

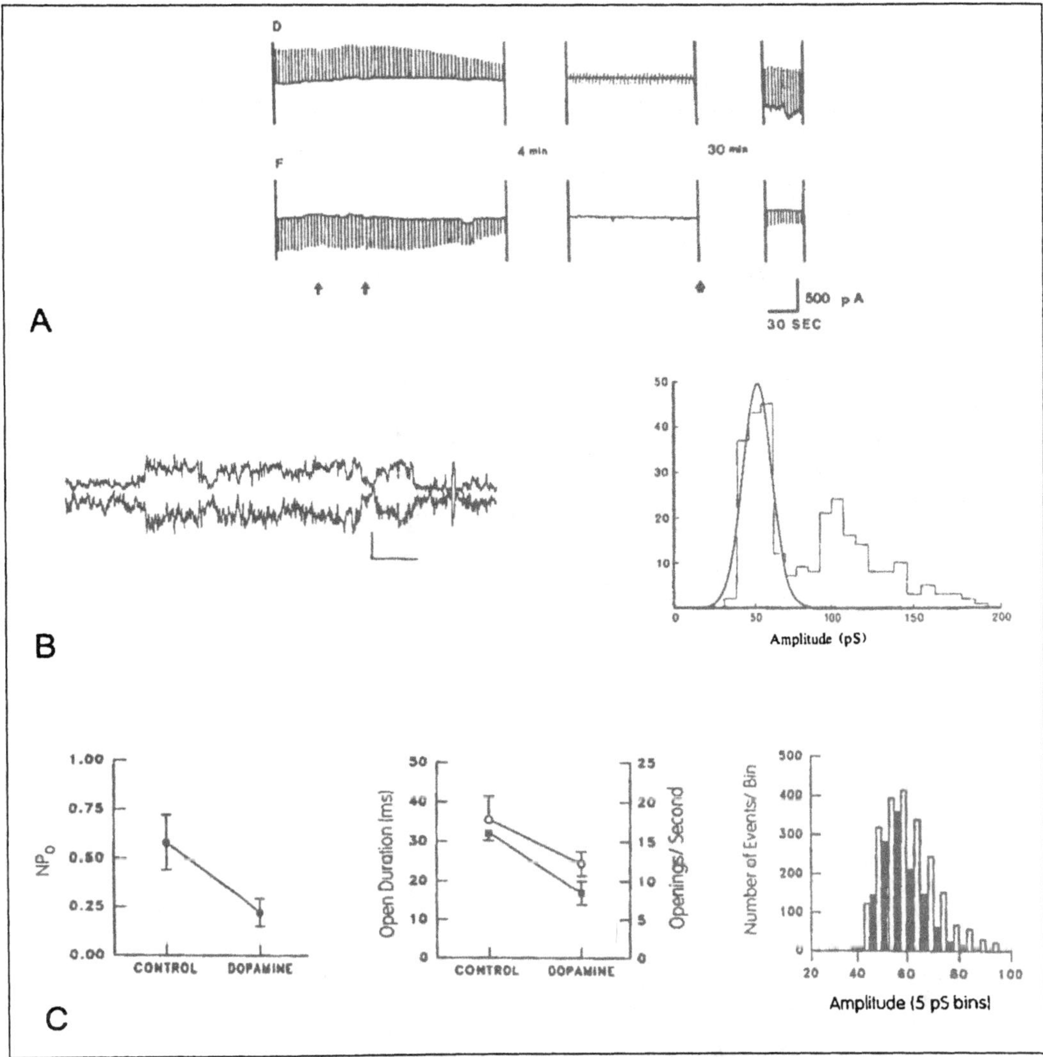

Fig. 6.3. Analysis of coupling between isolated pairs of horizontal cells from the fish retina using double whole-cell patch-clamp techniques. (A) Uncoupling effect of the catalytic subunit of cAMP-dependent protein kinase (PKA). At the time marked by the thin arrows, PKA was pressure injected into the follower cell (F) and uncoupling was complete after about 4.5 min. The thick arrow indicates the point where the pipette was withdrawn. After 30 min the pair was clamped with a new set of pipettes without PKA and coupling recovered. (B) Activity of gap junction channels recorded in a poorly coupled cell pair. Openings are represented as the traces moving apart (calibration: 5pA = 100 pS, 250 msec). Gaussian fit of first peak in the amplitude histogram on the right gives a unitary conductance of 51.8 +/-9.2 pS. (C) Modulation of channel-gating kinetics by dopamine. Mean reduction in the overall open probability (NP_o, left); reductions in both the mean duration of channel openings (open circles) and the mean frequency of channel openings (filled square, middle); amplitude histograms of channel openings at the unitary conductance level in the absence (open bars) and presence (filled bars) of dopamine, showing the relative stability of the unitary conductance during dopaminergic modulation. Reprinted with permission from: (A) Lasater EM, Proc Natl Acad Sci USA 1987; 84:7319-23; (B) McMahon DG et al, Proc Natl Acad Sci USA 1989; 86:7639-43; (C) McMahon DG et al, J Neurophysiol 1994; 72:2257-68.

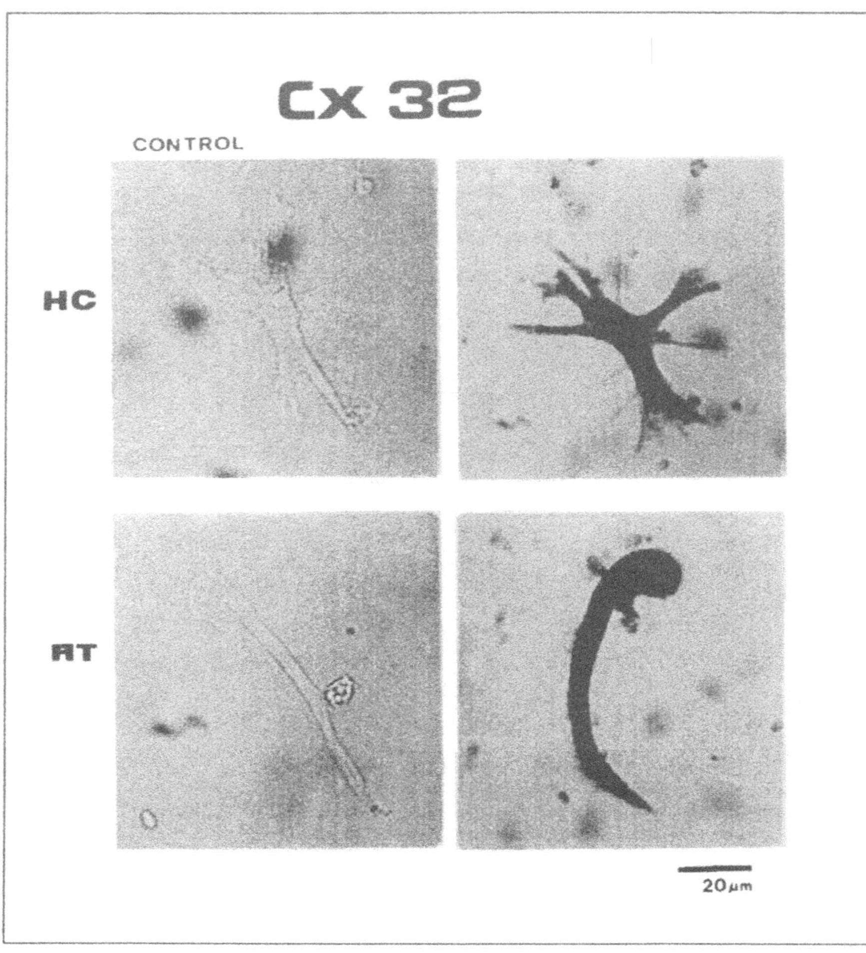
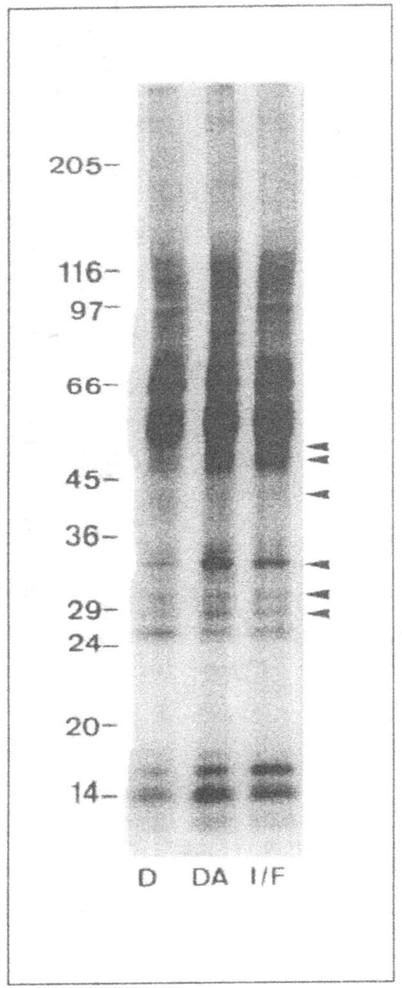

Fig. 6.4. (A) Identification of dopamine-regulated phosphoproteins from horizontal cells of the carp retina. Fractions[29] of horizontal cells from dark adapted (D) retinae were incubated either in dopamine (DA) or in a mixture of IBMX/forskolin (I/F). In vitro phosphorylation and subsequent separation on a gradient gel revealed several phosphoproteins (arrow heads) with an enhanced phosphorylation. (B) Antibodies directed against a specific sequence of connexin32 label isolated horizontal cells (HC) and their axon terminals (AT) obtained from the carp retina.

the resolution of single gap junction channels. Furthermore, the coupling is modulated by dopamine. The unitary conductance was 50 pS and the mean open time about 30 msec. Again, dopamine did not affect the unitary conductance but reduced the open probability from 0.58 to 0.22 by affecting both the mean open duration and the frequency of channel openings. Analysis of the distribution of channel open durations revealed the presence of two time constants, a fast component of about 13 msec and a slower component of about 56 msec. Both time constants were reduced by dopamine. The frequency of channel openings dropped from about 16/sec to about 8.5/sec after the addition of dopamine (Fig. 6.3C).

The unitary conductance measured in the two fish species matches that found in cardiac gap junctions[35] and it will be of great interest to determine how these functional similarities are reflected by homologies at the molecular level.

Recently, whole cell recordings from solitary, isolated horizontal cells from the catfish and skate retina revealed currents that were attributed to the opening of hemi-gap junctional channels.[36,37] The current was voltage-dependent with a reversal potential close to 0 mV. Activation of the current permitted the influx of Lucifer Yellow, and dopamine suppressed the current through activation of cAMP-dependent protein kinase. These studies suggest that hemi-gap junctional channels can be used as models for functional gap junctions between horizontal cells. This will facilitate the disclosure of the detailed molecular events underlying the modulation of the open probability by dopamine.

Dopamine does not only affect the open probability of gap junctional channels, but it also affects the connexon density within a given gap junctional area.[38-42] In the fish and in the turtle retina, the density was decreased by about a third. It has not yet been established whether this decrease results from a cAMP-dependent phosphorylation process. Also, we do not know how the density of connexons is re-

lated to the modulation of the gap junctional conductance. The data obtained in these studies are reviewed in chapter 7.

2.3. THE RELATIONSHIP BETWEEN LIGHT AND DOPAMINERGIC MODULATION OF GAP JUNCTIONAL CONDUCTANCE IN HORIZONTAL CELLS

The size of the receptive field is not only affected by dopamine but also by the ambient light conditions. The effect of light or darkness on the receptive field has been linked to a change of the coupling conductance using both an electrophysiological protocol based on a comparison of center and surround responses and in dye-coupling experiments with Lucifer Yellow. In the fish retina, a steady background light and a low-frequency flickering light both reduced the receptive field size (Fig. 6.5A) and restricted the transfer of Lucifer Yellow into neighboring horizontal cells.[43-45] Obviously, a uniform white background light reduced the gap junctional conductance. Such a decrease in coupling between horizontal cells due to an adapting light was also found in the mudpuppy retina[11,46] and in the turtle retina (Fig. 6.5B).[47] The idea that background light affects the size of the receptive field of horizontal cells by affecting g_j was recently challenged by the notion that the center/surround ratio generally used to demonstrate an effect on the coupling resistance also depends on the intensity of the stimuli used to construct it. Background light certainly affects the synaptic membrane resistance of horizontal cells due to its impact on transmitter release from the presynaptic photoreceptors and subsequently the length constant of the horizontal cell network is affected.[48] Taking these dependencies into account, the authors could not confirm a direct effect of background light on the junctional conductance. On the other hand, the effects on dye coupling by an adapting light strongly points to a modulatory role of light on the coupling conductance.

Interestingly, the coupling between horizontal cells appears to be affected not

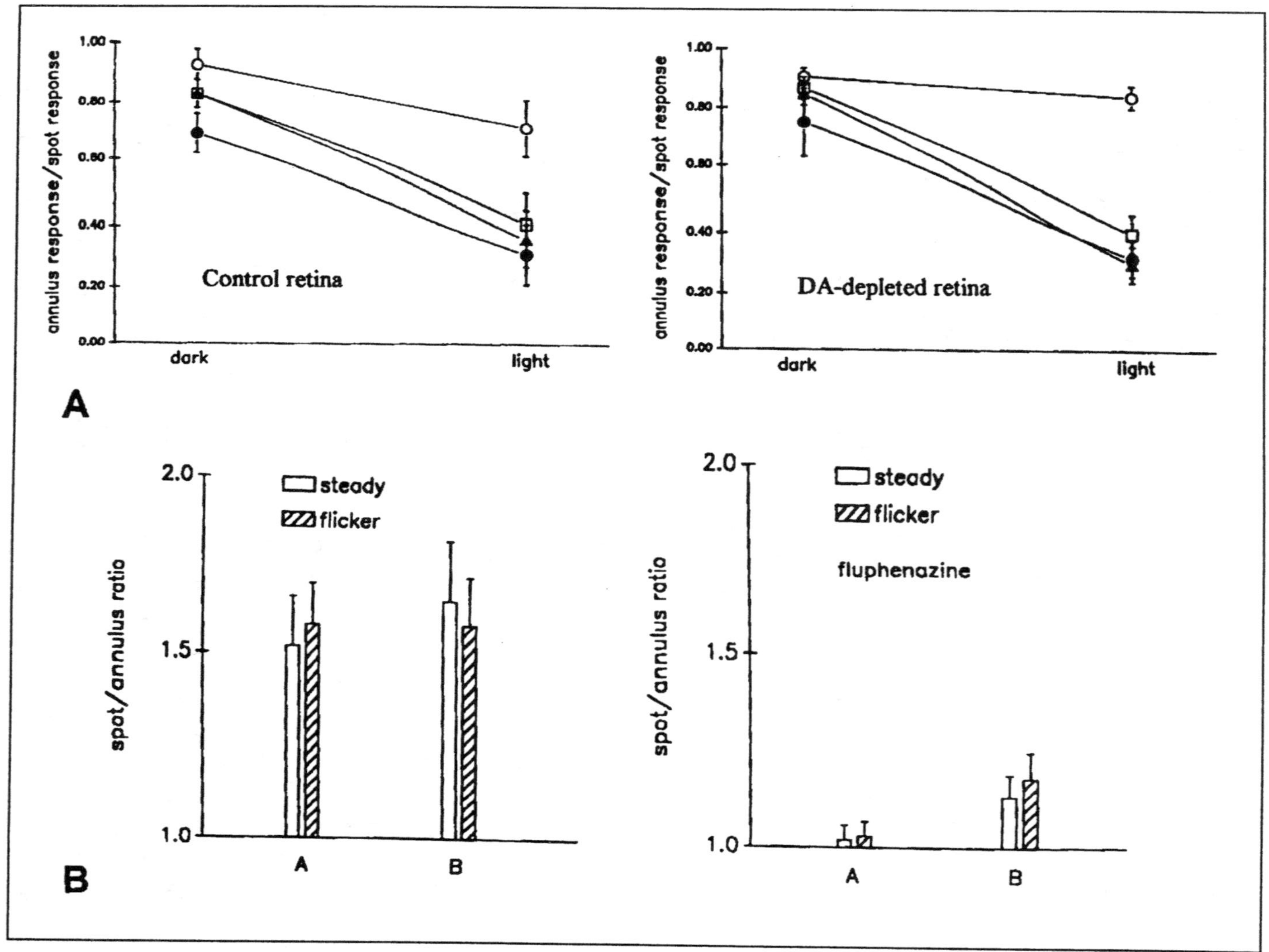

Fig. 6.5. The effect of background light on the coupling between horizontal cells in the fish and turtle retina. (A) A steady background light does affect the mean annulus/spot ratio of the light responses recorded intracellularly from fish horizontal cells in a control retina (left) and in a dopamine depleted retina (right). Four different spot/annulus combination were tested. (B) Steady and flickering background lights do affect the spot/annulus ratio also in the turtle retina. Two intensities were tested. There is an increase relative to the control before the background light was switched on. This increase is not detectable in the presence of fluphenazine, a dopamine antagonist. Reprinted from Weiler et al, Neuroscience Letters 1992; 140:121-124. Copyright 1992 Elsevier Science Ireland Ltd, Bay 15K, Shannon Industrial Estate, Co. Clare, Ireland.; and Baldridge et al, Visual Neuroscience 1991;7:411-450. Copy-

only by light but also by darkness. After about 2 hr in the dark, center responses from horizontal cells of the carp were increased and surround responses decreased.[49] Similar results were obtained from the white perch retina, where in addition a decrease of dye coupling after prolonged darkness was demonstrated.[50]

It is an obvious question to ask, whether ambient light and dopamine affect the gap junctional conductance in a correlated and interdependent way. This question is especially pertinent because dopamine release in the retina appears to be regulated by the ambient light conditions.[51-58]

The interdependence of the two stimuli was addressed by either testing the effects of different adaptational states in dopamine-depleted retinas or in the presence of dopamine antagonists. In the turtle and mudpuppy retina, the effects of both steady and flickering light (flickering light was used because it stimulated dopamine release more powerfully than steady light) were blocked by dopamine antagonists (Fig. 6.5B).[46,47] In the fish retina, only the effect of flickering light was blocked by dopamine antagonists, whereas the effect of steady light persisted in a dopamine-depleted retina (Fig. 6.5A) or in the presence of a D1 antagonist.[44,45] This suggests that regulating dopamine release is not the only mechanism by which light can affect the coupling conductance between horizontal cells.

2.4. NONDOPAMINERGIC, cAMP-INDEPENDENT MODULATION MECHANISMS OF GAP JUNCTIONAL CONDUCTANCE BETWEEN HORIZONTAL CELLS

Arginine, Nitric oxide (NO) and cGMP

NO is liberated in the L-arginine-L-citrulline metabolic pathway and activates soluble guanylate cyclase which results in the synthesis of cyclic GMP that is implicated in various neuronal functions.[59] Nitric oxide synthase is present in horizontal cells of the carp retina,[60] suggesting a pos-sible effect of the NO-cGMP pathway on the coupling conductance. Pairs of isolated horizontal cells from the catfish are indeed decoupled by about 50% if nitroprusside, an activator of guanylate cyclase, is added to the bathing medium.[28] Introduction of cGMP directly into one cell of the pair also decreased coupling by about 40%. Similar results were obtained in the turtle retina. Injection of cGMP or L-arginine, the precursor of NO, both increased the input resistance of horizontal cells in an eyecup preparation without affecting the dark adapted resting membrane potential. This suggests that the observed increase of the input resistance mainly results from an increase of the coupling resistance and not from the nonjunctional membrane resistance. At the same time, the light response to center stimulation was increased and that to surround stimulation was decreased.[19,61] Superfusion of the carp retina with nitroprusside yielded very similar results (Fig. 6.6A).[62] It is therefore likely that at least in the nonmammalian retina the conductance between horizontal cells is modulated through a NO-cGMP mediated pathway. At present we do not know through which signal such a pathway could be activated in the retina. Of course, glutamate, the neurotransmitter of the photoreceptors, is a likely candidate, since it is known to liberate NO.[63]

Arachidonic acid

Intracellular injection of arachidonic acid into horizontal cells of the turtle retina increased the coupling resistance and blocked dye coupling.[64] Its effect was blocked by a guanylate cyclase inhibitor and by a lipoxygenase inhibitor. These findings suggest that arachidonic acid blocks gap junctions via the activation of guanylate cyclase. This opens the possibility that activation of phospholipase A2 and subsequent production of arachidonic acid affects the coupling. The neurotransmitter glutamate again could be the putative signal, since glutamate can activate phospholipase A2. In this case, glutamate released from the photoreceptors could modulate

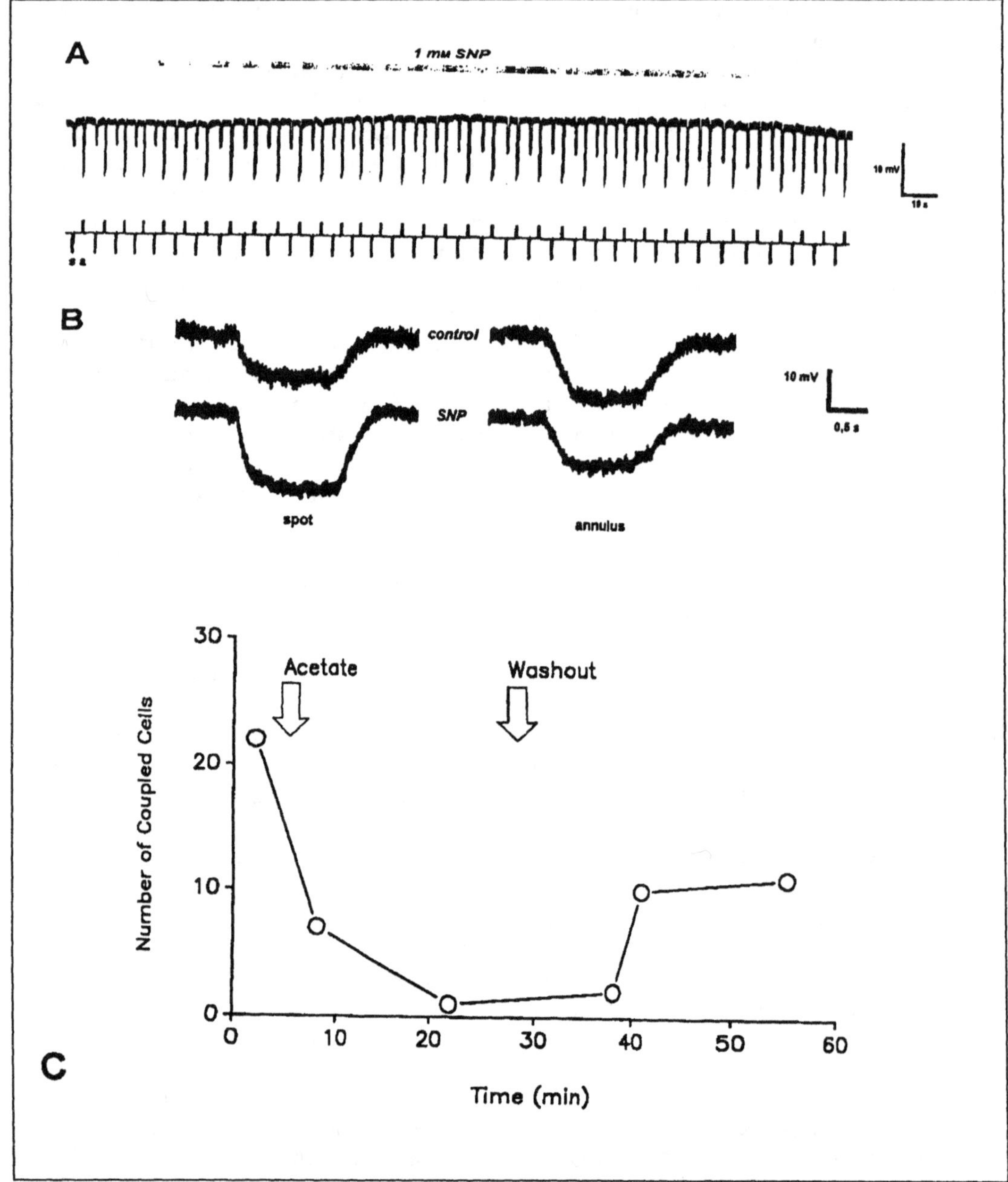

Fig. 6.6. *The effect of nitric oxide and pH on coupling of horizontal cells. (A, B) Electrophysiological recordings from horizontal cells of the carp retina. Application of sodium nitroprusside, an activator of NO, affects the annulus/spot ratio (s,a). It increases the responses to spot stimulation but decreases the response to annular stimulation. (C) Time course for the action of low intracellular pH on dye-coupling between A-type horizontal cells in the rabbit retina. The extracellular pH was kept at pH 7.2 during this experiment. Reprinted with permission from Hampson ECGM et al, Proc R Soc Lond [Biol] 1994; 255:67-72.*

junctional conductance by activating both the NO and the arachidonic acid pathways, both acting then through cGMP.

Calcium

Isolated pairs of coupled catfish horizontal cells were not decoupled when the internal Ca^{2+} concentration was seven times elevated.[28] Similar observations were made in the rabbit retina. Release of Ca^{2+} from internal stores did not affect the number of dye coupled A-type horizontal cells (Weiler & Hampson, unpublished data).

pH, CO$_2$

A reduction of the intracellular pH by substituting of sodium chloride in the bathing Ringer solution with sodium acetate decreased the coupling of isolated cell pairs of catfish horizontal cells.[28] The same procedure decreased dye coupling of A-type horizontal cells in the rabbit retina (Fig. 6.6B).[12] The weak acid acetate is membrane permeable and releases a proton once inside the cell resulting in an internal pH drop of some tenths of a pH unit. Decoupling can also result from an acidification of the external milieu with CO_2 or HCl,[12,65] which most likely acts through a subsequent acidification of the internal cell milieu. The pH effect in fish retina is unrelated to the dopaminergic uncoupling effect. Dopamine neither acts through internal acidification, nor does a change of external pH affect dopamine release.[12,28,65] By contrast, as already mentioned, the dopaminergic uncoupling effect in the rabbit retina with respect to horizontal cells is pH-gated.[12]

2.5. CONCLUSIONS

Taken together, the available data regarding the modulation of junctional conductance between horizontal cells in the retina appears to indicate that dopaminergic modulation might be a universal principle. Dopamine modulates g_j not only in retinas where dopamine-containing neurons (interplexiform cells) contact synaptically horizontal cells, but also via volume transmission from neurons (amacrine cells) of the inner retina.

The modulatory properties of messengers linked to glutamatergic activation point to a possible involvement also of glutamate, the principal transmitter of the photoreceptors which synapse onto horizontal cells.

The release of both neurotransmitters, dopamine and glutamate, has a tonic component that depends on the ambient light condition. It is therefore not surprising that coupling depends on the adaptational level of the retina. Furthermore, activity-dependent metabolic products might also affect the coupling or gating the effects of neurotransmitters on coupling.

Modulation of the coupling conductance affects the length constant of the horizontal cell network and thus the spread of activity. A functional key feature of horizontal cells is lateral inhibition that is essential for the control of the working range and the early processing of contrast. Fine tuning of this functional role seems to be essential for proper vision under different light conditions, and modulation of coupling conductance appears to be a powerful tool to achieve it.

3. MODULATION OF ROD-CONE COUPLING BY LIGHT

Photoreceptor coupling, at first glance a counter-intuitive feature for a sensory system producing high spatial resolution, is quite common among all vertebrate retinas. Functional electrical coupling between adjacent cones was first demonstrated in the turtle retina and subsequently in many other nonmammalian retinas.[66] In mammals functional cone-cone coupling is very likely, even between cones in the fovea of macaques, based on anatomical studies.[67,68] Surprisingly, this coupling is nonselective with regard to the spectral properties of cones and could therefore partially degrade not only the spatial but also chromatic resolution. Rods and cones are also electrically coupled[69-71] blurring the separation of scotopic and photopic pathways at the earliest stage. The finding that the coupling between horizontal cells can be modulated by the ambient light conditions prompted the search for a similar mechanism for the coupling between photoreceptors. Indeed, the coupling between rods and cones in the salamander retina is modulated by light.[72] Background light increased the voltage responses of rods to current injections into neighboring cones. Therefore cone to rod signals are enhanced under light adapted conditions. Under dark adapted conditions, rod-cone coupling is weak and does not permit the shunting of small rod signals. In the fish retina, a-light-dependent crosstalk through gap junctions between rods and cones has been suggested based on the chromatic properties of horizontal cell light responses under different adaptational conditions.[73]

At present nothing is known about the cellular and molecular mechanisms underlying the modulation of photoreceptor coupling.

4. MODULATION OF COUPLING BETWEEN AMACRINE CELLS

Many amacrine cell types in the mammalian and nonmammalian retina show dye coupling. There is not only intensive coupling between homologous types of amacrine cells, but also heterologous coupling with other neurons of the retina. A particularly well studied amacrine cell in this respect is the AII amacrine cell in the mammalian retina. This amacrine cell is a third order neuron in the rod pathway that receives signals from the rod bipolar cells and transmits signals to the cone bipolars through gap junctions and through chemical synapses.[74] Neighboring AII amacrine cells are also connected by gap junctions.[75] These gap junctions, however, are not permeable to Lucifer Yellow but only to the smaller tracer molecule neurobiotin.[1]

Dye coupling with neurobiotin was used to address the question of whether the AII amacrine cell coupling can be modulated by dopamine. Dopaminergic amacrine cells form with their processes a dense network in the rabbit retina that encapsulates the cell bodies of AII amacrine cells.[76] Quantitative analysis of the number of coupled AII cells after one cell had been injected with neurobiotin in the isolated and superfused rabbit retina clearly revealed an uncoupling effect of dopamine[77] (Fig. 6.7A). This uncoupling effect was not only achieved when dopamine was added to the superfusion medium, but also when the release of endogenous dopamine was stimulated. The effect was mediated through D1 receptor activation and subsequent increase of cAMP (Fig. 6.7B). Interestingly, the coupling between AII amacrine cells is not sensitive to pH. Acidification of the external milieu to a pH of 6.5 did not affect the AII coupling, although the coupling between horizontal cells in the same preparation was blocked. These functional differences, and the fact that the dye transfer properties of horizontal cell gap junctions and AII amacrine cell gap junctions differ, makes it likely that the molecular identity of the connexins in these cells might be different.

What might be the functional role of the coupling modulation of AII gap junctions? Direct measurements suggest that dopamine in the rabbit retina is preferentially released in the light.[57] Consequently, AII amacrine cells are uncoupled in the light adapted state. Since AII cells form a link between the rod and the cone pathway with the help of their electrical synapses, it would follow that the rod circuit in the inner retina is disconnected from the cone circuit in the photopic state. This would prevent a degrading of spatial acuity of the cone circuit by a signal flow from cone bipolars into AII cells.

Again, the data on the dopaminergic modulation of coupling between amacrine cells points to a role of coupling modulation in the course of light adaptation. Modulation of coupling conductances is obviously an important mechanism to support the important tuning of the working range of the retina. The example of the AII

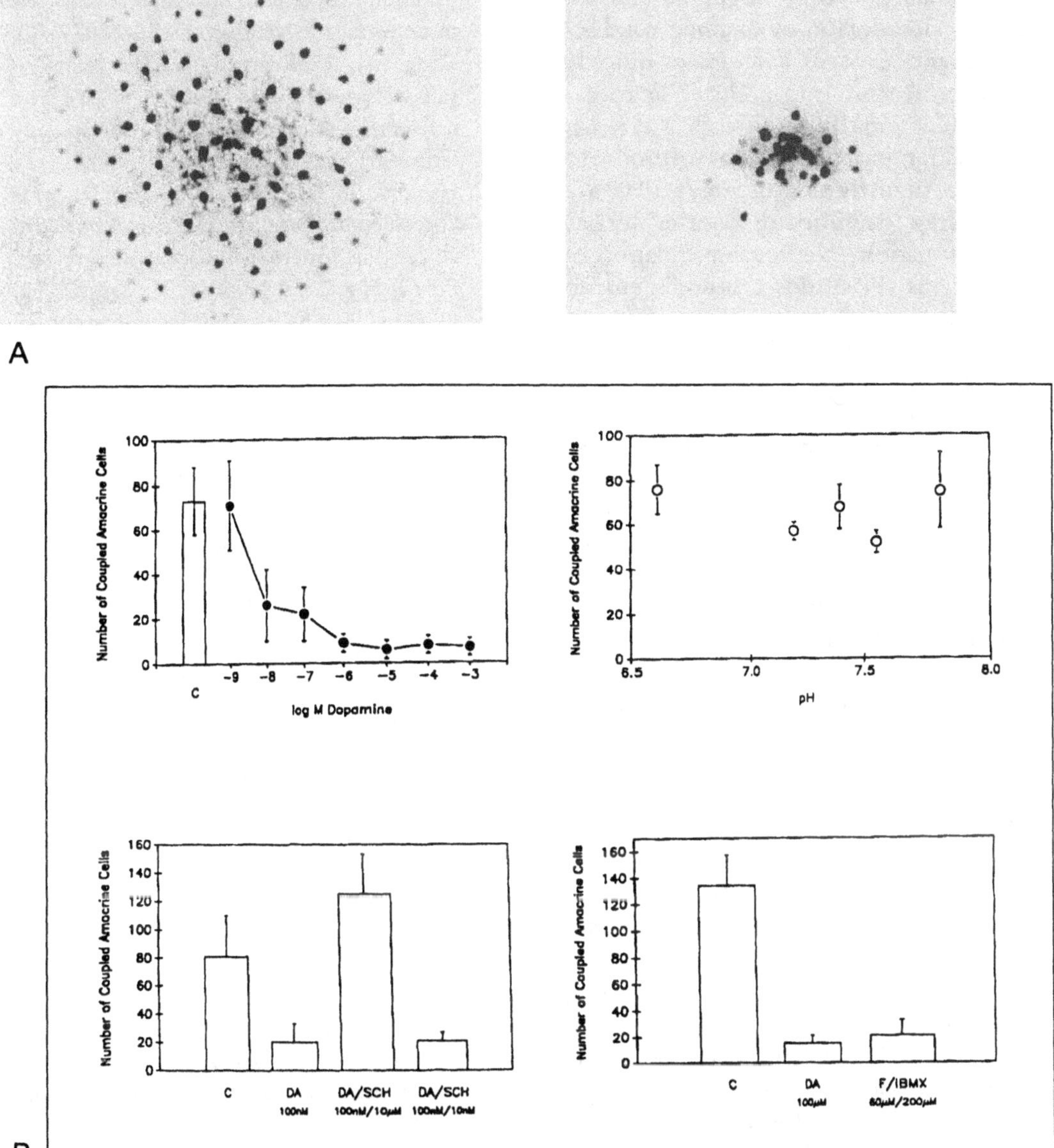

Fig. 6.7. Dopaminergic modulation of gap junction permeability between AII amacrine cells in the rabbit retina. (A) Micrographs showing the extent of tracer-coupling following injection of Neurobiotin into a single AII amacrine cell in a control (left) retina and in a retina superfused with a Ringer solution containing dopamine (right). (B) Dose-response plot of the number of tracercoupled amacrine cells against the concentration of dopamine (upper left). Acidification of the extracellular pH does not affect the coupling (upper right). The specific dopamine antagonist SCH-23390 blocks the uncoupling effect of dopamine (DA) (lower left). The combined effect of forskolin and IBMX (F/IBMX) was similar to the dopamine effect. Reprinted with permission from Hampson ECGM et al, J Neurosci 1992; 12:4911-22. Copyright 1992 Oxford University Press.

cells points to an additional aspect of coupling modulation that might be considered. The modulation of coupling conductances might serve as a means to quickly integrate a neuron into a circuit or to remove it functionally from such a circuit. It would, for example, allow neurons to participate in different networks that are configurated depending on the actual stimulus situation. Neuronal modulation of gap junctions thus adds a new dimension to the plasticity of neuronal networks.

REFERENCES

1. Vaney DI. Many diverse types of retinal neurons show tracer coupling when injected with biocytin or Neurobiotin. Neurosci Lett 1991; 125:187-90.
2. Hedden WL, Dowling JE. The interplexiform cell system II. Effects of dopamine on goldfish retinal neurons. Proc Royal Soc Lond 1978; B201:407-12.
3. Negishi K, Drujan BD. Effects of catecholamine on the horizontal cell membrane potential in the fish retina. Sens Proc 1978; 2:388-95.
4. Negishi K, Drujan BD. Reciprocal changes in center and surrounding S-potentials of fish retina in response to dopamine. Neurochem Res 1979; 4:313-8.
5. Negishi K, Drujan BD. Effects of catecholamines and related compounds on horizontal cells in the fish retina. J Neurosci Res 1979; 4:311-34.
6. Negishi K, Drujan BD. Similarities in effects of acetylcholine and dopamine on horizontal cells in the fish retina. J Neurosci Res 1979; 4:335-49.
7. Laufer M, Salas R. Intercellular coupling and retinal horizontal cell receptive field. Neurosci Lett 1981; 7:339
8. Teranishi T, Negishi K, Kato S. Dopamine modulates S-Potential amphitude and dye-coupling between external horizontal cells in carp retina. Nature 1983; 301:243-6.
9. Teranishi T, Negishi K, Kato S. Regulatory effect of dopamine on spatial properties of horizontal cells in the carp retina. J Neurosci 1984; 4:1271-80.
10. Piccolino M, Neyton J, Gerschenfeld HM. Decrease of the gap junction permeability induced by dopamine and cyclic 3-5-adenosine miniphosphate in horizontal cells of the turtle retina. J Neurosci 1984; 4:2477-88.
11. Dong C-J, McReynolds JS. The relationship between light, dopamine release and horizontal cell coupling in the mudpuppy retina. J Physiol 1991; 440:291-309.
12. Hampson ECGM, Weiler R, Vaney DI. pH-Gated dopaminergic modulation of horizontal cell gap junctions in mammalian retina. Proc R Soc Lond [Biol] 1994; 255:67-72.
13. Piccolino M, Witkovsky P, Trimarchi C. Dopaminergic mechanisms underlying the reduction of electrical coupling between horizontal cells of the turtle retina induced by d-amphetamine, bicuculline, and veratridine. J Neurosci 1987; 7:2273-84.
14. Dowling JE, Ehinger B. Synaptic organization of the amine-containing interplexiform cells of the goldfish and cebus monkey retinas. Science 1975; 188:270-3.
15. Witkovsky P, Eldred WD, Karten HJ. Catecholamine and indoleamine-containing neurons in the turtle retina. J Comp Neurol 1984; 228:217-25.
16. Witkovsky P, Nicholson C, Rice ME et al. Extracellular dopamine concentration in the retina of the clawed frog, *Xenopus laevis*. Proc Natl Acad Sci USA 1993; 90:5667-71.
17. Fuortes MGF, Simon EJ. Interactions leading to horizontal cell responses in the turtle retina. J Physiol (Lond) 1974; 240:177-98.
18. Ammermüller J, Weiler R, Perlman I. Short-term effects of dopamine on photoreceptors, luminosity- and chromaticity-horizontal cells in the turtle retina. Visual Neurosci 1995; 12:403-12.
19. Miyachi E-I, Murakami M. Decoupling of horizontal cells in the carp and turtle retinae by intracellular injection of cyclic AMP. J Physiol 1989; 419:213-24.
20. Dowling JE, Ehinger B. Synaptic organization of the dopaminergic neurons in the rabbit retina. J Comp Neurol 1978; 180:203-20.
21. Yamada M, Shigematsu Y, Umetani Y et al. Dopamine decreases receptive field size of rod driven horizontal cells in carp retina. Vision Res 1992; 32:1801-7.
22. Qian H, Malchow RP, Ripps H. Gap-junctional properties of electrically coupled skate

horizontal cells in culture. Vis Neurosci 1993; 10:287-95.

23. Yamamoto F, Borgula GA, Steinberg RH. Effects of light and darkness on pH outside rod photoreceptors in the cat retina. Expl Eye Res 1992; 54:685-97.

24. Watling KJ, Dowling JE, Iversen LL. Dopamine receptors in the retina may all be linked to adenylate cyclase. Nature 1979; 281:578-80.

25. Lasater EM, Dowling JE. Dopamine decreases conductance of the electrical junctions between cultured retinal horizontal cells. Proc Natl Acad Sci USA 1985; 82:3025-9.

26. Lasater EM, Dowling JE. Electrical coupling between pairs of isolated fish horizontal cells is modulated by dopamine and cAMP. In: Bennett, ed. Gap Junctions. Cold Spring Harb Symp Quant 1985:393-404.

27. Lasater EM. Retinal horizontal cell gap junctional conductance is modulated by dopamine through a cyclic AMP-dependent protein kinase. Proc Natl Acad Sci USA 1987; 84:7319-23.

28. DeVries SH, Schwartz EA. Modulation of an electrical synapse between solitary pairs of catfish horizontal cells by dopamine and second messengers. J Physiol 1989; 414: 351-75.

29. Janssen-Bienhold U, Nagel H, Weiler R. In vitro phosphorylation in isolated horizontal cells of the fish retina: Effects of the state of light adaptation. Eur J Neurosci 1993; 5:584-93.

30. Janssen-Bienhold U, Dermietzel R, Weiler R. Neuronal connexins in the retina. Europ J Neurosci 1995; (in press).

31. McMahon DG, Rischert JC, Dowling JE. Protein content and cAMP-dependent phosphorylation of fractionated white perch retina. Brain Res 1994; 659:110-6.

32. Janssen-Bienhold U, Schöpe B, Buschmann-Gebhardt B et al. Identification of neuronal connexin proteins in the retina. In: Elsner N, Menzel R, eds. Proceedings of the 23rd Göttingen Neurobiology Conference 1995, Volume II. Stuttgart, New York: Georg Thieme Verlag, 1995:458.

33. McMahon DG, Knapp AG, Dowling JE. Horizontal cell gap junctions: single-chan-

nel conductance and modulation by dopamine. Proc Natl Acad Sci USA 1989; 86:7639-43.

34. McMahon DG, Brown DR. Modulation of gap-junction channel gating at zebrafish retinal electrical synapses. J Neurophysiol 1994; 72:2257-68.

35. Burt JM, Spray DC. Single channel events and gating behavior of the cardiac gap junction channel. Proc Natl Acad Sci USA 1988; 85:3431-4.

36. DeVries SH, Schwartz EA. Hemi-gap-junction channels in solitary horizontal cells of the catfish retina. J Physiol 1992; 445: 201-30.

37. Malchow RP, Qian H, Ripps H. Evidence for hemi-gap junctional channels in isolated horizontal cells of the skate retina. J Neurosci Res 1993; 35:237-45.

38. Kurz-Isler G, Voigt T, Wolburg H. Modulation of connexon densities in gap junctions of horizontal cell perikarya and axon terminals in fish retina: effects of light/dark cycles, interruption of the optic nerve and application of dopamine. Cell Tissue Res 1992; 268:267-75.

39. Washioka H, Watanabe H, Negishi K et al. Horizontal cell gap junction in the goldfish retina: Area and density of particles as revealed by complementary freeze replicas. Arch Histol Cytol 1991; 54:181-8.

40. Baldridge WH, Ball AK, Miller RG. Gap Junction particle density of horizontal cells in goldfish retinas lesioned with 6-OHDA. J Comp Neurol 1989; 287:238-46.

41. Kohler K, Kolbinger W, Kurz-Isler G et al. Endogenous dopamine and cyclic events in the fish retina, II: Correlation of retinomotor movement, spinule formation, and connexon density of gap junctions with dopamine activity during light/dark cycles. Vis Neurosci 1990; 5:417-28.

42. Weiler R, Kohler K, Kolbinger W et al. Dopaminergic neuromodulation in the retinas of lower vertebrates. Neurosci Res 1988; Suppl. 8:S183-96.

43. Shigematsu Y, Yamada M. Effects of dopamine on spatial properties of horizontal cell responses in the carp retina. Neurosci Res 1988; 8:69-80.

44. Umino O, Lee Y, Dowling JE. Effects of

light stimuli on the release of dopamine from interplexiform cells in the white perch reetina. Vis Neurosci 1991; 7:451-8.

45. Baldridge WH, Ball AK. Background illumination reduces horizontal cell receptive-field size in both, normal and 6-hydroxy-dopamine-lesioned goldfish retinas. Vis Neurosci 1991; 7:441-50.

46. Dong C-J, McReynolds JS. Comparison of the effects of flickering and steady light on dopamine release and horizontal cell coupling in the mudpuppy retina. J Neurophysiol 1992; 67:364-72.

47. Akopian A, Weiler R. Effects of background illuminations on the receptive field size of horizontal cells in the turtle retina are mediated by dopamine. Neurosci Lett 1992; 140:121-4.

48. Perlman I, Ammermüller J. Receptive-field size of L1 horizontal cells in the turtle retina: Effects of dopamine and background light. J Neurophysiol 1994; 72:2786-95.

49. Mangel SC, Dowling JE. Responsiveness and receptive field size of carp horizontal cells are reduced by prolonged darkness and dopamine. Science 1985; 229:1107-9.

50. Tornqvist K, Yang X-L, Dowling JE. Modulation of cone horizontal cell activity in the teleost fish retina. III. Effects of prolonged darkness and dopamine on electrical coupling between horizontal cells. J Neurosci 1988; 8:2279-88.

51. Kirsch M, Wagner H-J. Release pattern of endogenous dopamine in teleost retinas during light adaptation and pharmacological stimulation. Vision Res 1989; 29:147-54.

52. Kolbinger W, Kohler K, Oetting H et al. Endogenous dopamine and cyclic events in the fish retina. I. HPLC assay of total content, release and metabolic turnover during different light/dark cycles. Vis Neurosci 1990; 5:143-9.

53. Weiler R, Kolbinger W, Kohler K. Reduced light responsiveness of the cone pathway during prolonged darkness does not result from an increase of dopaminergic activity in the fish retina. Neurosci Lett 1989; 99:214-8.

54. Boatright JH, Hoel MJ, Iuvone PM. Stimulation of endogenous dopamine release and metabolism in amphibian retina by light- and K⁺-evoked depolarization. Brain Res 1989; 482:164-8.

55. Boatright JH, Rubim NM, Iuvone PM. Regulation of endogenous dopamine release in amphibian retina by gamma-amino-butyric acid and glycine. Visual Neurosci 1994; 11:1003-12.

56. Dearry A, Burnside B. Light-induced dopamine release from teleost retinas acts as a light-adaptive signal to the retinal pigment epithelium. J Neurochem 1989; 53:870-8.

57. Godley BF, Wurtman RJ. Release of endogenous dopamine from the superfused rabbit retina in vitro: effect of light stimulation. Brain Res 1988; 452:393-5.

58. Kolbinger W, Weiler R. Modulation of endogenous dopamine release in the turtle retina: Effects of light, calcium, and neurotransmitters. Visual Neurosci 1993; 10:1035-41.

59. Garthwaite J, Boulton CL. Nitric oxide signaling in the central nervous system. Annu Rev Physiol 1995; 57:683-706.

60. Weiler R, Kewitz B. The marker for nitric oxide synthase, NADPH-diaphorase, colocalizes with GABA in horizontal cells and cells of the inner retina in the carp retina. Neurosci Lett 1993; 158:151-4.

61. Miyachi E-I, Murakami M, Nakaki T. Arginine blocks gap junctions between retinal horizontal cells. Neuro Rep 1990; 1(2):107-10.

62. Pottek M, Weiler R. Modulatory effects of nitric oxide on horizontal cells of the carp retina. In: Elsner N, Breer H, eds. Proceedings of the 22nd Göttingen Neurobiology Conference 1994; Volume I. Stuttgart, New York: Georg Thieme Verlag, 1994:55.

63. Vincent SR. Nitric oxide: A radical neurotransmitter in the central nervous system. Prog Neurobiol 1994; 42:129-60.

64. Miyachi E, Kato C, Nakaki T. Arachidonic acid blocks gap junctions between retinal horizontal cells. Neuroreport 1994; 5:485-8.

65. Negishi K, Teranishi T, Kato S. Opposite effects of ammonia and carbon dioxide on dye coupling between horizontal cells in the carp retina. Brain Res 1985; 342:330-40.

66. Attwell D. Ion channels and signal process-

ing in the outer retina. J Exp Physiol 1986; 71:497-536.

67. Raviola E, Gilula NB. Gap junctions between photoreceptor cells in the vertebrate retina. Proc Natl Acad Sci USA 1973; 70:1677-81.

68. Tsukamoto Y, Masarachia P, Schein SJ et al. Gap junctions between the pedicles of macaque foveal cones. Vision Res 1992; 32:1809-15.

69. Schwartz EA. Cones excite rods in the retina of the turtle. J Physiol 1975; 246:639-51.

70. Kolb H. Organization of the outer plexiform layer of the primate retina: EM Golgi impregnated cells. Phil Trans R Soc Lond 1970; 258:261-83.

71. Raviola E, Gilula NB. Intermembrane organization of specialized contacts in the outer plexiform layer of the retina: a freeze fracture study in monkeys and rabbits. J Cell Biol 1975; 65:192-222.

72. Yang X-L, Wu SM. Modulation of rod-cone coupling by light. Science 1989; 244: 352-254.

73. Mangel SC, Baldridge WH, Weiler R et al. Threshold and chromatic sensitivity changes in fish cone horizontal cells following prolonged darkness. Brain Res 1994; 659: 55-61.

74. Vaney DI, Young HM, Gynther IC. The rod circuit in the rabbit retina. Vis Neurosci 1991; 7:141-54.

75. Famiglietti EV, Kolb H. A bistratified amacrine cell and synaptic circuitry in the inner plexiform layer of the retina. Brain Res 1975; 84:293-300.

76. Voigt T, Wässle H. Dopaminergic innervation of A III amacrine cells in mammalian retina. J Neurosci 1987; 7(12):4115-28.

77. Hampson ECGM, Vaney DI, Weiler R. Dopaminergic modulation of gap junction permeability between amacrine cells in mammalian retina. J Neurosci 1992; 12:4911-22.

CHAPTER 7

MODULATION OF CONNEXON DENSITY IN GAP JUNCTIONS OF FISH HORIZONTAL CELLS

Hartwig Wolburg and Gertrud Kurz-Isler

1. INTRODUCTION

Gap junctions in the vertebrate retina are among those most widely investigated in the central nervous system. Many methods have been applied to measure receptive field sizes of retinal neurons in order to estimate the activity and strength of interneuronal coupling or to describe the morphology of the underlying gap junctions. In this chapter, the results gotten by application of the freeze-fracture method to problems of gap junctional modulation in the retina will be summarized. The advantage of this method is the two-dimensional visualization of planar lipidic monolayers of membranes showing inserted particles or—as in gap junctions—connexons. In conventional ultrathin section electron microscopy, gap junctions are linear domains with variable lengths; their ultrastructural peculiarities such as size, shape or the arrangement of connexons cannot be recognized. However, using standard freeze-fracture techniques the linkage of the gap junctional domain with underlying cytoskeletal elements cannot be observed. Surprisingly, the observation of cytoplasmic surfaces in the area of gap junctions as described, for example, in liver,[1] cardiac[2] and lens gap junctions[3] was not performed in gap junctions of the retina.

Most gap junctions are clusters of densely packed connexons which are—at least in vertebrate tissue—associated consistently with the protoplasmic fracture face (P-face). The P-face association is independent of whether or not cytoskeletal components are present at the cytoplasmic surface of gap junctions. Many past studies described an increase in the connexon density including a slight decrease of the channel extension parallel to the membrane plane following an uncoupling procedure (for an overview see ref. 4). However, most data on coupling/uncoupling-related connexon spacing describe subtle reorganizations of cardiac and liver connexon arrays on the order of 1 nm. It may be questionable whether interconnexon distances of less than 1 nm can be reliably measured in carbon/platinum freeze-fracture replicas whose thickness usually is in the magnitude of 2-3 nm. An exceptional situation was found initially by Wolburg and Kurz-Isler[5] in the gap junctions of the goldfish retina outer horizontal cells where the connexons differ largely in their density. The averaged interconnexon spacing varied between 20 and 40 nm or even more and was dependent on illumination conditions and was controlled by the pH and the dopamine content.

2. GAP JUNCTIONS IN THE VERTEBRATE RETINA

For many years the retina has been the subject of numerous morphological and physiological investigations many of which have studied the processing of spatiotemporal information in the neuronal network. The electrotonic coupling via gap junctions of cells in the retina including photoreceptors,[6-8] horizontal cells[9-11] (see below), bipolar cells,[12-14] amacrine cells,[15-17] ganglion cells[18-20] and glial cells.[21-25] is well known (for reviews of cell coupling in the retina see refs. 20, 26, 27). During development of the retina, gap junctions were recognized and described.[18,28-31] Connexin32 and connexin43 were found to be distributed in a distinct manner in retinal neurons and glial cells.[32-34] Hitherto, the main interest has been focused on the spread of electrotonic signals and dyes such as lucifer yellow, biocytin or neurobiotin.[12,17,19,35-43] Compared with the number of physiological studies performed on electrical synapses, morphological studies on gap junctions in the retina are less numerous (for examples see refs. 6, 44-50). Among all retina cells coupled by gap junctions, the outer horizontal cells have been studied most fre-

quently and their gap junction morphology correlated to physiological parameters.

3. GAP JUNCTIONS IN HORIZONTAL CELLS

As described in more detail in other chapters of this book, horizontal cells (HCs) are second order neurons in the outer part of the inner nuclear layer that are involved in the generation of the antagonistic surround of the receptive fields of bipolar and ganglion cells. In various species, some subtypes possess enlarged axon terminals extending into the inner nuclear layer; these axon terminals form their own network coupled by gap junctions.[11,19,51-54] Horizontal cells generate a different type of light-induced response (S-potential), the amplitude of which increases with the area of illumination. The spread of S-potentials is mediated by gap junctions[35] and can be decreased by dopamine.[39] In the light, dopamine is released from the interplexiform cells and binds to D1 and D2 receptors of HCs, which were shown to be involved in HC uncoupling.[55-57] The dopaminergic interplexiform cells are under the control of retinopetal fibers.[58] In the dark, γ-aminobutyric acid (GABA) is released from the HCs and binds to receptors of the interplexiform cell, thus maintaining a low rate of dopamine release. GABA antagonists have an uncoupling effect on HC gap junctions[38] by allowing an increase in dopamine release.[59,60] The effect of dopamine to decrease gap junctional conductance is achieved through a cyclic AMP-dependent protein kinase, which in turn is believed to phosphorylate the gap junction protein(s).[61-63] A recent review on the regulation of HC gap junction conductance by second messengers such as cyclic AMP, cyclic GMP, nitric oxide and Ca^{2+} has been published.[27]

4. GAP JUNCTIONS RESPOND ON DIFFERENT ILLUMINATION CONDITIONS

In freeze-fracture studies, the connexon density of HC gap junctions in fish retina was found to be dependent on illumination conditions.[5,64,65] In the dark, many small gap junctions on the order of 0.1 µm² revealed a high connexon density (about 8000/µm²; Fig. 7.1); after red light adaptation, the gap junctions became larger (up to 3 µm²) and the connexon density decreased (to about 3000/µm²; Figs. 7.2, 7.3). It should be pointed out that the connexon density that we termed as "high" is low in comparison with what in other systems was described as crystalline or densely packed. This dense, hexagonal package of connexons corresponding to a connexon density of 12000-14000/µm², has been hypothesized by several authors to reflect the uncoupled state and was not observed in HC gap junctions. In contrast, slight derangement of the crystalline connexon pattern corre-

sponding to an increase of the interconnexonal spacing on the order of one or two nanometers was interpreted as correlating with the coupled state and termed high connexon density in HC gap junctions.[65] Inversely, our "low" density of connexons obtained as a short-term and reversible physiological phenomenon has not been observed in other systems except under developmental conditions (for literature, see ref. 4). The distance between single small gap junctions appearing a few minutes after exposure to darkness seemed to be too large to be attributed to pure dissociation of a formerly large, light-adapted gap junctions. In contrast, the dramatic reduction in size of most gap junctions in the dark-adapted state cannot be explained solely by

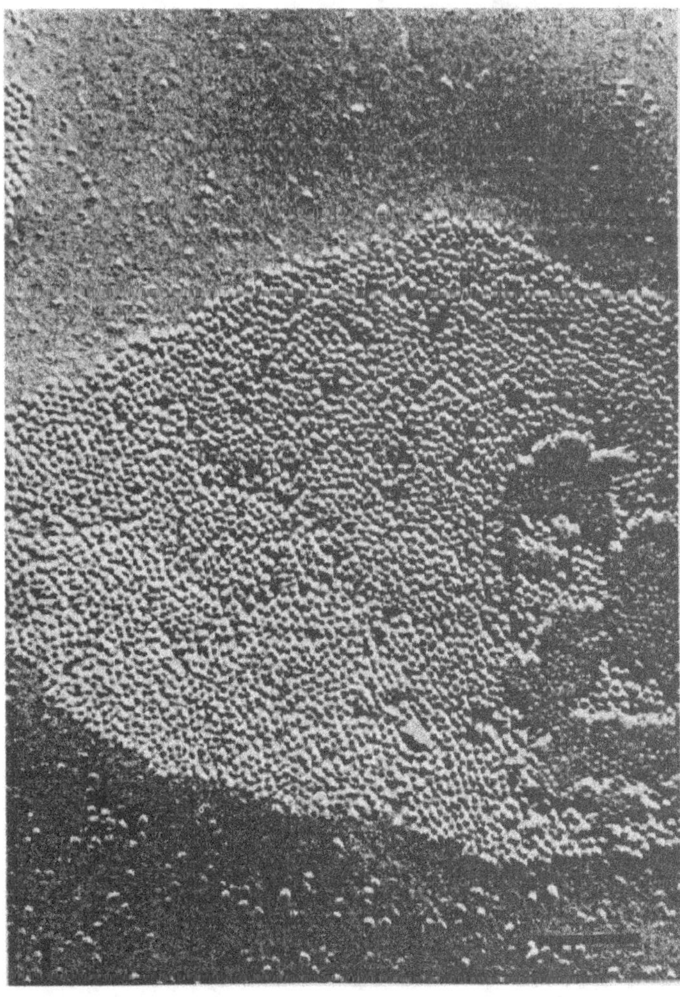

Fig. 7.1. *Goldfish outer HC axon terminal gap junction, 60 min after dark adaptation. The connexons are densely packed. x120,000. Bar: 0.1 µm.*

simple aggregation of the formerly dispersed connexons via lateral shift. An adequate explanation may be the de novo formation of small gap junctional plaques from a pool of formerly unidentified single connexons. Since in the dark the HC receptive field size is larger than in the light, Kurz-Isler and Wolburg[64,65] concluded from their freeze-fracture observations that the gap junctions with high connexon densities may be in the coupled state, and the gap junctions with low connexon densities in the uncoupled state. During dark adaptation, most small gap junctions were undercoated by an electron-dense material. Frequently, this material was associated with microtubules and microfilaments.[65]

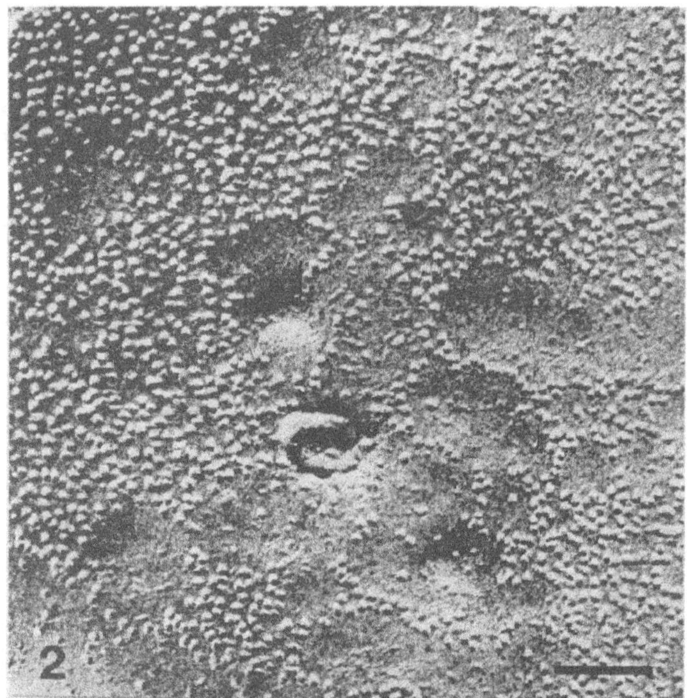

Fig. 7.2. Goldfish outer HC perikaryal gap junction with "fenestrations", 60 min after red light adaptation. The connexons are loosely scattered. x120,000. Bar: 0.1 μm.

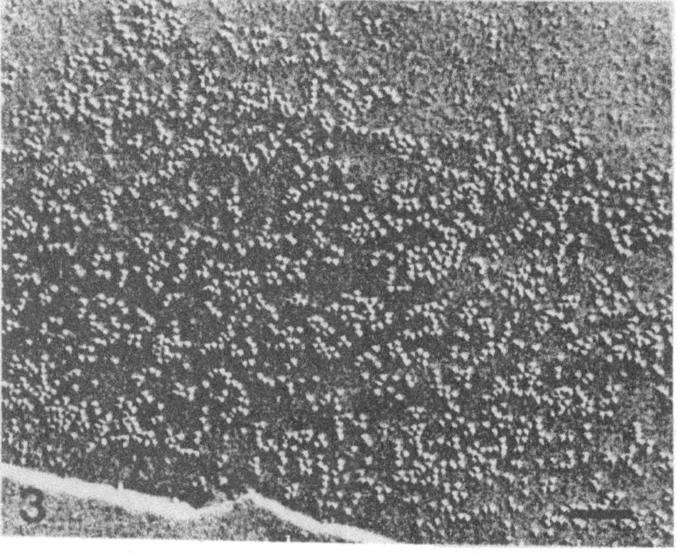

Fig. 7.3. Goldfish outer HC axon terminal gap junction, 30 min after light adaptation. The connexons are loosely scattered. x80,000. Bar: 0.1 μm.

The association of osmiophilic material beneath densified gap junctions during dark adaptation speaks in favor of a functional interaction between cytoskeleton and gap junctions. At places, ultrathin sections of red-light adapted HCs revealed patches of electron-dense material which were on the same order of magnitude as connexon clusters in gap junction replicas of the identical material. However, since the membrane areas between the connexon clusters are also as large as the connexon clusters themselves, it is not possible to distinguish which part of the gap junction—the connexon cluster or the membrane between—may be connected with cytoskeleton-associated plaque material.

As has been frequently stressed, outer HCs do not only form gap junctions between cell bodies (Figs. 7.2, 7.5) but also between axon terminals (Figs. 7.1, 7.3, 7.4, 7.6, 7.7; see above). As a result, the receptive fields of the axon terminals are larger than the dendritic spread of the HC perikarya.[35,36,53] In terms of freeze-facturing, perikaryal gap junctions of outer HCs of the fish retina frequently reveal a fenestrated shaping (Figs. 7.2, 7.5) whereas axon terminal gap junctions mostly are of oval shape[34,64] (Fig. 7.1). The increase of connexon density evoked by dark adaptation was more pronounced in the axon terminal than in perikaryal gap junctions.[54]

5. GAP JUNCTIONS RESPOND TO OPTIC NERVE CRUSH AND/OR DOPAMINE TREATMENT

Since retinopetal fibers running in the optic nerve are known to supply input to the dopaminergic interplexiform cells,[58] it seemed interesting to investigate the influence of crushing the optic nerve on the connexon density of the outer HC gap junctions. Transection of the optic nerve is the stereotypic operation performed in most electrophysiological studies using an open eye-cup preparation. Connexon density shifted from about 3000-4000/μm^2 to 6000-7000 connexons/μm^2 30-60 min after crushing of the optic nerve (Figs. 7.4, 7.5). The axon terminal gap junctions showed a 84% increase in density, but the perikaryal gap junctions only a 44% increase. Intraocular injection of dopamine is known to produce a low connexon density but in light-adapted retinas this decrease is less pronounced than in dark-adapted ones[66] (Fig. 7.6). Accordingly, administration of the dopamine antagonist haloperidol resulted in an increase of the connexon density.[66] Interestingly, this neutralizing effect of haloperidol was restricted to HC

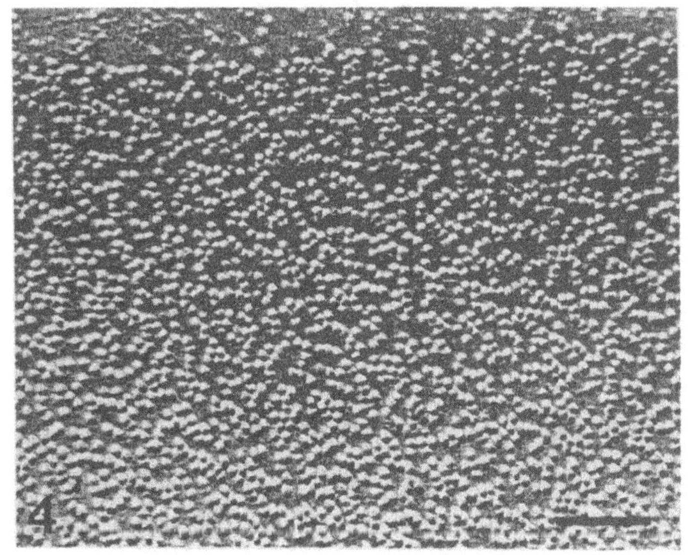

Fig. 7.4. Goldfish outer HC axon terminal gap junction, 120 min after optic nerve crush during light adaptation. The connexon density is increased. x120,000. Bar: 0.1 μm.

perikaryal gap junctions and not seen in axon terminal gap junctions,[66] giving rise to the suggestion that dopamine was acting on HC axon terminal gap junctions via an as yet unidentified transmitter. Dopamine injected immediately after optic nerve crush into the eye bulb prevented the increase of particle density.[54] An interesting experiment is demonstrated in Figure 7.7 which shows an axon terminal gap junction 30 min after intraocular dopamine injection and 60 min after crush of the cor-

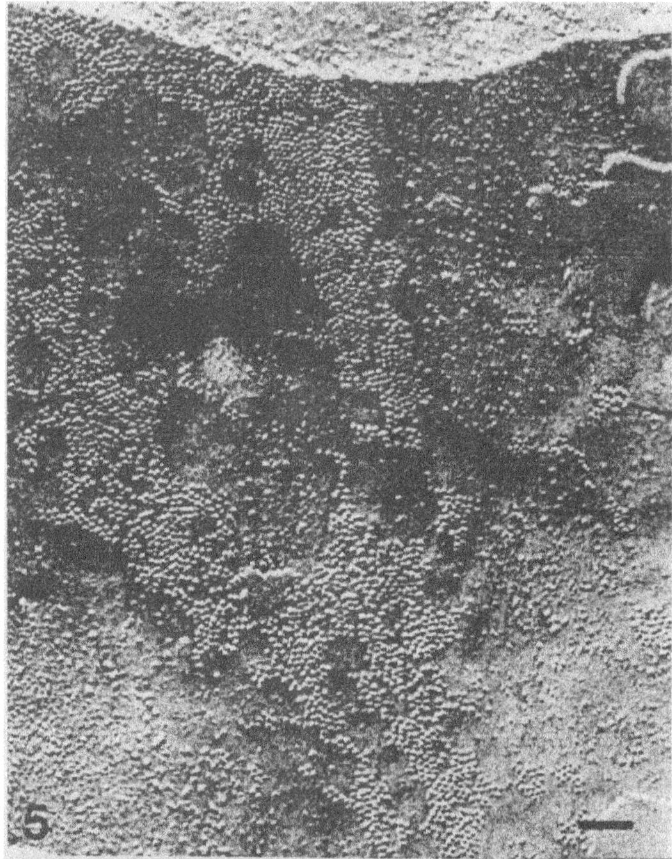

Fig. 7.5. Goldfish outer HC perikaryal gap junction with "fenestrations", 30 min after optic nerve crush during light adaptation. The connexon density is increased. x80,000. Bar: 0.1 µm.

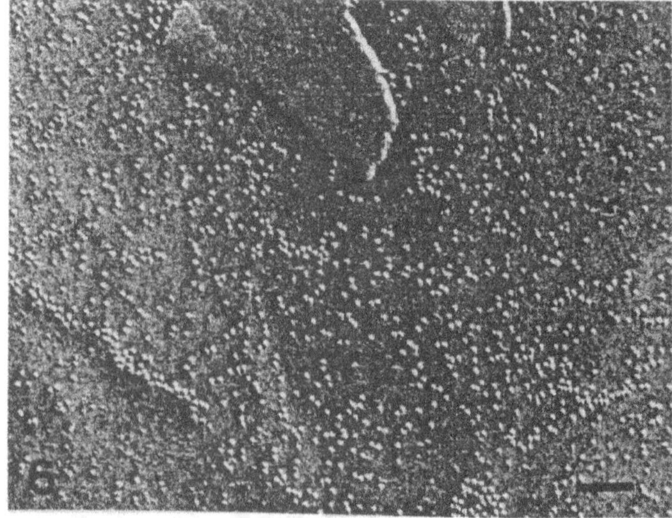

Fig. 7.6. Goldfish outer HC axon terminal gap junction, 30 min after intraocular dopamine injection (5 µl, 1 mM) and dark adaptation. Dopamine has antagonized the effect of dark adaptation on connexons. x80,000: Bar: 0.1 µm.

responding optic nerve. Most connexons are densely packed, presumably as a result of the primary optic nerve crush, but a number of connexons are scattered laterally, presumably as result of the dopamine injection. Prolonged interruption of the optic nerve that was nevertheless too short to allow axonal regeneration of the retinopetal fibers led to connexon densities that were similar in the treated and untreated eyes. Hence, since dopamine is able to antagonize the effect of optic nerve crush, it seems possible that the dopamine content available for gap junction modulation may be diminished after the interruption of the retinopetal fibers, so that dopamine cannot be synthesized and/or released. This suggests that dopamine synthesis and/or release is controlled by retinopetal projections. In contrast to the findings of Baldridge et al,[66] Kurz-Isler et al[65] described axon terminal gap junctions to be clearly sensitive to dopamine (Figs. 7.6, 7.7). The reaction of connexons in axon terminal gap junctions was even more marked than in perikaryal gap junctions. The increase of connexon density as a result of optic nerve crush and its prevention by dopamine treatment speaks in favor of a dopaminergic regulation of connexon density. This was furthermore substantiated by experiments that eliminated dopaminergic interplexiform cells by administration

of 6-OH-dopamine. The retinal depletion of dopamine by loss of dopaminergic neurons resulted in an increase of connexon density.[67,68] This again corroborated the hypothesis that the uncoupled state of HC gap junctions is characterized by a low, and the coupled state of HC gap junctions by a high connexon density.

6. GAP JUNCTIONS RESPOND TO VARIATIONS OF pH

Schmitz and Wolburg[69] reported that under conditions of decreased intracellular pH the connexon density of outer HC gap junctions was decreased. In the goldfish retina explanted in vitro, the connexon density at pH 6.5 was on the order of 8000-9000 connexons/μm^2, at pH 7.1 it was about 6000-7000 connexons/μm^2, and at pH 7.5 it was about 4000 connexons/μm^2. Density alterations in perikaryal and axon terminal gap junctions did not differ significantly. The highest connexon density was correlated with the appearance of membrane-associated plaque material, which disappeared following an increase in the pH. Regardless of the physiological coupling state that might be represented by different connexon densities in HCs, the results of Kurz-Isler and Wolburg[65] and Schmitz and Wolburg[69] consistently showed the coincidence of high connexon density with cytoplasm/cytoskeleton-GJ

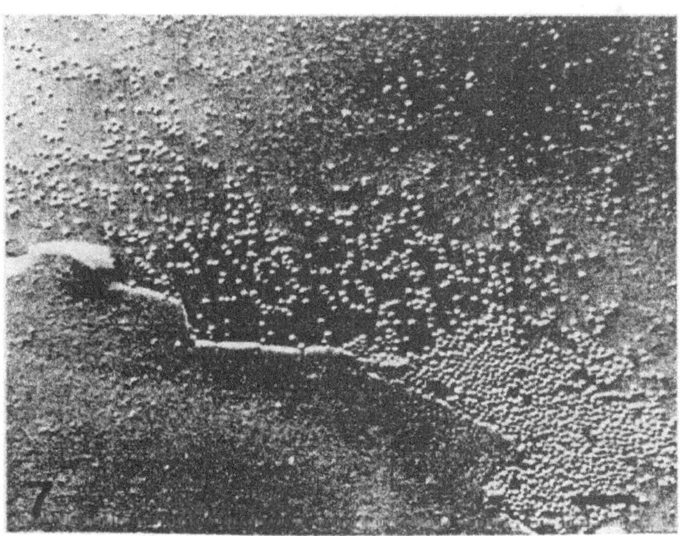

Fig. 7.7. Goldfish outer HC axon terminal gap junction, 30 min after intraocular dopamine injection (5 μl, 1 mM) and 60 min after optic nerve crush during light adaptation. Most connexons are densely packed as a result of the primary optic nerve crush, but some connexons have dispersed laterally within the membrane, presumably as a result of dopamine administration. x80,000. Bar: 0.1 μm.

association. A high connexon density seems to require the presence of cytoplasmic material with high osmium-binding capacity. On the other hand, it might be interesting that Vaughan and Lasater[70] have not been able to electrically uncouple fish HCs by exposing them to F-actin-disrupting cytochalasin D. If coupling in this cellular system occurs at a high connexon density as has been suggested by Belliveau et al[54] and Davidson et al,[65] F-actin obviously is not necessary for the maintenance of a high connexon density. Unfortunately, there are no immunocytochemical or freeze-fracture data available to demonstrate the effect of cytochalasin D on the fate and composition of the cytoplasmic plaques and the connexon positioning during and after microfilament disruption. In cultured HCs of the catfish retina, Hidaka et al[71] observed osmiophilic densifications beneath certain gap junctions; however, the authors were not able to correlate their appearance to the coupling state in a given cell pair.

The problem of whether a high connexon density in HC gap junctions may correlate to the coupled or uncoupled state has to be more closely considered. Since dark/light adaptation is associated with an increase/decrease of the HC receptive field size, we have argued that the increase/decrease of the connexon density reflects the coupling/uncoupling of the gap junctions, respectively. On the other side, pH decrease results either in uncoupling of HCs[63,72-74] and an increase of the connexon density.[69] Hampson et al[75] reported that in the rabbit retina dopamine uncouples the HCs at pH 7.2. However, dopamine did not exert any effect on the number of lucifer yellow-coupled cells at pH 7.4. The uncoupling effect of low extracellular pH was not reversed by dopaminergic antagonists, suggesting that dopaminergic uncoupling is pH gated. If the results on rabbits apply to the fish retina, they may explain the contradictory effects of both uncoupling agents, pH and dopamine, on connexon density. Lowering the (intra- or extracellular) pH results in a stereotypical increase of the connexon density; however,

dopamine seems to decrease the connexon density only when pH is between 7.2 and 7.4. Therefore, Schmitz and Wolburg[69] possibly were not able to neutralize the effect of pH values 7.1, 6.8, and 6.5 with dopamine, and to observe any effect of the dopamine antagonist haloperidol at pH 7.5. This would mean that below a certain pH threshold the uncoupling of HC gap junctions is indicated by a high connexon density; only within a small pH window dopamine has a negative effect on coupling that is characterized by a decreased connexon density.

REFERENCES

1. Hirokawa N, Heuser J. The inside and outside of gap junction membranes visualized by deep etching. Cell 1982; 30: 395-406.
2. Shibata Y, Yamamoto T. Cytoplasmic surface ultrastructures of cardiac gap junctions as revealed by quick-freeze, deep-etch replicas. Anat Record 1986; 214:107-12.
3. Hatae T, Iida H, Kuraoka A et al. Cytplasmic surface ultrastructures of gap junctions in bovine lens fibers. Invest Ophthalmol Vis Sci 1993; 34:2164-73.
4. Wolburg H, Rohlmann A. Structure-function relationships in gap junctions. Int Rev Cytol 1995; 157:315-73.
5. Wolburg H, Kurz-Isler G. Dynamics of gap junctions between horizontal cells in the goldfish retina. Exp Brain Res 1985; 60:397-401.
6. Raviola E, Gilula NB. Gap junctions between photoreceptor cells in the vertebrate retina. Proc Natl Acad Sci USA 1973; 70:1677-81.
7. Copenhagen DR, Owen WG. Coupling between rod photoreceptors in a vertebrate retina. Nature 1976; 260:57-59.
8. Tsukamoto Y, Masarachia P, Schein SJ et al. Gap junctions between the pedicles of macaque foveal cones. Vis Res 1992; 32:1809-15.
9. Dacheux RF, Raviola E. Horizontal cells in the retina of the rabbit. J Neurosci 1982; 2:1486-93.
10. Bloomfield SA, Miller RF. A physiological and morphological study of the horizontal

cell types of the rabbit retina. J Comp Neurol 1982; 208:288-303.

11. Vaney DI. The coupling pattern of axon-bearing horizontal cells in the mammalian retina. Proc R Soc Lond B 1993; 252: 93-101.

12. Kujiraoka T, Saito T. Electrical coupling between bipolar cells in the carp retina. Proc Natl Acad Sci USA 1986; 83:4063-66.

13. Saito T, Kujiraoka T. Characteristics of bipolar-bipolar coupling in the carp retina. J Gen Physiol 1988; 91:275-87.

14. Umino O, Maehara M, Hidaka S et al. Electrical coupling between bipolar cells in the retina. Invest Ophthalmol Vis Sci 1993; 34:984.

15. Bloomfield SA. Relationship between receptive and dendritic field size of amacrine cells in the rabbit retina. J Neurophysiol 1992; 68:711-25.

16. Hampson ECGM, Vaney DI, Weiler R. Dopaminergic modulation of gap junction permeability between amacrine cells in mammalian retina. J Neurosci 1992; 12:4911-22.

17. Hidaka S, Maehara M, Umino O et al. Lateral gap junction connections between retinal amacrine cells summating sustained responses. NeuroReport 1993; 5:29-32.

18. Hitchcock PF. Neurobiotin coupling between developing ganglion cells in the retina of the goldfish. Invest Ophthalmol Vis Sci 1993; 34:878.

19. Vaney DI. Many diverse types of retinal neurons show tracer coupling when injected with biocytin or neurobiotin. Neurosci Lett 1991; 125:187-90.

20. Vaney DI. Patterns of neuronal coupling in the retina. Progr Ret Eye Res 1994; 13:301-55.

21. Uga S, Smelser GK. Comparative study of the fine structure of retinal Müller cells in various vertebrates. Invest Ophthalmol 1973; 12:434-48.

22. Burns MS, Tyler NK. Interglial cell gap junctions increase in urethane-induced photoreceptor degeneration in rats. Invest Ophthalmol Vis Sci 1990; 31:1690-1701.

23. Wolburg H, Reichelt, W, Stolzenburg J-U et al. Rabbit retinal Müller cells in cell culture show gap and tight junctions which they do not express in situ. Neurosci Lett 1990;111:58-63.

24. Holländer H, Makarov F, Dreher Z et al. Structure of the macroglia of the retina: sharing and division of labour between astrocytes and Müller cells. J Comp Neurol 1991; 313:587-603.

25. Robinson SR, Hampson ECGM, Munro MN et al. Unidirectional coupling of gap junctions between neuroglia. Science 1993; 262:1072-74.

26. Dowling JE, Boycott BB. Organization of the primate retina: electron microscopy. Proc R Soc Lond B 1966; 166:80-111.

27. Murakami M, Miyachi E-I, Takahashi K-I. Modulation of gap junctions between horizontal cells by second messengers. Progr Ret Eye Res 1995; 14:197-221.

28. Dixon JS, Cronly-Dillon JR. The fine structure of the developing retina in Xenopus laevis. J Embryol Exp Morph 1972; 28: 659-66.

29. Fusisawa H, Morioka H, Watanabe K et al. A decay of gap junctions in association with cell differentiation of neural retina in chick embryonic development. J Cell Sci 1976; 22:585-96.

30. Hayes BP. The distribution of intercellular gap junctions in the developing retina and pigment epithelium of Xenopus laevis. Anat Embryol 1976; 150:99-111.

31. Hayes BP. Intercellular gap junctions in the developing retina and pigment epithelium of the chick. Anat Embryol 1977; 151: 325-34.

32. Vardi N, Hertzberg E, Sterling P. Gap junction distribution in cat and monkey retina visualized with monoclonal antibody to connexin32. Soc Neurosci Abstr 1990; 16:1076.

33. Jones CR, Becker DL, Cook JE. Gap junctions in the rat retina: immunoreactivity with antisera raised to oligopeptides of connexins Cx32 and Cx43. Neurosci Letters 1992; Suppl 42: S34.

34. Janssen-Bienhold U, Schöpe B, Buschmann-Gebhardt B, Dermietzel R, Weiler R. Identification of neuronal connexin proteins in the retina. In: Elsner N, Menzel R, ed. Proc 23rd Göttingen Neurobiol Conf. Stuttgart, New York: Georg Thieme Verlag, 1995:458.

35. Naka K-I, Rushton WAH. The generation and spread of S-potentials in fish (Cyprinidae). J Physiol 1967; 192:437-61.

36. Kaneko A. Electrical connexins between horizontal cells in the dogfish retina. J Physiol 1971; 213:95-105.

37. Kaneko A, Stuart AE. Coupling between horizontal cells in the carp retina revealed by diffusion of lucifer yellow. Neurosci Lett 1984; 47:1-7.

38. Piccolino M, Neyton J, Witkovsky P et al. Gamma-aminobutyric acaid antagonists decrease junctional communication between horizontal cells of the retina. Proc Natl Acad Sci USA 1982; 79:3671-75.

39. Teranishi T, Negishi K, Kato S. Regulatory effect of dopamine on spatial properties of horizontal cells in carp retina. J Neurosci 1984; 4:1271-80.

40. Marc RE, Liu WLS, Muller JF. Gap junctions in the inner plexiform layer of the goldfish retina. Vision Res 1988; 28:9-24.

41. Goddard JC, Behrens UD, Wagner HJ, Djamgoz MBA. Biocytin: intracellular staining, dye-coupling and immunocytochemistry in carp retina. Neuro Report 1991; 2:755-58.

42. Cuenca N, Fernández E, Garciá M, De Juan J. Dendrites of rod dominant ON-bipolar cells are coupled by gap junctions in carp retina. Neurosci Lett 1993; 162:34-38.

43. Teranishi T, Negishi K. Double-staining of horizontal and amacrine cells by intracellular injection with Lucifer Yellow and biocytin in carp retina. Neuroscience 1994; 59:217-26.

44. Lasansky A. Interactions between horizontal cells of the salamander retina. Invest Ophthalmol 1976; 15:909-16.

45. Raviola E. Intercellular junctions in the outer plexiform layer of the retina. Invest Ophthalmol Vis Sci 1976; 15:881-95.

46. Raviola E, Gilula NB. Intramembrane organization of specialized contacts in the outer plexiform layer of the retina. J Cell Biol 1975; 65:192-222.

47. Cooper NGF, Mc Laughlin BJ. Gap junctions in the outer plexiform layer of the chick retina: Thin section and freeze-fracture studies. J Neurocytol 1981; 10:515-29.

48. Zimmerman RP. Bar synapses and gap junctions in the inner plexiform layer: synaptic relationships of the interstitial amacrine cell of the retina of the cichlid fish, Astronotus ocellatus. J Comp Neurol 1983; 218:471-79.

49. Witkovsky P, Owen WG, Woodworth M. Gap junctions among the perikarya, dendrites, and axon terminals of the luminosity-type horizontal cell of the turtle retina. J Comp Neurol 1983; 261:359-68.

50. Tonosaki A, Washioka H, Nakamura H et al. Complementary freeze-fracture replication: an example of its use in the study of horizontal cell gap junctions of the carp retina. J Electr Microsc Tech 1985; 2:187-92.

51. Kouyama N, Watanabe K. Gap junctional contacts of luminosity-type horizontal cells in the carp retina. A novel pathway of signal conduction from the cell body to the axon terminal. J Comp Neurol 1986; 249:404-10.

52. Marshak DW, Dowling JE. Synapses of cone horizontal cell axons in goldfish retina. J Comp Neurol 1987; 256:430-43.

53. Kamermans M, van Dijk BW, Spekreijse H. Interaction between the soma and the axon terminal of horizontal cells in carp retina. Vision Res 1990; 30:1011-16.

54. Kurz-Isler G, Voigt T, Wolburg H. Modulation of connexon densities in gap junctions of horizontal cell perikarya and axon terminals in fish retina: effects of light/dark cycles, interruption of the optic nerve and application of dopamine. Cell Tiss Res 1992; 268:267-75.

55. Piccolino M, Neyton J, Gerschenfeld HM. Decrease of of the gap junction permeability induced by dopamine and cyclic 3'-5' adenosine-monophosphate in horizontal cells of the turtle retina. J Neurosci 1984; 4:2477-88.

56. Piccolino M, Demontis G, Witkovsky P et al. Involvement of D1 and D2 dopamine receptors in the control of horizontal cell electrical coupling in the turtle retina. Europ. J Neurosci 1989; 1:247-57.

57. McMahon DG, Knapp AG, Dowling JE. Horizontal cell gap junctions: single channel conductance and modulation by dopamine. Proc Natl Acad Sci USA 1989;

86:7639-43.

58. Zucker CL, Dowling JE. Centrifugal fibers synapse on dopaminergic interplexiform cells in the teleost retina. Nature 1987; 300: 166-68.

59. Negishi K, Teranishi T, Kato S. A GABA antagonist, bicuculline, exerts its uncoupling action on external horizontal cells through dopamine cells in carp retina. Neurosci Lett 1983; 37:261-66.

60. Piccolino M, Witkovsky P, Trimarchi C. Dopaminergic mechanisms underlying the reduction of electrical coupling between horizontal cells of the turtle retina induced by d-amphetamine, bicuculline, and veratridine. J Neurosci 1987; 7:2273-84.

61. Lasater EM. Retinal horizontal cell gap junctional conductance is modulated by dopamine through a cyclic AMP-dependent protein kinase. Proc Natl Acad Sci USA 1987; 84:7319-23.

62. Lasater EM, Dowling JE. Dopamine decreases conductance of the electrical junctions between cultured retinal horizontal cells. Proc Natl Acad Sci USA 1985; 82:3025-29.

63. Laufer M, Salas R, Medina R et al. Cyclic adenosine monophosphate as a second messenger in horizontal cell uncoupling in the teleost retina. J Neurosci Res 1989; 24:299-310.

64. Kurz-Isler G, Wolburg H. Gap junctions between horizontal cells in the cyprinid fish alter rapidly their structure during light and dark adaptation. Neurosci Lett 1986; 67:7-12.

65. Kurz-Isler G, Wolburg H. Light-dependent dynamics of gap junctions between horizontal cells in the retina of the crucian carp. Cell Tiss Res 1988; 251:641-49.

66. Baldridge WH, Ball AK, Miller RG. Dopaminergic regulation of horizontal cell gap junction particle density in goldfish retina. J Comp Neurol 1987; 265:428-36.

67. Baldridge WH, Ball AK, Miller RG. Gap junction particle density of horizontal cells in goldfish retinas lesioned with 6-OHDA. J Comp Neurol 1989; 287:238-46.

68. Weiler R Kohler K, Kolbinger W et al. Dopaminergic neuromodulation in the retina of lower vertebrates. Neurosci Res 1988; Suppl 8: S183-96.

69. Schmitz Y, Wolburg H. Gap junction morphology of retinal horizontal cells is sensitive to pH alterations in vitro. Cell Tiss Res 1991; 263:303-10.

70. Vaughan DK, Lasater EM. Distribution of F-actin in bipolar and horizontal cells of bass retinas. Am J Physiol 1990; 259: C205-14.

71. Hidaka S, Shingai R, Dowling JE et al. Junctions form between catfish horizontal cells in culture. Brain Res 1989; 498:53-63.

72. Negishi K, Teranishi T, Kato S. Opposite effects of ammonia and carbon dioxide on dye coupling between horizontal cells in the carp retina. Brain Res 1985; 342:330-39.

73. DeVries SH, Schwartz EA. Modulation of an electrical synapse between solitary pairs of catfish horizontal cells by dopamine and second messengers. J Physiol (Lond) 1989; 414:351-75.

74. Qian H, Malchow RP, Ripps H. Gap-junctional properties of electrically coupled skate horizontal cells in culture. Vis Neurosci 1993; 10:287-95.

75. Hampson ECGM, Weiler R, Vaney DI. pH-gated dopaminergic modulation of horizontal cell gap junctions in mammalian retina. Proc R Soc Lond B 1994; 255:67-72.

CHAPTER 8

CHARACTERIZATION AND REGULATION OF GAP JUNCTION CHANNELS IN CULTURED ASTROCYTES

Christian Giaume and Laurent Venance

For decades neurons have been considered as the unique active constituents of the brain while glial cells were viewed solely as bystanders. However, now there is increasing evidence to support that glial cells are more than simple supporting cells since they appear to nourish, protect and interact with neurons.[1-3] In particular the star-shaped cells called astrocytes constitute a major class of brain glial cells which possesses all the tools to receive, integrate and transmit signals to neurons. In the last few years, several reviews have focused on ionic channels in glia,[4-7] but none of these have taken into consideration the channels forming gap junctions. This omission is rather surprising since astrocytes, in culture systems as well as in vivo, are presumed to be the most widely coupled cell population in the central nervous system.

The purpose of this chapter is to summarize data which have contributed to the identification of the proteins forming junctional channels in astrocytes and the demonstration that intercellular communication mediated by gap junctions is controlled by endogenous compounds produced by several brain cell types, including neurons. Since most of this work has been carried out in vitro, the first part of this report will be focused on data obtained from primary cultures of rodent astrocytes. In the second part, attempts will be made to define the potential roles for gap junction communication in astrocytes and to identify the most intriguing questions to address in the future.

Gap Junctions in the Nervous System, edited by David C. Spray and Rolf Dermietzel.
© 1996 R.G. Landes Company.

1. INTRODUCTION

Glial cells, which outnumber neurons and occupy more than half of the brain volume,[8] are composed of different subtypes, including: (1) oligodendrocytes which produce the myelin sheath that wrap axons; (2) microglial cells which serve as the brain's immune cells; (3) ependymal cells which form an "epithelial" layer lining cerebral ventricles;[9] and (4) astrocytes which have distinctive, highly specialized contacts with a variety of brain elements. Due to their close relationship with other cell types astrocytes behave as "cross roads" cells, likely involved in multiple brain functions.[10] For instance, they appose neurons at synapses, cell bodies and nodes of Ranvier.[10,11] In conjunction with endothelial cells, they may participate in the formation and maintenance of the blood brain barrier.[12] Also, astrocytic processes line the subpial and ependymal surfaces of the brain. Moreover, astrocytes elaborate numerous contacts with themselves and with oligodendrocytes. At such contacts, the presence of gap junctions has led to the proposal of a syncytial organization of glial cells.[13-15]

1.1. HOMOTYPIC GAP JUNCTIONS IN GLIAL CELLS

At present, morphological, immunological and functional studies, performed in vivo as well as in vitro, indicate that brain glial cells represent the cell population that is most extensively and abundantly coupled by gap junctions.[16-18] Ultrastructure and freeze-fracture studies provide strong evidence for the presence of gap junctions in astrocytes. These homotypic junctions formed by astrocytes are abundant as compared to those observed between oligodendrocytes[15,19] and are localized joining cell bodies, joining processes and cell bodies,[13] and between astrocytic end-feet surrounding blood vessels, where they exhibit a typical "honeycomb" pattern.[20] Localization of homotypic junctions between oligodendrocytes is less well documented. Immunostaining with anticon-

nexin antibodies have shown some reactive sites at the cell body level in situ[21] and between processes in primary culture.[22] Functional confirmation of these morphological observations was achieved using electrophysiological techniques to test the occurrence of ionic or dye coupling. Thus, transfer of Lucifer yellow was demonstrated between astrocytes in cortical[23,24] and hippocampal[25] slices, and between oligodendrocytes in slices of the optic nerve.[26] Evidence of functional gap junctions were also obtained from primary cultures,[27] which have opened the way to use this simple model to investigate basic biophysical and pharmacological properties of glial gap junctions.

1.2. HETEROTYPIC GAP JUNCTIONS IN GLIA

Astrocyte-to-oligodendrocyte junctions were identified between cell bodies, between cell bodies and processes, and between astrocyte processes and the outer surface of the myelin sheath.[13] These heterotypic gap junctions are functional since these two types of glial cells are electrically[28] and dye[29] coupled when cocultured, and they exhibit a certain asymmetrical dye permeability when studied in the isolated rabbit retina.[30] Such heterotypic coupling between astrocytes and other brain cell types seems to be limited to oligodendrocytes.[29] However, there are some data indicating that gap junctions may mediate exchange between cocultured astrocytes and neurons[31,32] (but see ref. 33), leaving open the possibility that transient expression of gap junctions could exist which have escaped morphological or immunological detection so far.

2. COUPLED ASTROCYTES COULD PARTICIPATE IN INFORMATION PROCESSING IN THE BRAIN

The possibility that astrocytes could participate actively in information processing is based on the two way dialogue they establish with neurons.[34,35] This is supported by observations indicating that astrocytes are targets and sources of biologically active molecules.

At their surface astrocytes express numerous different receptors to neurotransmitters, peptides, growth factors and hormones which are coupled to intracellular signal transduction pathways.[36-39] Table 8.1 is a nonexhaustive list of membrane receptors to neurotransmitters which have been identified in astrocytes in vivo as well as in culture. It indicates that astrocytes can sense most signals released by neurons. In addition to this, the existence of regional heterogeneity[40] and of pharmacological subsets for receptors to neuroligands[41] suggests that astrocytes, like neurons, may exhibit some degree of specialization. This variety could be amplified, as these receptors are coupled via G proteins to all major second messenger pathways (PLC, PLA_2, PLD, adenylyl cyclase, tyrosine kinases, etc.) which can interact by crosstalk[42] and thus provide some plasticity to astrocyte responses.

Astrocytes can also transmit messages to surrounding cells either by direct release of signaling molecules, such as the excitatory neurotransmitter glutamate which activates neurons,[43] or by production of "transcellular" messengers including eicosanoids[44] and nitric oxide,[45] or as a consequence of a reduction of uptake mechanisms of neurotransmitters and ions which are highly active in these cells.[46,47]

Altogether, these observations have led to the proposal of a loop of neuro-glial interactions in which gap junction-mediated propagation of intercellular calcium waves is integrated.[34,48] According to this model, neuronal release of neurotransmitter could trigger rises in intracellular calcium concentration in nearby astrocytes, which then propagate to surrounding coupled astrocytes that in turn may activate adjacent neurons by a mechanism which remains to be clarified.[35] Consequently, in this loop interaction, the regulation of astrocyte gap junctions may define

Table 8.1. List of agonists which in astrocytes activate membrane receptors, transduction pathways and changes in membrane potential

Receptor	Adenylyl cyclase	Phosphoinosital breakdown	Arachidonic acid release	Intracellular calcium	Membrane potential
Noradrenaline	+ (β1, β2)	+ (α1, α2)	+	+ (α1, β1)	depol. (α1, α2) hyperpol. (β1, β2)
Histamine	+ (H2)	+ (H1)		+ (H1)	hyperpol.
Dopamine	+ (D1, D2)				hyperpol.
Seratonin	+ (5HT1, 5HT2)	+ (5HT1, 5HT2)	+ (5HT1, 5HT2)	+ (5HT1, 5HT2)	hyperpol.
Acetylcholine					
Muscarinic	+	+	+	+	hyperpol.
ATP	- (A1), + (A2)	+ (P2Y)	+ (A1)	+ (A1,A2)	hyperpol. (A1, A2)
Glutamate	- (metabo.)	+ (metabo.)	+ (metabo.)	+ (metabo., kainate)	depol. (kainate)
VIP	+				hyperpol.
Endothelins	- (ET1, ET3)	+ (ET1, ET3)	+ (ET1, ET3)	+ (ET1, ET3)	depol. (ET1, ET3)
Bradykinin (B2)		+	+	+	depol.
Substance P (NK1)		+	+	+	depol.
Adenosine (A2)	-				
Anandamide	-		+		

Most of these data are taken from previous reviews.[37-39,140]

glial networks which determine the size and shape of activated neuronal groups.

3. CONNEXIN43 IS THE MAJOR JUNCTIONAL PROTEIN IN ASTROCYTES

Characterization of connexins expressed in astrocytes was achieved by combining biochemical and electrophysiological studies using homogeneous cell populations obtained from primary cultures. After 3 weeks in culture, 95% of the cells are identified as astrocytes using anti-GFAP antibodies (Fig. 8.1C). Northern blot analysis of extracted RNA was achieved using specific cDNA probes for the three main connexins identified in mammal brain, Cx26, Cx32 and Cx43. At high stringency, the Cx43 cDNA probe detects a message in RNA extracted from rat heart, rat and mouse brain and 3-week-old cultured astrocytes (Fig. 8.1A). Using similar conditions, rat Cx26 and Cx32 cDNA probes detect messages in extracts from rat liver and rat and mouse brain. However, no message is detected in cultured astrocyte RNA with either of these two probes.[49,50] Replicas of whole fractions of rat heart and 3-week-old cultured astrocytes were probed according to two procedures, either by an enzymatic or a radiolabeling technique. In both approaches, a major band of 43 kDa was detected on replicas of the two preparations (Fig. 8.1B). Immunolabeling performed with the same antibody resulted in epitope recognition typically located at contact areas of astrocytes with some additional intracytoplasmic staining (Fig. 8.1C). Finally, immunogold labeling and electron microscope examination of ultrathin sections demonstrated that the antigenic sites determined for Cx43 were associated with characteristic gap junction structures.[49,50]

Besides these studies performed with primary cultures, both in situ hybridization and immunohistochemical staining have confirmed that Cx43 is expressed in vivo in astrocytes. These studies have also provided evidence that Cx43 expression is regionally[51-53] and developmentally[21,54] regulated. Since, in brain slices[21] and primary cultures,[22] oligodendrocytes are negative for Cx43 antibodies, it is likely that gap junctions between astrocytes and oligodendrocytes are made of different connexins. This possibility of heterologous channels is supported by the electron microscopic observation made in vivo that Cx43 is restricted to the astrocytic side at gap junctions between astrocytes and oligodendrocytes.[20] Although Cx32 is expressed by oligodendrocytes,[21,22] it is unlikely that heterologous gap junctions between these two types of glia are composed of Cx43 and Cx32 since these two connexins do not form functional gap junctions when expressed in Xenopus oocytes or transfected in HeLa cells.[55,56] Accordingly, it is likely that oligodendrocytes, in addition to Cx32, may express other unidentified connexins, as already suggested.[18]

Measurement of unitary conductance using the double whole-cell recording technique[57] represents another way to characterize gap junction channels expressed in defined cell types. Recordings of junctional currents were obtained from pairs of cells selected either within 48 hr after the beginning of the culture or from 3 week old replated astrocytes. From weakly coupled cell pairs, single-channel events can be resolved without pharmacological treatment. When a constant or pulsed transjunctional voltage difference is imposed on the junctions, typical "mirror images" of quantal events are recorded. Amplitude histograms of these recordings indicate that the main elementary conductance of junctional channels in astrocytes range around 60 pS. In addition to multiples of this value, smaller conductance changes (20-40 pS) are observed.[49,50] This main conductance state is not dependent upon the sign or the amplitude of transjunctional voltage (Fig. 8.2A). When recording of junctional currents is achieved by applying long lasting voltage pulses, junctional current inactivates (Fig. 8.2C). Analysis of these voltage-dependent properties (Fig. 8.2B) indicates that astrocyte gap junctions behave similarly to those present in cardiac myocytes.

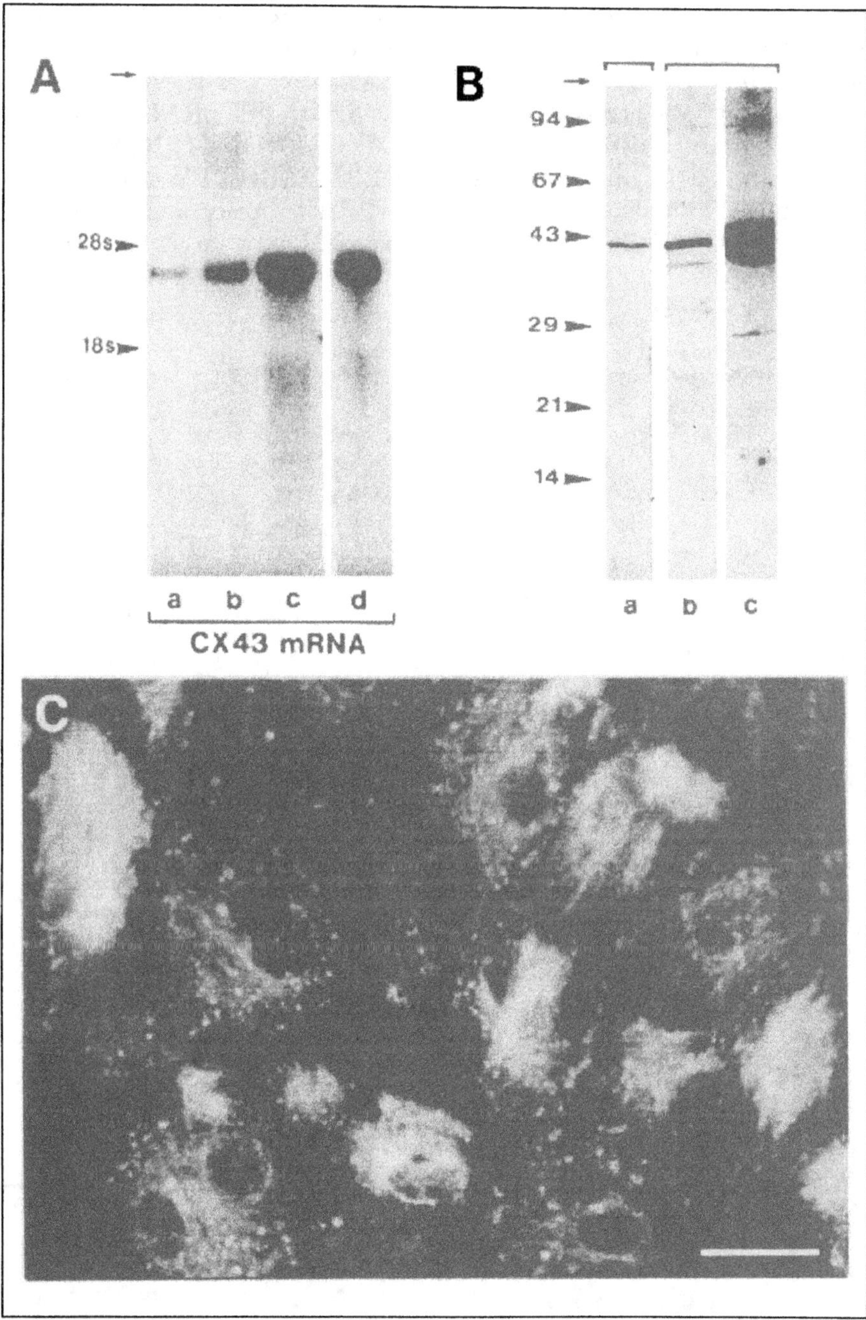

Fig. 8.1. Immunological detection and localization of Cx43 in cultured astrocytes. (A) Northern blot analysis of total RNA extracts from cultured mouse astrocytes (lane a), mouse (lane b) and rat (lane c) brains and rat heart (lane d). The arrow indicates the top of the blot, while arrowheads show the position of 28S and 18S mRNA subunits. (B) immunoblotting of total fractions of rat heart (lane a) and cultured mouse astrocytes (lane b and c). Anti-peptide antibodies were revealed using either secondary antibodies and peroxidase-labeled streptavidin (lane a and b) or ^{125}I- labeled protein A (lane c). The position of the molecular weight standards are indicated by arrowheads in kilodaltons. The arrow indicates the top of the immunoreplicas. (C) double immunofluorescence labeling of 3-week-old cultured astrocytes with antibodies specific to GFAP labeled with fluoresceine and to Cx43 labeled with rhodamine. Adapted with permission from Giaume et al, 1991.

These results suggest that astrocytic gap junctions are mainly composed of Cx43 and that channels composed of this protein have a unitary conductance of 60pS. Such an association of a defined channel-forming protein with single channel properties (unitary conductance, voltage dependence) is similar to that found in cardiac tissue[58,59] and in cell lines transfected with Cx43.[60]

4. SHORT-TERM REGULATION OF GAP JUNCTION PERMEABILITY IN ASTROCYTES

Besides heart and brain, Cx43 is also expressed in a large number of tissues in

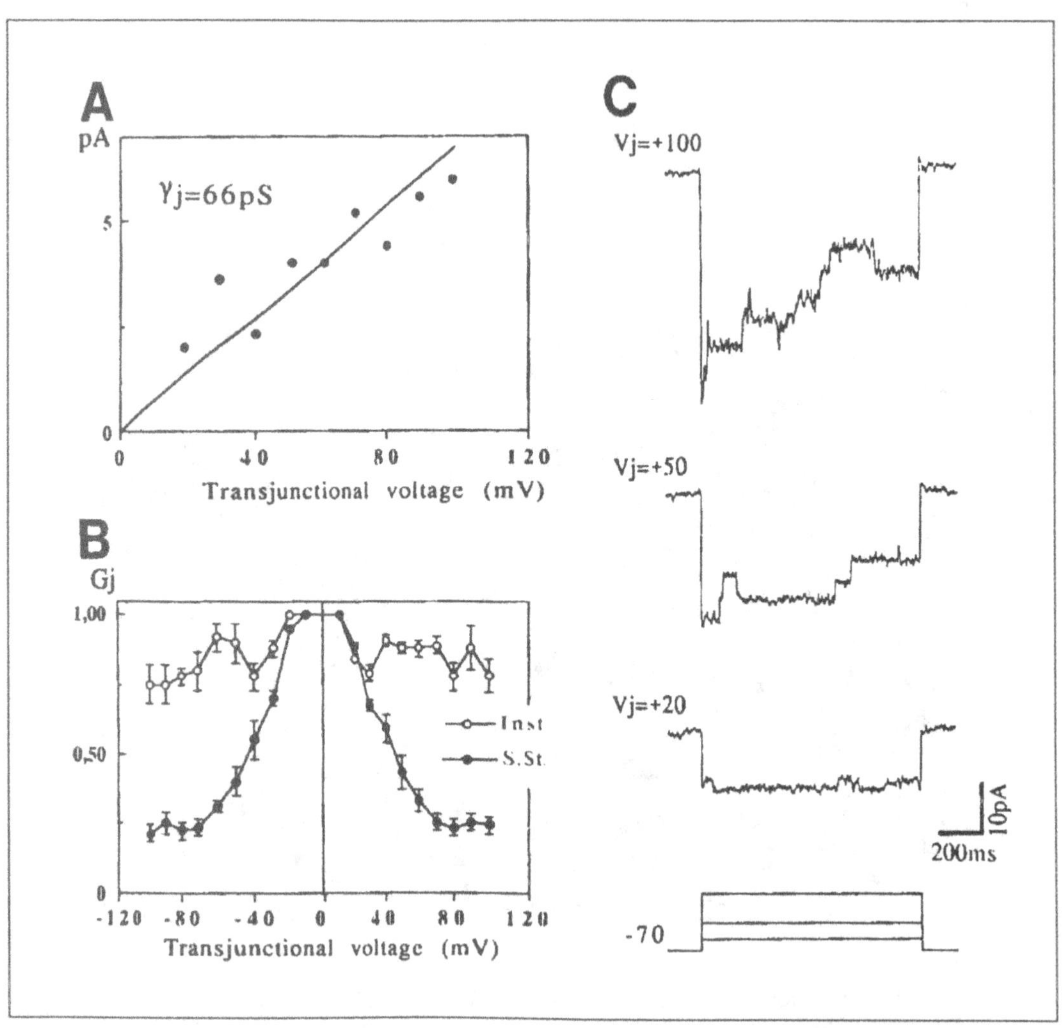

Fig. 8.2. Gating properties of gap junction channels in cultured astrocytes. (A) Plot of single junctional channel currents (ordinate) recorded, using the double whole-cell recording technique, from a pair of replated astrocytes at various transjunctional voltages (absissa). The averaged value of the unitary conductance (gj) was deduced from linear regression. (B) Voltage dependence of Cx43 channels in astrocytes. Normalized junctional conductances (Gj) measured at the beginning (Inst.) and at the end (S. St.) of the voltage pulse applied in one cell, are plotted versus transjunctional voltages. (C) Three examples of junctional currents recorded for the indicated values of transjunctional potentials (Vj).

mammals.[61] One of the characteristics of channels composed by this connexin—which is a target for multiple signaling pathways—is that regulation of junctional permeability does not operate in the same way. For example, PKA or PKC can have opposite effects depending on the system studied.[62] This suggests that the expression of a defined connexin will not per se be sufficient to determine how intercellular communication will be regulated. This will depend also on the cellular environment of the junctions. Short-term regulation of junctional permeability—which will be detailed below—likely results from a change in the balance between the open and closed states of individual channels. Here, several factors involved in this regulation will be considered in order to define the profile of Cx43 regulation in astrocytes.

4.1. NONRECEPTOR-MEDIATED EFFECTS

Gap junction permeability in astrocytes can be blocked by lipophilic agents such as alcohols (heptanol, octanol) and also volatile anesthetics (halothane, enflurane, isoflurane) but not by intravenous anesthetics (Table 8.2). Interestingly, the effect of octanol depends on the brain region from which astrocytes are cultured, with a higher potency associated with cells derived from spinal cord than from hippocampus or cerebellum.[63] However, in other systems these uncoupling agents may act on other ionic conductances[64] and so they should not be considered as specific agents in astrocytes.

Table 8.2. Receptor- and nonreceptor-mediated short-term regulation of gap junction permeability in cultured astrocytes

Effector	Effect	Technique	Reference
Membrane receptors			
α1-adrenergic (+adensoine)	Inhibition	SL/DT	Giaume et al., 1991
Adenosine triphosphate	Inhibition	IDI	Enkvist and MacCarthy, 1992
		ICW	Enkvist and MacCarthy, 1992
Endothelins (ET1, ET3)	Inhibition	SL/DT	Giaume et al., 1992
		IDI	(this chapter)
		DWCR	(this chapter)
Anandamide	Inhibition	DWCR	Venance et al., 1995
		SL/DT	Venance et al., 1995
		ICW	Venance et al., 1995
Glutamate (kainate)	Inhibition	DWCR (brain slices)	Müller and Kettenmann, 1995
Glutamate (Kainate, quisqualate)	Increase	IDI	Enkvist and MacCarthy, 1994
β-adrenergic (+IBMX)	Increase	SL/DT	Giaume et al., 1991
Non-receptors			
Volatile anesthetics	Inhibition	DWCR	Dermietzel et al., 1991
		FRAP	Finkbeiner, 1992
		SL/DT	Mantz et al., 1993
Alcohols	Inhibition	SL/DT	Giaume et al., 1991
		FRAP	Lee et al., 1994
		DWCR	(unpublished)
a-glycyrrhenitic acid	Inhibition	SL/DT	Venance et al., 1995
		ICW	Venance et al., 1995
		DWCR	(unpublished)
		PM	Goldberg et al., 1995
High potassium solution	Increase	IDI	Enkvist and MacCarthy, 1994

Abbreviations: SL/DT: scrape-loading/dye transfer technique; DWCR: double whole-cell recording; FRAP: fluorescence

Another class of uncoupling agent is represented by aglycones of glycyrrhizic acid, a saponin obtained from the licorice root, which also has anti-inflammatory and mineralocorticoid-like properties.[65] While glycyrrhizic acid has no effect by itself, its two derivatives, 18-α and 18-β glycyrrhetinic acid, in micromolar concentration induced reversible uncoupling of cultured astrocytes[66] and glioma cell lines transfected with connexins.[67] Specificity of these compounds has not been determined yet, but in contrast to the other uncoupling agents mentioned above, glycyrrhetinic acid does not affect the shape and the amplitude of agonist induced calcium responses in astrocytes.[66] Accordingly, due to its reversibility, nontoxicity and the absence of effect on calcium levels, these compounds appear to be promising tools to study intercellular calcium signaling in glial cells.

Finally, an increase of dye diffusion was observed in the presence of high potassium solutions (10-55 mM) and was interpreted as the consequence of cell depolarization which enhances junctional permeability.[68] Such voltage dependent behavior is opposite to that generally described for Cx43 channels sensitivity to inside-outside potential[69,70] and was not observed in pairs of astrocytes under voltage clamp conditions when junctional conductances were compared at -70 mV and 0 mV holding potentials (L. Venance and C. Giaume, unpublished data).

4.2. RECEPTOR-MEDIATED REGULATION

As indicated in Table 8.1, astrocytes express a large number of receptors for neurotransmitters which are known to be coupled to several transduction pathways, including phospholipases C and A2 and adenylyl cyclase. Several inhibitory effects are induced by neurotransmitters which provoke breakdown of phosphoinositides and release of arachidonic acid, as for example noradrenaline[71] and ATP.[72] Since the α₁-adrenergic and ATP receptors are heterogeneously distributed in astrocytes,[41] this could lead to the partial uncoupling and isolation of subpopulations of cells in the glial syncytium.[73] In contrast, increased junctional permeability was associated with a large accumulation of cAMP, as suggested by the effect of isoproterenol in the presence of the phosphodiesterase inhibitor IBMX.[71] The up- or downregulation of junctional permeability after stimulation of adrenergic or purinergic receptors suggests that products secreted by neurons can affect the functional state of the astrocytic network.

So far, the effect of the excitatory amino acid glutamate on astrocyte gap junctions is rather controversial because activation of kainate receptors has been reported to have opposite effects.[68,74] Moreover, the mechanism of action of glutamate has been interpreted differently. Inhibition is reported to be due to calcium influx through kainate receptors while increase was attributed to cell depolarization. However, since these studies have been performed using cultured cortical astrocytes in one case and cerebellar slices in the second, this discrepancy could be due to the nature of the preparation (culture versus slice) or the region from which the cells originate.

Endothelins form a family of vasoconstrictor peptides produced in the brain by different cell types such as endothelial cells, astrocytes and certain neurons.[75] Two isoforms of endothelins, ET1 and ET3, have potent, rapid and rather long lasting inhibitory effects on gap junction permeability in cultured astrocytes observed using several approaches (Fig. 8.3). In astrocytes, second messenger pathways mediating endothelin responses are now well documented.[76-79] Furthermore, as these peptides are powerful gap junction inhibitors, endothelins provide a useful tool for the study of intracellular mechanisms involved in the inhibition of Cx43 channels in astrocytes.[80] For example, inhibitory effect of

ET1 is largely diminished in the absence of extracellular calcium[80] or by pretreating the cells with pertussis toxin.[81] More detailed analysis has indicated that inhibitors of protein kinase C (staurosporine) or tyrosine kinase (genistein) do not prevent ET1-induced uncoupling, although these treatments were found to be effective in

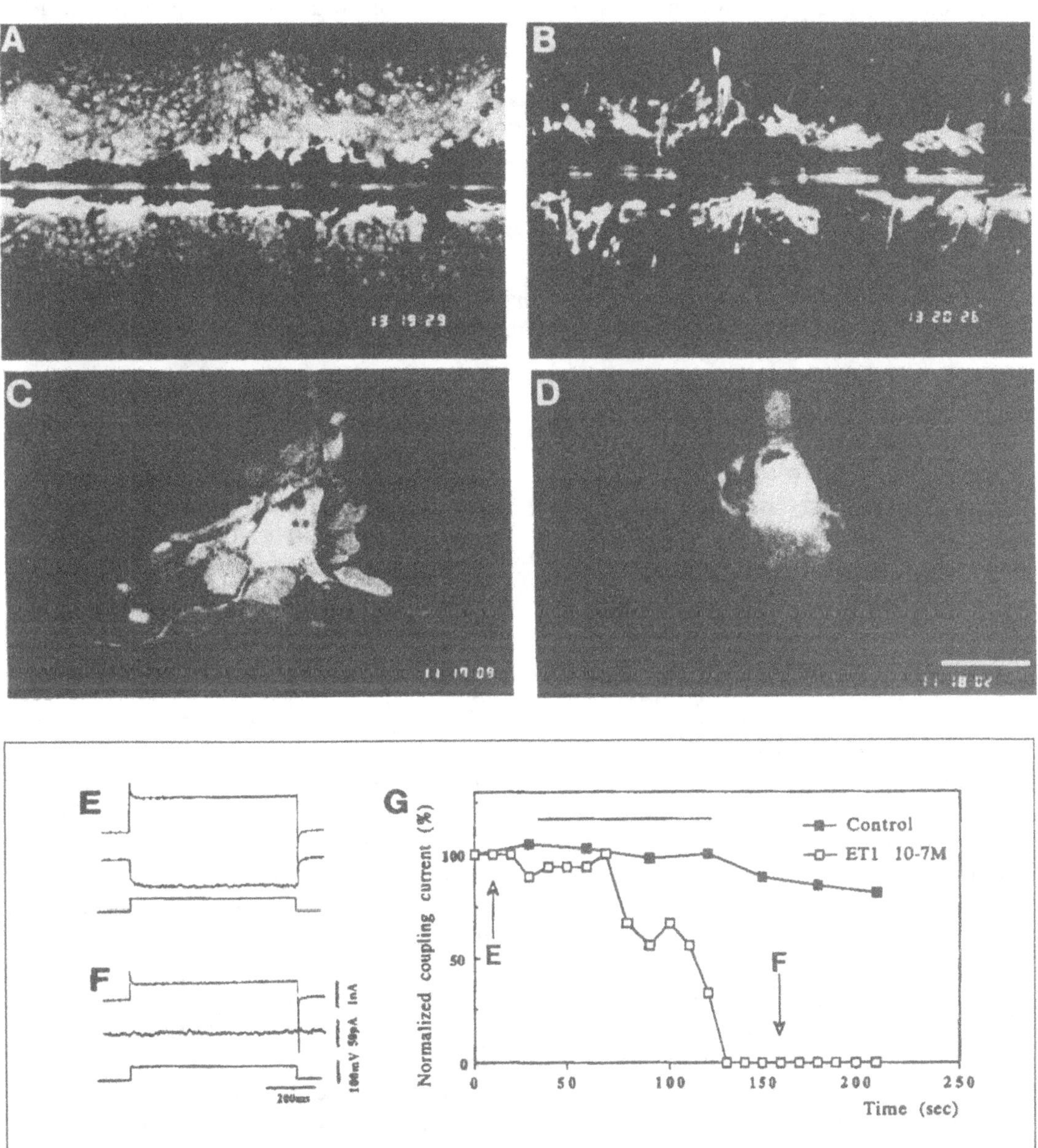

Fig. 8.3. Endothelin-induced inhibition of dye and electrical coupling in confluent astrocytes in culture. In control condition, dye coupling between confluent astrocytes was observed using either the scrape-loading/dye-transfer technique (A) or unitary loading of an astrocytes by a patch-clamp pipette filled with Lucifer yellow (C). When astrocytes were exposed to endothelin 1 (10^{-7}M), diffusion of the dye was restricted to the loaded cells (B and D). E-G: whole-cell recording of two adjacent cells within a layer of confluent astrocytes indicated that the inhibitory effect of ET1 is rapid and complete. Recordings shown in E and F correspond to the empty squares indicated by arrows in G. Reproduced with permission from Giaume and McCarthy, 1996 (Ref. 142).

blocking phosphorylation of identified protein substracts for these kinases in astrocytes.[82] In contrast, ET1 inhibition is blocked in the presence of a phospholipase A2 inhibitor, 4 bromophenalcylbromide, which suggests the participation of arachidonic acid or derivates.[81] However, more experiments are necessary to determine which steps are involved in the uncoupling mechanism.

4.3. SIGNAL TRANSDUCTION PATHWAYS

Up to now, there are converging data which confirm the initial observation that treatment with phorbol esters (PMA) induced a reduction of dye diffusion.[71] This has been reported in culture using different techniques and different dyes,[72,83] and also has been extended to brain slices.[25] Moreover, this inhibitory effect has been observed when intercellular calcium waves are generated by mechanical stimulation.[72] Although this inhibitory effect should be confirmed using electrophysiological techniques, these observations suggest that junctional permeability in astrocytes is inhibited by the activation of protein kinase C. Nevertheless, this finding contradicts somewhat the observation that endothelin uncoupling resists to treatments which block PKC activation.[82] An interpretation of this discrepancy could be that PMA and stimulation of endothelin receptors do not result in the activation of the same PKC isoforms which could be differently sensitive to staurosporine.[84,85]

Direct or indirect modulation of gap junctions by arachidonic acid or metabolites have been reported in several systems.[86-89] In astrocytes, exogenous application of arachidonic acid inhibits dye diffusion in a dose-dependent manner with an EC50 of 10^{-5}M.[71,90] In the brain, inhibition mediated by arachidonic acid is of particular interest, since eicosanoids can be produced by both neurons and astrocytes, and may act either as second messengers or as transcellular messengers due to their lipophilicity.[44] As illustrated in Figure 8.4, this fatty acid is also able to block propagation of intercellular calcium waves induced by mechanical stimulation of single astrocytes. Accordingly, due to its sources of production, arachidonic acid regulation of intercellular calcium signaling[90] is an example of neuroglial interaction and offers a means by which neurons may control the extent and the shape of the glial syncytium.

From studies performed in various systems—including transformed cells—inhibition of Cx43 junctions was correlated with tyrosine phosphorylation or MAP kinase activation,[91,92] increase in cGMP concentration[93] and lysophosphatidic acid application.[94] In cultured astrocytes gap junction permeability was not found to be sensitive to these transduction pathways. Indeed, it was not affected by direct applications of lysophosphatidic acid or by an inhibitor of tyrosine phosphatases. In addition, treatments which produce either nitric oxide and associated accumulation of cGMP or free radicals during the oxidation of xanthine by xanthine oxidase were also found to be ineffective (Table 8.3).

As mentioned above, the β-adrenergic agonist, isoproterenol, has no significant effect on junctional permeability when applied alone. Its action becomes significant in the presence of the phosphodiesterase inhibitor IBMX. Isoproterenol-induced accumulation of [³H] cAMP is large and sustained when applied in the presence of IBMX, while it is transient and more than 20 times smaller when the agonist was applied alone. Consequently, the increase of junctional permeability induced by this agonist associated with IBMX was interpreted as the consequence of large increase in cAMP concentration.[71] However, this observation was not confirmed when the permeant cAMP analog, 8 bromo-cAMP, was used alone.[68] To account for this discrepancy, the explanation advanced by the authors was the absence of IBMX in their trials. In agreement with that, in cardiac cells treatment with IBMX enhances the increase in coupling induced by 8 bromo-cAMP[95] or isoproterenol.[96]

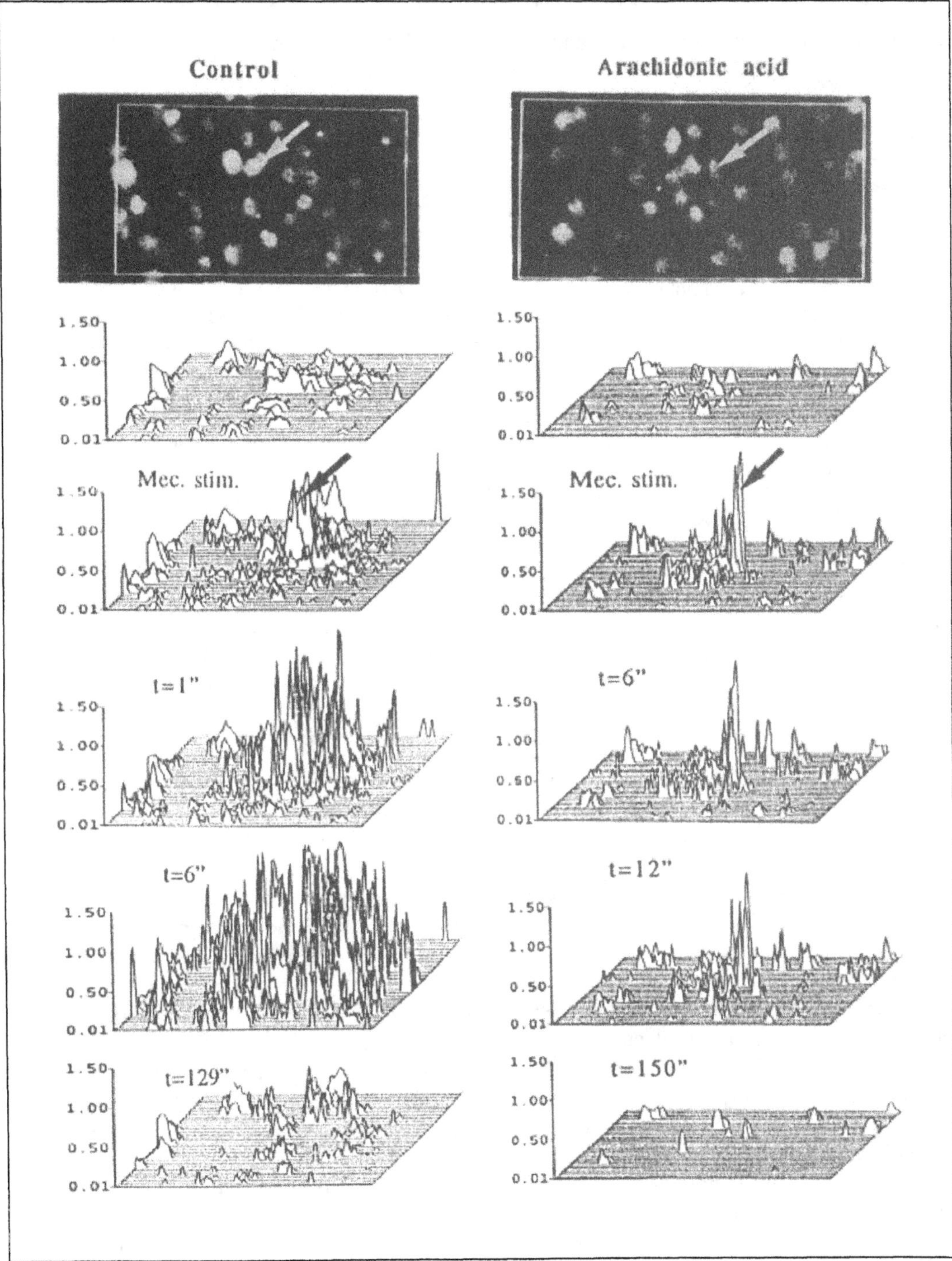

Fig. 8.4. Block of intercellular calcium waves by arachidonic acid. Mechanical stimulation of individual astrocytes (arrows) was carried out in the absence (left column) and in the presence of arachidonic acid $10^{-5}M$ (right column). The three-dimensional representation of changes in the ratio of Indo1 emissions indicates that, in control condition mechanical stimulation induced calcium rises in all the astrocytes of the field investigated (white rectangles). In contrast, in the presence of the fatty acid, only the stimulated cell and direct neighbors exhibit a change in $[Ca^{2+}]_i$.

Table 8.3. Control of junctional permeability by signal transduction pathways in cultured astrocytes

Transduction Pathway	Drug	Effect	Technique	Reference
Intracellular calcuim	Calcium ions	Inhibition	IDI	Enkvist and Maccarthy, 1994
	Ionomycin	Inhibition	SL/DT	(unpublished)
Low pH	Lactic acid	Inhibition	FRAP	Anders, 1988
	CO_2	Inhibition	IDI (brain slices)	Connors et al., 1982
Protein kinase C	Phobol ester	Inhibition	SL/DT	Giaume et al., 1991
			IDI	Enkvist and MacCarthy, 1992
			ICW	Enkvist and MacCarthy, 1992
			IDI (brain slices)	Konietzko and Müller, 1994
			IDI	Hossain et al., 1995
Eicosanoids	Arachidonic adid	Inhibition	SL/DT	Giaume et al., 1991
			ICW	(this chapter)
Tyrosine kinases	Vanadate + H_2O_2	No effect	SL/DT	(unpublished)
MAP kinase	Phenyl arsine oxide	No effect	SL/DT	(unpublished)
Nitric oxide (cGMP)	SIN -1	No effect	SL/DT	(unpublished)
Superoxide ions	xanthine + xanthine oxidase	No effect	SL/DT	(unpublished)
Lysophosphatidic adic	LPA	No effect	SL/DT	(unpublished)
cAMP	8-br-cAMP	No effect	IDI	Enkvist and MacCarthy, 1992
		No effect	ICW	Enkvist and MacCarthy, 1992

4.4. Ions

Initially, rise in intracellular free calcium ($[Ca^{2+}]_i$) was implicated in blockage of junctional permeability.[97] More recently, it has been established that the blocking effect of Ca^{2+} ions occurred at rather high concentrations unlikely to be physiological, $10^{-6}M$ and $10^{-5}M$.[98,99] Indeed, the most frequently observed action of Ca^{2+} ions at physiological concentrations ($10^{-8}M$ to $10^{-6}M$) is that they pass through gap junctions instead of closing them.[100] However, this is contradicted by a more recent study performed on Novikoff hepatoma cells which showed that $[Ca^{2+}]_i$ in the physiological range ($<5 \times 10^{-7}M$) Ca^{2+} affects junctional conductance.[101] Similarly, in astrocytes the intracellular injection of Lucifer yellow in buffer containing Ca^{2+} ions demonstrated a decrease in the number of coupled cells for $[Ca^{2+}]_i$ around $5 \times 10^{-7}M$.[68] Using the scrape-loading technique similar results were found with ionomycin (5 mM), while calcium measurements performed with

Indol indicated that $[Ca^{2+}]_i$ reaches $5.10^{-7}M$ in the presence of this calcium ionophore (L. Venance and C. Giaume, unpublished data). However, a direct action of calcium ions is far from being demonstrated. The uncoupling observed during increases in $[Ca^{2+}]_i$ induced by ionomyin could be due to mobilization of calcium-dependent signaling pathways. Interestingly in cultured astrocytes, at similar concentration ionomycin is known to induce accumulation of inositol-phosphates[102] and to release of arachidonic acid.[103]

Finally, as in many other preparations,[61] sensitivity to intracellular acidification have also been reported for astrocyte gap junctions[104] suggesting that protons may play a regulatory role. In glial cells, this could be physiologically relevant since intracellular pH is modulated by neuronal activity.[105] However, the titration curve of pH-sensitivity and the possibility of direct action of protons have not yet been determined in this system.

5. PUTATIVE ROLES FOR GAP JUNCTION REGULATION IN ASTROCYTES

5.1. CONTROL OF POTASSIUM MOVEMENTS

During neuronal activity, the K^+ concentration increases in the extracellular space.[106] In order to maintain neuronal excitability, rapid removal of excess potassium is necessary. As glial cells are permeable to potassium ions and an increase in their intracellular K^+ is observed during neuronal activity,[107] a role for astrocytes in the dissipation of potassium has been proposed.[108] In order to participate in the redistribution of K^+ ions, glial cells could either shunt the ions distally (spatial buffering) or store it locally (accumulation).[5] Although gap junctions are likely to contribute to spatial buffering performed by Müller glial cells,[109] this has not been demonstrated in other type of astrocytes. However, as gap junctions are likely to provide a direct pathway between sites where potassium ions are taken up and compartments where it is distributed, there are several models which have integrated junctional communication to account for K^+ movements within the brain.[12,110,111]

5.2. SUPPLY OF GLUCOSE METABOLITES TO NEURONS

Traditionally, astrocytes have been assigned a role in the metabolic and trophic support of active neurons.[112] These glial cells are ideally suited for this metabolic role: astrocytic end-feet surround intraparenchymal blood vessels—the source of glucose—while other astrocyte processes ensheath synaptic terminals. Thus, astrocytes may sense synaptic activity and deliver energetic metabolites to neurons. This metabolic role could be played in situ by single astrocytes, but more likely by functionally coupled astrocytes.[113] Demonstration that glucose, its phosphorylated form glucose-6-phosphate and the nonmetabolized derivate 3-ortho-methyl-glucose, are permeable to astrocyte gap junctions was achieved recently by adapting the scrape-loading technique to radiolabeled compounds. This permeability to glucose was shown to be blocked by various treatments including 18-α glycyrrhetinic acid, octanol, arachidonic acid and endothelins.[114] Moreover, these agents were shown to increase the uptake of 2-deoxy-glucose.[114] Altogether, these observations indicate that gap junctions in astrocytes may provide a pathway for the distribution of glucose metabolites within the brain. As inhibition of intercellular trafficking of energetic metabolites is associated with an increase in their uptake, this suggests that uncoupling of astrocytes may result in the formation of "hot-spots" of stored glucose and derivatives.

5.3. CONTROL OF CELL PROLIFERATION

A correlation between a reduction of intercellular communication and loss of growth control has been proposed for many years now.[115] In glia, this statement has been supported by experiments on connexin expressions in C6 glioma cell lines. Indeed, these cells are usually weakly coupled and expression of Cx43 is low. When expression of Cx43 was carried out in C6 cells, an inverse correlation was found between the rate of proliferation and the level of cell communication.[116] In addition, coculture of C6 cells with clones overexpressing Cx43 have led to the suggestion that decrease in cell proliferation is due to the secretion of a growth inhibitory factor and that the secretion of this factor may be linked to the level of gap junction permeability.[117] However, this relationship between rate of proliferation and level of cell coupling does not seem to occur in the opposite way, since sustained inhibition of gap junction permeability does not trigger glial proliferation in C6 cells transfected with Cx43.[67]

5.4. REGULATION OF CELL VOLUME

Astrocytes have been shown to swell in a number of pathological situations.[118] Cultured astrocytes, exposed to hypoosmotic stress, swell and subsequently reduce almost to their initial size. When electrical coupling was monitored in hypoosmolar medium, a decrease in cell-cell communication was found to occur.[119] This indicates that permeability of astrocyte gap junctions may participate in the regulation of cell volume, as proposed for pancreatic acinar cells.[120] In agreement with this, several agonists which induce cell swelling such as endothelins, noradrenaline and ATP,[121] also have an inhibitory effect on gap junction permeability in astrocytes (Table 8.2).

5.5. INTERCELLULAR CALCIUM SIGNALING

The initial observation made by A. Cornell-Bell and her colleagues that glutamate application initiated rises in $[Ca^{2+}]_i$ which propaged as intercellular waves between confluent astrocytes[122] has provoked interest in many laboratories. This finding was followed by the demonstration that cell-to-cell propagation of calcium signaling is mediated through gap junctions, either by inhibiting junctional permeability with uncoupling agents[18,72,90,123] or by comparing nontransfected and Cx43 tranfected C6 cells.[124] Furthermore, calcium waves were recorded in astrocytes from hippocampal slices, in which they were triggered by the firing of glutamatergic neuronal afferents.[125]

The ability of coupled astrocytes to respond to neuronal stimulation and to dispatch calcium signals to neighbors, has led to the proposal that astrocyte networks may constitute a long-range signaling system within the brain with its own kinetics and with a spatial organization distinct from neuronal circuits. Recently, a further step in understanding neuro-glial interactions has been taken by demonstrating direct signaling from astrocytes to neurons.[32,43] Although, the final mechanism which underlies this interaction remains to be elucidated, the existence of such glia-to-neuron interaction supports the participation of astrocytes in information processing in the central nervous system.[34,48] In the case of a communicating glial syncytium, when a presynaptic neuron is activated, release of glutamate will generate a response in the postsynaptic neuron which will be associated with a calcium increase in the astrocytic processes surrounding the synapse. This rise in $[Ca^{2+}]_i$ will then propagate between coupled astrocytes, which in turn should activate distant neurons (Fig. 8.5, upper diagram). When astrocytic gap junctions are closed by an uncoupler, neuronal activation would be limited to neurons which are directly connected to activated presynaptic elements (Fig. 8.5, lower diagram). Such control of neuro-glial interactions could be particularly relevant in the case of inhibition of astrocyte gap junctions by the endogeneous compound anandamide. Indeed, this arachidonic acid derivative uncouples astrocytes via the activation of a receptor without any rise in $[Ca^{2+}]_i$, which is not the case for other neurotransmitter-mediated uncoupling (Tables 8.1, 8.2). Thus by regulating the extent of the coupled astrocyte network in a calcium-independent manner, anandamide may participate in determining the size of activated neuronal groups.[90] This could be physiologically relevant in the striatum, where anandamide uncoupling is prominent and the existence of functional compartments of neuronal groups receiving somatotopic glutamatergic cortical afferents has been demonstrated.[127,128]

6. CONCLUSIONS AND PERSPECTIVES

During the last few years, identification and pharmacology of gap junctions in brain glial cells have progressed greatly. The finding that the type of connexin is distinctive to glial subtype and that their expression is developmentally and regionally regulated suggests that junctional communication exhibits some degree of specificity in glia. Moreover, the demonstration that pharmacological treatments control gap junction permeability indicates that regulation and plasticity of glial networks may occur. From these findings several directions for future investigations can be drawn.

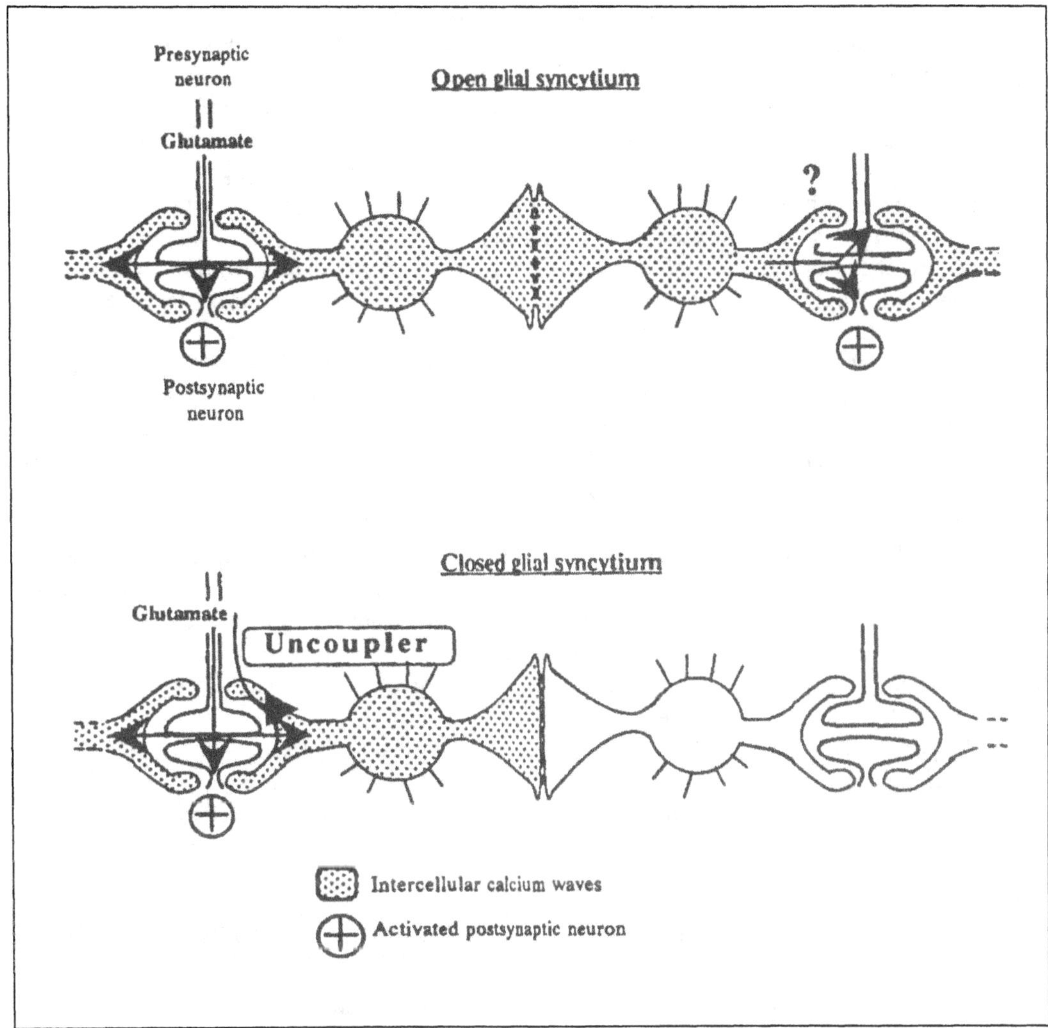

Fig. 8.5. Participation of astrocytic gap junctions in neuro-glial interaction through calcium signaling. Predictable consequences on neuronal activity of two functional states of the glial syncytium was considered in the drawings. Upper diagram: firing of a glutamatergic neuron (left) induces a postsynaptic response and a rise in [Ca²⁺]ᵢ in astrocytes which propagates through open gap junctions and activates distant neurons (right). The interrogation point indicates that the mechanism of astrocyte-to-neuron signal is not identified yet. Lower diagram: when gap junctions are inhibited, the activation of postsynaptic neurons is limited to the firing synapse.

6.1. FROM CULTURE TO IN VIVO STUDIES

Up to now, most of investigations concerning characterization and short term regulation of astrocyte gap junctions have been carried out in primary cultures (Tables 8.2, 8.3). Therefore, the next step will be to investigate whether data collected in vitro can be related to preparations which are closer to the in vivo situation. Such an approach has already been initiated for the identification of glial connexins with in situ hybridization and immunohistochemistry in the whole brain (see above), and studies of the regulation of junctional permeability in brain slices have begun (see Table 8.3). While in excitable cells gap junctions may play a role in impulse transmission, the expression of connexins in glial cells is more likely to provide a route for the exchange and distribution of metabolites or second messengers within the brain. Accordingly, functional assays of gap junction communication in glial cells should take into account this point and, when possible, be focused on the intercellular diffusion of physiologically relevant molecules.

What are the mechanisms involved in the dialogue between astrocytes and neurons? Does it occur in vivo? These are questions which will likely be addressed in the near future. If such an astrocyte-to-neuron signaling pathway is confirmed in vivo, one intriguing question will be to determine whether in brain structures, known to be composed by neuronal compartments, there are underlying glial compartments defined by coupled astrocyte networks.

As during neuronal development transient glial boundaries have been described to surround functional groups of neurons;[129] accordingly, dye coupling experiments should be performed to determine whether these boundaries are associated with restriction in the extent of dye coupling. Although dye injections into astrocytes from brain slices have not provided evidence for such limitation in glial connections either in the developing cortex[24] or hippocampus,[25] these studies should be extended to other structures such as striatum or cerebellum, while similar approaches should be carried out in anaesthetized whole animals.

6.2. LONG-TERM REGULATION OF Cx43 IN ASTROCYTES

Besides rapid changes in junctional permeability which are attributed to gating channel processes, slower changes can result from new formation of channels due to a precursor connexin pool or to a modification in transcription rate, mRNA stability or translation. In addition, because turn-over of connexin is rather fast, changes in the number of channels or junctions could involve protein degradation.[61,62] In astrocytes, brain injury and lesion have been associated with changes in the pattern of Cx43 expression.[130-133] However the nature and origin of signals involved in the

down- or upregulation of Cx43 have not yet been identified. In vitro models of reactive gliosis which reproduce some conditions of brain injuries are now available[134] and could contribute to the analysis of the mechanisms regulating expression and distribution of Cx43 in astrocytes during injury.

During circadian rhythms, changes in morphology and GFAP immunoreactivity of astrocytes from the suprachiasmatic nucleus have been associated with modification in the pattern of Cx43 distribution.[135] If these observations are correlated with changes in coupling strength, they could be of particular interest in demonstrating plasticity in glial networks. Such a relationship could be also investigated in other situations in which changes in astrocytes morphology have been observed, as for example following induction of long-term potentiation in the hippocampus.[136]

6.3. NEURO-GLIAL INTERACTIONS

The participation of astrocyte gap junctions in interactions between neurons and glial cells could occur at least at two levels. They could provide a pathway for direct calcium signaling between astrocytes and neurons if their presence, which is deduced solely from sensitivity to octanol,[32] is confirmed using morphological and electrophysiological approaches. Recently, combined patch-clamp technique and RT-PCR method performed in brain slice indicates expression of both Cx32 and Cx43 in neurons.[137] While Cx43/Cx32 heterotypic junctions are unlikely to be functional,[55,56] homotypic Cx43/Cx43 junctions could be involved in such communication between neurons and astrocytes.

Another aspect of neuro-glial interactions concerns the incidence of neurons on the expression and the functional state of astrocytic gap junctions. There is already evidence that kainic acid induced depletion of neurons results in a molecular alteration of Cx43 associated with loss of astrocytic gap junctions.[132] In addition, when glial coupling is compared using pure cultures of astrocytes versus cocultures with neurons, the percentage of coupled astrocytes is increased in cocultures[138,139] This up-regulation seems to be mediated by a soluble factor produced by neurons.[138] Thus, neurons may have the ability to control the level of junctional communication in astrocyte either by short-term regulation due to synaptic release of neurotransmitters (Table 8.2) or by long-term regulation through more complex mechanisms which remain to be understood.

Note added in proofs

Recently, Cx40 mRNAs have been detected in hippocampal and cerebelar astrocytes[143] and evidence for the expression of another connexin have been reported in astrocytes cultured from Cx43 knock out mice,[144] although intercellular communication are drastically reduced in these cells.[145]

Acknowledgments

We wish to thank Prof K.D Peusner, Dr B. Hamon and N. Stella for helpful comments on this manuscript, as well as Prof J. Glowinski for his enthusiasm and constant support.

References

1. Pope A. Neuroglia: quantitative aspect. In: Schoffeniels E, Franck G, Hertz L, Tower DB, eds. Dynamic Properties of Glial Cells. Oxford: Pergamon, 1978:13-20.
2. Kimelberg HK, Norenberg MD. Astrocytes. Scientific American 1989; 260:44-52.
3. Travis J. Glia: the brain's other cells. Science 1994; 266:970-972.
4. Barres BA, Chun LLY, Corey DP. Ion channels in vertebrate glia. Ann Rev Neurosci 1990; 13:441-474.
5. Barres BA. New roles for glia. J Neurosci 1991; 11:3685-3694.
6. Duffy S, MacVicar BA. Voltage-dependent ionic channels in astrocytes. In: Murphy S ed. Astrocytes: Pharmacology and Function. San Diego: Academic Press, 1993:137-169.
7. Sontheimer H. Voltage-dependent ion channels in glial cells. Glia 1994; 11: 156-172.
8. Ranck JB. Specific impedence of rabbit cerebral cortex. Exp Neurol 1963; 7: 144-174.
9. Oksche A. Circumventricular structures and pituitary functions. In Scow RO, ed. Endocrimology. Amsterdam: Excerpta Medica, 1973:73-79.
10. Palay SL, Chan-Palay V. General morphology of neurons and neuroglia. In: Brookhart JM, Mountcastle VB, eds. Handbook of Physiology The Nervous System, Vol. 1. Bethesda: American Physiological Society, 1977:5-37.
11. Black JA, Waxman SG. The perinodal astrocyte. Glia 1988; 1:169-183.
12. Risau W, Wolburg H. Development of the blood-brain barrier. Trends Neurosci 1990; 13: 174-178.
13. Massa PT, Mugnaini E. Cell junctions and intramembrane particles of astrocytes and oligodendrocytes a freeze-fracture study. Neuroscience 1982; 7:523-538.
14. Massa PT, Mugnaini E. Cell-cell junctional interactions and characteristic plasma membrane fractures of cultured rat glial cells. Neuroscience 1985; 14: 695-709.
15. Mugnaini E. Cell junctions of astrocytes, ependymal and related cells in the mammalian central nervous system, with emphasis on the hypothesis of a generalized functional syncytium of supporting cells. In: Fedoroff S, Vernadakis A, eds. Astrocytes. New York: Academic Press, 1986:329-371.
16. Peters A. Plasma membrane contacts in the central nervous system. J Anat (London) 1962; 96:237-248.
17. Ransom BR, Carlini WG. Electrophysiological properties of astrocytes. In: Fedorff S, Vernadakis A eds. Astrocytes. New York: Academic Press, 1986:1-49.
18. Dermietzel R, Spray DC. Gap junctions in the brain: where, what type, how many and why? Trends Neurosci 1993; 16:186-192.
19. Gonatas N, Hiryama M, Stieber A, Silberberg. The ultrastructure of isolated rat oligodendroglial cultures. J Neurocytol 1982; 11:997-1008.
20. Yamamoto T, Ochalski A, Hertzberg EL, Nagy JI. On the organization of astrocytes gap junctions in rat brain as suggested by LM and EM immunohistochemistry of connexin43 expression. J Comp Neurol 1990; 302: 853-883.

21. Dermietzel R, Traub O, Hwang TK, Beyer E, Bennett MVL, Spray DC. Differential expression of three gap junction proteins in developing and mature brain tissues. Proc Natl Acad Sci USA 1989; 87:1328-1331.

22. Giaume C, Venance L. Gap junctions in brain glial cells and development. Persp Develop Neurobiol 1995; 2:335-345.

23. Connors BW, Benardo LS, Prince DA. Carbon dioxide sensitivity of dye coupling among glia and neurons of the neocortex. J Neurosci 1984; 4:1324-1330.

24. Binmöller F-J, Müller C. Postnatal development of dye-coupling among astrocytes in rat visual cortex. Glia 1992; 6:127-137.

25. Konietzko U, Müller C. Astrocytic dye coupling in rat hippocampus: topography, developmental onset, and modulation by protein kinase C. Hippocampus 1994; 4: 297-306.

26. Butt AM, Ransom BR. Visualization of oligodendrocytes and astrocytes in the intact rat optic nerve by intracellular injection of Lucifer yellow and horseradish peroxidase. Glia 1989; 2:470-475.

27. Kettenmann H, Orkand RK, Schachner M. Coupling among identified cells in mammalian nervous system cultures. J Neurosci 1983; 3:506-516.

28. Ransom BR, Kettenmann H. Electrical coupling, without dye coupling, between mammalian astrocytes and oligodendrocytes in cell culture. Glia 1990; 3:258-266.

29. Venance L, Cordier J, Monge M, Zalc B, Glowinski J, Giaume C. Homotypic and heterotypic coupling mediated by gap junctions during glial cell differentiation in vitro. Eur J Neurosci 1995; 7:451-461.

30. Robinson SR, Hampson EC, Munro MN, Vaney DI. Unidirectional coupling of gap junctions between neuroglia. Science 1993; 262:1072-1074.

31. Kadle R, Cohan CS, Nicholson BJ. Cx43 mediated neuronal-neuronal and neuronal-glial coupling. International Meeting on Gap Junctions, Asilomar 1991:121 (abstract).

32. Nedergaard M. Direct signaling from astrocytes to neurons in cultures of mammalian brain cells. Science 1994; 263: 1768-1771.

33. Schneider SP. Demonstration of artifactual coupling between spinal neurons and glial cells during intracellular recording with micropipette electrodes. Brain Res 1992; 585:343-348.

34. Smith S. Neuromodulatory astrocytes. Cur Biol 1994; 4:807-810.

35. Attwell D. Glia and neurons in dialogue. Nature 1994; 369:707-708.

36. DeGeorge JJ, Morell P, MacCarthy KD, Lapetina EG. Adrenergic and cholinergic stimulation of arachidonate and phosphatidate metabolism in cultured astroglial cells. Neurochem Res 1986; 11:1061-1071.

37. Murphy S, Pearce B. Functional receptors for neurotransmitters on astroglial cells. Neuroscience 1987; 22:381-394.

38. Hösli E, Hösli L. Receptors for neurotransmitters on astrocytes in the mammalian central nervous system. Prog Neurobiol 1993; 40:477-506.

39. Finkbeiner SM. Glial calcium. Glia 1994; 9:83-104.

40. Wilkin GP, Marriott DR, Cholewinski AJ. Astrocyte heterogeneity. Trends Neurosci 1990; 13:43-46.

41. McCarthy KD, Slam AK. Pharmacological-distinct subsets of astroglia can be identified by their calcium response to neuroligands. Neuroscience 1991; 41:325-333.

42. Liscovitch M. Crosstalk among multiple signal-activated phospholipases. Trends Biol Sci 1992; 17:393-399.

43. Parpura V, Basarsky TA, Liu F, Jeftinija K, Jeftinija S, Haydon PG. Glutamate-mediated astrocyte-neuron signalling. Nature 1994; 369:744-747.

44. Piomelli D. Eicosanoids in synaptic transmission. Critical Rev Neurobiol 1994; 8:65-83.

45. Murphy S, Simmons ML, Agullo, Garcia A, Feinstein DL, Galea E, Reis DJ, Minc-Golomb D, Schwartz JP. Synthesis of nitric oxide in CNS glial cells. Trends Neurosci 1993; 16:323-328.

46. Barbour B, Szatkowski M, Ingledew N, Attwell D. Arachidonic acid induces a prolonged inhibition of glutamate uptake into glial cells. Nature 1989; 342:918-920.

47. Kimelberg HK, Jalonen T, Walz W. Regulation of the brain microenvironment: transmitters and ions. In: Murphy S, ed. Astro-

cytes: Pharmacology and Function. San Diego: Academic Press, 1993:193-228.

48. Smith SJ. Do astrocytes process neural information? In Yu ACH, Hertz L, Norenberg MD, Sykovà E, Waxman SG, eds. Progress in Brain Research, Vol. 94. 1992:119-136.

49. Dermietzel R, Hertzberg H, Kessler JA, Spray DC. Gap junction between cultured astrocytes: immunocytochemical, molecular, and electrophysiological analysis. J Neurosci 1991; 11:1421-1432.

50. Giaume C, Fromaget C, El Aoumari A, Cordier J, Glowinski J, Gros D. Gap junctions in cultured astrocytes: single-channel currents and characterization of channel-forming protein. Neuron 1991; 6:133-143.

51. Naus CCG, Belliveau DJ, Bechberger JF. Regional differences in connexin32 and connexin43 in the rat brain. Neurosci Lett 1990; 111:297-302.

52. Matsumoto A, Tamusa A, Urano A, Hyodo S. Cellular localization of gap junction mRNA in the neonatal rat brain. Neurosci Lett 1991; 124:225-228.

53. Micevytch PE, Abelson L. Distribution of mRNAs coding for liver and heart gap junction proteins in the rat central nervous system. J Comp Neurol 1991; 305:96-118.

54. Belliveau DJ, Kidder GM, Naus CCG. Expression of gap junction genes during postnatal neural development. Dev Genet 1991; 12:308-317.

55. White TW, Paul DL, Goodenough DA, Bruzzone R. Functional analysis of selective interactions among rodent connexins. Mol Biol Cell 1995; 6:459-470.

56. Elfgang C, Eckert R, Lichtenberg-Fraté H, Butterweck A, Traub O, Klein RA, Hülser DF, Willecke K. Specific permeability and selective formation of gap junction channels in connexin-transfected HeLa cells. J Cell Biol 1995; 129:805-817.

57. Neyton J, Trautmann A. Single-channel currents of an intercellular junction. Nature 1985; 317:331-335.

58. Rook MB, Jongsma HJ, van Ginneken. Properties of single gap junctional channels between isolated neonatal rat heart cells. Am J Physiol 1988; 255:H770-H782.

59. Burt J, Spray DC. Single-channel events and gating behavior of the cardiac gap junction channel. Proc Natl Acad Sci USA 1998; 85:3431-3434.

60. Fishman GI, Spray DC, Leinwand LA. Molecular characterization and functional expression of the human cardiac gap junction channel. J Cell Biol 1990; 111:589-598.

61. Bennett MVL, Barrio LC, Bargiello TA, Spray DC, Hertzberg EL, Saez JC. Gap junctions: new tools, new answers, new questions. Neuron 1991; 6: 305-320.

62. Saez JC, Berthoud VM, Moreno AP, Spray DC. Gap Junctions: Multiplicity of controls in differentiated and undifferentiated cells and possible functional implications. In: Shenolikar S and Nairn AC, eds. Advances in Second Messengers and Phosphoprotein Research, Vol. 27. New York, Raven Press 1993:163-198.

63. Lee SH, Kim WT, Cornell-Bell AH, Sontheimer H. Astrocytes exhibit regional specificity in gap-junction coupling. Glia 1994; 11:315-325.

64. Jongsma HJ, Rook M. Morphology and electrophysiology of cardiac gap junction channels. In: Zipes DP, Jalife J, eds. Cardiac Physiology: From Cell to Bedside. 2nd ed. Philadelphia: WB Saunders Company 1995:115-126.

65. Davidson JS, Baumgarten IM, Harley EH. Reversible inhibition of intercellular junctional communication by glycyrrhetinic acid. Biochem Biophys Res Com 1986; 134:29-36.

66. Venance L, Giaume C. 18-&-glycyrrhetinic acid inhibits junctional permeability in cultured astrocytes. 1995 Gap Junction Conference, Les Embiez, (abstract).

67. Goldberg GS, Bechberger JF, Naus CCG. A preloading method of evaluating gap junctional communication by fluorescent dye transfer. Biotechnologies 1995; 18:490-496.

68. Enkvist MOK, McCarthy KD. Astroglial gap junction communication is increased by treatment with either galutamate or high K^+ concentration. J Neurochem 1994; 62:489-495.

69. Barrio LC, Handler A, Bennett MVL. Inside-outside and transjunctional voltage dependence of rat connexin43 channels expressed in pairs of *Xenopus* oocytes. Biophys

J 1991; 64:A191 (abstract).

70. White TW, Bruzzone R, Wolfram S, Paul DL, Goodenough DA. Selective interactions among the multiple expressed in the vertebrate lens: the second extracellular domain is a determinant of compatibility between connexins. J Cell Biol 1994; 125:879-892.

71. Giaume C, Marin P, Cordier J, Glowinski J, Prémont J. Adrenergic regulation of intercellular communication between cultured striatal astrocytes from the mouse. Proc Natl Acad Sci USA 1991; 88:5577-5581.

72. Enkvist MOK, McCarthy KD. Activation of protein kinase C blocks astroglial gap junction communication and inhibit the spread of calcium waves. J Neurochem 1992; 59:519-526.

73. Giaume C. Noradrenergic control of gap junction permeability in cultured striatal astrocytes. In: Briley M, Marien M, eds. Noradrenergic Mechanisms in Parkingson's Disease. Boca Raton, CRC Press 1994: 205-223.

74. Müller T, Kettenmann H. Physiology of Bergmann glial cells. Int Rev Neurosci 1995; 38:341-359.

75. MacCumber MW, Ross CA, Snyder SH. Endothelin in the brain: receptors, mitogenesis, and biosynthesis in glial cells. Proc Natl Acad Sci USA 1990; 87:2359-2363.

76. Marsault, Vigne P, Breittmayer JP, Frelin C. Astrocytes are target cells for endothelins and sarafotoxin. J Neurochem 1990; 54:2142-2144.

77. Marin P, Delumeau JC, Durieu-Trautmann O, Le Nguyen D, Prémont J, Strosberg AD, Couraud PO. Are several G proteins involved in the different effects of endothelin-1 in mouse striatal astrocytes? Eur J Neurosci 1991; 56:1270-1275.

78. Tencé M, Cordier J, Glowinski J, Prémont J. Endothelin-evoked release of arachidonic acid from mouse astrocytes in primary culture. Eur J Neurosci. 4:993-999.

79. Cazaudon S, Parker PJ, Strosberg AD, Couraud PO. Endothelins stimulate tyrosine phosphorylation and activity of p42/mitogen-activated protein kinase in astrocytes. Biochem J 1993; 293:381-386.

80. Giaume C, Cordier J, Glowinski J. Endothelins inhibit junctional permeability in cultured mouse astrocytes. Eur J Neurosci 1992; 4:877-881.

81. Giaume C, Venance L, Cordier J, Glowinski J. Endothelins inhibit intercellular communication in brain glial cells. IVth International Conference on Endothelin, London 1995; P28 (abstract).

82. Venance L, Siciliano JC, Yokoyama M, cordier J, Glowinski J, Giaume C. Inhibition of astrocyte gap junctions by endothelins. In: Kanno Y, Kataoka K, Shiba Y, Shibata Y, Shimazu T, eds. Intercellular Communication Through Gap Junctions. Amsterdam: Elsevier, 1995:245-249.

83. Hossain MZ, Ernst LA, Nagy JI. Utility of intensely fluorescent cyanine dyes (CY3) for assay of gap junctional communication by dye-transfer. Neurosci Lett 1995; 184: 71-74.

84. Orr N, Yavin E, Lester DS. Identification of two distinct populations of protein kinase C in rat brain membranes. J Neurochem 1992; 58:461-470.

85. Seynaeve CM, Kazanietz MG, Blumberg PM, Sausville EA, Worland PJ. Differential inhibition of protein kinase C isozymes by UCN-01, a staurosporine analogue. Mol Pharmacol 1994; 45:1207-1214.

86. Giaume C, Randriamampita C, Trautmann A. Arachidonic acid closes gap junction channels in rat lacrimal glands. Pflügers Arch 1989; 413:273-279.

87. Burt JM, Massey KD, Minnich BN. Uncoupling of cardiac cells by fatty acids: structure-activity relationship. Am J Physiol 1991; 260:C439-C448.

88. Miyachi E, Kato C, Nakaki T. Arachidonic acid blocks gap junctions between retinal horizontal cells. NeuroReport 1994; 5:485-488.

89. Lazrak A, Peres A, Giovannardi S, Peracchia C. Ca-mediated and independent effects of arachidonic acid on gap junctions and Ca-independent effects of oleic acid and halothane. Biophys J 1994; 67:1052-1059.

90. Venance L, Piomelli D, Glowinski J, Giaume C. Inhibition by anandamide of gap junctions and intercellular signalling in striatal astrocytes. Nature 1995, 376: 590-594.

91. Filson AJ, Azarnia R, Beyer EC, Loewenstein WR, Brugge JS. Tyrosine phosphorylation

of a gap junction protein correlates with inhibition of cell-to-cell communication. Cell Growth Diff 1990; 1:661-668.

92. Kanemitsu MY, Lau AF. Epidermal growth factor stimulates the disruption of gap junctional communication and connexin43 phosphorylation independent of 12-0-tetracanoylphorbol 13-acetate-sensitive protein kinase C: the possible involvement of mitogen-activated protein kinase. Mol Biol Cell 1993; 4:837-848.

93. Takens-Kwak BR, Jongsma HJ. Cardiac gap junctions: three distinct single channel conductances and their modulation by phosphorylating treatments. Pflügers Arch 1992; 422:198-200.

94. Hii CST, Oh S-Y, Schmidt SA, Clark KJ, Murray AW. Lysophosphatidic acid inhibits gap-junctional communication and stimulates phosphorylation of connexin43 in WB cells: possible involvement of the mitogen-activated protein kinase cascade. Biochem J 1994; 303:475-479.

95. Burt JM, Spray DC. Ionotropic agents modulate gap junctional conductance between cardiac myocytes. Am J Physiol 1988; 254:H1206-H1210.

96. De Mello WC. Effect of isoproterenol and 3-isobutyl-methylxanthine on junctional conductance in heart cell pairs. Biochem Biophys Acta 1989; 1012: 291-298.

97. Loewenstein WR. Junctional intercellular communication: the cell-to-cell membrane channel. Physiol Rev 1981:61;829-913.

98. Neyton J, Trautmann A. Acetylcholine modulation of the conductance of intercellular junctions between rat lacrimal cells. J Physiol (London) 1986; 377:283-295.

99. Noma A, Stubio N. Dependence of junctional conductance on protons, calcium and magnesium ions in cardiac paired cells of guinea-pig. J Physiol (London) 1987; 382:193-211.

100. Saez JC, Connor JA, Spary DC, Bennett MVl. Hepatocyte gap junctions are permeable to second messenger, inositol 1,4,5-trisphosphate, and to calcium ions. Proc Natl Acad Sci USA 1989; 86:2708-2712.

101. Lazrak A, Peracchia C. Gap junction gating sensitivity to physiological internal calcium regardless of pH in Novikoff hepatoma cells. Biophys J 1993; 2002-2012.

102. Venance L, Stella N, Glowinski J, Giaume C. Activation of phospholipase C is a necessary step for propagation of intercellular calcium signaling in cultured astrocytes. 25th Neuroscience Meeting San Diego 1995; (abstract).

103. Oomagari K, Buisson B, Dumuis A, Bockaert J, Pin JP. Effect of glutamate and ionomycin on the release of arachidonic acid, prostaglandins and HETEs from cultured neurons and astrocytes. Eur J Neurosci 1991; 3:928-939.

104. Anders JJ. Lactic acid inhibition of junctional intercellular communication in in vitro asytrocytes as measured by fluorescence recovery after laser photobleaching. Glia 1988; 1:371-379.

105. Chesler M, Kraig R. Intracellular pH transients of mammalian astrocytes. J Neurosci 1989; 9:2011-2019.

106. Orkand RK, Nicholls JG, Kuffler SW. Effect of nerve impulses on the membrane potential of glial cells in the central nervous system of amphibia. J Neurophysiol 1966; 29:788-806.

107. Ballanyi K, Grafe P, Ten Brugencate G. Ion activities and potassium uptake mechanisms of glial cells in guinea-pig olfactory cortex slices. J Physiol (London) 1987; 382:159-174.

108. Newman EA. Regulation of extracellular potassium by glial cells in the retina. TINS 1985; 8:156-159.

109. Mobbs P, Brew H, Attwell D. A quantitative analysis of glial cell coupling in the retina of the axolotl (*Ambystoma Mexicanum*). Brain Res 1988; 460:235-245.

110. Somjen G. Glial cells functions. In: Adelman, ed. Encyclopedia of Neuroscience, Vol 1. Boston: Birkhäuser, 1987:465-466.

111. Reichenbach A. Glial K^+ permeability and CNS K^+ clearance by diffusion and spacial buffering. Ann NY Acad Sci 1991; 633: 272-286.

112. Somjen G. Nervenkitt: notes on the history of the concept of neuroglia. Glia 1988; 1:2-9.

113. Magistretti PJ, Pellerin L, Martin JL. Brain energy metabolism: an integrated cellular perspective. In: Bloom FE, Kupfer DJ eds, Psychopharmacology. New York: Raven

Press, 1995:657-670.

114. Tabernero A, Giaume C, Medina JM. Endothelin-1 regulates glucose utilization in cultured astrocytes by controling intercellular communication through gap junctions. Glia 1996; 16:187-195.

115. Loewenstein WR. Junctional intercellular communication and the control of growth. Biochem Biophys Acta Cancer Rev 1978; 560:1-65.

116. Zhu D, Caveney S, Kidder GM, Naus CCG. Transfection of C6 glioma cells with connexin43 cDNA: analysis of expression, intercellular coupling and cell proliferation. Proc Natl Acad Sci USA 1991; 88: 1883-1887.

117. Zhu D, Kidder GM, Caveney S, Naus CCG. Growth retardation in glioma cells cocultured with cells overexpressing a gap junction protein. Proc Natl Acad Sci USA 1992; 89:10218-10221.

118. Kimelberg HK, Ransom BR. Physiological and pathological aspect of astrocyte swelling. In: Fedoroff S, Vernadakis A, eds. Astrocytes. New York: Academic Press, 1986:129-166.

119. Kimelberg HK, Kettenmann H. Swelling-induced changes in electrophysiological properties of cultured astrocytes and oligodendrocytes. I. Effects on membrane potentials, input impedance and cell-cell coupling. Brain Res 1990; 529:255-261.

120. Ngezahayo A, Kolb H-A. Gap junctional permeability is affected by cell volume changes and modulates volume regulation. FEBS Lett 1990; 276:6-8.

121. Bender AS, Neary JT, Norenberg MD. Role of phosphoinositide hydrolysis in astrocyte volume regulation. J Neurochem 1993; 61:1506-1514.

122. Cornell-Bell AH, Finkbeiner SM, Cooper MS, Smith SJ. Glutamate induces calcium waves in cultured astrocytes: long-range glial signaling. Science 1990; 247:470-473.

123. Finkbeiner SM. Calcium waves in astrocytes-filling the gaps. Neuron 1992; 8:1101-1108.

124. Charles AC, Naus CCG, Kidder GM, Dirksen ER, Sanderson MJ. Intercellular calcium signaling via gap junctions in glioma cells. J Cell Biol 1992; 118:195-201.

125. Dani JW, Chernjavsky A, Smith SJ. Neuronal activity triggers calcium waves in hippocampal astrocytes networks. Neuron 1992; 8:429-440.

126. Charles AC. Glia-neuron intercellular calcium signaling. Dev Neurosci 1994; 16:196-206.

127. Gerfen CR. The neostriatal mosaic: multiple levels of compartemental organization. Trends Neurosci 1992; 15:133-139.

128. Graybiel AM, Aosaki T, Flaherty AW, Kimura M. The basal ganglia and adaptative motor control. Science 1994; 265: 1826-1831.

129. Steindler DA. Glial boundaries in the developing nervous system. Annu Rev Neurosci 1993; 16:445-470.

130. Vukelic JI, Yamamoto T, Hertzberg EL, Nagy JI. Depletion of connexin43-immunoreactivity in astrocytes after kainic acid-induced lesions in the rat brain. Neurosci Lett 1991; 130:120-124.

131. Rohlmann A, Laskawi R, Hofer A, Dobo A, Dermietzel R, Wolff J. Neurosci Lett 1993; 154:206-208.

132. Hossain MZ, Sawchuk, MA, Murphy LJ, Hertzberg EL. Kainic acid induced alterations in antiboy recognition of connexin43 and loss of astrocytic gap junctions in rat brain. Glia 1994; 10:250-265.

133. Dermietzel R, Hofer A, Rohlmann and Müller C. Functional plasticity and cell specific expression of connexins in normal and pathological glial tissues. In: Kanno Y, Kataoka K, Shiba Y, Shibata Y, Shimazu T, eds. Intercellular Communication Through Gap Junctions. Amsterdam: Elsevier, 1995:235-237.

134. McMillian MK, Thai L, Hong J-S, O'Callaghan JP, Pennypacker KR. Brain injury in a dish: a model for reactive gliosis. Trends Neurosci 1994; 17:137-141.

135. Lavialle M, Serviere J. Are the astrocyte gap junctions implicated in the neuronal synchronization in the circadian clock? Neuroscience Meeting, Miami, 1994:294-17.

135a. Serviere J, Lavialle M. Astrocytes in the mammalian circadian clock: putative roles. Prog Brain Res 1996 (in press).

136. Wenzel J, Lammert G, Meyer U, Krug M. The influence of long-term potentiation on the spatial relationship between astrocyte processes and potentiated synapses in the

gyrus neuropil of the rat brain. Brain Res 1991; 560:122-131.

137. Dermietzel R, Kirchhoff F, Dichtel B, and Kremer M. Application of patch recordings from brain slice preparations for single cell detection of connexin gene transcripts. 1995 Gap Junction Conference, Les Embiez, (abstract).

138. Fisher G, Kettenmann H. Cultured astrocytes form a syncytium after maturation. Exp Cell Res 1985; 159:273-279.

139. Anders JJ, Woolery S. Microbeam laser-injured neurons increase in vitro astrocytic gap junctional communication as measured by fluorescence recovery after laser photobleaching. Lasers Surg Med 1992; 12:51-62.

140. Glowinski J, Marin P, Tence M, Stella N, Giaume C, Prémont J. Glial receptors and their intervention in astrocyto-astrocytic and astrocyto-neuronal interactions. Glia 1994; 11:201-208.

141. Mantz J, Cordier J, Giaume C. Effects of general anesthetics on intercellular communications mediated by gap junctions between astrocytes in primary culture. Anesthesiology 1993; 78:892-901.

142. Giaume C, McCarthy KD. Control of gap junctional communication in astrocytic networks. TINS 1996 (in press).

143. Zheng X, Friedman L, Fan D, Zhang L, Aronica DH, Hall R, Dermietzel R, Zukin S, Bennett MVL. Cx40 mRNA in astrocytes and neurons of rat brain. (1995) Soc Neurosci Abstr 25:563.

144. Spray D, Viera D, El-Sabban ME, Gao Y, Bennett MVL, (1995) Gap junction properties in astrocytes from connexin43 (Cx43) knock-out (KO) mice. (1995) Soc Neurosci Abstr 25:563.

145. Bechberger JF, Naus CCG, Giaume C, Venance L, Juneja SC, Kidder GM. Functional characterization of astrocytes deficient in connexin43. (1995) Soc Neurosci Abstr 25:563.

CHAPTER 9

Do Glial Gap Junctions Play a Role in Extracellular Ion Homeostasis?

Bruce R. Ransom

1. INTRODUCTION

In the adult mammalian brain, glial cells, to a much greater extent than neurons, express the intriguing form of intercellular communication that is mediated by gap junctions[1,2] (see also chapters 8, 11). The function(s) served by glial gap junctions is not clearly established. One long held notion is that electrical coupling between glial cells, mediated by these junctions, helps to redistribute K^+ that accumulates with neural activity, the so-called spatial buffer hypothesis[3] (see below). Although there is strong evidence that K^+ release associated with intense neural activity would quickly overwhelm diffusion-based K^+ removal leading to disruptive increases in extracellular $[K^+]$ ($[K^+]_o$),[4-6] there is no evidence-based consensus about how the brain prevents this from happening. This fact is not widely appreciated. The theories about $[K^+]_o$ homeostasis, especially spatial buffering, have been around for so long that they are sometimes mistakenly assumed to be proven. I will briefly review what is known about control of brain $[K^+]_o$ and discuss possible roles of glial gap junctions in these processes.

2. $[K^+]_O$ WOULD INCREASE TO DISRUPTIVE LEVELS IF SPECIAL CONTROL MECHANISMS DID NOT OPERATE

Neuronal excitability and synaptic transmission are both influenced by the concentration of extracellular ions, especially $[K^+]_o$.[6-8] Not surprisingly, therefore, powerful physiological mechanisms exist to stabilize the concentration of ions in brain extracellular space (ECS). Some of these operate slowly and set the resting levels of extracellular ions while

Gap Junctions in the Nervous System, edited by David C. Spray and Rolf Dermietzel.
© 1996 R.G. Landes Company.

others are designed to deal with rapid changes in extracellular ions generated by neural activity.

The choroid plexus, which forms cerebrospinal fluid (CSF), and the blood-brain-barrier (BBB) are the two principal mechanisms that set resting extracellular levels of $[K^+]$. CSF in the ventricles and subarachnoid space is in slow equilibrium with brain extracellular fluid which accounts for the ionic similarity of the two.[9] The chemical composition of CSF is tightly controlled by an active secretory process in the choroid plexus. Freshly produced CSF has a $[K^+]$ of about 3.3 mM. The other specialized mechanism contributing to steady state control of the ionic content of extracellular fluid is the BBB. The BBB prevents ions and other solutes in blood from freely exchanging with brain extracellular fluid. As CSF circulates though the ventricles and subarachnoid space its $[K^+]$ drops slightly to about 3.0 mM, probably because some K^+ is actively transferred to blood by capillary endothelial cells.[9]

Together, the BBB and CSF establish and maintain a steady baseline level of brain $[K^+]_o$. Blood $[K^+]$ can fluctuate widely with diet, activity and renal function, but this has very little effect on brain $[K^+]_o$ because of these mechanisms.[10]

Although baseline $[K^+]_o$ is stable, the moment to moment concentration of extracellular K^+, and other ions as well, fluctuates with neural activity.[11] The anatomic basis for this is the restrictive extracellular space (ECS). The average width of the space between brain cells, that is the ECS, is only about 0.02 μm (Fig. 9.1). The vast web of infolding neuronal and glial membranes results in the ECS being a sizable fraction, about 20% on average, of total brain volume.[12] The volume of brain occupied by ECS varies somewhat in different areas of the brain, and rapidly (i.e., within seconds) and reversibly decreases by as much as 15% with intense neural activity.[13,14] Although brain ECS is extremely small, it does not greatly impede ionic diffusion.[12] However, a diffusing particle in the ECS encounters

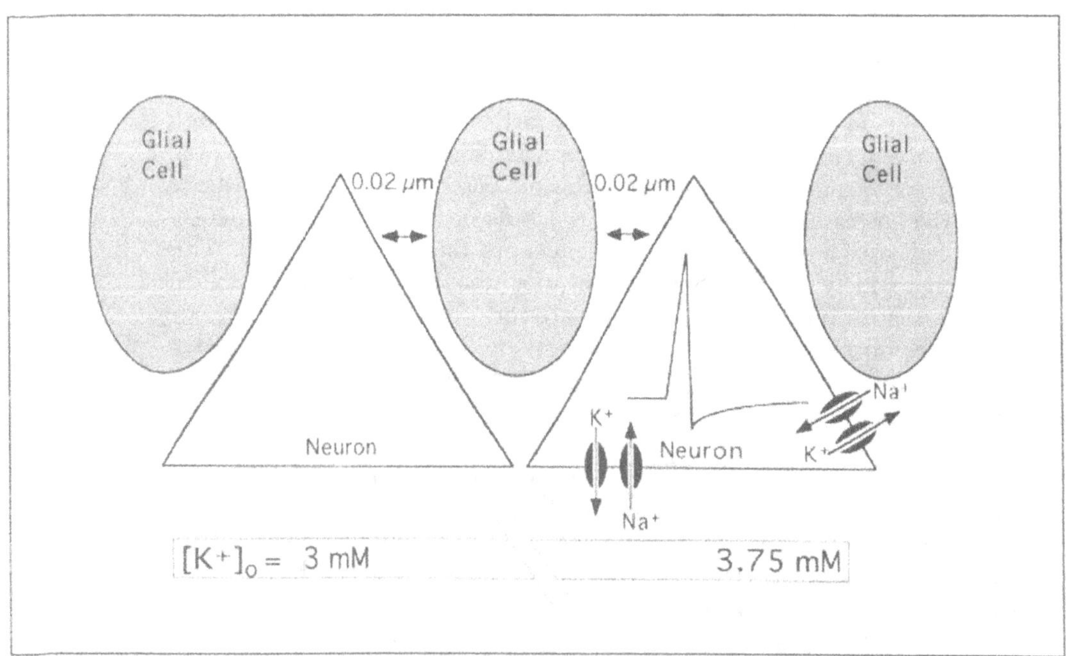

Fig. 9.1. Anatomic relationship between glial cells, neurons and brain extracellular space (ECS). The average distance between cells in the mammalian brain is only about 0.02 μm, producing a very restricted ECS. Single action potentials in neurons cause sufficient ionic flux to increase $[K^+]_o$ by ~0.75 mM. Communication between glial cells and neurons takes place by way of substances moving in the ECS.

obstacles in the form of cell membranes that limit free movement from one point in the brain to another. The circuitous route that a diffusing molecule must take in moving around cells strongly affects ECS diffusion; the degree to which a diffusional path from one point to another point is lengthened is described by the term tortuosity. The tortuosity in rat brain reduces ionic diffusion in ECS by 60% compared to movement in free solution.[12]

Ionic fluctuations occur continuously in the ECS with neuronal signaling. Brain activity is associated with significant decreases in neuronal $[K^+]_i$ leading to increases in $[K^+]_o$.[15-18] Single action potentials, for example, can increase $[K^+]_o$ by about 0.75 mM (Fig. 9.1).[19] More intense activity causes larger perturbations in $[K^+]_o$. Changes in the concentration of extracellular ions and molecules can serve a signaling purpose, but unregulated changes will disrupt brain function. Even small increases in $[K^+]_o$ (e.g., > 5mM) can affect synaptic transmission in the hippocampus.[7] Changes in brain $[K^+]_o$ can also influence cerebral blood flow,[20] ECS volume,[13,14] glucose metabolism[21] and neuronal activity.[17,22-26] The increase in $[K^+]_o$ tends to act as a positive feedback signal increasing excitability (e.g., ref. 27). Several observations indicate that control mechanisms beyond simple ionic diffusion have evolved to deal with rapid fluxes of K^+ into the ECS. (1) Calculations indicate that $[K^+]_o$ would increase to much higher levels and dissipate much more slowly than is seen experimentally.[28-33] (2) Even with intense stimulation, the level of $[K^+]_o$ never exceeds ~12 mM in the CNS of mammals, referred to as the K^+ ceiling level.[34,35] (3) When K^+ is injected into brain extracellular space it accumulates to a lessor extent than injected inert cations and, unlike the inert cations, its disappearance differs from what is predicted by the diffusion equation.[32]

Glial cells, especially astrocytes, are strategically placed in the brain to be able to sense changes in the neuronal microenvironment and help correct alterations that might arise in that environment due to neuronal discharge, metabolism or excessive neurotransmitter release. Glia, neurons,[16] and perhaps blood vessels[36] are likely to contribute to K^+ homeostasis, but glial mechanisms appear to predominate for rapid control.[4,6,18,37-41] Two types of mechanisms of K^+ removal have been proposed for glia: (1) K^+ accumulation by transport mechanisms (i.e., Na^+/ K^+-ATPase, KCl cotransport) and/or Donnan forces, and (2) K^+ redistribution by a spatial buffering mechanism.

The ability of glial cells to accumulate net amounts of K^+ in response to increases in $[K^+]_o$ has been directly demonstrated by measuring increases in glial $[K^+]_i$.[18,38,41-45] Uptake of K^+ can occur regardless of the spatial pattern of $[K^+]_o$ increase and therefore is referred to as space-independent (Fig. 9.2).[46] A portion of this net K^+ uptake is blocked by ouabain, indicating participation of the glial Na^+ pump.[40,41] The isoform of the Na^+ pump expressed in astrocytes is ideally suited for this purpose because it is activated by increases in $[K^+]_o$ as small as 1 mM.[47,48] In cultured astrocytes and glioma cells, KCl influx via an anion transporter that cotransports K^+ and Na^+ with Cl^-,[49] is activated by increases in $[K^+]_o$.[40,50] Neurons must reaccumulate K^+ lost during activity and appear to do so by the Na^+ pump and KCl cotransport.[16,49] Neuronal uptake, however, occurs slowly compared to glial uptake, probably over minutes following an abrupt increase in $[K^+]_o$.[16,51,52] Both neurons and glia contribute to K^+ removal in the intact CNS, therefore, but only glial cells show rapid net accumulation of K^+.

Another space-independent type of K^+ uptake by glial cells is mediated by Donnan forces which drive KCl influx in the face of elevated $[K^+]_o$. This type of KCl uptake is supported by the observations that K^+ accumulation is accompanied by an increase in $[Cl^-]_i$ and is partially blocked by the K^+ channel blocker Ba^{2+}.[41,53]

The concept of spatial buffering of focal increases in $[K^+]_o$ was introduced by Kuffler and his colleagues.[31,54] They reasoned that the high K^+ permeability of glia

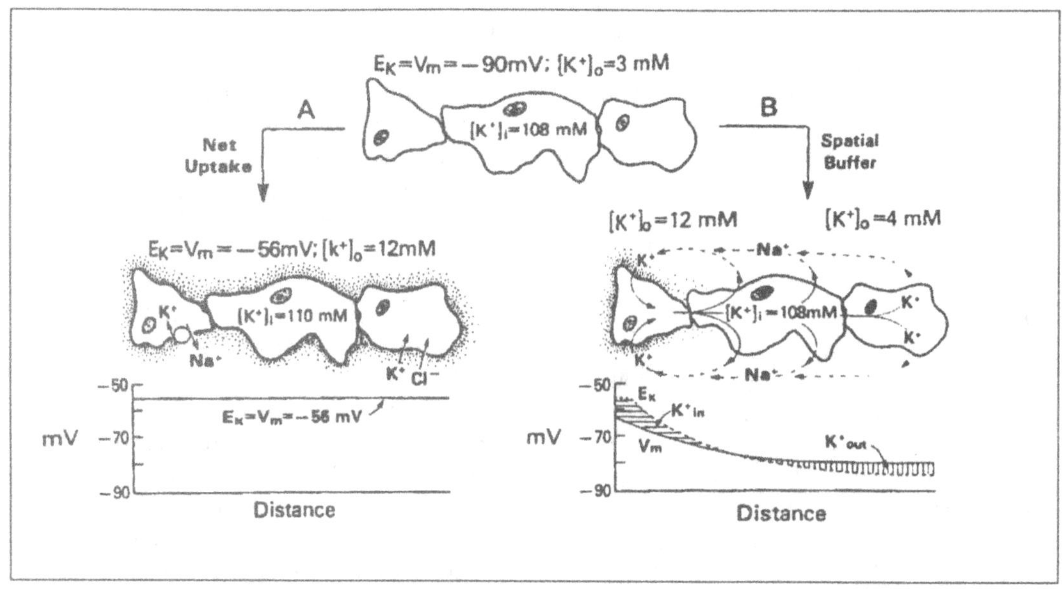

Fig. 9.2. Schematic diagram of the role of glial cells in K^+ homeostasis. Top: Three glial cells coupled by gap junctions are depicted in their resting state. The membrane potentials (V_m) are at the K^+ equilibrium potential (E_K). (A) Space-independent net uptake of K^+. K^+ released from active nerve cells has produced a spatially uniform increase in $[K^+]_o$ (shaded area) to 12 mM. The rise in K^+ increases the rate of the Na^+ pump and K^+ and Cl^- uptake (either by cotransport or Donnan effect). This results in an increase in $[K^+]_i$. V_m remains at approximately E_K (graph). (B) Space-dependent spatial buffering. Focal neural activity produces a local increase in $[K^+]_o$ to 12 mM (shaded area left side). This produces a glial depolarization that spreads passively to the coupled glial cells. The difference between V_m and E_K drives K^+ inward at the region where it is raised and outward at distant regions. Because of membrane selectivity, the transmembrane current is carried by K^+. The extracellular current is carried by the primary extracellular ions Na^+ and Cl^-. The result is a net flux of K^+ away from where it has accumulated. Average $[K^+]_i$ is not affected. The graph shows the distribution of E_K and V_m as a function of distance along the glial syncytium. During recovery, as the neurons take up K^+, these processes are reversed. Reprinted with permission from Orkand RK, Ann NY Acad Sci 1986; 481:269-272.

coupled with their low-resistance intercellular connections would permit them to transport K^+ from focal areas of high concentration, where a portion of the glial syncytium would be depolarized, to areas of normal $[K^+]_o$, where the glial syncytium would be more normally polarized (Fig. 9.2).[46] This transport depends on a current loop in which K^+ enters glial cells at the point of high $[K^+]_o$ and leaves them at sites of normal $[K^+]_o$, with intracellular and extracellular ionic current flow completing this circuit; i.e., the process is "space-dependent" (Fig. 9.2).[55,56] Experimental and theoretical analysis indicates that under conditions of focal increases in $[K^+]_o$ five times as much K^+ flux moves by way of glial cells as through the ECS, except where only very localized K^+ gradients are involved.[56-58] Studies on retinal Müller cells, a specialized astrocyte, have revealed a special form of K^+ spatial buffering (Fig. 9.3).[6] The end-foot of the Müller cell, which abuts the vitreous humor of the eye, contains much higher K^+ permeability than the remainder of the cell's membrane which causes accumulated ECS K^+ to be preferentially transported to the vitreous humor, which acts as a disposal site. This process has been termed K^+ siphoning. The generality of topographic inhomogeneity of membrane K^+ channels in astrocytes remains to be established but astrocytes of the salamander optic nerve have higher K^+ channel densities on their endfeet.[59] These processes contact blood vessels and the brain surface and should facilitate transport of accumulated K^+ from the parenchyma of the brain to areas that could reasonable serve as large K^+ sinks.[36]

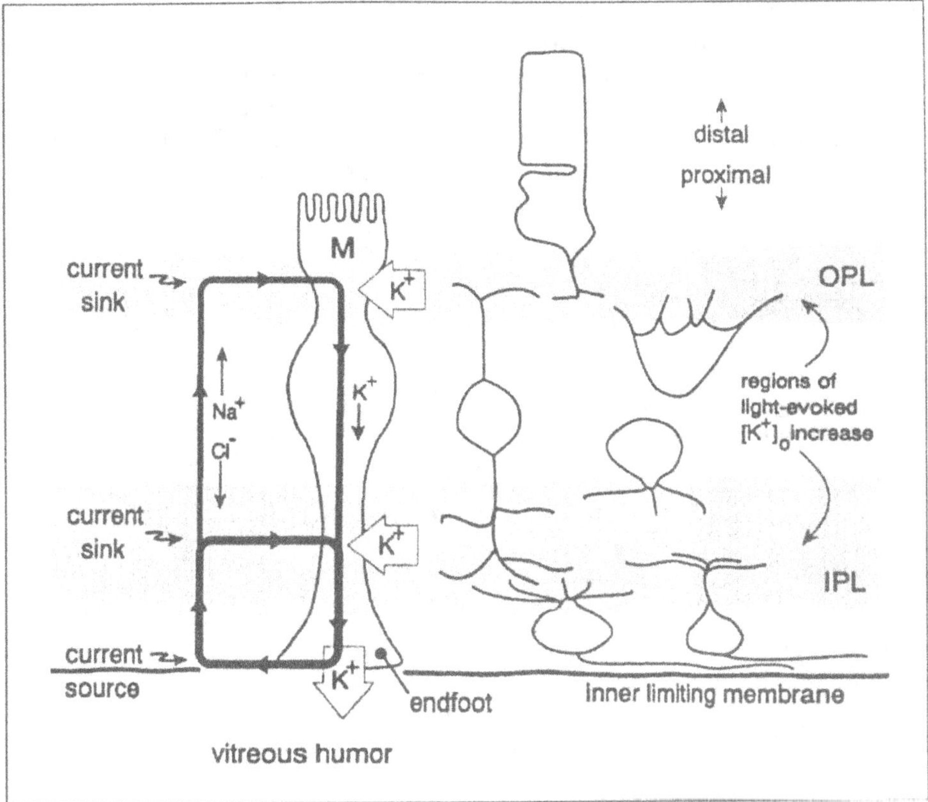

Fig. 9.3. Potassium siphoning, a form of K+ spatial buffering, helps regulate [K+]o in the retina. Increases in [K+]o are evoked by light stimulation in the inner and outer plexiform layers of the amphibian retina. The [K+]o increase causes K+ to enter Müller cells at the sites of maximal accumulation and leave the cell at the endfoot which is located at the vitreous humor of the eye where [K+]o is normal. This transfer of K+ from the plexiform layers to the vitreous humor is facilitated by a high density of K+ channels on the endfoot membrane. Reprinted with permission from Newman EA. In: Kettenmann H and Ransom B. Neuroglia. New York: Oxford University Press 1995:717-731.

The K+ channels implicated in K+ siphoning[6] and K+ spatial buffering[59a] have biophysical properties that are well suited to participate in K+ redistribution.

To summarize, neural activity leads to K+ accumulation in the ECS and this could cause disruption of CNS function if passive diffusion away from the sites of focal increase were the only means of K+ removal. Glial cells can assist with K+ removal by taking up net amounts of this ion or by simply transporting it away from sites of increase. The relative contribution of these glial mechanisms remains unclear and may depend on the magnitude and spatial distribution of the K+ increase within the tissue, tissue geometry, brain region and the species studied.

3. ARE GLIAL GAP JUNCTIONS NECESSARY FOR K+ SPATIAL BUFFERING?

The K+ spatial buffer hypothesis remains attractive but its quantitative significance has been difficult to establish.[5,6,17,60] As originally conceived, K+ spatial buffering depended on the presence of strong gap junctional communication between adjacent glial cells (Fig. 9.4A).[3] It was argued that strong glial coupling would extend the distance over which this K+ distribution system could operate and, perhaps, increase its efficiency[46,61] (Fig. 9.4A). A few indirect observations support this idea but the importance of gap junctional communication between glial cells for K[+] spatial buffering has never been critically tested.

If the theory is to work at all there must be strongly coupled glial cells in the brain (see reviews by Dermietzel, Spray[62] and Ransom[2]). The submammalian preparations favored by Kuffler and his coworkers contained glial cells that were strongly coupled to each

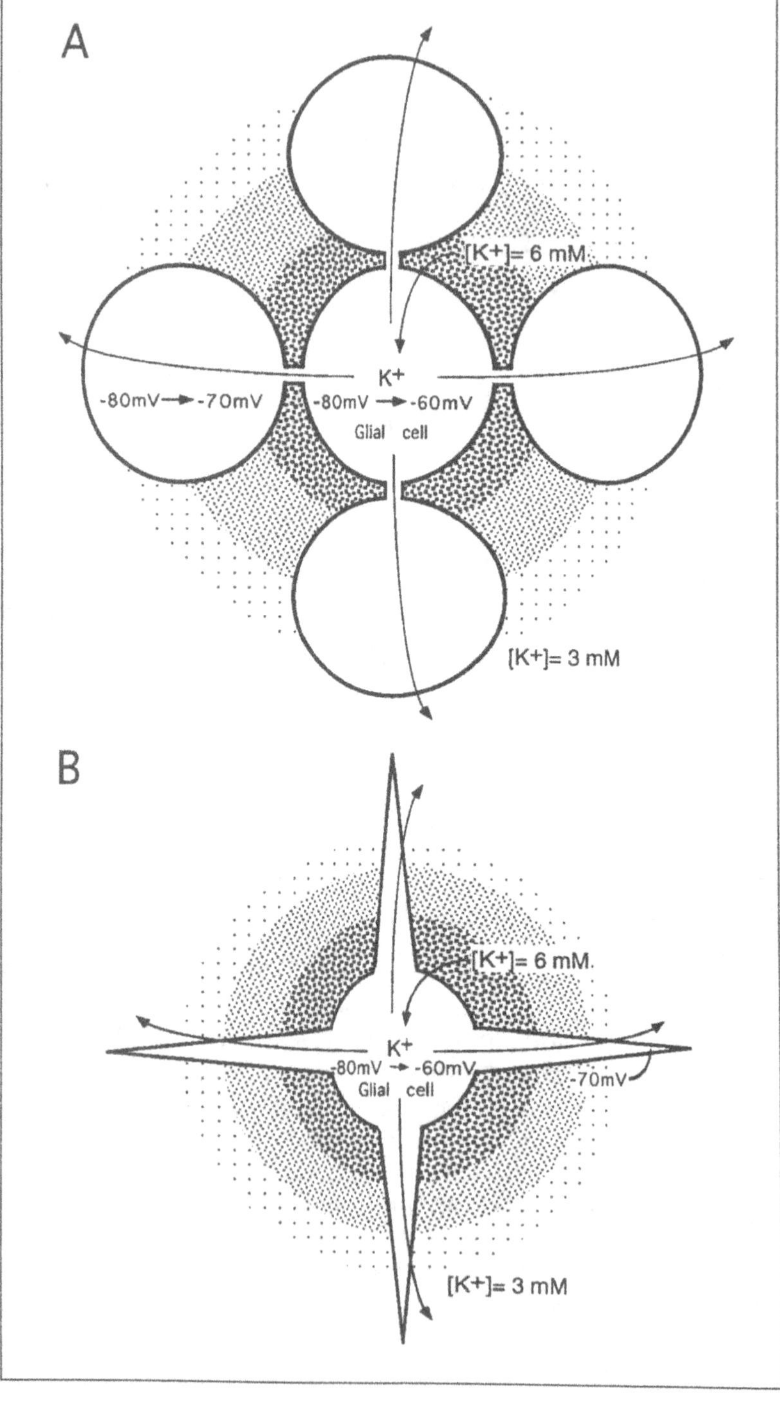

Fig. 9.4. Schematic diagrams showing how astrocytes might participate in K[+] removal by the spatial buffer mechanism. (A) Gap junctions between astrocytes create a highly coupled network of cells that cover a large area. This arrangement would assist spatial buffering of K[+] because it would insure that cells within the electrically coupled aggregate reached beyond the area of [K[+]]o increase, setting up the space-dependent type of K[+] movement shown in Fig. 9.2. It has seemed intuitive that a coupled network of cells would be able to convey a greater amount of K[+], a greater distance from the epicenter of accumulation than would noncoupled cells. (B) Individual astrocytes, however, could also convey K[+] from sites of accumulation to areas where [K[+]]o is normal. This would only require that they had processes that extended from the area of increased [K[+]]o to areas of lower or normal [K[+]]o.

other[62a] (e.g., see Cohen[63]). Mammalian astrocytes usually, but not always, show wide spread dye- and electrical-coupling. When Lucifer Yellow (LY) is injected into individual cortical astrocytes as many as 100 adjacent cells are stained.[34,64,65] The coupled cells form spherical arrays suggesting that LY diffuses from the injected astrocyte into uniformly coupled surrounding cells. In cultured white matter astrocytes, derived from optic nerve, dye coupling is restricted to type-1 astrocytes, with type-2 astrocytes devoid of coupling.[66] The type-2 astrocyte that has been defined in vitro, however, may not exist in vivo.[67] Thus far, studies of astrocyte dye-coupling in situ have failed to detect subpopulations of noncoupled astrocytes.[68]

The mammalian CNS is an anatomically and functionally diverse structure and cellular differences between regions can be anticipated. Studies in vitro suggest that regional differences in astrocyte coupling may exist. A maximum of ~50% of hippocampal astrocytes[69] and ~80% of optic nerve astrocytes[66] are coupled. Connexin43, the protein that forms gap junctions in astrocytes,[70,71] appears to be widely distributed throughout the mammalian brain, although some areas show a higher concentration of the protein than others.[72,73] These findings suggest that the strength of astrocyte intercellular communication varies with brain region (see also Lee et al[74]) and that this difference is controlled at the level of gene expression. It is not clear if these qualitative differences in coupling strength would dictate variations in the regional efficacy of spatial buffering.

Electrophysiological studies provide a more quantitative assessment of glial coupling. Rat cortical astrocytes in confluent monolayer cultures are uniformly and strongly coupled electrically.[75] The average coupling ratio (i.e., voltage change in the current-injected cell divided by the voltage change in a nearby cell) of astrocytes within 100 μm of each other is 0.44. Coupling is also seen between oligodendrocytes but is much weaker and less consistent than between astrocytes (coupling ratio = 0.11). Although heterologous coupling occurs between astrocytes and oligodendro-

cytes, it is extremely weak (coupling ratio = 0.03).[76] Astrocytes, therefore, are likely to participate more vigorously than oligodendrocytes in most of the functions related to intercellular coupling including ion redistribution.

The cortical distribution of glial membrane potentials and extracellular current flow in relationship to focal accumulations of extracellular K^+ provide support for the spatial buffer hypothesis. Futamachi and Pedley[77] recorded glial membrane potentials and $[K^+]_o$ at various depths in the cortex of a cat with experimentally induced epilepsy. During interictal discharges, $[K^+]_o$ increased up to 10 mM and the increase was mainly seen in the middle layers of the cortex, falling off sharply near the surface and at the deepest layers. Glial cells, however, were uniformly depolarized throughout the cortex.[77] This observation can be explained if the cortical glial cells were coupled to each other causing the glial cells distal to the K^+-depolarized cells in the middle layers to be depolarized as well. The depolarized cells away from the areas of high $[K^+]_o$ would have an outward driving force on K^+, a necessary condition for spatial buffering (Fig. 9.2). As expected, this arrangement results in a closed field-type distribution of extracellular current flow.[78] Ionic current in the form of K^+ enters the cells at the center of the $[K^+]_o$ increase causing extracellular negativity, and leaves from the distal cells causing extracellular positivity.

If coupling plays a significant role in the redistribution of focal K^+ accumulation seen with neural activity, the time course and/or magnitude of recorded increases in $[K^+]_o$ should be altered by blocking gap junctions. Technical factors, including the fact that uncoupling agents can have direct effects on various aspects of excitability, make this a difficult experiment. In the isolated frog spinal cord, massive increases in pCO_2, which presumably uncoupled glial cells due to intracellular acidification (but see Kettenmann et al[79]), slowed the clearance of neural activity-induced increases in K^+,[80] consistent with gap junction involvement in K^+ clearance.

It is important to consider that the theoretical advantage provided by gap junctions might be achieved more simply by the intrinsic anatomy of individual astrocytes. As far as spatial buffering of K^+ is concerned, the main advantage of gap junctions is to allow the membrane potential of distant glial cells to be influenced by cells in the vicinity of $[K^+]_o$ increases (Fig. 9.4A). Individual astrocytes, especially in white matter, have processes that can extend for 200-300 μm.[68] In the course of physiological activity, focal areas of an astrocyte would be exposed to increases in $[K^+]_o$ as shown in Figure 9.4B. This is in essence the situation necessary for the movement of K^+ from areas of high concentration to areas of low concentration via glial cytoplasm; the glial membrane near the local increase in $[K^+]_o$ would be depolarized relative to areas of the same cell that were adjacent to lower or baseline $[K^+]_o$. Noncoupled astrocytes could work in a cooperative manner to remove K^+ from areas of focal accumulation in this manner (Fig. 9.5).

The effectiveness of the model shown in Figure 9.4B (see also Fig. 9.5) has not been tested, but there is a precedent for this kind of K^+ transport in the retina (Fig. 9.3). Müller cells remove light-evoked increases in extracellular K^+ by transporting it to the vitreous humor of the eye which acts as a temporary storage site.[6] The efflux is facilitated by a higher density of K^+ channels in the endfoot compared to membrane elsewhere. In the cat, blockade of this removal system causes a 3-fold increase in the light-evoked $[K^+]_o$ increase.[81] This spatial buffering system in the retina, called K^+ siphoning, can operate effectively without coupling. In fact, coupling may not exist in mammalian Müller cells (see review by Ransom[2]).

4. ARE GLIAL GAP JUNCTIONS NECESSARY FOR DIRECT K^+ SEQUESTRATION?

The other mechanism by which glial cells participate in $[K^+]_o$ homeostasis is by direct uptake and sequestration of K^+ (see above).[5,18,41] This process might also benefit from glial coupling because the coupled glial aggregate would have a larger intracellular volume to buffer against changes in $[K^+]_i$ that would occur with K^+ uptake (Fig. 9.6). This would be desirable because the K^+ uptake system (e.g., passive KCl uptake and/or KCl cotransport and/or Na^+-K^+ ATPase) would tend to function better against a lower K^+ concentration gradient.

Our recent studies on the homeostasis of $[Na^+]_i$ in astrocytes indicate that the astrocyte Na^+ pump is ideally designed to participate in direct K^+ uptake because it is stimulated by very small increases in $[K^+]_o$.[47] To operate efficiently, the Na^+ pump requires a certain minimum $[Na^+]_i$ (see Sontheimer et al[82]). If the K^+ uptake process were intense, it might deplete astrocyte $[Na^+]_i$ to the point where the pump would fail. Other ionic mechanisms, including Na^+-K^+-Cl^- cotransport and Na^+-HCO_3^- cotransport, work in conjunction with the K^+-stimulated Na^+ pump to maintain a relatively constant $[Na^+]_i$. When gap junctions are uncoupled, the intracellular $[Na^+]_i$ of these astrocytes, which tends to be stable and uniform within a coupled aggregate, drifts in an unpredictable manner suggesting that the mechanisms setting the level of this critical intracellular ion vary from cell to cell (C. Rose and B. Ransom, unpublished observations). Maintaining stable cytoplasmic levels of glial $[Na^+]_i$ at rest and during heightened activity could be another important role of gap junctions related to regulation of $[K^+]_o$.

5. IS THE PARTICIPATION OF GAP JUNCTIONS IN $[K^+]_O$ HOMEOSTASIS UNDER DYNAMIC CONTROL?

The fact that the conductance of glial gap junctions can be modulated by a variety of circumstances including nearby neural discharge[83,84] adds an interesting dimension to how they might operate in the context of K^+ homeostasis. Norepinephrine, apparently acting through a second messenger pathway, downregulates functional

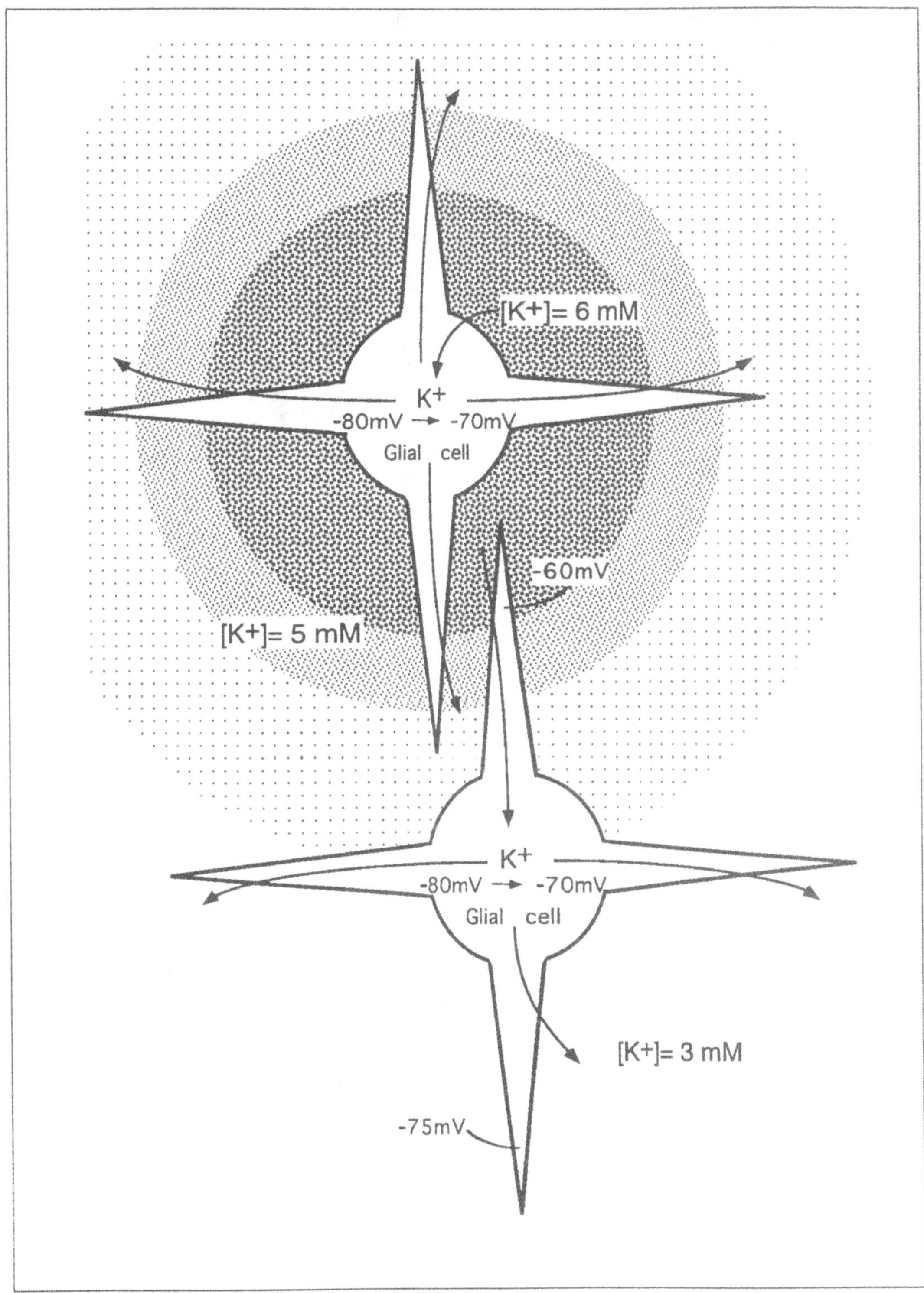

Fig. 9.5. Individual astrocytes could work in tandem to move K^+ away from areas of focal $[K^+]_o$ increase. This would require only that participating cells 'see' a gradient of $[K^+]_o$ along their processes which would provide the driving force for K^+ influx at points of maximal $[K^+]_o$ increase and efflux at points where $[K^+]_o$ is relatively lower. These sites would also correspond to points within the individual glial cells that were more and less depolarized, respectively.

gap junctional communication between striatal astrocytes.[85] On the other hand, the β-adrenergic agonist isoproterenol enhances dye-coupling by elevating cAMP.[85] Astrocyte dye-coupling appears to increase with application of glutamate or high [K⁺] solutions,[86] although brief applications of high [K⁺] solutions did not increase coupling between oligodendrocytes and astrocytes.[76] Glutamate or high [K⁺] solution

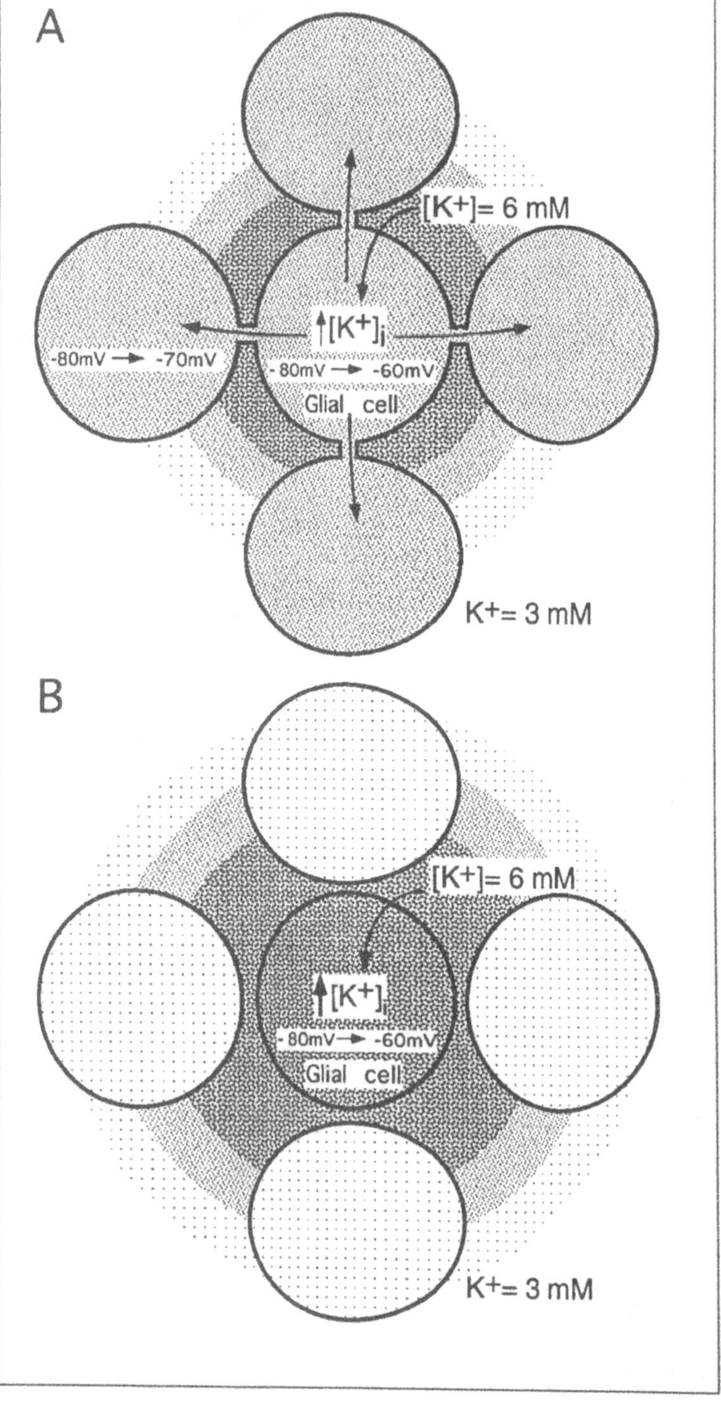

Fig. 9.6. Diagram indicating the possible role of glial gap junctions in extracellular K⁺ homeostasis mediated by net accumulation of K⁺. (A) Glial cells exposed to high [K⁺]ₒ will increase their intracellular [K⁺] ([K+]ᵢ) by direct uptake of K⁺ (see text). If strong coupling exists between adjacent cells the net increase in [K⁺]ᵢ will be modest (thin arrow) because the K⁺ which enters can distribute itself throughout the glial network by diffusion. This may prevent the development of high levels of [K⁺]ᵢ that could retard the uptake process. The levels of [K⁺] are qualitatively indicated by the density of shading. (B) In the absence of coupling, the uptake process causes much higher levels of [K⁺]ᵢ (thick arrow) to occur in the glial cells at the center of the [K⁺]ₒ increase. The suggestion that this will result in less effective control of [K⁺]ₒ is implied by the larger area of dense shading outside the cell, compared to that shown in panel A.

cause membrane depolarization and this might be linked to changes in coupling via changes in intracellular pH.[8,87] Gap junction conductance is increased by alkaline shifts in intracellular pH and depolarization produces an alkaline shift in astrocytes that could enhance coupling.[2,87,88] It was recently reported that dye coupling between glial cells of the frog optic nerve was increased if the dye injection occurred during electrical stimulation of the nerve.[84] This effect appears to be mediated at least in part by the associated increase in $[K^+]_o$.

If gap junctions are involved in any of the K^+ removal mechanisms discussed here, the capacity of these junctions to be functionally altered by conditions which might be induced by neural activity is fascinating. It suggests that K^+ removal or transfer may be a more dynamic process than previously thought. Glutamate and K^+, two factors that can affect glial gap junctional conductance, are released into the extracellular space in a fashion that is graded according to neural activity.[8,11] The demands on a K^+ removal system would also vary with the intensity of neural activity and it is attractive to imagine that a critical element of the removal system (perhaps rate-limiting?) could be upregulated to meet increased demand. The requirements of such a modulated system would be that the effects on gap junction conductance be rapid in onset and reversible. These important questions can be tested experimentally.

6. CONCLUSIONS

Gap junctions are abundantly expressed in the adult mammalian brain and are primarily located in glial cells, especially in astrocytes. Glial gap junctions have been hypothesized to participate in redistribution of neural activity-dependent increases in $[K^+]_o$ by the spatial buffering mechanism. Though theoretically attractive, the necessity of gap junctions for this mechanism to operate has not been proven. Individual astrocytes devoid of gap junctions could transport K^+ away from sites of accumulation if the $[K^+]_o$ increase was nonuniformly distributed along their pro-

cesses. K^+ siphoning by Müller cells in the retina exemplifies how this could work. Glial cells also assist in removal of excess extracellular K^+ by direct uptake of this ion. Coupling among glial cells could facilitate this process by creating a functionally larger intracellular compartment into which K^+ could be accumulated. This would prevent massive increases in $[K^+]_i$ which would limit the effectiveness of this process.

It is surprising that one of the most conspicuous physiological features of glial cells, the strong gap junction coupling between them, is so poorly understood in terms of its function. The importance of controlling $[K^+]_o$ in the brain and the central role of glial cells in this process are established facts. What needs experimental verification is whether glial gap junctions play a critical part in this function. We have hypotheses, we must now test them.

ACKNOWLEDGMENTS

This work was supported by NIH grants NS 15589 and NS 06208. I thank Chris Ransom for helpful comments about the manuscript.

REFERENCES

1. Loewenstein WR. Junctional intercellular communication: The cell-to-cell membrane channel. Physiol Rev 1981; 61:829-913.
2. Ransom BR. Gap junctions. In: Kettenmann H, Ransom BR, eds. Neuroglia. New York: Oxford University Press, 1995:299-318.
3. Kuffler SW, Nicholls JG. The physiology of neuroglial cells. Ergeb Physiol 1966; 57:1-90.
4. Nicholson C. Dynamics of brain cell microenvironment. Neurosci Res Program Bull 1980; 18:183-322.
5. Ransom BR, Carlini WG. Electrophysiological properties of astrocytes. In: Fedoroff S, Vernadakis A, eds. Astrocytes, Vol. 2. New York: Academic Press, 1986:1-49.
6. Newman EA. Glial cell regulation of extracelluar potassium. In: Kettenmann H, Ransom BR, eds. Neuroglia. New York: Oxford University Press, 1995:717-731.

7. Rausche G, Igelmund P, Heinemann U. Effects of changes in extracellular potassium, magnesium, calcium concentration on synaptic transmission in area CA1 and the dentate gyrus of rat hippocampal slices. Pflugers Arch 1990; 415:588-593.

8. Ransom BR. Glial modulation of neural excitability mediated by extracellular pH: a hypothesis. Prog Brain Res 1992; 94:37-46.

9. Johanson CE. Ventricles and cerebrospinal fluid. In: Conn PM, ed. Neuroscience in Medicine. Philadelphia: J.B. Lippincott, 1995.

10. Ames A, Higashi K, Nesbett FB. Relation of potassium concentration in choroid-plexus fluid to that in plasma. J Physiol (London) 1965; 181:506-515.

11. Somjen GG. Extracellular potassium in the mammalian contral nervous system. Ann Rev Physiol 1979; 41:159-177.

12. Nicholson C. Extracellular space as the pathway for neuron-glial interaction. In: Kettenmann H, Ransom BR, eds. Neuroglia. New York: Oxford University Press, 1995:387-397.

13. Dietzel I, Heinemann U, Hofmeier G, Lux HD. Transient changes in the size of the extracellular space in the sensorimotor cortex of cat. Exp Brain Res 1980; 40:432-439.

14. Ransom BR, Yamate CL, Connors BW. Activity-dependent shrinkage of extracellular space: A developmental study. J Neurosci 1985; 5:532-535.

15. Grafe P, Rimpel J, Reddy MM, Ten Bruggencate G. Changes to intracellular sodium and potassium ion concentrations in frog spinal motoneurons induced by repetitive synaptic stimulation. Neurosci 1982; 7:3213-3220.

16. Ballanyi K, Grafe P, Reddy MM, ten Bruggencate G. Different types of potassium transport linked to carbachol and y-aminobutyric acid actions in rat sympathetic neurons. Neuroscience 1984; 2:917-927.

17. Sykova E. Extracellular potassium accumulation in the central nervous system. Prog Biophys Mol Biol 1983; 42:135-189.

18. Coles JA, Tsacopoulos M. Potassium activity in photoreceptors, glial cells and extracellular space in the drone retina: Changes during photostimulation. J Physiol (London) 1979; 290:525-549.

19. Adelman WJ, Fitzhugh R. Solutions of the Hodgkin-Huxley equations modified for potassium accumulation in periaxonal spaces. Fedn Proc 1975; 34:1322-1329.

20. Kontos HA. Regulation of the cerebral circulation. Annu Rev Physiol 1981; 43:497-507.

21. Salem RD, Hammerschlag R, Bracho H, Orkand RK. Influence of potassium ions on accumulation and metabolism of [^{14}C] glucose by glial cells. Brain Res 1975; 86:499-503.

22. Baylor DA, Nicholls JG. Changes in extracellular potassium concentration produced by neuronal activity in the central nervous system of the leech. J Physiol (London) 1969a; 203:555-569.

23. Baylor DA, Nicholls JG. After effects of nerve impulses on signalling in the central nervous system of the leech. J Physiol (London) 1969b; 203:571-589.

24. Sykova E, Orkand RK. Extracellular potassium accumulation and transmission in frog spinal cord. Neuroscience 1980; 5:1421-1428.

25. Malenka RC, Kocsis JD, Ransom BR, Waxman SG. Modulation of parallel fiber excitability by postsynaptically mediated changes in extracellular potassium. Science.1981; 214:339-341.

26. Yarom Y, Spira ME. Extracellular potassium ions mediate specific neuronal interaction. Science 1982; 216:80-82.

27. Traynelis S, Dingledine R. Role of extracellular space in hyperosmotic suppression of potassium-induced electrographic seizures. J Neurophys 1989; 61:927-938.

28. Krnjevic K, Morris ME. Factors determining the decay of K^+ potential and focal potentials in the central nervous system. Can J Physiol Pharmacol 1975; 53:923-934.

29. Vern BA, Schuette WH, Thibault LE. [K^+]$_0$ clearance in cortex: A new analytical method. J Neurophysiol 1977; 40:1015-1023.

30. Cordingley GE, Somjen GG. The clearing of excess potassium from extracellular space in spinal cord and cerebral cortex. Brain Res 1978; 151:291-306.

31. Lewis DV, Schuette WH. NADH fluorescence and [K^+]$_0$ changes during hippocam-

pal electrical stimulation. J Neurophysiol 1975; 38:405-417.

32. Nicholson C, Phillips JM. Ion diffusion by tortuosity and volume fraction in the extracellular microenvironment of the rat cerebellum. J Physiol (London) 1981; 321: 225-257.

33. Coles JA, Poulain DA Extracellular K$^+$ in the supraoptic nucleus of the rat during reflex bursting activity by oxytocin neurones. J Physiol 1991; 439:383-409.

34. Connors BW, Ransom BR, Kunis DM, Gutnick MJ. Activity-dependent K$^+$ accumulation in the developing rat optic nerve. Science 1982; 216:1341-1343.

35. Heinemann U, Lux HD. Ceiling of stimulus-induced rises in extracellular potassium concentration in the cerebral cortex of cat. Brain Res 1977; 120:231-249.

36. Goldstein GW, Betz AL. Recent advances in understanding brain capillary function. Ann Neurol 1983; 14:389-395.

37. Henn FA, Haljame H, Hamberger A. Glial cell function: Active control of extracellular K$^+$ concentration. Brain Res 1972; 43: 437-443.

38. Hertz L. An intense potassium uptake into astrocytes, its further enhancement by high concentrations of potassium, and its possible involvement in potassium homeostasis at the cellular level. Brain Res 1978; 145:202-208.

39. Gardner-Medwin AR, Coles JA, Tsacopoulos M. Clearance of extracellular potassium: Evidence for spatial buffering by glial cells in the retina of the drone. Brain Res 1981; 209:452-457.

40. Walz W, Hinks EC. Carrier-mediated KCl accumulation accompanied by water movements is involved in the control of physiologicial K$^+$ levels by astrocytes. Brain Res 1985; 343:44-51.

41. Ballanyi K, Grafe P, ten Bruggencate G. Ion activities and potassium uptake mechanisms of glial cells in guinea-pig olfactory cortex slices. J Physiol (London) 1987; 382:159-174.

42. Kettenmann H, Sonnhof U, Schachner M. Exclusive potassium dependence of the membrane potential in cultured mouse oligodendrocytes. J Neurosci 1983; 3:500-505.

43. Schlue WR, Wuttke W. Posassium activity in leech neuropil glial cells changes with external potassium concentration. Brain Res 1983; 270:368-372.

44. Walz W, Hertz L. Intracellular ion changes of astrocytes in response to extracellular potassium. J Neurosci Res 1983; 10:411-423.

45. Walz W. Role of glial cells in the regulation of the brain microenvironment. Progr in Neurobiol 1989; 33:309-333.

46. Orkand RK. Introductory remarks: Glial-interstitial fluid exchange. Ann NY Acad Sci 1986; 481:269-272.

47. Rose CR, Ransom BR. Intracellular sodium homeostasis in rat hippocampal astrocytes. J Physiol 1996; 491:291-305.

48. Sweadner KJ. Na, K-ATPase and its Isoforms. In: Kettenmann H, Ransom BR, eds. Neuroglia. New York: Oxford University Press, 1995:259-272.

49. Russell JM. Cation-coupled chloride influx in squid axon. J Gen Physiol 1983; 81:909-925.

50. Winter-Wolpaw E, Martin DL. Cl$^-$ transport in a glioma cell line: Evidence for two transport mechanisms. Brain Res 1984; 297:317-327.

51. Heinemann U, Lux HD. Undershoots following stimulus induced rises of extra-cellular potassium concentration in cerebral cortex of cat. Brain Res 1975; 93:63-67.

52. Walz W, Hertz L. Ouabain-sensitive and ouabain-resistant net uptake of potassium into astrocytes and neurons in primary cultures. J Neurochem 1982; 39:70-77.

53. Boyle PJ, Conway EJ. Potassium accumulation in muscle and associated changes. J Physiol 1941; 100:1-63.

54. Orkand RK, Nicholls JG, Kuffler SG. The effect of nerve impulses on the membrane potential of glial cell in the central nervous system of amphibia. J Neurophysiol 1966; 29:788-806.

55. Dietzel I, Heinemann U, Hofmeier G, Lux HD. Stimulus-induced changes in extracellular Na$^+$ and Cl$^-$ concentration in relation to changes in the size of the extracellular space. Exp Brain Res 1982; 46:73-84.

56. Gardner-Medwin AR. Annalysis of potassium dynamics in mammalian brain tissue.

J Physiol (London) 1983b; 335:393-426.

57. Gardner-Medwin AR. A study of the mechanisms by which potassium moves through brain tissue in rat. J Physiol (London) 1983a; 335:353-374.

58. Gardner-Medwin AR, Nicholson C. Changes of extracellular potassium activity induced by electric current through brain tissue in the rat. J Physiol (London) 1983; 335:375-392.

59. Newman EA. High potassium conductance in astrocyte endfeet. Science 1986; 233:453-454.

59a. Ransom CB, Sontheimer H. Biophysical and pharmacological characterization of inwardly rectifying potassium currents in rat spinal cord astrocytes. J Neurophysiol 1995; 73:333-346.

60. Karwoski CJ, Lu H-K, Newman EA. Spatial buffering of light-evoked potassium increases by retinal Müller (glial) cells. Science 1989; 244:578-580.

61. Mobbs P, Brew H, Attwell D A quantitative analysis of glial cell coupling in the retina of the axolotl (Ambystoma mexicanum). Brain Res 1988; 460:235-245.

62. Dermietzel R, Spray DC. Gap junctions in the brain: where, what type, how many and why? Trends Neurosci 1993; 16:186-192.

62a. Kuffler SW, Nicholls JG, Orkand RK. Physiological properties of glial cells in the central nervous system of amphibia. J Neurophysiol 1966; 29:768-787.

63. Cohen MW. The contribution of glial cells to surface recordings from the optic nerve of an amphibian. J Physiol (London) 1970; 210:565-580.

64. Binmöller FJ, Müller CM. Postnatal development of dye-coupling among astrocytes in rat visual cortex. Glia 1992; 6:127-137.

65. Gutnick MJ, Connors BW, Ransom BR. Dye-coupling between glial cells in the guinea pig neocortical slice. Brain Res 1981; 213:486-492.

66. Sontheimer H, Minturn JE, Black JA, Waxman SG, Ransom BR. Specificity of cell-cell coupling in rat optic nerve astrocytes in vitro. Proc Natl Acad Sci USA 1990; 87:9833-9837.

67. Fulton BP, Burne JF, Raff MC. Glial cells in the rat optic nerve: The search for the type-2 astrocyte. Ann N Y Acad Sci 1991; 663:27-34.

68. Butt AM, Ransom BR. Morphology of astrocytes and oligodendrocytes during development in the intact rat optic nerve. J Comp Neurol 1993; 338:141-158.

69. Sontheimer H, Waxman SG, Ransom BR. Relationship between Na^+ current expression and cell-cell coupling in astrocytes cultured from rat hippocampus. J Neurophysiol 1991; 65:989-1002.

70. Dermietzel R, Hertzberg EL, Kessler JA et al. Gap junctions between cultured astrocytes: Immunocytochemical, molecular, and electrophysiological analysis. J Neurosci 1991; 11:1421-1432.

71. Giaume C, Fromaget C, El Aoumari A, CordierJ, Glowinski J, Gros D. Gap junctions in cultured astrocytes: Single-channel currents and characterization of channel-forming protein. Neuron 1991a; 6:133-143.

72. Batter DK, Corpina RA, Roy C, Spray DC, Hertzberg EL, Kessler JA. Heterogeneity in gap junctional expression in astrocytes cultured from different brain regions. Glia 1992; 6:213-221.

73. Nagy JI, Yamamoto T, Sawchuk MA, Nance DM, Hertzberg EL. Quantitative immunohistochemical and biochemical correlates of connexin 43 localization in rat brain. Glia 1992; 5:1-9.

74. Lee SH, Kim WT, Cornell-Bell AH, Sontheimer H. Astrocytes exhibit regional specifity in gap-junction coupling. Glia 1994; 11:315-325.

75. Kettenmann H. Ransom BR. Electrical coupling between astrocytes and between oligodendrocytes studied in mammalian cell cultures. Glia 1988; 1:64-73.

76. Ransom BR, Kettenmann H. Electrical coupling, without dye coupling, between mammalian astrocytes and oligodendrocytes in cell culture. Glia 1990; 3:258-266.

77. Futamachi KJ, Pedley TA. Glial cells and extracellular potassium: Their relationship in mammalian cortex. Brain Res 1976; 109:311-322.

78. Ransom BR. The behavior of presumed glial cells during seizure discharge in cat cerebral cortex. Brain Res 1974; 69:83-99.

79. Kettenmann H, Ransom BR, Schlue WR. Intracellular pH shifts capable of uncoupling cultured oligodendrocytes are seen only in low HCO_3^- solution. Glia 1990; 3:110-117.

80. Syková E, Orkand RK, Chvátal, Hájek, Kriz N. Effects of carbon dioxide on extracellular potassium accumulation and volume in isolated frog spinal cord. Pflügers Arch 1988; 412:183-187.

81. Frishman LJ, Yamaoto F, Bogucka J, Steinberg RH. Light-evoked changes in $[K+]o$ in proximal portion of light-adapted cat retina. J Neurophysiol 1992; 67: 1201-1212.

82. Sontheimer H, Fernandez-Marques E, Ulrich N, Pappas CJ, Waxman SG. Astrocyte Na^+ channels are required for maintenance of Na^+/K^+-ATPase activity. J Neurosci 1994; 14:2464-2475.

83. Spray DC, Bennett MVL. Physiology and pharmacology of gap junctions. Ann Rev Physiol 1985; 47:281-303.

84. Marrero H, Orkand RK Nerve impulses increase glial intercellular permeability. Glia 1996; 16:285-289.

85. Giaume C, Marin P, Cordier J, Glowinski J, Premont J. Adrenergic regulation of intercellular communications between cultured striatal astrocytes from the mouse. Proc Natl Acad Sci USA 1991b; 88: 5577-5581.

86. Enkvist K, McCarthy KD. Astroglial gap junction communication is increased by treatment with either glutamate of high K^+ concentration. J Neurochem 1994; 62: 489-495.

87. Pappas CA, Ransom BR. Depolarization-induced alkalinization (DIA) in rat hippocampal astrocytes. J Neurophysiol 1994; 72:2816-2826.

88. Spray DC, Harris AL, Bennett MVL. Gap junctional conductance is a simple and sensitive function of intracellular pH. Science 1981; 211:712-715.

CHAPTER 10

SUBCELLULAR TOPOGRAPHY AND PLASTICITY OF GAP JUNCTION DISTRIBUTION ON ASTROCYTES

Astrid Rohlmann and J. R. Wolff

1. INTRODUCTION

The aim of this chapter is to characterize the cellular and subcellular morphology of the astrocytic network and its plasticity with respect to cellular connectivity. As will be shown, the plasticity of gap junction coupling is a dynamic process, which enables astrocytes to alter the junctional-coupling pattern with a very short latency on neural induction. This was studied by immunohistochemistry using antibodies against connexin43, the major astrocytic gap junction protein.[1] Morphological characterization of astrocytes allows analysis of subcellular gap junction distribution on astrocytic surfaces. Surface-to-volume analysis of astrocytes reveals that predominantly lamellae, flat sheet-like or finger-like protrusions of astrocytic processes and cell body form gap junctions and frequently build coupled ensheathments around synapses.[2] Data will be shown that gap junctions exist between different cells, but that they also couple processes and lamellae of *one* cell. Hence, we conclude that astrocytic gap junctions do not only form a system of intercellular junctions. Rather, gap junctions serve to build astrocytic compartments regardless of whether processes and lamellae of different cells are connected or parts of the same cell are coupled.

Quantitative data reveal an enormous coupling capacity within the astrocytic network providing each astrocyte with a mean of 30,000 gap junctions. Following neuronal axotomy, rapid changes in the number of

Gap Junctions in the Nervous System, edited by David C. Spray and Rolf Dermietzel.
© 1996 R.G. Landes Company.

astrocytic gap junctions occur in the vicinity of the reorganizing neurons and their synapses. This may indicate that astrocytic ensheathment and its degree of coupling may have a role in the maintenance and remodeling of synapses.

2. CHARACTERIZATION OF ASTROCYTIC MORPHOLOGY: DETERMINANTS OF THEIR CELL BIOLOGICAL FUNCTION

Recent data reveal an increasing list of probable astroglial functions, which by far surpass the traditional role as "glue" substance between neurons.[3] The astrocytic network is involved in modulating synaptic transmission by buffering the extracellular concentration of ions and through specific uptake and release mechanisms for transmitters.[4,5] Astrocytes are also involved in neuronal energy supply and may serve as carbohydrate storage.[4,5] They may provide trophic signals for neurons either through transmitter-related effects or by trophic signals.[4,5] Astroglia are involved in the regulation of the cerebrovascular supply, and generate electrical fields in the extracellular space which in turn may influence neurons.[4,5] They also regulate the size of the extracellular space because of the capability of astrocytic lamellae to swell.[6-8]

All these presumptive functions of astrocytes depend on ion pumps, carriers, receptor molecules etc., which are localized in their plasma membranes. To fully understand how these functions are served it is essential to know how astrocytic surface membranes are distributed in the neuropil. In addition most of the astrocytic functions require an intimate interaction between the astroglial and neuronal networks, e.g., during excitation and integration of synaptic input on neurons. This interaction is apparently not diffuse but requires focally organized sites of interaction between neurons and astrocytes. Regarding those neuron-glia relations, the smallest possible interaction site is comprised by a synapse and a perisynaptic meshwork composed of coupled astroglial cell processes and/or lamellae. A functional model for coupled astrocytic processes regulating potassium and glutamate in the extracellular space surrounding synapses has been proposed.[9] To morphologically describe those units on the subcellular level we must characterize the structure of astrocytes, the distribution of astrocytic processes and lamellae in the neuropil, and the arrangement of astrocytic plasma membranes around synapses before we can describe and discuss astrocytic gap junction coupling in general.

2.1. NUMBER AND SIZE OF ASTROCYTES AND ARRANGEMENT OF ASTROCYTIC COMPONENTS IN THE NEUROPIL

Astrocytes are "dendritic cells" penetrating a tissue space with their cell processes and lamellae.[10] This complex architecture has probably prevented morphometric approaches to astrocytic cell shape in vivo. We tried to evaluate the shape and spatial organization of astrocytes in the occipital cortex of rats including the interdigitation of cell processes originating from neighboring astrocytes.

Astrocytes were investigated on the light microscopic (LM) level using glial fibrillary acidic protein (GFAP)-immunocytochemistry and selective filling with horseradish peroxidase (HRP) of single astrocytes. Sections of rat cortex were stained with GFAP-antibodies following standard protocols. Diaminobenzidine (DAB) was used as a chromogen which was intensified by heavy metals.[11] GFAP is a widely accepted, specific marker of intermediate filaments in astrocytes. In adult rat brains GFAP selectively labels astrocytes and other astrocytic cell types, e.g., radial astrocytes, tanycytes and ependymal cells. In protoplasmatic astrocytes predominantly located in the grey matter of the cerebral cortex, GFAP-positive intermediate filaments are clustered in virtually all cell bodies while being loosely distributed in processes and, important for morphological studies, specifically sparing their tips.[12] In addition, brains of rats in which HRP crystals had been implanted were immunohistochemically studied. Single astrocytes accumulating HRP were found to be located in the terminal fields of anterogradely loaded axon terminals (mechanism of transport unknown) and, in contrast to the GFAP-staining, showed a Golgi-like complete staining of all their processes and lamellae.

Photographs of GFAP- and HRP-stained astrocytes were then used to evaluate two parameters necessary to describe the spatial organization of the astrocytic network: Firstly, the numerical *density* of astrocytes in one mm³ cortex tissue as obtained by GFAP-staining and, secondly, the *diameter* of individual astrocytes after HRP-filling. In contrast to GFAP-, HRP-staining allows more accurate measurements of cell diameters because it does not spare process tips and even lamellae are filled with the dye.

GFAP-stained astrocytic cell bodies were counted and the astrocytic cell densities (numbers per mm³) were calculated and corrected for double counts.[13] To control against the possibility of sectioning an astrocyte cell body (diameter approximately 6 µm) several times and thereby overestimating the number of cell bodies, the section had to be rather thick. Hence, we cut 40 µm sections. Results revealed a numerical density of 17,575 ± 2,936 astrocytes/mm³ (±S.D.) in laminae V and VI of rat visual cortex. This value is within the range of lamina variations in astroglial cell density and is in agreement with previous results described for the visual cortex laminae of albino rats of 16,400-66,900/mm³.[14]

The contours of HRP-filled astrocytes were drawn with a camera-lucida microscope. Both the projection areas and the polygons connecting the tips of cell processes were converted to circles of a constant area. The diameter of each circle represents the size of the tissue space penetrated by processes and lamellae of the cell under investigation. Analysis resulted in a mean diameter of HRP-filled cells of 78.0 ± 26.0 µm (±S.D.).

The *density* and *size* of astrocytes may be utilized to estimate the average amount of *interdigitation* between neighboring as-

trocytes: A mean number of astrocytes of 17,575 cells per mm³ of tissue suggests that each cell on average should occupy a tissue volume of 56,899 µm³, if astrocytic processes and lamellae did not interdigitate. The mean diameter of such a theoretically calculated sphere would be approximately 48 µm. However, HRP-filled processes and lamellae of single astrocytes show a larger *size* which results in a mean diameter of 78.0 µm. The actual diameter of an astrocytic territory then exceeds the diameter calculated from cell density by about 30 µm for each cell. Hence, processes and lamellae of adjacent cells radially interdigitate over a mean distance of 30 µm. This overlapping tissue space will be termed *heterocontrol space* in contrast to the *autocontrol space*. Both spaces represent tissue which is monitored and controlled by astrocytes. The autocontrol space (about 18 µm) comprises the cell body and a few micrometer tissue space surrounding it and is penetrated by processes and lamellae of one cell only. This autocontrol space (with a volume of about 3,000 µm³), the "private territory" of an astrocyte, is very small compared to the average tissue volume (about 248,500 µm³) occupied by one cell. Figure 10.1 summarizes schematically the organization of interdigitating astrocytes.

Why do we present these data in such detail? For the gap junction distribution the autocellularly- and heterocellularly-controlled tissue organization predicts that gap junctions should occur mainly in the heterocontrol space, if regarded as intercellular contacts. This raises the question of whether the autocontrol space of astrocytes is devoid of gap junctions. This question will be addressed in more detail in subheading 3.2.

2.2. SURFACE AND VOLUME ANALYSIS OF ASTROCYTIC COMPARTMENTS

The size and contribution of astrocytic compartments (processes and lamellae) were estimated by morphometric and stereological techniques in relation to the whole cell surface and to the tissue volume. These data are necessary to characterize the distribution

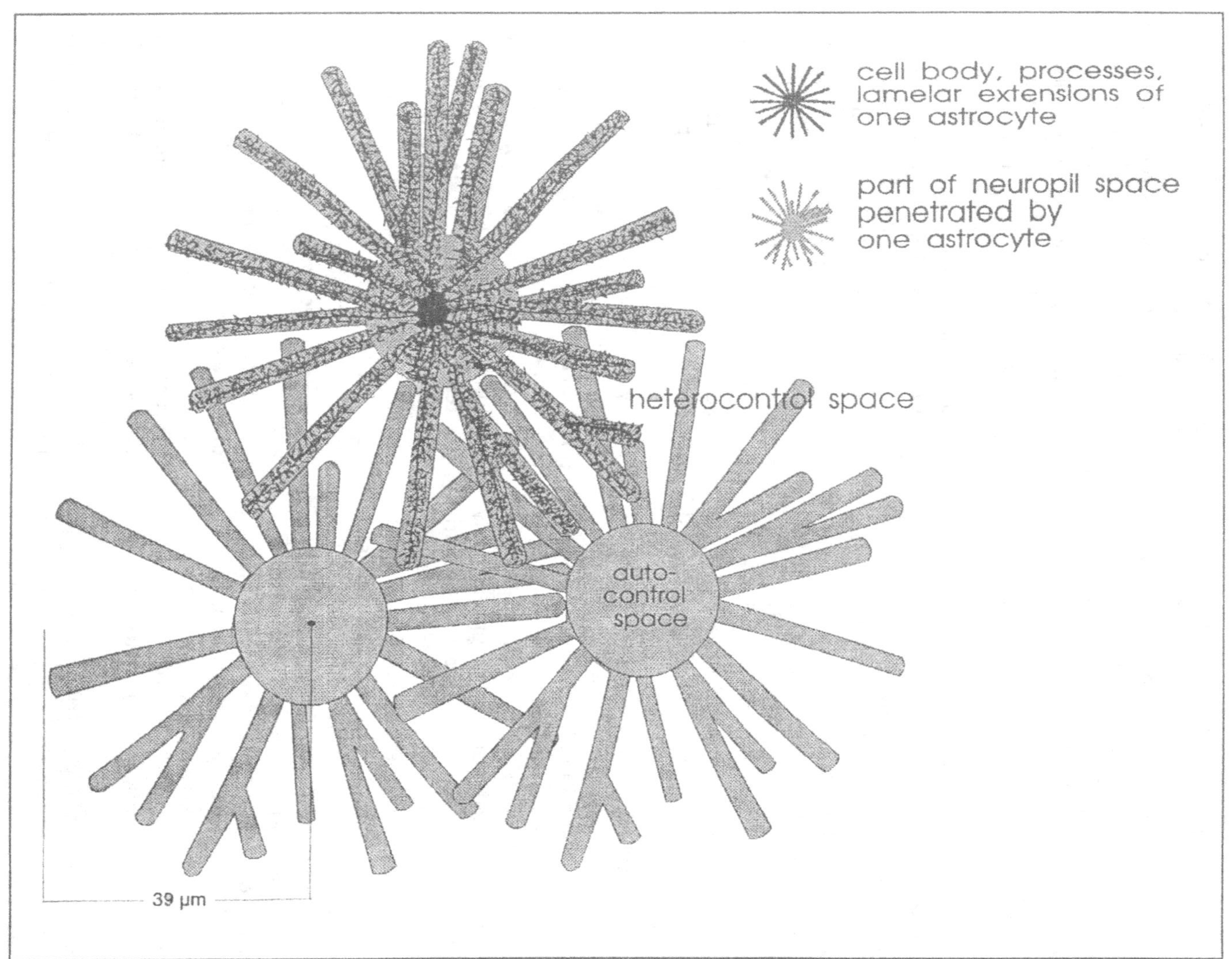

Fig. 10.1. Schematic overview of three interdigitating astrocytes which on average interdigitate over a distance of 30 µm in their periphery. The center comprises the "private territory" of each cell which we termed autocontrol space with a radius of 9 µm. This is in contrast to the heterocontrol space where interdigitation of astrocytes occurs.

Table 10.1. Elements of astrocytic compartments

	Surface/Neurophil Volume	Volume/Neuropil Volume
Cell body	0.002 µm²/m³ (0.1 %)	0.043 µm³/µm³ (34 %)
Processes	0.29 µm²/µm³ (19 %)	0.047 µm³/µm³ (38 %)
Lamellae	1.21 µm²/µm³ (81 %)	0.035 µm³/µm³ (28 %)

of gap junctions on these cellular compartments. We used HRP-filled astrocytes which we completely stained up to the very small tips of their lamellae and analyzed sections of these cells using electron micrographs with the point-counting-method.[15,16]

A lattice consisting of test points connected by half-circles was randomly superimposed on electron micrographs containing section profiles of labeled astrocytes. For counting irregularly shaped and inhomogeneously distributed profiles such as astrocytic lamellae, the minimal distance of points has to be equal to the square root of the area of the smallest profile to be evaluated. The *ratio of points* lying on stained astrocytic profiles to all reference points is according to the point-counting approach an estimate of the *volume* density or the volume fraction occupied by these tissue components measured as the astrocytic volume (µm³)/neuropil volume (µm³). *Surface* area densities were estimated by counting *intercepts* of marked glial plasma membranes with test lines of the grid as astrocytic surface (µm²)/neuropil volume (µm³).

We estimated the contribution of astrocytic compartments with their surface and volume values (see Tab. 10.1).

Volume data revealed that astrocytic soma, processes and lamellae contribute almost equally to the cell volume. This is in contrast to the cell surface which is predominantly built by astrocytic lamellae contributing about 80%. These data show that lamellae contain most of the astrocytic surface membrane. As mentioned above lamellae and finger-like protrusions are found on processes as well as on the cell body and form irregularly shaped thin sheets which invade even very small extracellular space clefts. They mostly represent membrane enlargements and include little cytoplasm. Their cytoplasm does not contain any organelles and is free of filament bundles. This membrane reservoir builds a huge surface for possible interactions with surrounding neuropil elements. Hence, lamellae probably serve local monitoring and focal exchange demands (also through gap junctions) but not rapid transport systems to the cell body.

Gap junctions are evenly distributed on the astrocytic cell surface (as will be shown on LM and EM level in section 3.1-3.3) and consequently about 80% of the gap junctions are located on lamellae. Lamellae may thus form larger coupled compartments which enables them to penetrate more extracellular space or to completely ensheath larger structures.

3. TOPOGRAPHY OF GAP JUNCTIONS IN CORTICAL TISSUE ON IDENTIFIED ASTROCYTES

All morphometrical studies of glial gap junctions in the nervous system done so far provide an idea of the cellular propensity to form gap junctions, but do not allow estimation of the amount of gap junction coupling of one astrocyte and gap junction distribution over its cell surface. Heterogeneity of gap junction expression has been described for cultured astrocytes from hypothalamus and striatum. These immunohistochemical-, mRNA- and dye-coupling studies showed region-specific differences, with a very high level of connexin43 expression in hypothalamic astrocytes.[17] In addition to region-specific coupling, regulation of glial coupling has been found during development, e.g., in olfactory tissue (see chapter 15) or postnatally in various parts of rat brain.[18] Because of the highly heterogeneous distribution of connexin43 immunoreactivity in adult brain[18,19] it was concluded that the diversity of coupling patterns may be related to the development of neuronal activity and establishment of functional circuits. Other studies[20] confirmed that the density of connexin43 immunostaining in brain sections shows a high correlation to relative amounts of connexin43 protein detected in Western blots. They hypothesized that immunohistochemical staining reflects the relative amounts of connexin43 and possibly the numbers of gap junctions present in different brain areas. On this background we tried to obtain data for single astrocytes and their gap junction coupling at both light and electron microscopical levels.

3.1. LIGHT MICROSCOPICAL ANALYSIS OF CONNEXIN43 STAINED OCCIPITAL CORTEX

We first checked connexin43 staining of cortical astrocytic gap junctions by comparing our staining pattern with data from the literature. Tissue of cerebral cortex of adult rats was processed as described for staining of facial nuclei, see chapter 4. Immunohistochemistry with antibodies against connexin43 resulted in the same type of punctate staining pattern as described in detail elsewhere.[19]

However, the laminar distribution pattern we obtained in the visual cortex (Fig. 10.2A) was slightly different from that previously shown for the parietal cortex region of adult rats.[19] In the occipital cortex staining was highest in lamina I and VIb, intermediate in layers II, III, Va and VIa, low in lamina Vb and lowest in lamina IV. This finding suggests that

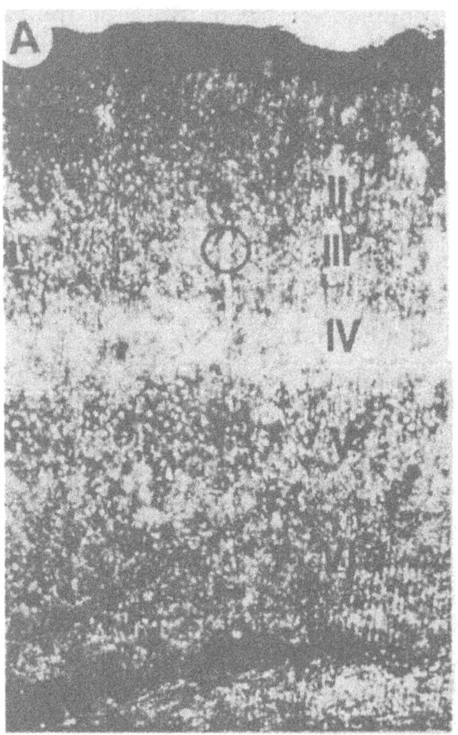

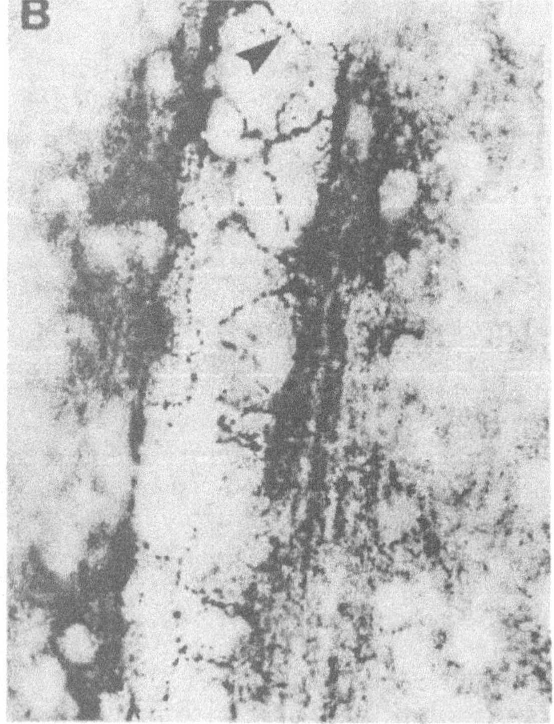

Fig. 10.2. (A) Connexin43 stained visual cortex of adult rats shows a laminar pattern of immunostaining. (B) Higher magnification reveals connexin43 positive gap junctions of astrocytic endfeet around blood vessels. (Magnification A: x30, B: x520)

there is not one uniform staining pattern throughout the rat neocortex, but rather connexin43 staining seems to vary slightly in different cortical areas. The main difference was found in lamina IV, the level of specific thalamic afferent input, which was consistently less stained compared to neighboring layers in the occipital cortex. In the parietal cortex a boundary was clearly demarcated between the superficial and the deeper layers by a lower staining intensity.[19] Irrespective of the laminar pattern, higher magnification (Fig. 10.2B) revealed a characteristic staining around blood vessels by astrocytic endfeet which are often coupled by numerous gap junctions.[19] Inter-estingly, we were not able to find this typical perivascular staining pattern around most capillaries. Thus, some variation in the occurrence and/or size or number of gap junctions may exist not only around neurons and synapses of certain layers of cortex but also around some blood vessels.

3.2. LIGHT MICROSCOPICAL ANALYSIS OF CONNEXIN43 STAINED GAP JUNCTIONS AND GFAP STAINED ASTROCYTES

Connexin43 and GFAP-immunohistochemistry were done as described above. Comparison of both stainings at equal magnification (Fig. 10.3) demonstrates

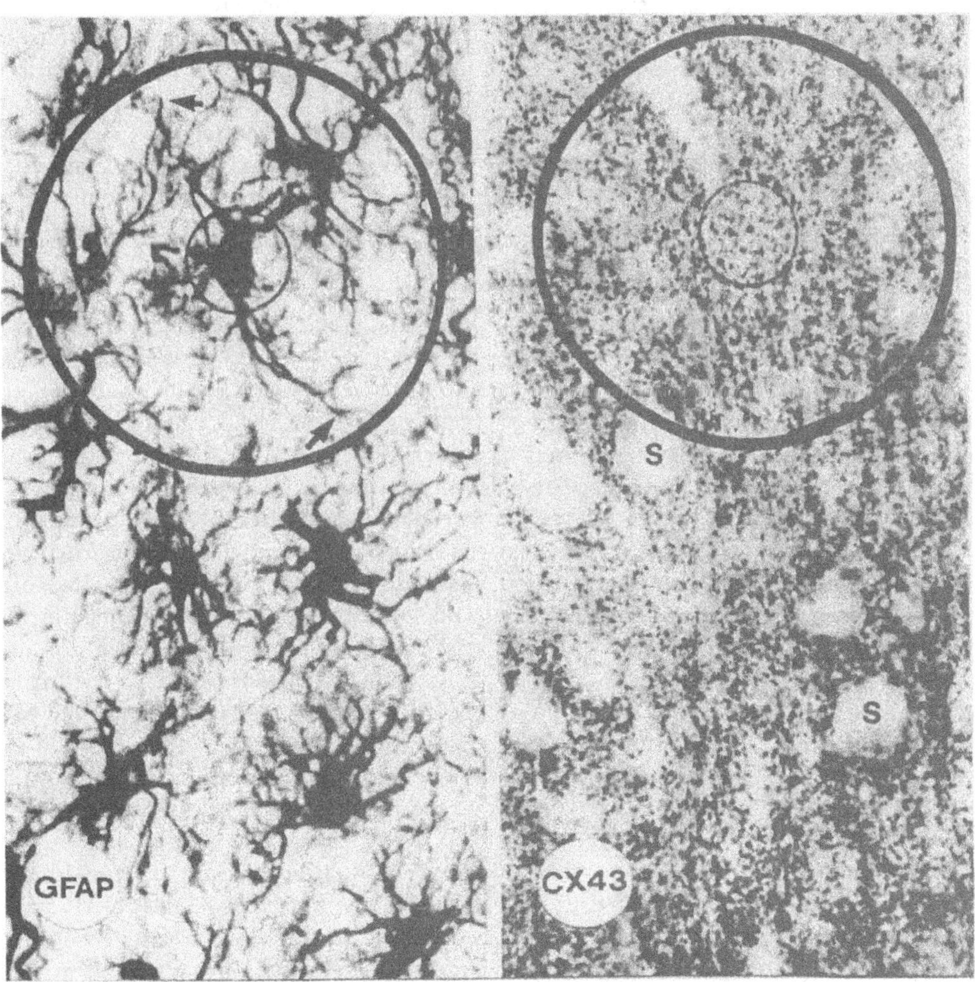

Fig. 10.3. Connexin43 staining reveals a homogeneous staining pattern. In contrast to this GFAP-staining discriminates single stained astrocytes. S = soma of neuron, arrows point to tips of astrocytic processes. (Magnification A,B: x400)

different cellular staining patterns for astrocytic gap junctions versus GFAP-positive intermediate filaments. While connexin43 immunohistochemistry shows homogeneously distributed punctate staining throughout the cortical tissue, GFAP-immunoreactivity is confined to discernible astrocytes. Cellular boundaries, however, are not detectable in gap junctions staining and arrangement of connexin43 immunoreactive puncta does not discriminate single cells. Correspondingly, autocontrol and heterocontrol spaces as morphologically described in Section 2.1 cannot be resolved in connexin43 stained sections. In Figure 10.3 an inner circle has been drawn around an astrocytic cell body to indicate the presumptive size of the autocontrol space. The outer circle indicates the diameter of the whole tissue space penetrated by the cell (about 78 μm). We never found the autocontrol space devoid of gap junction-puncta. Connexin43 immunopositive aggregates of different sizes are rather evenly distributed everywhere in cortical tissue except for neuronal perikarya and blood vessels. If gap junctions coupled only *between* different cells, they should occur in the periphery of astrocytes where interdigitation takes place. In contrast, connexin43 immunoreactivity is evenly distributed, suggesting that gap junctions may occur everywhere within the cell territory of an astrocyte, including the auto- as well as the heterocontrol space. This implicates that to some extent *intra*cellular coupling may occur at least in the autocontrol space. This hypothesis has been analyzed in more detail at the electron microscopical level.

3.3. Electron Microscopical Identification of Astrocytic Gap Junctions in Unstained, HRP-Stained and Connexin43 Stained Cortical Tissue

For quantitative studies it is crucial that gap junctions can be unequivocally identified by structural criteria. In accord with the literature ultrastructural analysis revealed a pentalaminar (without narrow extracellular gap) or heptalaminar (outer layers of the plasma membranes are separated by about 2 nm) structure of gap junctions. These differences between the stripe patterns depend on preparative methods, e.g., steps of fixation and contrasting procedures of the tissue,[21] and on electron irradiation conditions.[22] Ultrathin sections of gap junctions exhibit a "periodic intercellular density"[23] which probably corresponds to sequences of channels interspaces within a gap junction. Subunits of channels become visible on freeze-fracturing images[24] and in negatively stained gap junction-sheets.[25] Gap junction channels have been further described in size, with a diameter of about 4-6 nm and a center-to-center spacing of 9.1 nm between channels.[26] This characteristic submicroscopic appearance of a gap junction in orthogonal sections enabled us to analyze the subcellular distribution and number of gap junctions on identified single astrocytes.

Astrocytic gap junctions examined in unstained cortical tissue were found to exhibit the ultrastructural criteria described above.[19,27,28]

Electron micrographs frequently reveal gap junctions on astrocytic cell bodies (Fig. 10.4A). One prediction of astrocytic organization (see chapter 2.1) would be that cell bodies should be devoid of *inter*cellular gap junctions, because astrocytes mostly interdigitate in their periphery. In contrast, we found cell bodies to couple as frequently as do processes and lamellae of astrocytes. In addition, processes and lamellae often couple with themselves as shown in Figure 10.4B where a pentalaminar gap junction builds a "mesaxon-like" structure and couples compartments of one cell. This phenomenon has already been described as "reflexive gap junctions" for smooth muscle cells[29] and decidual cells of the ovary.[30]

To further investigate coupling between compartments of the same astrocyte in situ, we analyzed single HRP-stained astrocytes at the EM level.

This HRP-filling allowed us to identify a single astrocyte at the LM level (Fig. 10.5A) followed by further processing for EM. Electron micrographs revealed

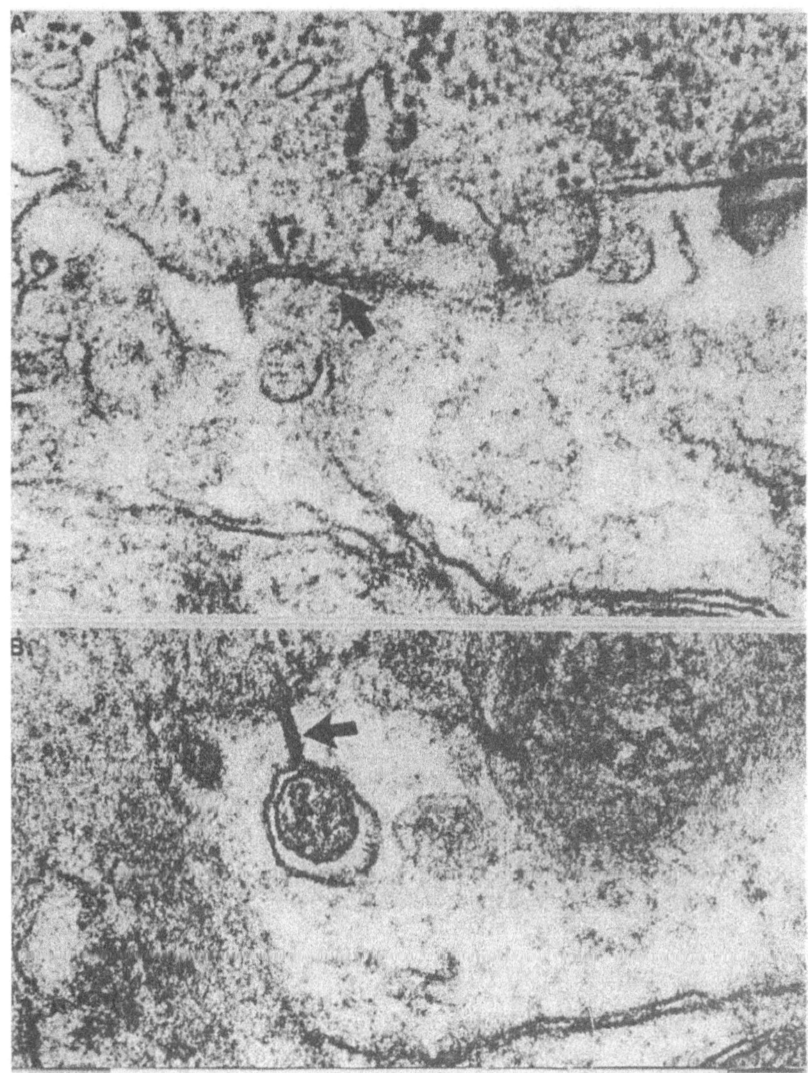

Fig. 10.4. (A) A gap junction couples between a cell body and a process. (B) A reflexive gap junction builds a "mesaxon-like" structure and couples a process with itself. Arrows point to gap junctions. (Magnification in A,B: x110,000)

numerous gap junctions between processes and lamellae of one identified cell (Fig. 10.5B,C). We found that single stained cells couple between their own stained compartments with about the same frequency as to unstained astrocytes. Although we cannot exclude that in vivo filling of astrocytes with HRP may modify the coupling pattern, we conclude that gap junctions are probably evenly distributed on astrocytic surfaces regardless of whether they couple different parts of one cell or two different cells. Section profiles of HRP-stained lamellae were characteristically located next to synaptic clefts. Ac-

cordingly, we found many gap junctions in the neighborhood of synapses.

We also examined connexin43 stained gap junctions at the EM level. The staining procedure included pretreatment of tissue with a $Na(BH_4)$ solution to facilitate antibody penetration. The immunohistochemical procedure was the same as for light microscopy except that we did not use any detergent.

These preparations confirmed that many gap junctions are located next to synapses where they frequently couple astrocytic lamellae (Fig. 10.6). Connexin43 pre-embedding immunoreaction stained

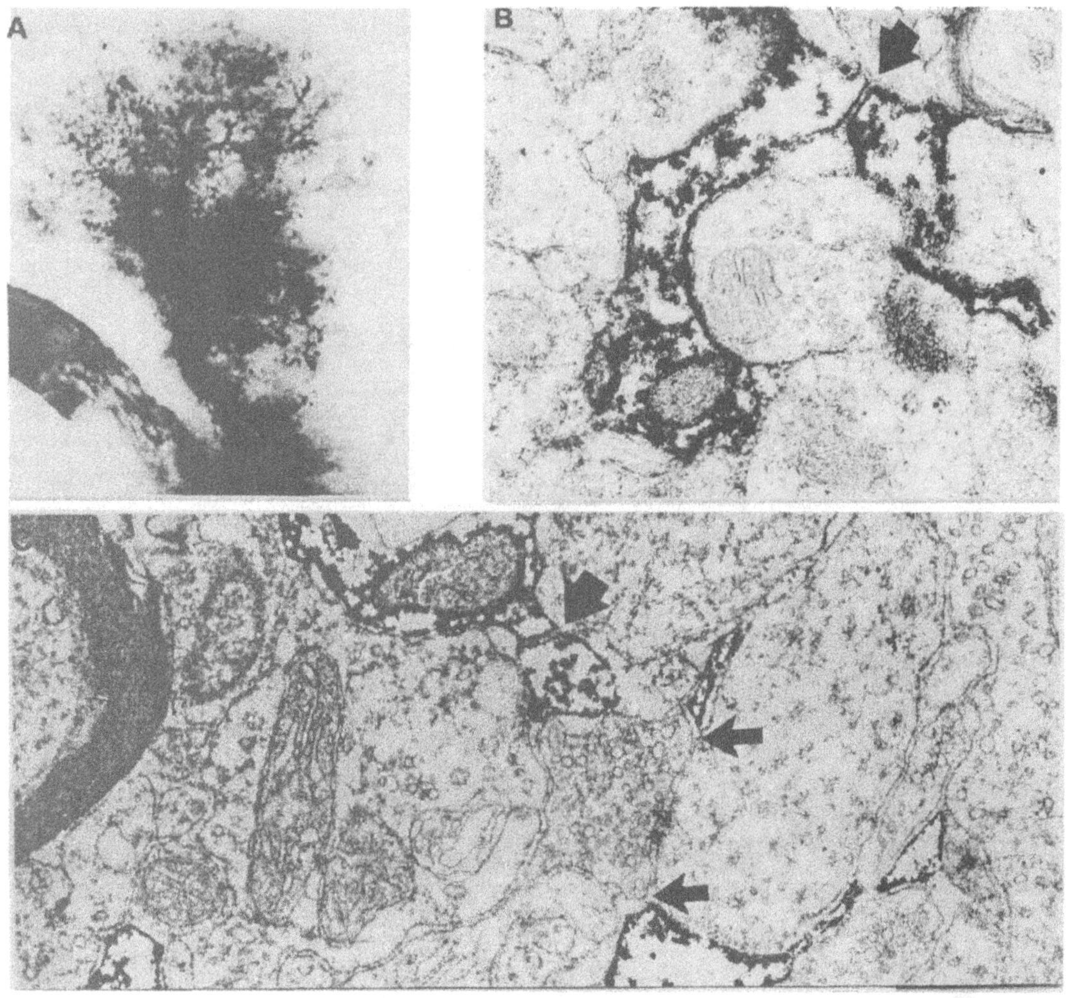

Fig. 10.5. (A) LM view of an HRP-filled astrocyte. (B,C) Electron micrographs of gap junctions (thick arrows) between stained processes and lamellae of the same cell are frequently located next to synapses. The thin arrows in C indicate stained astrocytic profiles next to the synaptic cleft. (Magnification A: x810, B,C: x43,000).

gap junctions together with the surrounding cytoplasm unlike staining with gold-labeled secondary antibodies. This is probably due to diffusion of the DAB-reaction product.[19,20]

3.4. QUANTIFICATION OF ASTROCYTIC GAP JUNCTIONS IN CORTICAL TISSUE

For an unbiased estimation of gap junctions we studied electron micrographs using the disector method.[31-34] We used tissue of occipital cortex prepared according to standard methods for EM analysis to yield optimal quality in tissue preservation. According to the terminology of the disector method the first of three consecutive sections is termed the "reference section" (RS), the second the "intermediate section" (IS) and the last the "look-up section" (LS). We made two series of photographs including all laminae of visual cortex. Gap junctions were scored positively, if found on RS but not on LS. The disector height (DH) is the sum of the section thicknesses of RS and IS, each determined with the fold technique.[35] The ratio between scored gap junctions and the total number of identified gap junctions on the RS has to be corrected for section thicknesses by dividing the quotient by DH to yield

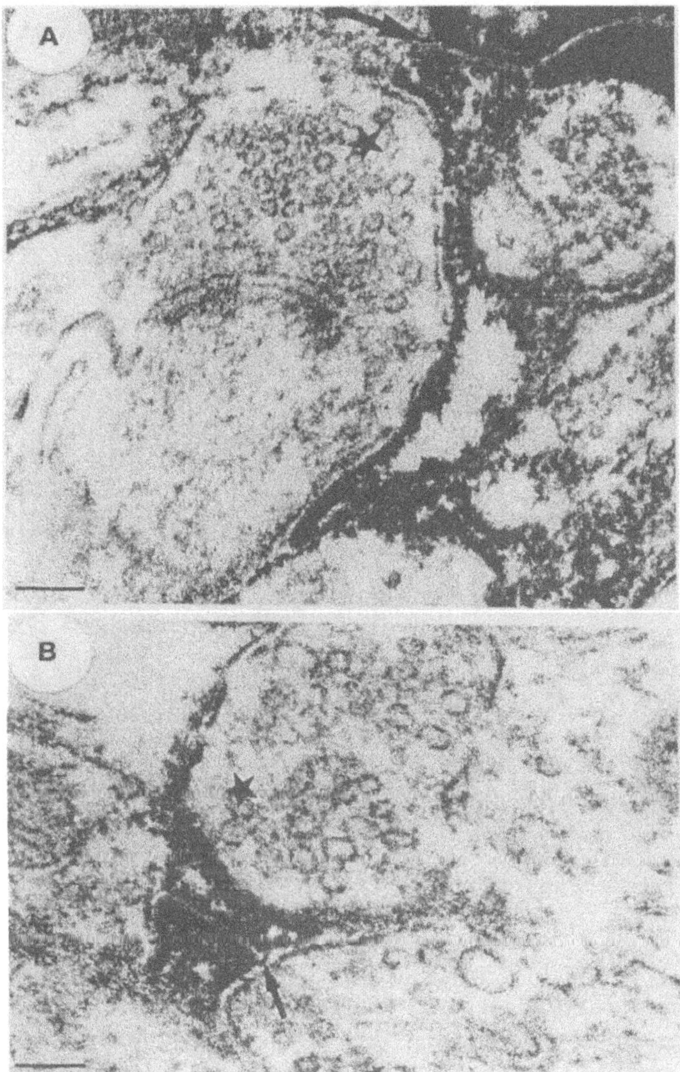

Fig. 10.6. (A,B) Connexin43 immunostained gap junctions (arrows) with surrounding cytoplasm located next to presynaptic terminals (asterisks). (Scale bar = 0.08 µm).

the relative density of gap junctions per µm of section thickness. The latter value multiplied by the number of gap junction profiles in the examined section area results in the number of gap junctions per volume tissue space. Since gap junctions are small in orthogonal diameter compared with the thickness of ultrathin sections, we tilted each section in the EM from -20° to +20°. We did not use the maximal tilting angle form -90° to +90°, because extremely tilted sections can hardly be focused. Results were corrected to this maximal angle of 180°.

Figure 10.7 shows the difficulty in recognizing the full length of a spirally bent gap junction, where as orthogonally cut parts are easy to identify, as in the upper part, but the image is more and more diffuse towards the lower part, which is obliquely cut. Tilting of any obliquely cut gap junction exhibits their pentalaminar structure, which is a basic criterion for their proper identification, if its contact

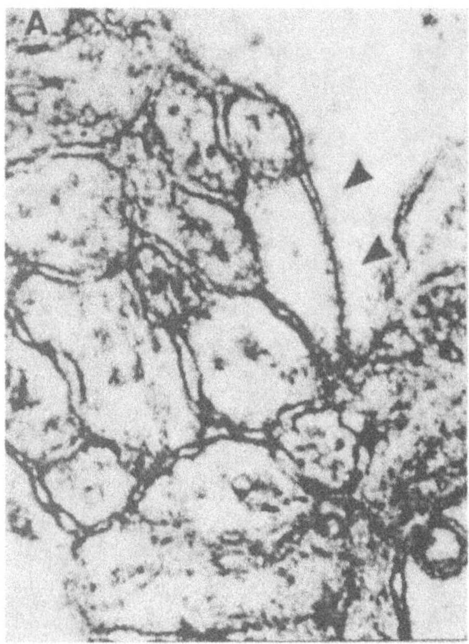

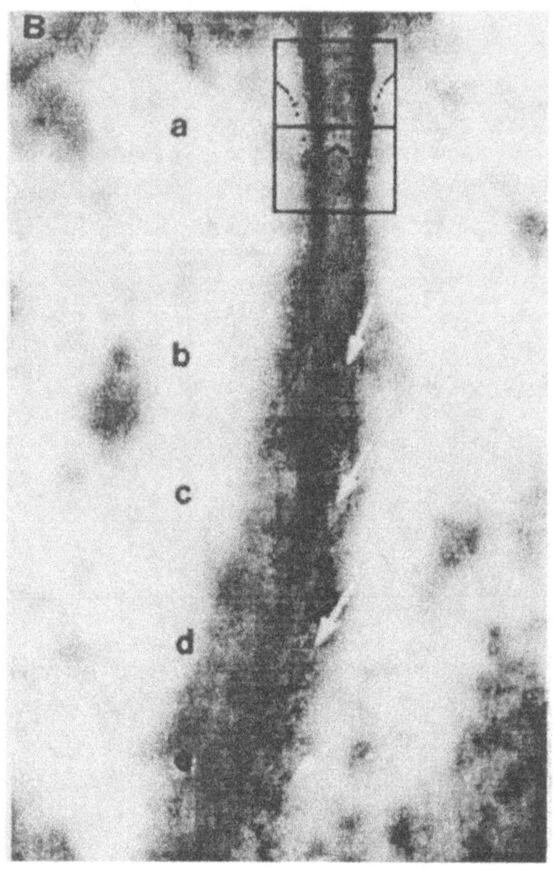

Fig. 10.7. (A) Electron micrograph of a gap junction (arrowheads), the same gap junction is scanned in B, which is orthogonally cut in the upper part (a in B), but bent towards the section plane in the lower part and not identifiable when obliquely cut (b-e in B). (Photograph magnification A: x80,000)

zone is perpendicularly oriented to the section plane.

Counting of gap junctions with the disector method and correcting for undetectable, obliquely cut junctions resulted in a mean value of 890 million gap junctions per mm³. This is a conservative measurement, because it probably does not include very small gap junctions and does not fully comprise strongly bent gap junctions forms. Assuming a regular distribution of gap junctions one can statistically estimate one gap junction per 1-2 µm³ tissue.

Taking into account the numerical density of astrocytes in visual cortex which may vary between 16,400 and 66,900 astrocytes/mm³,[14] a conservative estimation of gap junction numbers reveals that 30,000-60,000 gap junctions on average may occur per astrocyte. This number is a measure for the huge cellular coupling capacity that characterizes the astrocytic network.

Taken together with the average area of a gap junction = 0.03 µm² (cortical gap junction profile diameter about 0.17 µm) a mean coupling area of 900-1,800 µm² of gap junctions exists per astrocyte. This value corresponds to a collective coupling area of about 15-55 x 10⁶ µm² gap junctions per mm³ of cortical tissue.

4. PLASTICITY OF THE ASTROCYTIC COUPLING

To analyze astrocytic coupling plasticity we used the in vivo model of facial nerve transection.[36,37] Astrocytes are known to react rapidly to peripheral facial nerve axotomy with an early induction of glial fibrillary acidic protein (GFAP) and S100 protein in the facial nucleus and facial motor cortex, respectively.[38]

We transected the right facial nerve trunk at its extracranial origin excluding the postauricular nerve in 14 anaesthetized rats. Loss of the right facial motor function was verified by absence of the eyelid reflex. The **facial nucleus** in the brainstem was studied after various survival times starting with 30 min and the latest being 4.5 days. The corresponding motor cortex area was analyzed after 3 hr. Brains of rats were stained against connexin43.[36,37] Non-operated control animals and the contralateral side confirmed the faint to moderate connexin43 immunoreactivity of facial nuclei. As early as 45 min postoperative, connexin43 staining was enhanced around the axotomized facial motoneurons. Differences in the intensity of staining between the right, i.e., the operated side, and the contralateral side were apparently based on differences in the numerical density of connexin43 immunoreactive puncta. These differences persisted at least for 4.5 days. In coronal sections connexin43 staining was not enhanced in the medial part of the right facial nucleus, where those motoneurons are located sending their axons into the postauricular nerve which was left intact. Hence, enhanced astrocytic connexin43 staining was restricted to the immediate environment of lesioned motoneurons. At high LM magnification, enhanced staining was characterized by aggregated puncta forming clusters distributed around axotomized motoneurons.

Facial nerve representation in the **motor cortex** of the left hemisphere (contralateral to nerve transection) was more connexin43 immunopositive compared to the unoperated control side.

Figure 10.8A shows a schematic overview of the left facial motor cortex. Connexin43 staining included the facial nerve representation area and additional staining was found to be elongated towards the occipital cortex (Fig. 10.8B), resulting in a striped pattern. Our data on facial nerve axotomy showed a very fast neural induction of astrocytic gap junction coupling. Peripheral nerve axotomy leaves the blood-brain-barrier intact. Thus, the inducing stimulus must include a retrograde signal transfer to the motoneuron somata inducing a modification of gap junction coupling in neighboring astrocytes. Although the stimulus is still unknown, the short latency of enhanced glial coupling (45 min) excludes signal transfer via rapid axonal transport. The latter would be expected to take at least one hour for the distance of approximately 18 mm from the operation site to the facial nucleus according to known transport velocities.[39]

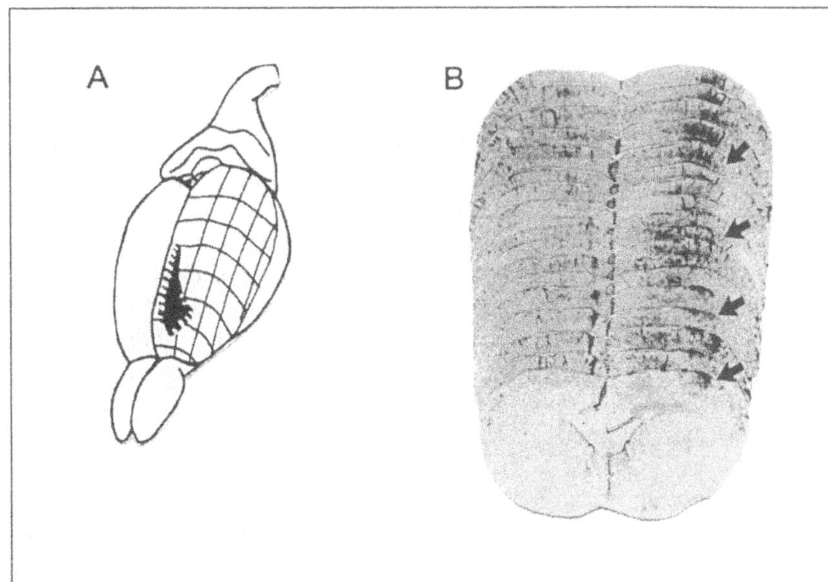

Fig. 10.8. (A) Schematic overview of the left facial nerve representation area in the motor cortex. (B) Connexin43 immunopositive stripe (arrows) in this area in the left hemisphere 3 hr after facial nerve transection. (Magnification B: x3.8)

The rapid reaction of astrocytic gap junction coupling is the earliest response to facial nerve transection known so far. Table 10.2 lists different cell types in the facial nucleus and their reported reactions during the first 24 hr after facial nerve transection. Biochemical and molecular alterations in the lesioned motoneurons could not be detected within less than 5 hr after axotomy (see Table 10.2 for references). Ornithine-decarboxylase, an essential enzyme in polyamine synthesis, is enhanced at this time. Polyamine is involved in DNA replication and RNA synthesis occurring in regenerating motoneurons. Also the immediate early genes *c-Jun*, *jun B* and TIS 11 are induced with a similar latency, and m-RNA expressions are altered for neurofilaments (NF), actin polypeptides and growth associated protein (GAP-43) even later. After 8 hr chromatolysis was reported to commence. In 50% of the neurons cytokines are induced, in particular calcitonin gene-related peptide (CGRP) within about 10 hr. This time course suggests that these rapid neuronal changes cannot induce modifications in astrocytic gap junction coupling and that the proper signal has not been found yet.

Besides these neuronal reactions microglial cells respond to the axotomy of motoneurons. Beta4A-amyloid precursor protein (APP) is synthesized de novo and levels of microglial growth factors are altered, including those of macrophage colony-stimulation factor receptor (MCSF) and C3 receptor type 3 (CR3). These receptors have functions in regulating, e.g., cell proliferation, phagocytosis, hormone interactions and cell adhesion. A cytokine is induced, interleukin 6 (IL-6), which is known to be produced by cultured astrocytes and microglia. Considering that the onset of microglial reactions occurs a couple of hours after astrocytic induction it seems to be unlikely that these cells induce the rapid changes of astrocytic gap junction coupling.

Enhanced connexin43 immunoreactivity is followed by increasing expression of GFAP in astrocytes with a delay of almost one day.

All these reactions of different cell types do not present a uniform response pattern but rather indicate that several different cascades of cellular changes are triggered by axotomy of motoneurons. The very early alteration in astrocytic connexin43 expression may be involved in a cascade that leads to rapid and transient synaptic remodeling on axotomized motoneurons. In addition for the facial nucleus "synaptic stripping" has been reported, in which afferent synaptic terminals detach and displace from the surface of regenerating motoneurons examined after 4 days.[51] Early synaptic reorganization events occur in the facial motor cortex 45 min to 3 hr after facial nerve lesion resulting in an expansion of the representation of neighboring cortical areas according to electrophysiological data.[52] Hence, stronger astrocytic coupling may be induced around reorganizing neurons and may allow more detailed interaction between neurons and specifically formed small astroglial meshes.

5. A NEW MODEL OF GAP JUNCTION COUPLING IN THE ASTROCYTIC NETWORK

Most of the examined membrane contacts were easy to identify as astrocytic gap junctions connecting astrocytic lamellae. Since lamellae are sheet-like protrusions of astrocytic membranes which bear about 80% of their cell surface, this high frequency of gap junctions on lamellae is not unexpected. Considering a more or less even distribution of gap junctions in tissue the question arises whether gap junctions represent intercellular contacts which should preferentially occur in the heterocontrol space of an astrocyte. The autocontrol space is predominantly penetrated by processes and lamellae of one cell which do not interdigitate with neighboring cells. Hence, one would not expect to find intercellular contacts in this compartment. Data presented here suggest that this is not true. Thus, we conclude that gap junctions form intercellular as well as autocellular contacts. This conclusion is based on the following findings: First, LM studies of

Table 10.2. Reaction of various cell types in the facial nucleus after facial nerve transection

Time Course	1.5 Hours	5-8 Hours	5-7 Hours	6 Hours	8 Hours	10 Hours	12-24 Hours	24 Hours
Enhanced cx-43 immunostaining[36,37]	astrocytes							
Ornithine decarboxylase enhanced[40]		mainly *moto-neurons*, (slight increase also in unoperated nucleus)						
c-Jun, jun B, TIS 11-mRNA induced[41]			*motoneurons* (*jun* B in glial cells?) (TIS 11 also in glial cells?)					
Upregulation of APP-mRNA[42]				*microglia*				
Induced (IL) 6-mRNA[43]					? (known to be produced by cultured astrocytes and microglia)			
Chromatolysis[44]					*motoneurons*			
CGRP-mRNA accumulation[45]						*50% of moto-neurons*, (slight increase also in unoperated nucleus)		
Increased GFAP-expression[46]							astrocytes	
Decreased NF and increased actin polypeptides[47]							*motoneurons*	
Cytoskeleton and GAP-43-mRNA increased[48]							*motoneurons*	
Upregulation of CR3 complement receptor[49]								*microglia*
Upregulation of MCSF-receptor[50]								microglia

connexin43 stained cortex did not reveal any zones around astrocytic cell bodies devoid of gap junctions. Second, electron micrographs also show gap junctions on astrocytic cell bodies or autocellular coupling through reflexive gap junctions. Third, single HRP-stained astrocytes at the EM level reveal abundant gap junctions between processes and lamellae of the same cell.

Therefore we propose the following model of gap junction coupling between astrocytes: Gap junction coupling is present in both the **autocontrol space**, where gap junctions couple between processes and lamellae of the same cell, and the **hetero-control space**, which predominantly corresponds to intercellular coupling. One may estimate that at least 25% of all gap junctions belong to autocellular coupling. This value may be underestimated because a considerable amount of autocoupling may also occur between processes and lamellae in the interdigitation territory of each astrocyte.

6. CONCLUSIONS AND OUTLOOK

Quantitative evaluation of the numerical density and distribution of astrocytes as well as connexin43 positive gap junctions in the visual cortex of adult rats revealed a degree of gap junction coupling which is unexpectedly large (30,000-60,000 gap junctions per cell, about 1,000 μm^2 gap junction area per astrocyte, approximately 900 million gap junctions per mm³ of tissue).

The distribution of gap junctions on various compartments of astrocytes and their plasma membrane suggests that gap junctions not only serve intercellular coupling. In addition, a considerable fraction of these junctions couple processes and lamellar extensions of the same cell. Thus, gap junction coupling cannot be regarded as establishing exclusively a system of connected astrocytes.

Rather, gap junctions establish an astrocytic network with meshes of subcellular size. These meshes apparently either surround individual synapses and small groups of synapses or separate synaptic units from each other.

Experimental evidence accumulates suggesting that after neuron injury transsynaptic and even transneuronal signals of unknown nature can rapidly (within less than one hour) induce expression of connexin43 and possibly the formation of aggregated gap junctions around specific sets of synapses on neurons with modified functions.

New types of experiments are required to clarify the nature of processes which underly rapid modifications in neuropil compartments remote from sites of neural lesion (and possibly other functional changes in neurons) that lead to changes in connexin43 immunoreactivity or number of astrocytic gap junctions as described here. At present, the cellular and molecular cascades on which these changes are based remain enigmatic.

Future research should be directed towards a better understanding of the physiological role of the subcellular meshwork built by astrocytic processes and lamellae connected by gap junctions and the related compartmentalization of neuropil detected in various brain regions.

ACKNOWLEDGMENTS

During the time of this work A. Rohlmann was supported by the Graduate College "Dynamics and organization of neuronal networks" (Goettingen, Germany). Connexin43 antibodies were provided by Prof. R. Dermietzel. Facial nerve transections were done by Dr. R. Laskawi.

REFERENCES

1. Dermietzel R, Hertzberg EL, Kessler JA et al. Gap junctions between cultured astrocytes: immunocytochemical, molecular, and electrophysiological analysis. J Neurosci 1991; 11:1421-32.
2. Wolff JR. Quantitative aspects of astroglia. In: Masson und Cie eds., Proc VIth Intern Congr Neuropath Paris, 1970:327-52.
3. Virchow R. Über das granulilrte Aussehen der Wandungen der Gehirnventrikel. Allg Z Psychiatr Psych-Gerichtl Med 1846; 3:424-50.

4. Barres BA. New roles for glia. J Neurosci 1991; 11: 3685-94.

5. Smith SJ. Do astrocytes process neural information? Prog Brain Res 1992; 94:119-36.

6. Güldner F-H, Wolff JR. Perisynaptic reactions of astroglia in the visual cortex after optic nerve stimulation. Exp Brain Res Suppl I 1977:343-47.

7. Wolff JR, Güldner F-H. Perisynaptic astroglial reactions to neuronal activity. In: E. Schoffeniels et al, eds. Dynamic properties of glia cells. Oxford and New York: Pergamon Press, 1978:115-18.

8. Kimelberg HK, Sankar P, O'Connor ER et al. Functional consequences of astrocytic swelling. Prog Brain Res 1992; 94:57-68.

9. Jensen AM, Chiu S-Y. Astrocyte networks. In: S. Murphy, ed. Astrocytes, pharmacology and function. San Diego, New York, Boston, London, Syndney, Tokyo, Toronto: Academic Press, 1993:309-30.

10. Lassmann G. Einige Bemerkungen über die dem peripherischen Nervensystem zugeordneten Dendritenzellen. Wien Klin Wochenschr 1966; 7:293-95.

11. Adams JC. Heavy metal intensification of DAB-based HRP reaction product. J Histochem Cytochem 1981; 29:775.

12. Connor JR, Berkowitz EM. A demonstration of glial filament distribution in astrocytes isolated from rat cerebral cortex. Neuroscience 1985; 16:33-44.

13. Floderus S. Untersuchungen ueber den Bau der menschlichen Hypophyse mit besonderer Beruecksichtigung der quantitativen mikromorphologischen Verhaeltnisse. Acta Pathol Microbiol Second Suppl 1944; 53:1-26.

14. Gabbott PLA, Stewart MG. Distribution of neurons and glia in the visual cortex (area 17) of the adult albino rat: a quantitative description. Neuroscience 1987; 21:833-45.

15. Glagolev AA. On the geometrical methods of quantitative mineralogic analysis of rocks. Trans Inst Econ Min Moscow 1933; 59:1.

16. Thomson E. Quantitative microscopic analysis. J Geol 1930; 38:193.

17. Batter DK, Corpina RA, Roy C et al. Heterogeneity in gap junction expression in astrocytes cultured from different brain regions. Glia 1992; 6:213-21.

18. Yamamoto T, Vukelic J, Hertzberg EL et al. Differential anatomical and cellular patterns of connexin43 expression during postnatal development of rat brain. Dev Brain Res 1992; 66:165-80.

19. Yamamoto T, Ochalski A, Hertzberg EL et al. On the organization of astrocytic gap junctions in rat brain as suggested by LM and EM immunohistochemistry of connexin43 expression. J Comp Neurol 1990; 302:853-83.

20. Nagy JI, Yamamoto T, Sawchuk MA et al. Quantitative immunohistochemical and biochemical correlates of connexin43 localization in rat brain. Glia 1992; 5:1-9.

21. Brightman MW, Reese TS. Junctions between intimately apposed cell membranes in the vertebrate brain. J Cell Biol 1969; 40:648-77.

22. Sosinsky GE, Jesior JC, Caspar DLD et al. Gap junction structures VIII. Membrane cross-sections. Biophys J 1988; 53:709-22.

23. Larsen WJ. Biological implications of gap junction structure, distribution and composition: a review. Tiss Cell 1983; 15:645-71.

24. Dermietzel R, Hwang TK, Spray DS. The gap junction family: structure, function and chemistry. Anat Embryol 1990; 182: 517-28.

25. Unwin PNT, Zampighi G. Structure of the junction between communicating cells. Nature 1980; 283:545-48.

26. Hoh JH, Lal R, John, SA et al. Atomic force microscopy and dissection of gap junctions. Science 1991; 253:1405-08.

27. Yamamoto T, Ochalski A, Hertzberg EL et al. LM and EM immunolocalization of the gap junctional protein connexin43 in rat brain. Brain Res 1990; 508:313-19.

28. Wolburg H, Rohlmann A. Structure-function relationships in gap junctions. Int Rev Cytol 1995; 157:315-56.

29. Iwayama T. Nexuses between areas of the surface membrane of the same arterial smooth muscle cell. J Cell Biol 1971; 49:521.

30. Herr JC. Reflexive gap junctions: gap junctions arising from the same ovarian decidual cell. J Cell Biol 1976; 69:495.

31. Sterio DC. The unbiased estimation of number and sizes of arbitrary particles using the disector. J Microsc 1984; 134:127-36.

32. Gundersen HJG, Jensen EB. The efficiency of systematic sampling in stereology and its prediction. J Microsc 1987; 147:229-63.

33. Siklós L, Párducz A, Halász N et al. An unbiased estimation of the total number of synapses in the superior cervical ganglion of adult rats established by the disector method. Lack of change after long-lasting sodium bromide administration. J Neurocytol 1990; 19:443-54.

34. Missler M, Wolff A, Merker HJ et al. Pre- and postnatal development of the primary visual cortex of the common marmoset. II. Formation, remodelling, and elimination of synapses as overlapping processes. J Comp Neurol 1993; 333:53-67.

35. deGroot DMG. Disk-like and complex-shaped synapses: number, size and dense projections. A critical note. Acta Stereol 1985; 4:147-51.

36. Rohlmann A, Laskawi R, Hofer A et al. Facial nerve lesions lead to increased immunostaining of the astrocytic gap junction protein (connexin43) in the corresponding facial nucleus of rats. Neurosci Lett 1993; 154:206-08.

37. Rohlmann A, Laskawi R, Hofer A et al. Astrocytes as rapid sensors of peripheral axotomy in the facial nucleus of rats. NeuroReport 1994; 5:409-12.

38. Laskawi R, Rickmann M, Böttcher H et al. Glial changes in the motorcortex of the rat following various types of facial nerve lesion. In: Elsner N, Heisenberg M, eds. Gene-Brain-Behaviour. New York; Georg Thieme Verlag Stuttgart, 1993:805.

39. Ochs S. Fast transport of materials in mammalian nerve fibers. Science 1972; 176:252-60.

40. Tetzlaff W, Kreutzberg GW. Ornithine decarboxylase in motoneurons during regeneration. Exp Neurol 1985; 89:679-88.

41. Haas CA, Donath C, Kreutzberg GW. Differential expression of immediate early genes after transection of the facial nerve. Neuroscience 1993; 53:91-99.

42. Banati RB, Gehrmann J, Czech C et al. Early and rapid de novo synthesis of Alzheimer βA4-amyloid precursor protein (APP) in activated microglia. Glia 1993; 9:199-210.

43. Kiefer R, Lindholm D, Kreutzberg GW. Interleukin-6 and transforming growth factor-β1 mRNAs are induced in rat facial nucleus following motoneuron axotomy. Europ J Neurosci 1993; 5:775-81.

44. Neiss WF, Schulte E, Guntinas-Lichius O et al. Quantification of chromatolysis in motoneurons of the Wistar rat. Ann Annat Suppl 1993; 175:6.

45. Haas CA, Streit WJ, Kreutzberg GW. Rat facial motoneurons express increased levels of calcitonin gene-related peptide mRNA in response to axotomy. J Neurosci Res 1990; 27:270-75.

46. Tetzlaff W, Graeber MB, Bisby MA et al. Increased glial fibrillary acidic protein synthesis in astrocytes during retrograde reaction of rat facial nucleus. Glia 1988; 1:90-95.

47. Tetzlaff W, Bisby MA, Kreutzberg W. Changes in cytoskeletal proteins in the rat facial nucleus following axotomy. J Neurosci 1988; 8:3181-89.

48. Tetzlaff W, Alexander SW, Miller FD et al. Response of facial and rubrospinal neurons to axotomy: changes in mRNA expression for cytoskeletal proteins and GAP-43. J Neurosci 1991; 11:2528-44.

49. Graeber MB, Streit WJ, Kreutzberg GW. Axotomy of the rat facial nerve leads to increased CR3 complement receptor expression by activated microglial cells. J Neurosci Res 1988; 21:18-24.

50. Raivich G, Kreutzberg GW. Peripheral nerve regeneration: role of growth factors and their receptors. Int J Devl Neurosci 1993; 11:311-24.

51. Blinzinger K, Kreutzberg G. Displacement of synaptic terminals from regenerating motoneurons by microglial cells. Z Zellforsch 1968; 85:145-57.

52. Sanes JN, Suner S, Lando JF et al. Rapid reorganization of adult rat motor cortex somatic representation patterns after motor nerve injury. Proc Natl Acad Sci USA 1988; 85:2003-07.

CHAPTER 11

EFFECT OF GAP JUNCTIONAL COMMUNICATION ON GLIOMA CELL FUNCTION

Christian C.G. Naus, John F. Becherger and Shari L. Bond

1. INTRODUCTION

The gap junction is the site of the intercellular membrane channels which provide for direct cytoplasmic continuity between adjacent cells.[1] The structural unit of the gap junction is the *connexon*, a proteinaceous cylinder with a hydrophilic channel; connexons spanning the plasma membranes of closely-apposed cells align end-to-end, forming intercellular channels. Gap junction channels provide for the exchange of small molecules (less than 1000 Daltons) between cells to allow metabolic cooperation;[2] they also transmit developmental signals involved in cell patterning.[3] Among the metabolites which pass through gap junction channels are second messengers such as inositol 1,4,5-trisphosphate and Ca^{2+}, agents involved in cellular regulation.[4-9] Due to the recent isolation of gap junction proteins, and the cloning of their mRNAs, it is now obvious that these proteins, the *connexins*, are encoded by a multigene family.[10] Each connexin exhibits a characteristic tissue distribution in the adult animal, implying functional differentiation among the different types of channel.

2. GAP JUNCTIONS IN ASTROCYTES AND GLIOMA CELLS

Gap junctions were previously considered prevalent in invertebrate and lower vertebrate nervous systems, presumably functioning as a "primitive" form of interneuronal communication via electrotonic synapses. However, recent evidence has shown this to be a prevalent phenomenon in the mammalian CNS (see other chapters in this book). In addition to

Gap Junctions in the Nervous System, edited by David C. Spray and Rolf Dermietzel.
© 1996 R.G. Landes Company.

intercellular coupling between neurons, morphological data support the existence of gap junctions between glial cells, including astrocytes, oligodendrocytes and ependymal cells.[11,12] These latter nonneuronal junctions presumably play a role in ion transfer, such as astrocytic spatial buffering of potassium ions generated through neuronal activity.[13] Recently, several studies have characterized a number of aspects of gap junctional expression and function in astrocytes[5,8,9,14-30] (also see chapters 8-10, 11). It thus appears that gap junctional communication is a fundamental property characteristic of astrocytes.

In contrast, glioma cells show deficits in gap junctional communication compared to that seen in astrocytes.[31-39] Indeed, this deficiency in intercellular communication via gap junctions has been implicated in the disturbance of the regulation of cell proliferation and subsequent differentiation. There is good evidence that an interruption of this communication pathway is one of the steps in malignant transformation.[40-42] A variety of chemical agents which transform cells in vitro cause a reduction of intercellular coupling. Recently, it was demonstrated that Cx43 is a direct target for phosphorylation by the product of the v-*src* oncogene, the result of which is the blockage of intercellular coupling.[43] Furthermore, in a report describing subtractive hybridization to identify tumor-suppressor genes, the gene encoding Cx26 was selected.[44] More recently, mutation of the Cx37 gene has been suggested to contribute to the malignancy of lung carcinomas.[45] Although the majority of studies support the hypothesis that a loss of gap junctional intercellular communication is associated with tumor phenotype, there are some exceptions.[46,47] In such cases it may be that a growth-regulatory step downstream of intercellular coupling has been perturbed.

We have used the C6 glioma cell line as a model system to investigate the role of gap junctions in the control of cell growth and differentiation. This cell line was derived from a rat glial tumor induced by n-nitrosomethylurea.[48] C6 glioma cells grow very rapidly in culture and are capable of forming large tumors in nude mice and aggressive neoplastic gliomas in the brains of rats.[49,50] The expression of Cx43 in C6 glioma cells has been shown to be relatively low in comparison to primary cortical astrocyte cells in culture[51] (Fig. 11.1A,B). Primary astrocytes in cul-

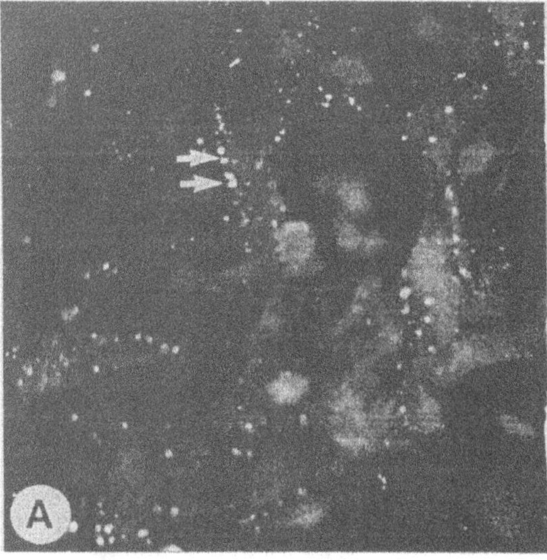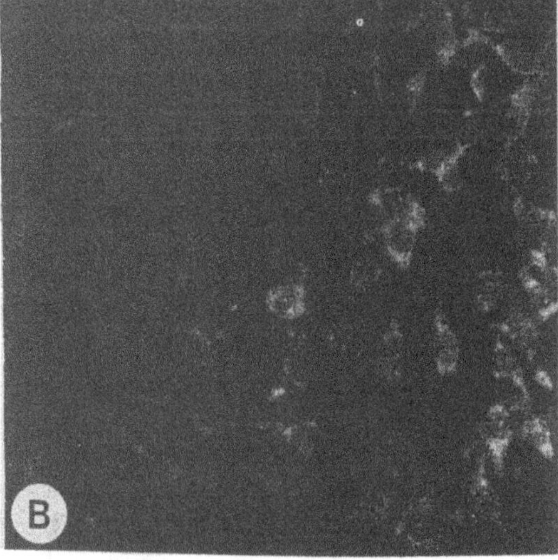

Fig. 11.1. Immunocytochemical labeling of primary astrocytes (A) and C6 glioma cells (B) with connexin43 antibody. While distinct punctate labeling is apparent in astrocytes, there is little labeling present in glioma cells. Mag. = 180X.

ture show extensive coupling when monitored by intracellular dye injection, whereas such coupling of C6 cells is very low.[51] In addition, while intercellular calcium waves readily pass through cultures of astrocytes,[5-9] such waves are severely limited in C6 glioma cells.[6] Several studies have provided evidence supporting a role for gap junctional communication in this aspect of calcium signalling.[5-9]

In order to determine if a similar deficit in gap junctional communication exists in human gliomas, we have examined connexin expression and intercellular coupling in a number of human glioma cell lines. There is some variability in the pattern of immunocytochemical localization (Fig. 11.2A-F). However, all lines show reduced staining relative to primary astrocytes. While some cell lines show more immunoreactivity than others, this appears to be confined primarily to the cytoplasm, with very little staining at areas of cell to cell contact. Furthermore, all cell lines examined showed limited coupling when evaluated for intercellular passage of the low molecular weight dye carboxyfluorescein. Western blot analysis confirms the presence of Cx43 protein in these cell lines (Fig. 11.2G). However, all cell lines show very little of the slower migrating, presumably phosphorylated, Cx43 protein which is prevalent in primary astrocytes. We have also noted that Cx43 expression is reduced in surgically resected human gliomas (unpublished observations). These data suggest that gap junctional communication is also compromised in human gliomas.

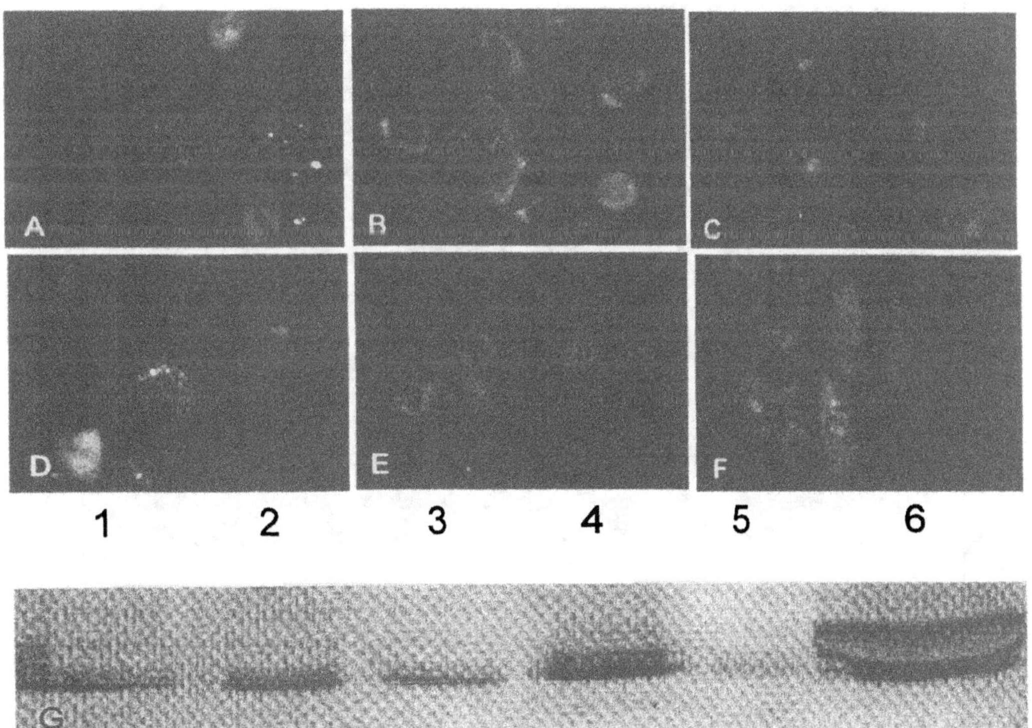

Fig. 11.2. Immunocytochemistry for connexin43 in several human glioma lines (A-F) reveals limited staining at areas of cell-cell contact. However, some cells contain variable cytoplasmic staining. The following cell lines are shown: A, SF126; B, SSF539; C, U251MG; D, SF188; E, U87MG; F, SF188. Western blotting for connexin43 in the same cell lines is shown in G. While primary astrocytes show several distinct connexin43 species, the glioma lines display only the faster migrating species. Lanes were loaded as follows: 1, SF188; 2, U251MG; 3, U87MG; 4, SF539; 5, C6 glioma cells; 6, primary astrocytes. A-F Mag. = 225X.

3. TRANSFECTION OF GLIOMA CELLS

One approach to specifically study the function of a protein is to induce its expression in a system where it is absent or very low. To isolate the effect of changing only cell coupling, specific methods for altering gap junction gene expression would be useful. Our laboratory,[49,50,52,53] as well as others,[54-57] have utilized transfection of gap junction cDNAs into communication-deficient cell lines to increase intercellular communication via gap junctions.

By transfecting C6 glioma cells with a plasmid construct constitutively expressing Cx43, it was possible to increase the expression of this gap junction gene, resulting in increased Cx43 mRNA and protein (Fig. 11.3A,B). With immunocytochemistry, protein could be localized at areas of cell-to-cell contact as well as intracellularly (Fig. 11.3C). The increased level of intercellular coupling resulted in dramatic effects on cell growth and differentiation in culture. In the course of these experiments various clones were iso-

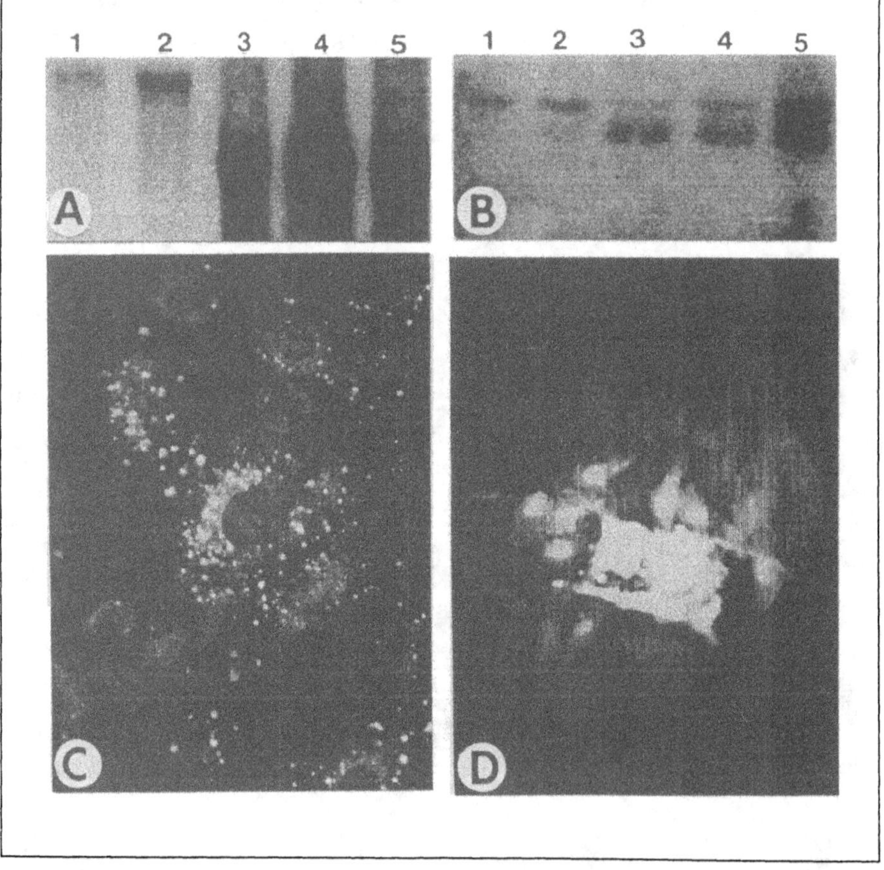

Fig. 11.3. (A) Northern blot analysis of RNA isolated from nontransfected C6 cells (lane 1), transfected control (lane 2) and three different connexin43-transfected C6 clones (lane 3, clone 12; lane 4, clone 13; lane 5, clone 14). The blot was hybridized with a radiolabeled connexin43 cDNA probe. The upper band is the endogenous 3 kb connexin43 mRNA, while the lower band is the mRNA transcribed from the transfected plasmid. (B) Western blot analysis for connexin43 in the same cell lines depicted in panel A (lane 1, nontransfected C6; lane 2, transfected control; lane 3, clone 12; lane 4, clone 14; lane 5, clone 13). (C) Connexin43-transfected clone 13 displays abundant connexin43 immunolocalization both at areas of cell-cell contact and intracellularly. Mag. = 405X. (D) Following intracellular injection of carboxyfluorescein into a single transfected (clone 13) cell, the low molecular weight dye quickly passes to many neighboring cells. Mag. = 180X.

lated which could be grouped into low, moderate, or high expressers of Cx43. The level of expression of this gene correlated well with the level of intercellular coupling (Fig. 11.3D) and the decreased rate of growth of the clones in culture. Expression of Cx43 also resulted in morphological changes consistent with the loss of tumor phenotype, as evidenced by the appearance of actin stress fibers in transfected cells[58] and the ability of those cells to display intercellular calcium signaling typical of astrocytes.[6]

In order to extend our characterization of the effect of increased Cx43 expression on cell growth, we used an in vivo rat glioma model.[49] We implanted these cells into the rat cerebrum and monitored tumor growth over two weeks. We observed a similar reduction in growth of tumors derived from glioma cells transfected with Cx43. Immunocytochemical staining for Cx43 confirmed that the transfected cells maintained their expression of this gap junction cDNA up to two weeks in vivo (Fig. 11.4). Nontransfected C6 glioma cells gave rise to large tumors in which very little Cx43 protein was detectable (Fig. 11.4A). The host/tumor interface was readily discerned due to the lack of Cx43 immunostaining in the C6 glioma cells. In contrast, C6 cells transfected with Cx43 cDNA grew more slowly, with Cx43 immunoreactivity being readily detected throughout the smaller tumor (Fig. 11.4B). The host/tumor interface could no longer be readily distinguished based on Cx43 immunoreactivity.

The mechanism by which increased gap junctional coupling leads to decreased cell growth remains to be clarified. In these C6 glioma cells transfected with Cx43, we have observed that conditioned medium from these cells has growth inhibitory activity.[59] This may be unique to glioma cells transfected with Cx43 since similar effects were not observed with Cx32 transfected glioma cells (unpublished observations) or other cell lines transfected with various connexins (personal communications, H. Yamasaki, R. Ruch, J. Trosko, G. Goldberg).

Another approach to address the role of gap junctions in cellular function is to downregulate expression of gap junction genes or interfere with gap junctional communication. Attempts have been made to selectively uncouple cells via intracellular injections of specific gap junction antibodies. Antibodies raised against gap junctions isolated from liver have been reported to reduce dye-coupling between astrocytes,[60] but the specificity of this effect is unclear. Such studies are technically demanding and

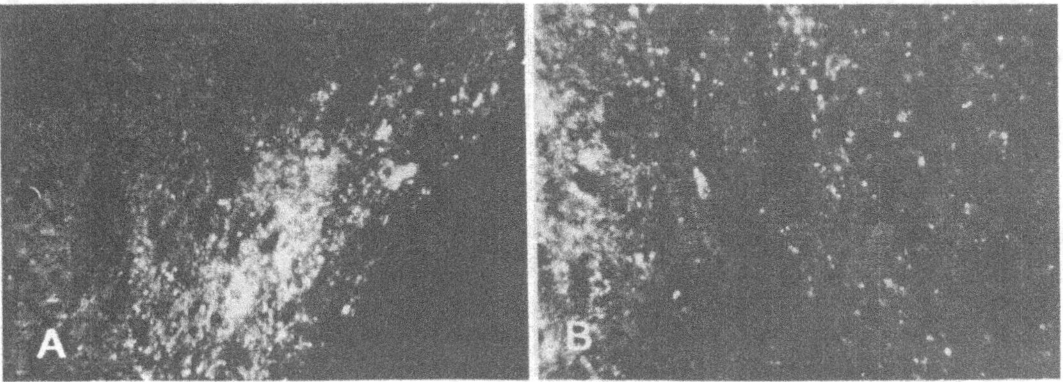

Fig. 11.4. (A) Following two weeks of intracerebral growth, immunocytochemical staining for connexin43 clearly demarcates the border between the host brain (connexin43 positive) and a tumor formed by nontransfected C6 cells (connexin43 negative). (B) In contrast, the smaller tumor formed by connexin43-transfected C6 cells displays connexin43 immunoreactivity, making the host brain (left) difficult to distinguish from the tumor (right). Mag. − 210X.

difficult to interpret since damaged cells will uncouple. The use of antibodies directed against,[61] or synthetic peptides homologous to,[62] extracellular epitopes of gap junction proteins has resulted in successful blocking of these channels in some cells in culture, and antisense strategies have also been used to interfere with Cx43 expression,[63,64] but such studies have not been conducted with astrocytes.

One of the most striking differences between C6 glioma cells and primary astrocytes is the distinct deficiency in gap junctional intercellular communication in the transformed cells. We now have available a source of primary astrocytes from

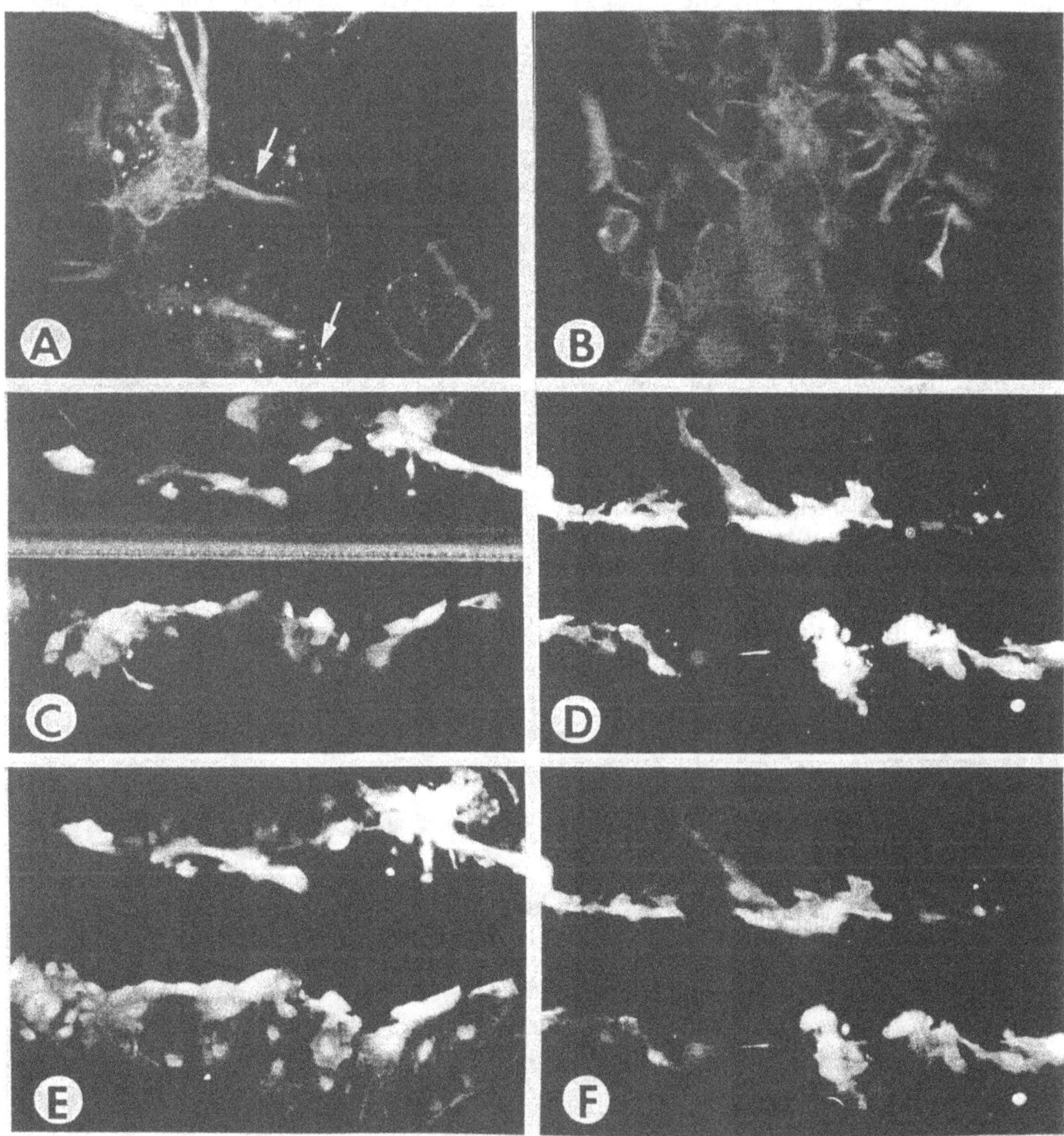

Fig. 11.5. Cultured astrocytes from fetal brains of wild type (A,C,E) and Cx43-null mutant (B,D,F) mice. Cells were double labeled with a monoclonal antibody for GFAP and a polyclonal antibody for Cx43. The figures were obtained following double exposure photomicrography. While both express GFAP (filamentous staining in A,B), only the wild type show punctate immunoreactivity for Cx43 (arrows in A). With the scrape loading technique, both cultures show equivalent labeling with rhodamine-dextran (MW - 10,000) at the cut surface (C,D); only the wild type astrocytes show spread of carboxyfluorescein (MW - 376) through gap junctions to cells away from the cut (E). A,B Mag. = 345X. C - F Mag. = 64X.

transgenic mice with a null mutation of the Cx43 gene.[65] In preliminary experiments, separate cultures of primary astrocytes were prepared from late gestation mouse fetal brains using established methods.[66] After several days, some cultures were prepared for immunocytochemistry and dye coupling assays as previously described.[30] The remaining cultures were set up for growth analysis. While primary astrocytes from wild type fetal brain exhibited clear Cx43 immunoreactivity and cell-to-cell passage of low molecular weight dye, similar cultures from homozygous null fetal mice showed no Cx43 immunoreactivity and negligible dye passage (Fig. 11.5). Expression of the astrocytic marker glial fibrillary acidic protein (GFAP) appeared unaffected. Analysis of growth over several days revealed a difference between the wild type and homozygous null astrocytes, the latter showing reduced growth.

4. SUMMARY

It is clear that gap junctional intercellular communication plays an important role in a number of astrocytic functions. It is also clear that this communication pathway is compromised in glioma cells. The abnormal growth characteristics of glioma cells can be partially reversed by transfection with Cx43 cDNA. However, the loss of gap junctional coupling, in and of itself, does not appear to have immediate effects on cell growth. Nevertheless, several aspects of astrocytic function will be disturbed in Cx43-deficient astrocytes.

Recently, major initiatives have begun into the evaluation of viral gene therapy for gliomas. Several laboratories have used retroviral vectors to transmit the herpes simplex virus thymidine kinase (HSV-tk) gene into glioma cells.[39,67] This enzyme catalyzes the rate-limiting reaction in the phosphorylation of deoxynucleosides, which subsequently become incorporated not only into the viral genome, but also into the replicating cellular genome. HSV-tk also phosphorylates deoxynucleoside analogs such as ganciclovir, resulting in disruption of DNA synthesis in replicating cells, and ultimately cell death. In some cases, gap junctional transfer of cytotoxic ganciclovir metabolites to surrounding cells has been implicated in the mechanism of tumor regression.[68] This so-called "bystander effect" provides a mechanism for extending cytotoxicity from cells expressing tk to non-tk expressing cells. While this provides an augmentation to the killing effect on tumor cells, it may also cause undesirable effects in endogenous cells if the tk-expressing cells form gap junctions with replicating glial or endothelial cells. These issues remain to be explored.

ACKNOWLEDGMENTS

The authors are grateful for collaborative interactions with S. Caveney, G. Kidder and D. Zhu, for technical assistance provided by L. Bechberger, N. Khoo and D. Belliveau, to A. Lau for Cx43 antibodies, to E. Beyer for Cx43 cDNA and to J. Rutka for human glioma cell lines. The research summarized in this report was made possible by research grants from the Medical Research Council of Canada and the Cancer Research Society Inc. S.L. Bond is a recipient of a Health Research Personnel Development Fellowship from the Ontario Ministry of Health.

REFERENCES

1. Beyer EC. Gap junctions. Int Rev Cytol 1993; 137C:1-37.
2. Gilula NB, Reeves OR, Steinbach A. Metabolic coupling, ionic coupling, and cell contacts. Nature 1972; 235:262-265.
3. Fraser SE, Green CR, Bode HR et al. Selective disruption of gap junctional communication interferes with a patterning process in hydra. Science 1987; 237:49-55.
4. Saez JC, Connor JA, Spray DC et al. Hepatocyte gap junctions are permeable to the second messenger, inositol 1,4,5-trisphosphate, and to calcium ions. Proc Natl Acad Sci USA 1989; 86:2708-2712.
5. Charles AC, Merrill JE, Dirksen ER et al. Intercellular signaling in glial cells: Calcium waves and oscillations in response to mechanical stimulation and glutamate.

Neuron 1991; 6:983-992.

6. Charles AC, Naus CCG, Zhu D et al. Intercellular calcium signaling via gap junctions in glioma cells. J Cell Biol 1992; 118: 195-201.

7. Finkbeiner S. Calcium waves in astrocytes-filling in the gaps. Neuron 1992; 8: 1101-1108.

8. Dani JW, Chernjavsky A, Smith SJ. Neuronal activity triggers calcium waves in hippocampal astrocyte networks. Neuron 1992; 8:429-440.

9. Cornell-Bell AH, Finkbeiner SM, Cooper MS et al. Glutamate induces calcium waves in cultured astrocytes: Long-range glial signalling. Science 1990; 247:470-473.

10. Kumar NM, Gilula NB. Molecular biology and genetics of gap junction channels. Seminars in Cell Biology 1992; 3:3-16.

11. Brightman MW, Reese TS. Junctions between intimately apposed cell membranes in the vertebrate brain. J Cell Biol 1969; 40:648-677.

12. Mugnaini E. Cell junctions of astrocytes, ependyma, and related cells in the mammalian central nervous system, with emphasis on the hypothesis of a generalized functional syncytium of supporting cells. In: Federoff S, Vernadakis A, eds. Development, Morphology, and Regional Specialization of Astrocytes, Vol. I. Orlando, Florida: Academic Press, 1986:329-371.

13. Walz W, Hertz L. Functional interactions between neurons and astrocytes. II. Potassium homeostasis at the cellular level. Prog Neurobiol 1983; 20:133-183.

14. Giaume C, Marin P, Cordier J et al. Adrenergic regulation of intercellular communications between cultured striatal astrocytes from the mouse. Proc Natl Acad Sci USA 1991; 88:5577-5581.

15. Giaume C, Fromaget C, el Aoumari A et al. Gap junctions in cultured astrocytes:single-channel currents and characterization of channel-forming protein. Neuron 1991; 6:133-143.

16. Giaume C, Cordier J, Glowinski J. Endothelins inhibit junctional permeability in cultured mouse astrocytes. Eur J Neurosci 1992; 4:877-881.

17. Mantz J, Cordier J, Giaume C. Effects of general anesthetics on intercellular communications mediated by gap junctions between astrocytes in primary culture. Anesthesiology 1993; 78:892-901.

18. Giaume C. Noradrenergic control of gap junction permeability in cultured striatal astrocytes. In: Noradrenergic Mechanisms in Parkinson's Disease. CRC Press, 1994; 205-224.

19. Yamamoto T, Ochalski A, Hertzberg EL et al. On the organization of astrocytic gap junctions in rat brain as suggested by LM and EM immunohistochemistry of connexin43 expression. J Comp Neurol 1990; 302:853-883.

20. Yamamoto T, Vukelic J, Hertzberg EL et al. Differential anatomical and cellular patterns of connexin43 expression during postnatal development of rat brain. Brain Res Dev Brain Res 1992; 66:165-180.

21. Vukelic JI, Yamamoto T, Hertzberg EL et al. Depletion of connexin43-immunoreactivity in astrocytes after kainic acid-induced lesions in rat brain. Neurosci Lett 1991; 130:120-124.

22. Nagy JI, Yamamoto T, Sawchuk MA et al. Quantitative immunohistochemical and biochemical correlates of connexin43 localization in rat brain. Glia 1992; 5:1-9.

23. Charles AC. Glia-neuron intercellular calcium signaling. Dev Neurosci 1994; 16:196-206.

24. Charles AC, Dirksen ER, Merrill JE et al. Mechanisms of intercellular calcium signaling in glial cells studied with dantrolene and thapsigargin. Glia 1993; 7:134-145.

25. Fischer G, Kettenmann H. Cultured astrocytes form a syncytium after maturation. Exp Cell Res 1985; 159:273-279.

26. Dermietzel R, Hertberg EL, Kessler JA et al. Gap junctions between cultured astrocytes: immunocytochemical, molecular, and electrophysiological analysis. J Neurosci 1991; 11:1421-1432.

27. Batter DK, Corpina RA, Roy C et al. Heterogeneity in gap junction expression in astrocytes cultured from different brain regions. Glia 1992; 6:213-221.

28. Kettenmann H, Ransom BR. Electrical coupling between astrocytes and between oligodendrocytes studied in mammalian cell

cultures. Glia 1988; 1:64-73.

29. Lee SH, Kim WT, Cornell-Bell AH et al. Astrocytes exhibit regional specificity in gap-junction coupling. Glia 1994; 11:315-325.

30. Belliveau DJ, Naus CCG. Cortical type 2 astrocytes are not dye coupled nor do they express the major gap junction genes found in the central nervous system. Glia 1994; 12:24-34.

31. Lin ZX, Chang CC, Trosko JE. Gap junctional intercellular communication (GJIC) in rat glioma cells—characterizations to detect inhibitors of metabolic cooperation. Sci China 1990; 33:435-443.

32. Knedlitschek G, Anderer U, Weibezahn KF et al. Radioresistance of rat glioma cell lines cultured as multicellular spheroids. Correlation with electrical cell-to-cell-coupling. Strahlenther Onkol 1990; 166:164-167.

33. Tiffany Castiglioni E, Neck KF, Caceci T. Glial culture on artificial capillaries: electron microscopic comparisons of C6 rat glioma cells and rat astroglia. J Neurosci Res 1986; 16:387-396.

34. Aslanidi KB, Boitsova LJ, Chailakhyan LM et al. Energetic cooperation via ion-permeable junctions in mixed animal cell cultures. FEBS Lett 1991; 283:295-297.

35. Ciment G, de Vellis J. Cell surface-mediated cellular interactions: effects of B104 neuroblastoma surface determinants on C6 glioma cellular properties. J Neurosci Res 1982; 7:371-386.

36. Hülser DF, Lauterwasser U. Membrane potential oscillations in homokaryons. An endogenous signal for detecting intercellular communication. Exp Cell Res 1982; 139:63-70.

37. Tani E, Higashi N. Freeze-fracture study of human brain tumors. Childs Brain 1975; 1:46-71.

38. Tani E, Nishiura M, Higashi N. Freeze-fracture studies of gap junctions of normal and neoplastic astrocytes. Acta Neuropathol (Berl) 1973; 26:127-138.

39. Martuza RL, Malick A, Markert JM et al. Experimental therapy of human glioma by means of genetically engineered virus mutant. Science 1991; 252:854-855.

40. Ruch RJ. The role of gap junctional intercellular communication in neoplasia. Ann Clin Lab Sci 1994; 24:216-231.

41. Holder JW, Elmore E, Barrett JC. Gap junction function and cancer. Cancer Res 1993; 53:3475-3485.

42. Yamasuki H, Naus CCG. Role of connexin genes in growth control. Carcinogenesis 1996; 17:1199-12

43. Swenson KI, Piwnica-Worms H, McNamee H et al. Tyrosine phosphorylation of the gap junction protein connexin43 is required for the pp60$^{v\text{-}src}$-induced inhibition of communication. Cell Regul 1990; 1:989-1002.

44. Lee SW, Tomasetto C, Sager R. Positive selection of candidate tumor-suppressor genes by subtractive hybridization. Proc Natl Acad Sci USA 1991; 88:2825-2829.

45. Mandelboim O, Berke G, Fridkin M et al. CTL induction by a tumor-associated antigen octapeptide derived from a murine lung carcinoma. Nature 1994; 369:67-71.

46. Yamasaki H. Gap junctional intercellular communication and carcinogenesis. Carcinogenesis 1990; 11:1051-1058.

47. Yamasaki H. Changes of gap junctional intercellular communication during multistage carcinogenesis. Prog Clin Biol Res 1990; 340D:153-164.

48. Benda P, Lightbody J, Sato G et al. Differentiated rat glial cell strain in tissue culture. Science 1968; 161:370-371.

49. Naus CCG, Elisevich K, Zhu D et al. In vivo growth of C6 glioma cells transfected with connexin43 cDNA. Cancer Res 1992; 52:4208-4213.

50. Bond SL, Bechberger JF, Khoo NKS et al. Transfection of C6 glioma cells with connexin32: the effects of expression of a nonendogenous gap junction protein. Cell Growth & Diff 1994; 5:179-186.

51. Naus CCG, Bechberger JF, Caveney S et al. Expression of gap junction genes in astrocytes and C6 glioma cells. Neurosci Lett 1991; 126:33-36.

52. Zhu D, Caveney S, Kidder GM et al. Transfection of C6 glioma cells with connexin43 cDNA: analysis of expression, intercellular coupling, and cell proliferation. Proc Natl Acad Sci USA 1991; 88:1883-1887.

53. Bechberger JF, Khoo N, Naus CCG. Invis-

ible expression of connexin43 in glioma cells. Cell Grown & Diff 1996; in press.

54. Eghbali B, Kessler JA, Reid LM et al. Involvement of gap junctions in tumorigenesis: Transfection of tumor cells with connexin32 cDNA retards growth in vivo. Proc Natl Acad Sci USA 1991; 88:10701-10705.

55. Tomasetto C, Neveu MJ, Daley J et al. Specificity of gap junction communication among human mammary cells and connexin transfectants in culture. J Cell Biol 1993; 122:157-167.

56. Jou YS, Matesic D, Dupont E et al. Restoration of gap-junctional intercellular communication in a communication-deficient rat liver cell mutant by transfection with connexin43 cDNA. Mol Carcinog 1993; 8:234-244.

57. Mehta PP, Hotz-Wagenblatt A, Rose B et al. Incorporation of the gene for a cell-cell channel protein into transformed cells leads to normalization of growth. J Membrane Biol 1991; 124:207-225.

58. Naus CCG, Zhu D, Todd SDL et al. Characterization of C6 glioma cells over-expressing gap junction protein. Cell Mol Neurobiol 1992; 12:163-175.

59. Zhu D, Kidder GM, Caveney S et al. Growth retardation in glioma cells co-cultured with cells overexpressing a gap junction protein. Proc Natl Acad Sci USA 1992; 89:10218-10221.

60. Dudek FE, Gribkoff VK, Olson JE et al. Reduction of dye coupling in glial cultures by microinjection of antibodies against the liver gap junction polypeptide. Brain Res 1988; 439:275-280.

61. Meyer RA, Laird DW, Revel JP et al. Inhibition of gap junction and adherens junction assembly by connexin and A-CAM antibodies. J Cell Biol 1992; 119:179-189.

62. Dahl G, Nonner W, Werner R. Attempts to define functional domains of gap junction proteins with synthetic peptides. Biophys J 1994; 67:1816-1822.

63. Moore LK, Burt JM. Selective block of gap junction channel expression with connexin-specific antisense oligodeoxynucleotides. Amer J Physiol-Cell Physiol 1994; 36: C1371-C1380.

64. Goldberg GS, Martyn KD, Lau AF. A connexin43 antisense vector reduces the ability of normal cells to inhibit the foci formation of transformed cells. Mol Carcinogen 1994; 11:106-114.

65. Reaume A, de Sousa P, Kulkarni S et al. Cardiac malformation in neonatal mice lacking connexin43. Science 1995; 267: 1831-1834.

66. McCarthy KD, de Vellis J. Preparation of separate astroglial and oligodendroglial cell cultures from rat cerebral tissue. J Cell Biol 1980; 85:890-902.

67. Culver KW, Ram Z, Wallbridge S et al. In vivo gene transfer with retroviral vector-producer cells for treatment of experimental brain tumors. Science 1992; 256: 1550-1552.

68. Bi WL, Parysek LM, Warnick R et al. In vitro evidence that metabolic cooperation is responsible for the bystander effect observed with HSV tk retroviral gene therapy. Human Gene Therapy 1993; 4:725-731.

CHAPTER 12

GAP-JUNCTIONAL COMMUNICATION IN MAMMALIAN CORTICAL ASTROCYTES: DEVELOPMENT, MODIFIABILITY AND POSSIBLE FUNCTIONS

Christian M. Müller

1. INTRODUCTION

Intercellular communication among cells in the central nervous system can occur via surface-surface interactions, the release of neuroactive substances which bind to specialized receptors on other cells, or via the direct exchange of electrical or chemical signals between cells. The latter means of communication utilizes gap junctions, pores comprised of two hemichannels localized in both of two communicating cells. Each hemichannel is formed by six connexin molecules, being integral membrane proteins. The pore formed by the associated hemichannels has a diameter of approximately 1.2 nm and allows the passage of electrical charge, as well as small molecules.[1]

In the mature central nervous system of mammals, gap-junctional communication is most pronounced among astrocytes, though gap junctions are also expressed between neurons[2,3] and oligodendrocytes.[4-6] Coupling is not exclusively confined to cells of the same type, but occurs also as heterologous coupling, e.g., between astrocytes and oligodendrocytes.[7,8] Unlike the obvious role in providing fast electrical transmission between some specialized neurons, little is known about the functional implications of gap-junctional communication between glial cells in the mature CNS. Possible functions of gap junctional communication among astrocytes may reside in the propagation of signals,[9] the redistribution of ions,[10] or metabolic interactions among coupled cells. These possibilities will be discussed in more detail below.

In the developing CNS gap junctional communication has been shown to be a widespread phenomenon in neurons.[11-13] Functional implications for developmental processes which have been proposed are the nonsynaptic propagation of neuronal activity[14] or the diffusion of molecules expressed only in a subpopulation of cells. The latter mechanism may be of importance for the establishment of molecular gradients underlying pattern formation in the brain.[15] The development of gap junctional communication among astrocytes in the mammalian cortex and its significance for putative functions will be addressed below.

2. DYE COUPLING AMONG ASTROCYTES

The existence of gap junctions between cells can be shown by visualizing them on the ultrastructural level[7,16] or by specific antibodies directed against individual members of the connexin family.[17,18] Functional gap junctional communication, however, can only be proven by the injection of low-molecular weight dyes into single cells and observing the dye-spread to adjacent cells,[19,20] by double cell recording to determine the transfer of electrical signals,[5,17] or by observing the spread of physiological signals, e.g., calcium elevations, between neighboring cells.[9] An advantage of the dye-transfer technique is that the gap-junctional communication can be visualized on the network level and that the technique can be combined with other anatomical techniques, e.g., immunocytochemical analyses of cell types involved in the communicating network.

Today, two low molecular weight dyes are commonly used for dye coupling stud-

ies, Lucifer Yellow and biocytin-derivatives.[19,21] Due to its lower molecular weight, biocytin and its derivatives allow visualization of dye coupling even in weakly coupled cells where the fluorescent compound Lucifer Yellow fails to pass gap junctions.[20] Comparison of the spread observed with dyes of different molecular weight can thereby reveal the strength of coupling. We injected Lucifer Yellow and Neurobiotin (Vector Laboratories) into electrophysiologically identified astrocytes in slices of rat visual cortex[19] and hippocampus.[20] Typically, the injections resulted in a circular area containing labeled cells (Fig. 12.1A) suggesting an omnidirectional and uniform spread of the dye within the functional syncythium of coupled cells. Depending on the injection time and the tracer such areas could reach a diameter of up to 1 mm. Quantification of the density of labeled cells revealed 10,000-40,000 cells/mm³. Following application of 2 mM heptanol which strongly reduces gap junctional communication, the dye accumulated in the injected cell (Fig. 12.1B) proving that dye spread occurred via gap junctions.

Counterstaining of the dye-labeled cells with an antiserum against the astroglial marker GFAP showed that the vast majority of cells were astrocytes (Fig. 12.2). The impression of a cell type-specific and quantitative coupling of astrocytes was partially supported by comparing the overall number of dye-coupled cells and the total number of astrocytes in the cortical layers and different subregions of the hippocampal formation. In the stratum radiatum of the hippocampus and in the visual cortex both values were very similar, the number of dye-filled cells slightly exceeding the number of GFAP-positive astrocytes. This finding could either reflect the inclusion of a few non-astrocytic cells in the functional syncytium, e.g., oligodendrocytes, or the presence of GFAP-negative astrocytes.[22] The data from gray matter areas therefore suggest a dominant cell type-specific interaction among astrocytes. A clearcut exception from this generalization emerged from observations in the *stratum lacunosum*

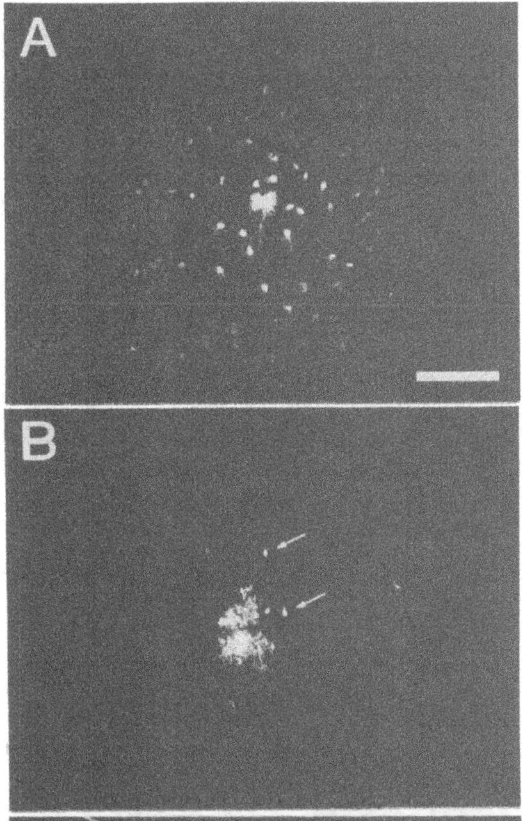

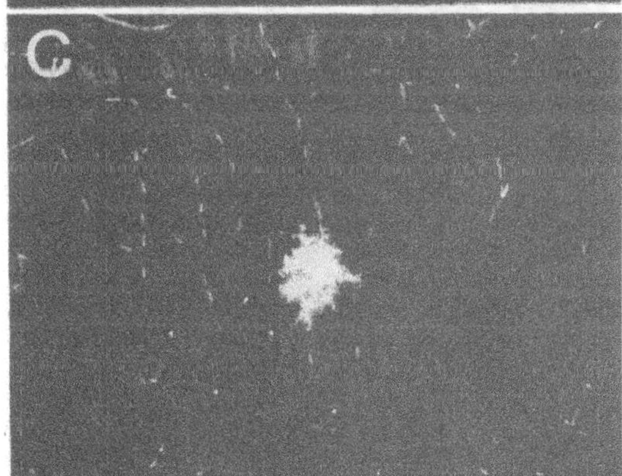

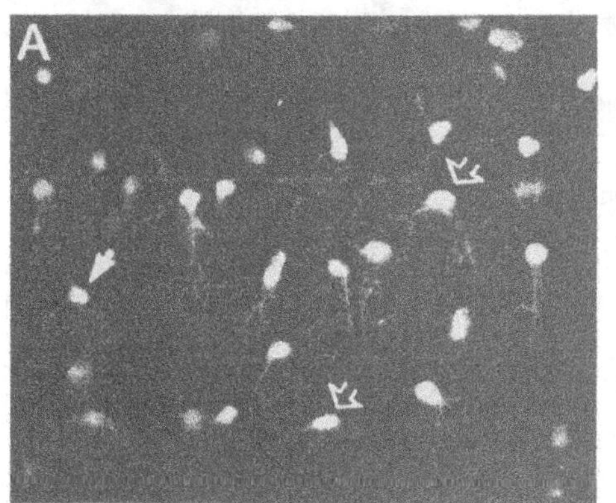

Fig. 12.1. Lucifer Yellow injections into electrophysiologically identified astrocytes in the CA1 region of rat hippocampus. (A) Dye-injection into one astrocyte results in labeling of cells in a circular region around the injected cell. At the end of the initial filling-period the gap-junction blocker heptanol was applied to the bathing solution leading to a stronger labeling of the impaled cell upon continuing of the dye-injection. (B) When 2 mM heptanol was present in the superfusate throughout the dye-injection, Lucifer Yellow is significantly retained in the injected cell with very little spread to adjacent cells (arrows). (C) Preincubation of the slice with the protein kinase C activator phorbol-12,13-dibutyrate also results in a significant reduction of dye coupling. Note the intense labeling even in very fine processes. The dotted background fluorescence is caused by autofluorescence of blood cells. Scaling bar: 100 μm.

Fig. 12.2 (below). Astrocytic identity of dye-coupled cells shown by double staining with the astroglial marker GFAP. Lucifer Yellow injection into an astrocyte in the CA1 region of rat hippocampus resulted in numerous labeled cells (A). The majority of cells (examples are labeled by open arrows) can also be identified when viewing the section with a filter combination specific for TRITC-labeling used to visualize GFAP-immunoreactivity. Few labeled cells (solid arrow) lack GFAP-positivity. Scaling bar: 50 μm. Reprinted with permission from Konietzko and Müller, Hippocampus 1994; 4:297-306.

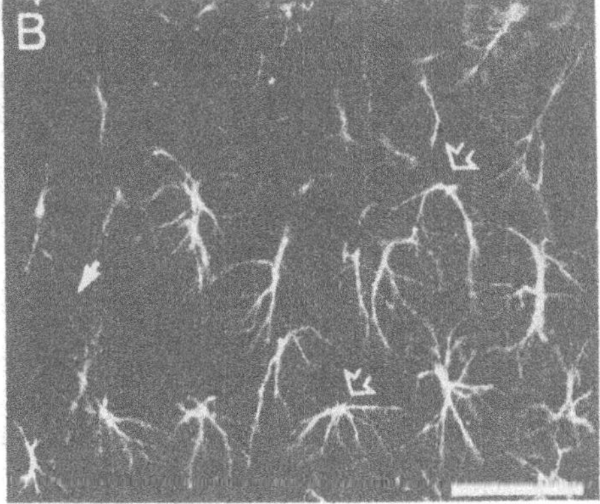

moleculare and alveus of the hippocampus. In these regions the number of dye-coupled cells significantly exceeded the number of GFAP-positive astrocytes.[20] Both regions contain a high proportion of oligodendrocytes, supporting the notion of heterologous coupling between the two types of macroglia.[23] In a few cases we injected oligodendrocytes with Lucifer Yellow and indeed observed dye coupling to neighboring cells with an astrocyte morphology (Fig. 12.3). These data confirm the presence of heterotypic coupling between macroglial cells.[7,8]

We conclude from our data that within neuropilar structures (e.g., cortical gray matter) astroglial cells are predominantly and most likely also quantitatively coupled to each other forming a global functional syncytium. In areas containing a significant proportion of oligodendrocytes (e.g., neuropilar structures containing many myelinated axons) also heterotypic coupling occurs.

3. TOPOGRAPHY OF DYE COUPLING

Earlier studies had proposed that astroglial coupling reveals a heterogeneity with respect to substructures of given brain regions. Using the immunocytochemical analysis of connexin43 expression, Yamamoto et al[18] showed that the expression of this astroglial gap-junction protein differs in the substructures of the hippocampus. The authors proposed that this may relate to differences in astroglial coupling. In our studies we could not observe differences in the degree of dye spread when the injected astrocyte was located in different subareas of the hippocampal formation[20] or the neocortex.[19] However, in line with the immunocytochemical data we observed a reduced density of astrocytes in areas shown to reveal low connexin43 immunoreactivity. Differences in the amount of connexin43 immunoreactivity appear to correlate with the astroglial density rather than with functional coupling efficacy. The lack of a functional separation of astroglial populations with respect to dye coupling

was further supported by the observation of dye-spread across the hippocampal fissure,[20] separating the CA1-region from the dentate gyrus. These data gave rise to the conclusion that astroglial cells show functional coupling even across well defined areal boundaries within the CNS. The only exception from this rule, as observed in our studies on visual cortex, was the border between gray and white matter where dye

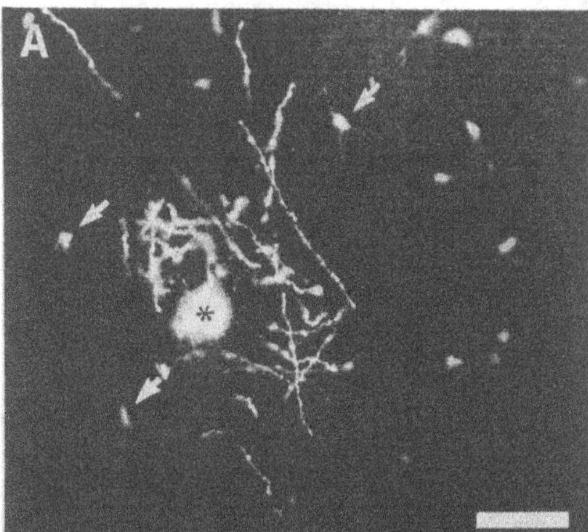
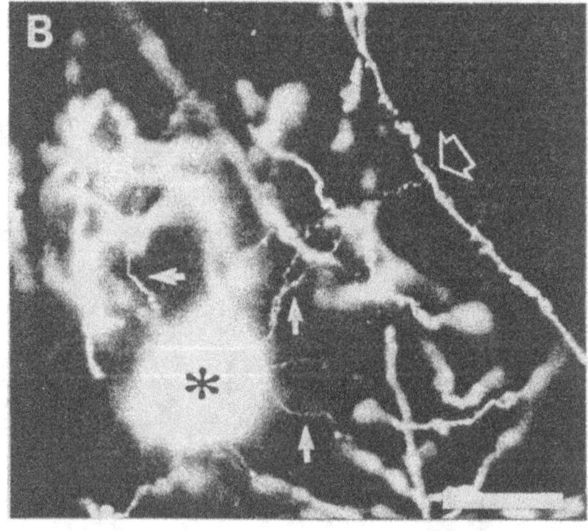

Fig. 12.3. Heterotypic coupling between an injected oligodendrocyte and astrocyte-like cells in rat hippocampus. The brightly fluorescent cell in A can be identified as an oligodendrocyte by its morphology. Note the presence of very fine processes emanating from the cell body (arrows in B) which can be followed to thick processes (open arrow in B) corresponding to myelinating processes. Numerous cells with smaller somata and an astrocytic morphology are labeled in the vicinity of the injected oligodendrocyte (arrows in A). Scaling bars: 50 µm (A), 25 µm (B).

spread was absent.[19] This is in line with earlier investigations, in vitro, showing that dye coupling among astrocytes is confined to A2B5-immunonegative cells.[24] A2B5 is an antigen predominant in white matter astrocytes.[25] Thus, astroglial coupling appears to be a characteristic of neuropilar astrocytes irrespective of the areal compartment, but appears to be absent or strongly reduced in white matter regions.

4. DEVELOPMENT OF DYE COUPLING

In order to assess the developmental time course of astroglial dye coupling we injected astrocytes in slices from rats aged 4-120 days. While dye coupling was routinely observed in tissue from animals older than two weeks, gap-junctional coupling appeared to be absent or strongly reduced in tissue from younger animals when Lucifer Yellow was used as tracer.[19,20] Because the membrane potential of immature astrocytes was usually very instable in immature cells we performed control experiments which revealed that astrocytic dye-coupling was robust in older animals when injections were done in the presence of high extracellular potassium, leading to a strong depolarization of the cells.[19] These

data gave rise to the hypothesis that gap-junctional communication between cortical astrocytes occurs only postnatally. However, when injections were done using neurobiotin in hippocampal slices dye coupling was readily observed even in four day old animals and included the entire local population of astrocytes.[20] These data show that gap-junctional communication is actually present very early in development

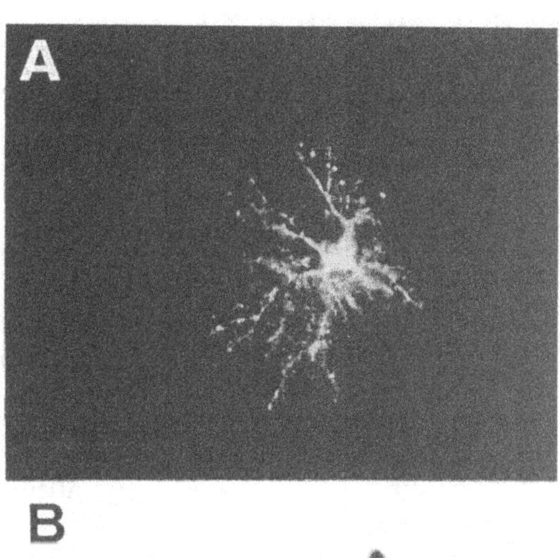

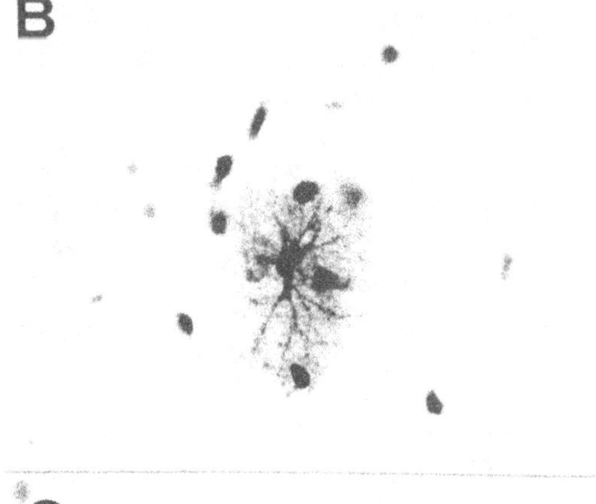

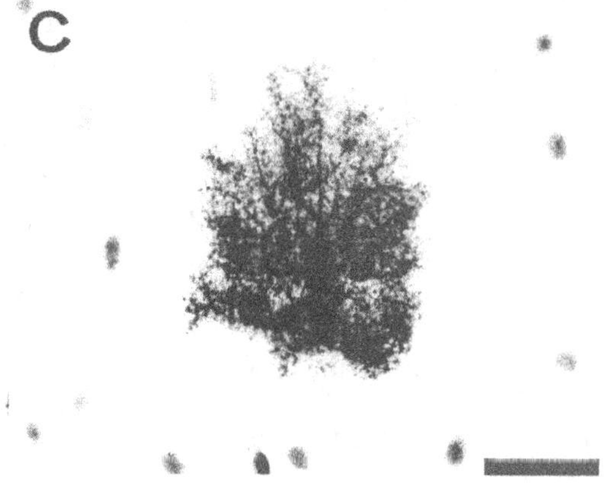

Fig. 12.4. Reduction of dye coupling in gliotic tissue one week after a stab wound was placed in rat visual cortex. (A) In the direct vicinity of the lesion Lucifer Yellow is fully retained in an injected cell. Note the stellate morphology of the cell, reminiscent of a reactive astrocyte. (B) Injection of Neurobiotin into an astrocyte situated close to a stab wound reveals some dye coupling to neighboring cells. Note the intense filling of the injected cell and its stellate morphology. (C) Neurobiotin injection into an astrocyte in control tissue after preincubation of the slice with the protein kinase C activator phorbol-12,13-dibutyrate. The accumulation of the dye in the injected cell indicates a significant reduction of gap-junctional communication (the exposure time during photographic printing was doubled for the tissue around the injected cell in order to visualize the weak labeling due to dye-coupling). In contrast to the stellate morphology of astrocytes in gliotic tissue (A/B), astrocytes in gray matter of control tissue reveal the bushy morphology of protoplasmic astrocytes. Scaling bar: 50 μm.

and includes most astroglial cells already at these early developmental periods. The observed difference in the degree of dye spread with Lucifer Yellow and Neurobiotin may indicate that gap junctional communication is lower in immature compared to mature astrocytes. This is in line with data obtained in cultured astrocytes[26] and the fact that the expression of connexin43 strongly increases in the CNS during postnatal development.[17,27,28]

5. MODIFICATION OF DYE-COUPLING BY PROTEIN KINASE C AND UPON CENTRAL NERVOUS SYSTEM DAMAGE

With respect to the apparent widespread communication of astrocytes suggested from our topographical data, it is of particular interest whether gap junctions are modifiable. We addressed this question by two different approaches. First we injected astrocytes during the application of the protein kinase C (PKC) activator phorbol-12,13-dibutyrate. Activation of PKC is a well known effect of several signaling cascades between neurons and glial cells. It was shown that gap junctional communication was significantly reduced by activation of PKC (Figs. 12.1C, 12.4C and ref. 20). These data are in line with studies in cultured astrocytes[29] and suggest that astroglial gap junctions are modifiable by second messenger cascades, in situ. This modification may be mediated by phosphorylation of connexins known to shift the unitary conductance of channels assembled by connexin43.[30] Such a change in the coupling efficacy may be of functional importance, as one messenger, inositol 1,4,5-triphosphate, known to pass gap junctions and underlying the spread of calcium signals[29,31] has a fairly high molecular weight. An overall reduction of gap junctional conductance by phosphorylating agents may therefore specifically impair the spread of calcium signals without affecting the exchange of smaller molecules, e.g., ions. The functional syncytium of astroglial cells may therefore be dynamically compartmentalized by physiological stimuli.

In a second approach we investigated the influence of central nervous system damage onto astroglial coupling (Müller, Konietzko and Dermietzel, in preparation). Damage was either induced by lowering a hypodermic needle into the cortex under deep anesthesia (stab wound), or by injecting the neurotoxic compound kainic acid into the white matter area of newborn animals to eliminate the transient population of neurons innervating layer IV.[32] Two to 30 days after the operation we prepared slices and investigated astrocytic dye coupling by intracellular injection of Lucifer Yellow or Neurobiotin.

Close to the mechanical damage by a stab wound reactive astrocytes revealed significantly reduced coupling. Following Lucifer Yellow injections the dye was always maintained in the injected cell (Fig. 12.4A) while Neurobiotin revealed some spread to neighboring cells (Fig. 12.4B). Similarly, astrocytes in layer IV, the axonal target of the neurons destroyed by prior kainate injection, also revealed no dye coupling after Lucifer Yellow injection. Again, Neurobiotin still revealed weak coupling of astrocytes to neighboring cells. We therefore conclude that astroglial coupling is strongly reduced following central nervous damage. Our experiments are in line with the severe reduction of connexin43 immunoreactivity observed after neuronal degeneration induced by neurotoxin injection.[33,34]

What are the mechanisms underlying the changes in astroglial coupling after central nervous system damage? In our experimental approach using kainic acid injection only the afferent axons were affected and the neurotoxin was injected remote from the studied location in the tissue. Therefore we exclude the possibility of a direct effect of the toxin onto the astrocytes, as well as a mere loss of neurons as a cause for the decoupling of astroglial cells. We rather suggest that both treatments induce an astrogliosis and that decoupling is a feature of reactive astrocytes. The astrogliosis is most likely induced by signals emanating from degenerating axons (remote kainate injection) or neurons (stab wound).

The available data indicate that gap junctional communication can be modified both by posttranslational mechanisms, i.e., by phosphorylation, as well as on the level of gene expression or degradation of gap junction proteins. Both mechanisms may interact, as connexin degradation has been shown to be influenced by the degree of phosphorylation.[35] Short-term effects of central nervous system damage on the expression of gap junction proteins are described in chapter 10 (see also ref. 36).

6. POSSIBLE FUNCTION(S) OF GAP-JUNCTIONAL COMMUNICATION AMONG ASTROCYTES

With respect to the functional significance of gap junctional communication among astrocytes experimental evidence is as yet virtually lacking. Three possible functions—being not necessarily exclusive—may be supported by experimental data. A well accepted role of astroglial cells lies in the homeostasis of the extracellular ionic composition. Especially potassium ions, which are released upon neuronal activation, are known to be buffered by astrocytes. Gap junctional coupling of astroglial cells could play a role in the spatial redistribution of potassium ions remote from the site of neuronal release.[5,10,37] The uniform coupling of astrocytes seen in our studies allows for a very efficient—because omnidirectional—redistribution. A similar hypothesis could also be valid for other mechanisms of homeostasis, e.g., with respect to extracellular pH.[38,39] A reduction of gap junctional communication by physiological signals (e.g., neurotransmitters released from neurons and acting on astrocytes via activation of PKC) may hinder the diffusion of excess ions to sites of previous activation. The observed strong reduction in gap junctional communication upon neuronal degeneration can be viewed as a protection mechanism of the intact tissue, as redistribution of pathologically high

concentrations of neuronally released ions from the damaged brain region is hindered.

A second avenue of thinking is based on the finding that gap junctional communication among astrocytes allows the active propagation of signals. It has been shown in astrocytic cultures that activation of transmitter receptors and mechanical stimulation leads to an increase in the intracellular calcium concentration which is propagated to neighboring cells.[9,40] This finding has been replicated in organotypic hippocampal slice cultures where the astroglial signaling can be induced by stimulation of afferent axon pathways.[41] The propagation is most likely mediated by the transfer of inositol 1,4,5-trisphosphate passing through gap junctions and eliciting the release of calcium from intracellular stores of neighboring cells.[31] The functional significance of this kind of neuro-glial communication inducing glia-glial signal propagation is far from understood. A possible relation to the above mentioned ion redistribution could lie in the possibility that gap junctional communication is reduced upon application of excitatory transmitters and intracellular elevation of calcium.[42] The slow propagation of calcium signals may support a unidirectional redistribution of ions within the functional syncytium of astrocytes. In addition, a recent study has shown that neurons are excited in the direct vicinity of astroglial cells revealing an increased intracellular calcium concentration.[43] Such a mechanism may result in nonsynaptic excitation of neuronal clusters. Indeed such activity is seen in the developing cortex where it is thought to play a role in the shaping of neural connections and the emergence of functional architectures.[14]

A third possibility for a functional implication of astroglial gap junctions could be the "synchronization" of the astroglial population. Gap junctions may 'clamp' neighboring cells to an identical resting potential which could be essential for the maintenance of a uniform extracellular ion concentration. In addition, gap junctions may simply signal the presence of neighboring astrocytes and could thereby control the spatial arrangement and density of astroglia. Interestingly, decoupling upon central nervous damage is usually paralleled by astroglial proliferation. Conversely, it has been shown that proliferative activity of cell lines is reduced by the presence of gap junctions.[44,45]

As yet, one can only speculate about the exact functional role of astroglial gap junctions. In order to address such questions experimentally it is essential to have tools to selectively uncouple cells without disturbing other physiological properties. The immense progress in the characterization of gap junction proteins and the genomic sequences over the last years makes it very likely that such tools, e.g., functional antibodies or knockout mutants, will be available soon and will allow experimental access to the proposed hypotheses.

Acknowledgment

Supported by the Deutsche Forschungsgemeinschaft (Mu 908/2-1) and the Bundesministerium für Forschung und Technologie (316902A).

References

1. Simpson J, Rose B, Loewenstein WR. Size limit of molecules permeating the junctional membrane channels. Science 1977; 195: 294-301.
2. Furshpan EJ. "Electrocal transmission" at an excitatory synapse in a vertebrate brain. Science 1964; 144:878-880.
3. Cobbett P, Hatton GI. Dye-coupling in hypothalamic slices: dependence on in vivo hydration state and osmolarity of incubation medium. J Neurosci 1984; 4:3034-3038.
4. Dermietzel R, Traub O, Hwang TK et al. Differential expression of three gap junction proteins in developing and mature brain tissues. Proc Natl Acad Sci USA 1989; 86:10148-10152.
5. Kettenmann H, Ransom BR. Electrical coupling between astrocytes and between oligodendrocytes studied in mammalian cell cultures. Glia 1988; 1:64-73.
6. von Blankenfeld G, Ransom BR, Kettenmann H. Development of cell-cell coupling among cells of the oligodendrocyte lineage. Glia 1993; 7:322-328.
7. Mugnaini E. Cell junctions of astrocytes, ependyma and related cells in the mammalian central nervous system, with emphasis on the hypothesis of a generalized syncythium of supporting cells. In: Fedoroff S, Vernadakis A, eds. Astrocytes. London: Academic Press, 1986:329-371.
8. Robinson SR, Hampson ECGM, Munro MN et al. Unidirectional coupling of gap junctions between neuroglia. Science 1993; 262:1072-1074.
9. Cornell-Bell AH, Finkbeiner SM, Cooper MS et al. Glutamate induces calcium waves in cultured astrocytes: Long-range glial signaling. Science 1990; 247:470-473.
10. Walz W. Role of glial cells in the regulation of the brain ion microenvironment. Prog Neurobiol 1989; 33:309-333.
11. Peinado A, Yuste R, Katz LC. Extensive dye coupling between rat neocortical neurons during the period of circuit formation. Neuron 1993; 10:103-114.
12. Penn AA, Wong ROL, Shatz CJ. Neuronal coupling in the developing mammalian retina. J Neurosci 1994; 14:3805-3815.
13. Paternostro MA, Reyher CKH, Brunjes PC. Intracellular injections of Lucifer Yellow into lightly fixed mitral cells reveal neuronal dye-coupling in the developing rat olfactory bulb. Dev Brain Res 1995; 84:1-10.
14. Yuste R, Nelson DA, Rubin WW, Katz LC. Neuronal domains in developing neocortex: mechanisms of coactivation. Neuron 1995; 14:7-17.
15. Gierer A, Müller CM. Development of layers, maps and modules. Curr Opinion Neurobiol 1995; 5:91-97.
16. Robertson JD. The occurrence of a subunit pattern in the unit membranes of club endings in Mauthner cell synapses in goldfish brains. J Cell Biol 1963; 19:201-221.
17. Dermietzel R, Hertzberg EL, Kessler JA et al. Gap junctions between cultured astrocytes: Immunocytochemical, molecular, and electrophysiological analysis. J Neurosci 1991; 11:1421-1432.
18. Yamamoto T, Ochalski A, Hertzberg EL et al. LM and EM immunolocalization of the

gap junctional protein connexin43 in rat brain. Brain Res 1990; 508:313-319.

19. Binmöller F-J, Müller CM. Postnatal development of dye-coupling among astrocytes in rat visual cortex. Glia 1992; 6:127-137.

20. Konietzko U, Müller CM. Astrocytic dye coupling in rat hippocampus: Topography, developmental onset, and modulation by protein kinase C. Hippocampus 1994; 4: 297-306.

21. Vaney DI. Many diverse types of retinal neurons show tracer coupling when injected with biocytin or Neurobiotin. Neurosci Lett 1991; 125:187-190.

22. Stichel CC, Müller CM, Zilles K. Distribution of glial fibrillary acidic protein and vimentin immunoreactivity during rat visual cortex development. J Neurocytol 1991; 20: 97-108.

23. Massa PT, Mugnaini E. Cell junctions and intramembrane particles of astrocytes and oligodendrocytes: a freeze-fracture study. Neuroscience 1982; 7:523-538.

24. Sontheimer H, Minturn JE, Black JA et al. Specificity of cell-cell coupling in rat optic nerve astrocytes in vitro. Proc Natl Acad Sci USA 1990; 87:9833-9837.

25. Raff MC, Abney ER, Cohen J et al. Two types of astrocytes in cultures of developing rat white matter. differences in morphology, surface gangliosides, and growth characteristics. J Neurosci 1983; 3:1289-1300.

26. Fischer G, Kettenmann H. Cultured astrocytes from a syncytium after maturation. Exp Cell Res 1985; 159:273-279.

27. Giaume C, Fromaget C, El Aoumari A et al. Gap junctions in cultured astrocytes: Single-channel currents and characterization of channel-forming protein. Neuron 1991; 6:133-143.

28. Belliveau DJ, Kidder GM, Naus CCG. Expression of gap junction genes during postnatal neural development. Dev Genet 1991; 12:308-317.

29. Enkvist MOK, McCarthy KD. Activation of protein kinase C blocks astroglial gap junction communication and inhibits the spread of calcium waves. J Neurochem 1992; 59:519-526.

30. Moreno AP, Fishman GI, Spray DC. Phosphorylation shifts unitary conductance and modifies voltage dependent kinetics of human connexin43 gap junction channels. Biophys J 1992; 62:51-53.

31. Charles AC, Dirksen ER, Merrill JE et al. Mechanisms of intercellular calcium signaling in glial cells studied with dantrolene and thapsigargin. Glia 1993; 7:134-145.

32. Ghosh A, Antonini A, McConnell SK et al. Requirement for subplate neurons in the formation of thalamocortical connections. Nature 1990; 347:179-181.

33. Vukelic JI, Yamamoto T, Hertzberg EL et al. Depletion of connexin43-immunoreactivity in astrocytes after kainic acid-induced lesions in rat brain. Neurosci Lett 1991; 130:120-124.

34. Hossain MZ, Sawchuk MA, Murphy LJ et al. Kainic acid induced alterations in antibody recognition of connexin43 and loss of astrocytic gap junctions in rat brain. Glia 1994; 10: 250-265.

35. Elvira M, Diez JA, Wang KKW et al. Phosphorylation of connexin32 by protein kinase C prevents its proteolysis by u-calpain and m-calpain. J Biol Chem 1993; 268: 14294-14300.

36. Rohlmann A, Laskawi R, Hofer A et al. Astrocytes as rapid sensors of peripheral axotomy in the facial nucleus of rats. NeuroReport 1994; 5:409-412.

37. Gardner-Medwin AR. Analysis of potassium dynamics in mammalian brain tissue. J Physiol 1983; 335:393-426.

38. Ransom BR. Glial modulation of neural excitability mediated by extracellular pH: a hypothesis. Progr Brain Res 1992; 94:37-46.

39. Rose CR, Deitmer JW. Evidence that glial cells modulate extracellular pH transients induced by neuronal activity in the leech central nervous system. J Physiol (Lond) 1994; 481:1-5.

40. Charles AC, Merrill JE, Dirksen ER et al. Intercellular signaling in glial cells: Calcium waves and oscillations in response to mechanical stimulation and glutamate. Neuron 1991; 6:983-992.

41. Dani JW, Chernjavsky A, Smith SJ. Neuronal activity triggers calcium waves in hippocampal astrocyte networks. Neuron 1992; 8:429-440.

42. Müller T, Möller T, Kirchhoff F et al.

Calcium influx through kainate receptors blocks gap junctions on Bergmann glia cells in cerebellar slices. In: Elsner N, Breer H, eds., Göttingen Neurobiology Report 1994, New York: Thieme, Stuttgart, 1995:206.

43. Nedergaard M. Direct signaling from astrocytes to neurons in cultures of mammalina brain cells. Science 1994; 263:1768-1771.

44. Zhu D, Kidder GM, Caveney S et al. Growth retardation in glioma cells cocultured with cells over expressinga gap junction protein. Proc Natl Acad Sci USA 1992; 89:10218-10221.

45. Naus CCG, Hearn S, Zhu D et al. Ultrastructural analysis of gap junctions in C6 glioma cells transfected with connexin43 cDNA. Exp Cell Res 1993; 206:72-84.

CHAPTER 13

CONNEXIN32 AND X-LINKED-CHARCOT-MARIE-TOOTH DISEASE

Suzanne M. Deschênes, Linda Jo Bone,
Kenneth H. Fischbeck and Steven S. Scherer

"... that the women of this family, themselves even uncommonly buxom and healthy, should be able to ... transmit to the males alone tissues unlike their own, and endowed with a regular form of weakness which they do not themselves possess, is ... marvelous. It seems as if the daughter of a diseased father carried from the beginning of her life ova of two sexes, the female healthy, the male containing within it the representation of the father's disease."

Herringham (1888)

1. SUMMARY

The discovery that X-linked Charcot-Marie-Tooth disease (CMTX) is caused by mutations in connexin32 (Cx32)[1] has united the fields of molecular genetics, neurology, and gap junction biology by demonstrating the importance of Cx32 in myelinating Schwann cells. The lack of overt clinical manifestations in other tissues that express Cx32 suggests the existence of compensatory mechanisms elsewhere that are absent in peripheral nerve. Determining the mechanism by which Cx32 mutations cause the phenotype of CMTX will contribute to the understanding of the function of the myelin sheath and elucidate the role of Cx32 in other tissues.

2. HISTORICAL BACKGROUND

Charcot-Marie-Tooth disease (CMT), or hereditary motor and sensory neuropathy, was first described in 1886[2,3] and is characterized by

Gap Junctions in the Nervous System, edited by David C. Spray and Rolf Dermietzel.
© 1996 R.G. Landes Company.

slowly progressive weakness and atrophy of distal limb muscles, *pes cavus*, sensory loss, and loss of reflexes. CMT is the most common form of inherited peripheral neuropathy, affecting about 1 in 3000 individuals,[4] and is classified into two groups according to electrophysiological and pathological findings: type I (demyelinating) and type II (axonal).

Demyelinating CMT is genetically heterogeneous, with similar clinical manifestations produced by gene defects on chromosomes 17 (CMT1A), 1 (CMT1B), and X (CMTX). In the past few years, mutations of the genes for peripheral myelin protein 22 kDa (PMP-22) and myelin protein zero (P_0) have been associated with CMT1A and CMT1B, respectively.[5-15]

In 1888, Herringham reported a family with clinical features the same as those first described by Charcot, Marie, and Tooth, but with asymptomatic carrier females.[16] This disorder, CMTX, first manifests in males in late childhood or adolescence and progresses to moderate disability by the third decade of life. Females usually have onset of symptoms later in life with a less severe course, but they may be asymptomatic. The less severe phenotype in females may be attributable to the random inactivation of the X chromosome carrying the wild-type copy of Cx32 in myelinating Schwann cells.

After CMT1A, CMTX is the second most common hereditary neuropathy.[4,17] The exact incidence of CMTX is difficult to calculate because of the extensive overlap in the clinical features of all forms of CMT and the existence of sporadic cases in which the genetic defect has not been determined. Ionasescu and colleagues[18,19] have described two types of X-linked recessive neuropathy with distinct clinical features (e.g., spasticity and cognitive deficits) that are not seen in typical CMTX.

CMTX is usually classified as a demyelinating neuropathy. However, nerve biopsies have been reported from two different kindreds, and show prominent axonal loss as well as variable amounts of demyelination and remyelination.[20,21] Biopsies from these two kindreds are shown in Figure 13.1. Nerve conduction velocities observed in CMTX patients (25-40 m/sec for males and 25-50 m/sec for females, with 50 m/sec being normal)[22] are not as slow as those seen in kindreds of CMT1A (10-40 m/sec).[23] Thus, there appears to be more axonal loss and less demyelination/remyelination in typical CMTX patients than in typical CMT1A and CMT1B patients. The correct interpretation of these differences depends on elucidating the pathophysiological consequences of mutations in PMP-22, P_0, and Cx32.

Until the 1980s, there was some question as to whether CMTX was due to sex-limited penetrance of an autosomally-inherited defect, in which males are more severely affected than females,[24] or to X-linked "recessive";[25-28] or X-linked "dominant"[29-32] inheritance. This issue was clarified by the reinvestigation of a large North Carolina family with X-linked inheritance and variable expressivity in males and females, and with no instances of male-to-male transmission.[20,25] Furthermore, these reports suggested that distinctions between dominant and recessive modes of inheritance in CMTX are attributable to incomplete penetrance and variable expressivity in heterozygous females.

The pattern of inheritance in CMTX kindreds also demonstrated that the genetic defect maps to a single locus, which was amenable to positional cloning. The earliest reports of linkage analysis excluded both the distal short arm[30] and the distal long arm[31] of the X chromosome. The use of restriction fragment length polymorphisms and short tandem repeat polymorphic markers restricted the CMTX critical region to the proximal segment of the long arm.[33-44] Further recombination analyses in several large families refined the localization of CMTX to an approximately 1.5 megabase interval in Xq13.1, between markers DXS106 and DXS559.[45]

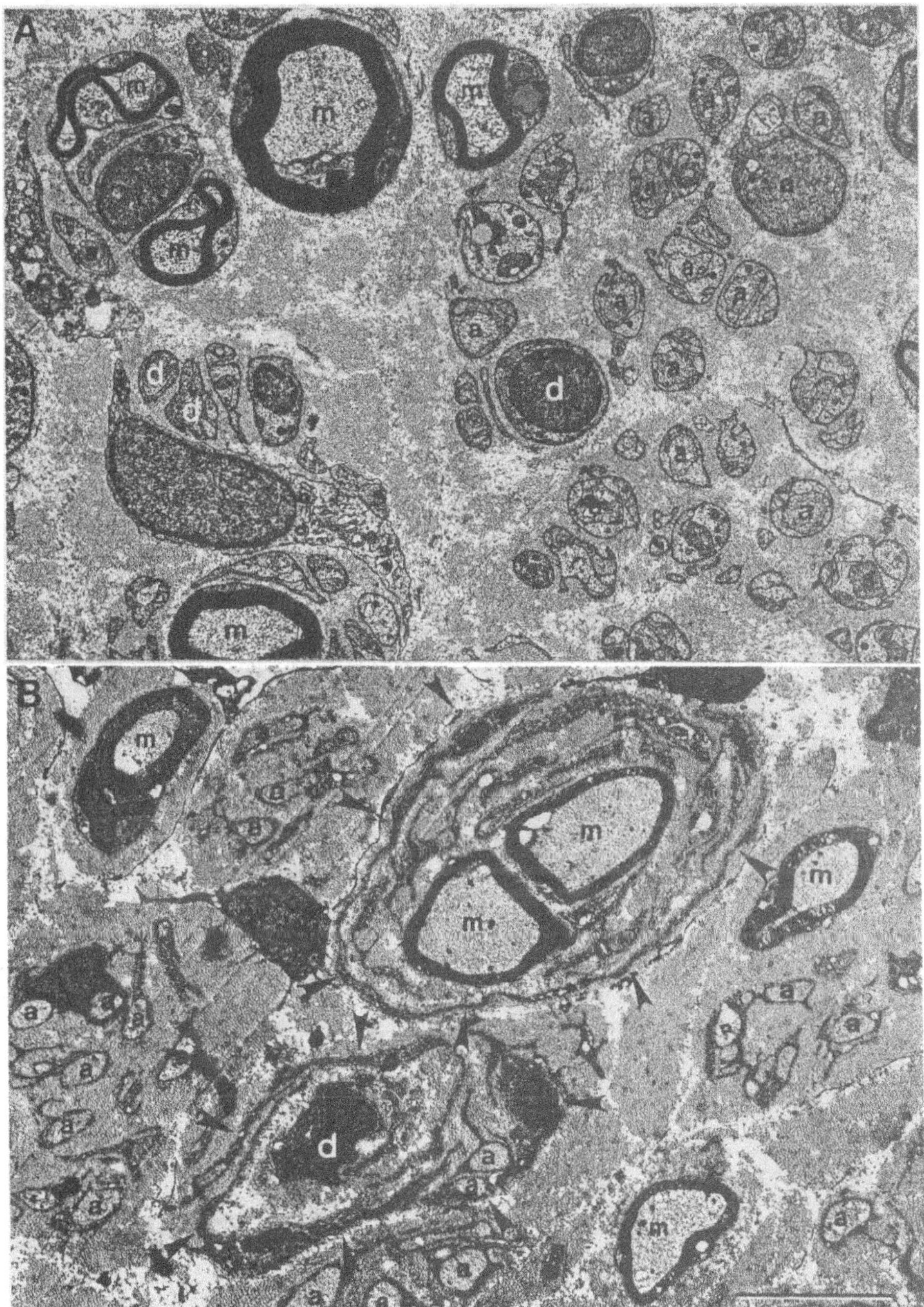

Fig. 13.1. Electron micrographs of sural nerve biopsies from two different CMTX kindreds: (A) a 61 year-old male reported by Hahn et al (1990), and (B) a 28 year-old male reported by Rozear et al (1987). Most of the myelinated fibers (m) in both biopsies have myelin sheaths that are inappropriately thin for the axonal caliber. The loss of myelinated axons is more pronounced in (A), while (B) exhibits rudimentary "onion bulbs" (arrowheads), which is histological evidence for previous episodes of demyelination and remyelination. Note the large number of unmyelinated axons (a), many of which are associated with Schwann cells in a 1:1 manner. Some Schwann cells are "denervated" (d), as they lack any association with an axon. Scale bar: 5 μm.

3. CONNEXIN32 MUTATIONS AND CMTX

Three genes were previously mapped to the same segment of Xq13.1: cell cycle gene-1 (CCG1),[46] the γ subunit of the interleukin-2 receptor (IL2RG)[47,48] and gap junction gene β1 or Cx32 (GJB1).[49,50] No DNA rearrangements were detected by Southern blot analysis, so these candidate genes were screened for expression in peripheral nerve by Northern blot analysis. Of these candidate genes, only Cx32 mRNA was expressed in rat peripheral nerve at significant levels. The level of Cx32 mRNA in nerve was comparable to brain and spleen, although lower than in liver.[1]

Direct sequence analysis of PCR fragments covering the entire Cx32 coding region, which is completely contained within exon 2 (see Fig. 13.2), initially revealed 7 different base changes in 8 independent families.[1] Subsequent reports confirmed these findings,[51-56] bringing the number of different mutations to 42 in 54 families (Table 13.1). Coding region mutations have not been found in seven X-linked families;[1,52,53,56] these families may have as yet unidentified mutations in promoter elements, at the splice sites, or in regulatory/enhancer elements involved in transcriptional control.

To understand how mutations in the noncoding regions of Cx32 could cause CMTX, one needs to consider the structure of the Cx32 gene. In mice, rats and humans, it is composed of two exons and a large intron which is about 6.1 kb in mice and rats,[57,59,60] as shown in Figure 13.2. Neuhaus et al[61] have recently shown that all Cx32 transcripts in peripheral nerve are initiated at a previously described potential promoter, termed P2, which is located in the last third of the intron.[59] Cx32 transcripts in the liver, brain, and spinal cord, on the other hand, are initiated from both the P1 and the P2 promoters.[61] These data suggest that different sets of transcription factors regulate expression of Cx32 in different tissues, and that peripheral nerve may be selectively vulnerable to mutations in P2.

4. FUNCTIONAL ANALYSIS OF CX32 MUTATIONS IN VITRO

CMTX mutations are found in every region of the Cx32 protein (Fig. 13.3). The phenotypic effects of some of these mutations can be anticipated. For example, mutations within the third transmembrane domain, which is thought to line the pore of the gap junction channel, might be expected to interrupt or alter the passage of ions and small molecules (e.g., second

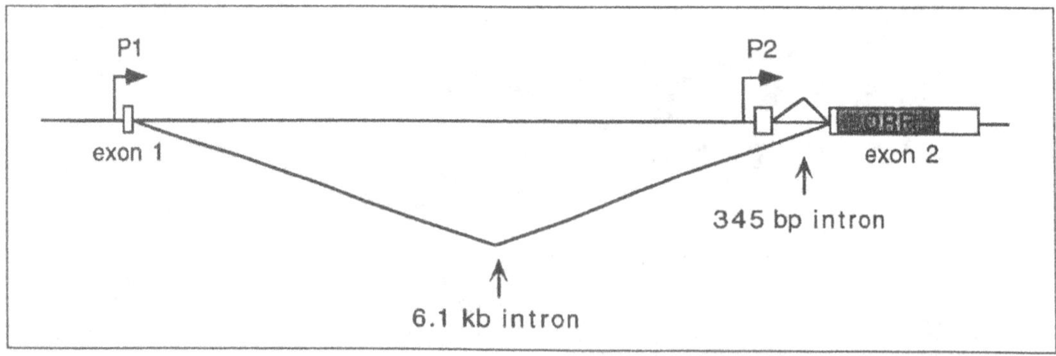

Fig. 13.2. Genomic structure of Cx32 and alternative splicing of the transcript. The entire coding region, designated ORF (open reading frame) is contained within exon 2. All nontranslated regions of the gene are represented by white boxes, and the two promoters are designated by P1 and P2. When the P2 promoter is active, an exon at the 3' end of the intron is alternatively spliced to exon 2, yielding a transcript with a 5' end that differs from that initiated from P1[61].

Table 13.1. Connexin32 mutations in CMTX patients

Codon affected	Nucleotide change	Amino acid change	Reference
Amino terminus			
12	GGC → AGC	Gly → Ser	Bergoffen et al., 1993
13	GTG → TTG	Val → Leu	Bone et al., in press
15	CGG → CAG	Arg → Gln	Fairweather et al., 1994
15	CGG → TGG	Arg → Trp	Janssen, in preparation
1st Transmembrane domain			
22	CGA → CAA	Arg → Gln	Kazazian, unreported
22	CGA → CAA	Arg → Gln	Ionasescu, 1995
22	CGA → TGA	Arg → stop	Ionasescu et al., 1994
22	CGA → TGA	Arg → stop	Janssen, in preparation
30	ATC → AAC	Ile → Asn	Bone et al., in press
34	ATG → ACG	Met → Thr	Tan & Ainsworth, 1994
35	GTG → ATG	Val → Met	Cherryson et al., 1994
38	GTG → ATG	Val → Met	Orth et al., 1994
1st Extracellular loop			
60	TGC → TTC	Cys → Phe	Fairweather et al., 1994
63	GTT → ATT	Val → Ile	Fairweather et al., 1994
63	GTT → ATT	Val → Ile	Janssen, in preparation
65	TAT → TGT	Tyr → Cys	Bone et al., in press
73→	CAT → AT	frameshift	Fairweather et al., 1994
2nd Transmembrane domain			
75	CGG → CAG	Arg → Gln	Tan & Ainsworth, 1994
80	CAG → CGG	Gln → Arg	Ionasescu, 1995
89	CTC → CCC	Leu → Pro	Janssen, in preparation
Intracellular loop			
95	GTG → ATG	Val → Met	Kant, unreported
95	GTG → ATG	Val → Met	Bone et al., in press
102	GAG → GGG	Glu → Gly	Ionasescu et al., 1994
111-116	18 bp deletion	6 aa deletion	Cherryson et al., 1994
111-116	18 bp deletion	6 aa deletion	Ionasescu, 1995
3rd Transmembrane domain			
133	TGG → CGG	Trp → Arg	Bone et al., in press
137→	ATC → AC	frameshift	Bone et al., in press
139	GTG → ATG	Val → Met	Bone et al., in press
139	GTG → ATG	Val → Met	Bergoffen et al., 1993
139	GTG → ATG	Val → Met	Bergoffen et al., 1993
142	CGG → TGG	Arg → Trp	Bergoffen et al., 1993
142	CGG → TGG	Arg → Trp	Ionasescu et al., 1994
143	TTG deletion	Leu deletion	Fairweather et al., 1994
2nd Extracellular loop			
156	CTC → CGC	Leu → Arg	Bergoffen et al., 1993
156	CTC → CGC	Leu → Arg	Bone et al., in press
158	CCT → GCT	Pro → Ala	Cherryson et al., 1994
164	CGG → TGG	Arg → Trp	Ionasescu, 1995
172	CCC → TCC	Pro → Ser	Bergoffen et al., 1993
175→	AAC → AAAC	frameshift	Bergoffen et al., 1993
182	TCC → ACC	Ser → Thr	Cherryson et al., 1994
185→	G insertion	frameshift	Ionasescu, 1995
186	GAG → TAG	Glu → stop	Ionasescu et al., 1994
Carboxy terminus			
208	GAG → AAG	Glu → Lys	Fairweather et al., 1994
211	TAC → TAA	Tyr → stop	Tan & Ainsworth, 1994
215	CGG → TGG	Arg → Trp	Fairweather et al., 1994
217	TGT → TGA	Cys → stop	Ionasescu et al., 1994
220	CGA → TGA	Arg → stop	Bone et al., in press
220	CGA → TGA	Arg → stop	Bone et al., in press
220	CGA → TGA	Arg → stop	Fairweather et al., 1994
220	CGA → TGA	Arg → stop	Ionasescu, 1995
264-272→	29 bp deletion	deletion/frameshift	Ionasescu, 1995

messengers) through the channel. Frame-shift mutations that change downstream amino acid residues and prematurely truncate the protein might alter Cx32 so that connexon assembly, insertion into the membrane, or interactions between apposed connexons would not occur normally. Moreover, changes in the highly conserved cysteine residues in the extracellular loops might alter intramolecular disulfide bonds so that proper folding would not occur (for reviews of gap junction channels and their

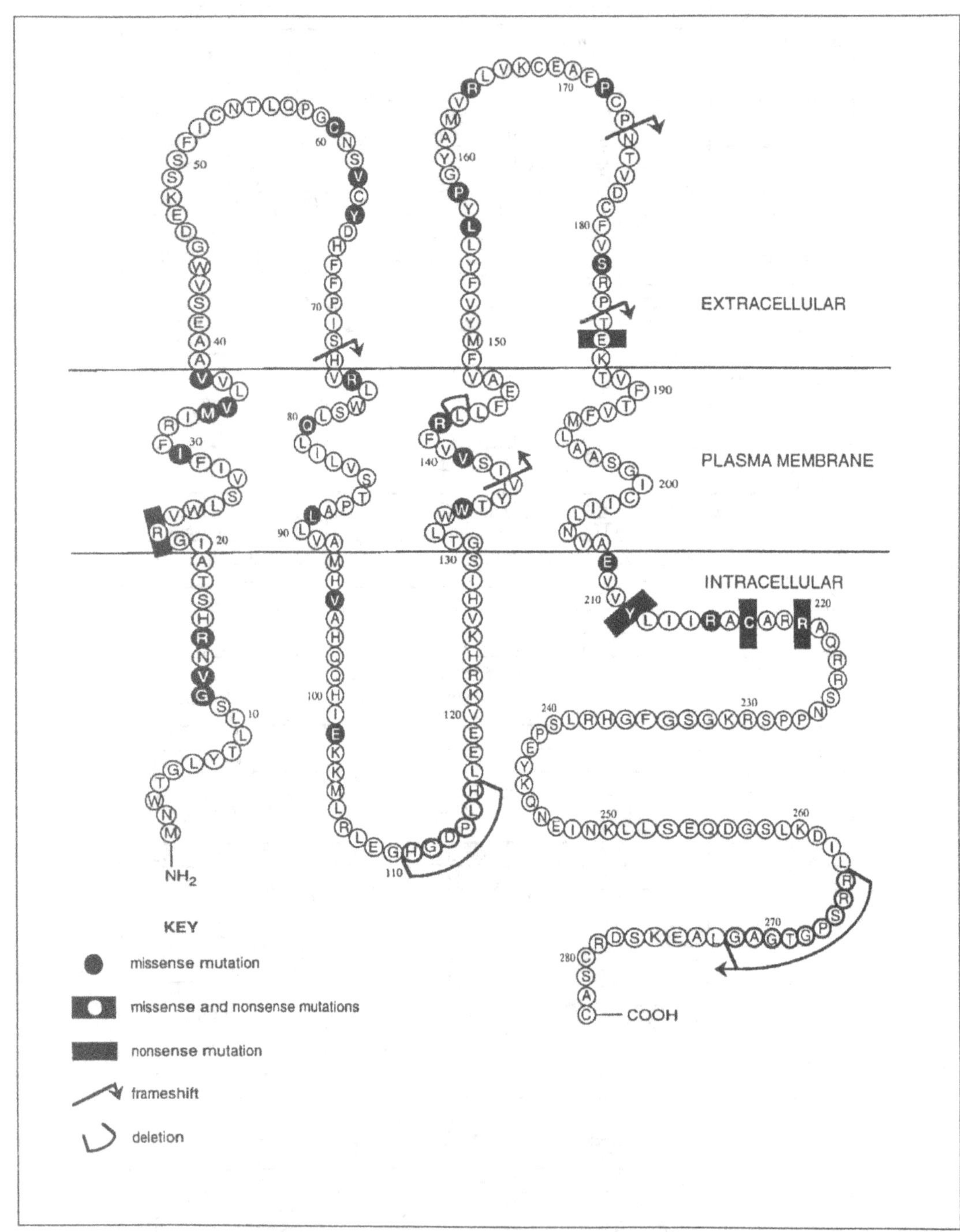

Fig. 13.3. Diagrammatic representation of Cx32 mutations in CMTX families.

structure and assembly see refs. 62,63). The effects of carboxy terminal mutations that prematurely truncate the protein are less clear, however. Nonsense mutations at residue 220 were found in four different CMTX families, and a deletion of amino acids 264-272 was found in another family (Table 13.1), yet truncations of up to 64 amino acids (up to and including residue 220) of the C-terminus are compatible with normal function in both *Xenopus* oocytes and mammalian cells.[64-66] Larger C-terminal truncations result in progressive loss of function in oocytes.[65] The apparent contradiction between in vitro and in vivo effects of nonsense mutations at residue 220 may reflect limitations in the experimental systems and methods used to assess the effects of the mutations.

To determine whether other CMTX mutations result in a loss of function, Bruzzone et al[67] expressed three different mutant genes (175 frameshift, Arg142Trp, and Glu186Lys), as well as the wild-type Cx32 gene, in *Xenopus* oocytes. Whereas wild-type Cx32 formed gap junctions, no oocytes expressing any of the three CMTX cDNAs formed functional gap junctions with oocytes expressing wild-type connexin26 (Cx26) or Cx32. To determine whether Cx32 mutations could also have a dominant-negative effect, the same three CMTX mutations were coinjected with wild-type Cx26, which is coexpressed with Cx32 in several tissues.[68,69] Oocytes that were coinjected with wild-type Cx26 and Cx32 cDNAs formed gap junctions when apposed to oocytes that expressed wild-type Cx26 cDNA, as previously demonstrated by Barrio et al.[70] Oocytes that were coinjected with wild-type Cx26 and Cx32 cDNA with CMTX mutations, however, showed decreased junctional conductance.[67] Similarly, other evidence suggests that three other CMTX mutations (Cys60Phe, Val139Met, Arg215Trp) cause loss of function and/or dominant-negative effects in cultured mammalian cells, as assayed by dye transfer.[66] These data indicate that CMTX mutations cause a loss of functional Cx32 protein, and that at least some

CMTX mutations can also have a dominant-negative effect in vitro.

5. Cx32 EXPRESSION IN MYELINATING CELLS

To understand how CMTX mutations cause peripheral neuropathy, one needs to consider the structure of the myelin sheath. Myelin is formed by spiral wrapping of the cell membrane of myelinating glia—oligodendrocytes in the central nervous system (CNS) and Schwann cells in the peripheral nervous system (PNS). The myelin sheaths produced by these two cell types are structurally similar, consisting mostly of compact myelin that is characterized by unique but overlapping sets of proteins.[71,72] Proteolipid protein (PLP) and myelin basic protein (MBP) are the major structural proteins in the CNS, while P_0, MBP, and PMP-22 are the major proteins in the PNS. Each of these proteins is essential for proper myelination, since mutations in PLP and MBP cause dysmyelination in the CNS and mutations in P_0 and PMP-22 cause dysmyelination in the PNS.[72-75]

The compact myelin in the PNS contains periodic interruptions called Schmidt-Lanterman incisures or clefts.[76] These incisures and the paranodal regions of the myelin sheath contain a distinct group of proteins, including myelin-associated glycoprotein, Cx32, E-cadherin, and oligodendrocyte-myelin glycoprotein.[1,77-79] Incisures have also been reported in the CNS,[76] but no molecules have yet been found to be preferentially located there. As shown in Figure 13.4a, Cx32-immunoreactivity is found in the incisures and paranodes and in the inner mesaxon.[1,80] This distribution differs from P_0 (Fig. 13.4b), which is found in compact myelin.

The location of Cx32 in the myelin sheath is similar to the location of putative gap junctions found by freeze-fracture electron microscopy.[81] As depicted in Figure 13.5, the gap junctions lie between rows of tight junctions at the paranodes, incisures, and inner mesaxon. Since CMTX mutations can disrupt the formation of functional gap junctions in vitro,[65-67] it is

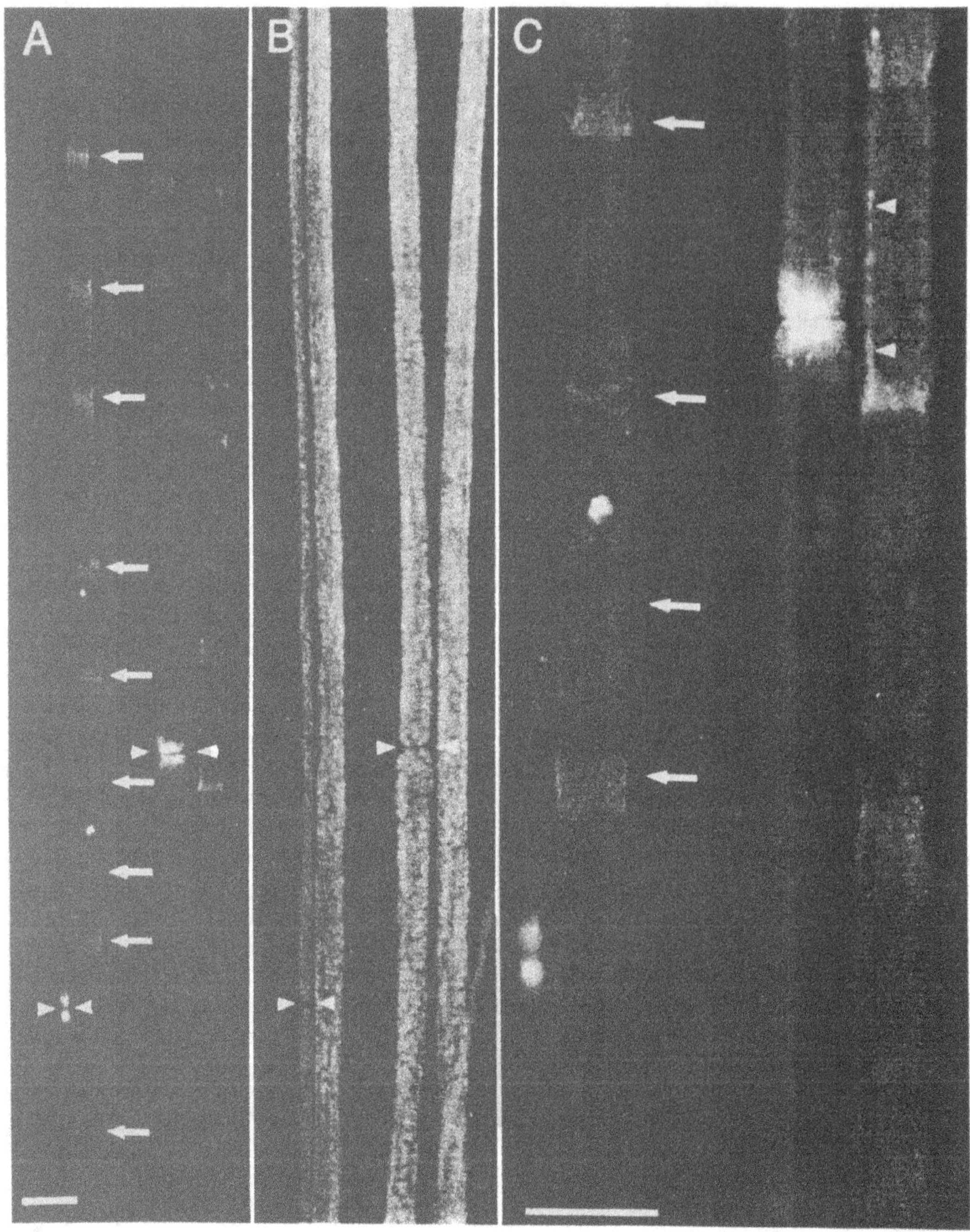

Fig. 13.4. Immunohistochemical localization of Cx32 in peripheral nerve. Panels (A) and (B) show the same field of four teased fibers from adult rat sciatic nerve, double-labeled with a monoclonal antibody against Cx32 (panel A) and a rabbit polyclonal antiserum against rat P_0 (panel B). Note in panel (A) that Cx32 is predominantly found at the incisures (arrows) and the paranodes (arrowheads). P_0, on the other hand, is localized to the compact myelin (panel B). Panel (C) is a 0.5 µm thick scanning laser confocal image of the fibers shown in panels (A) and (B). Scale bars: 22.5 µm in panels (A) and (B); 10 µm in panel (C). Reprinted with permission from Scherer S, J Neuro 1995; 15:8281-8294.

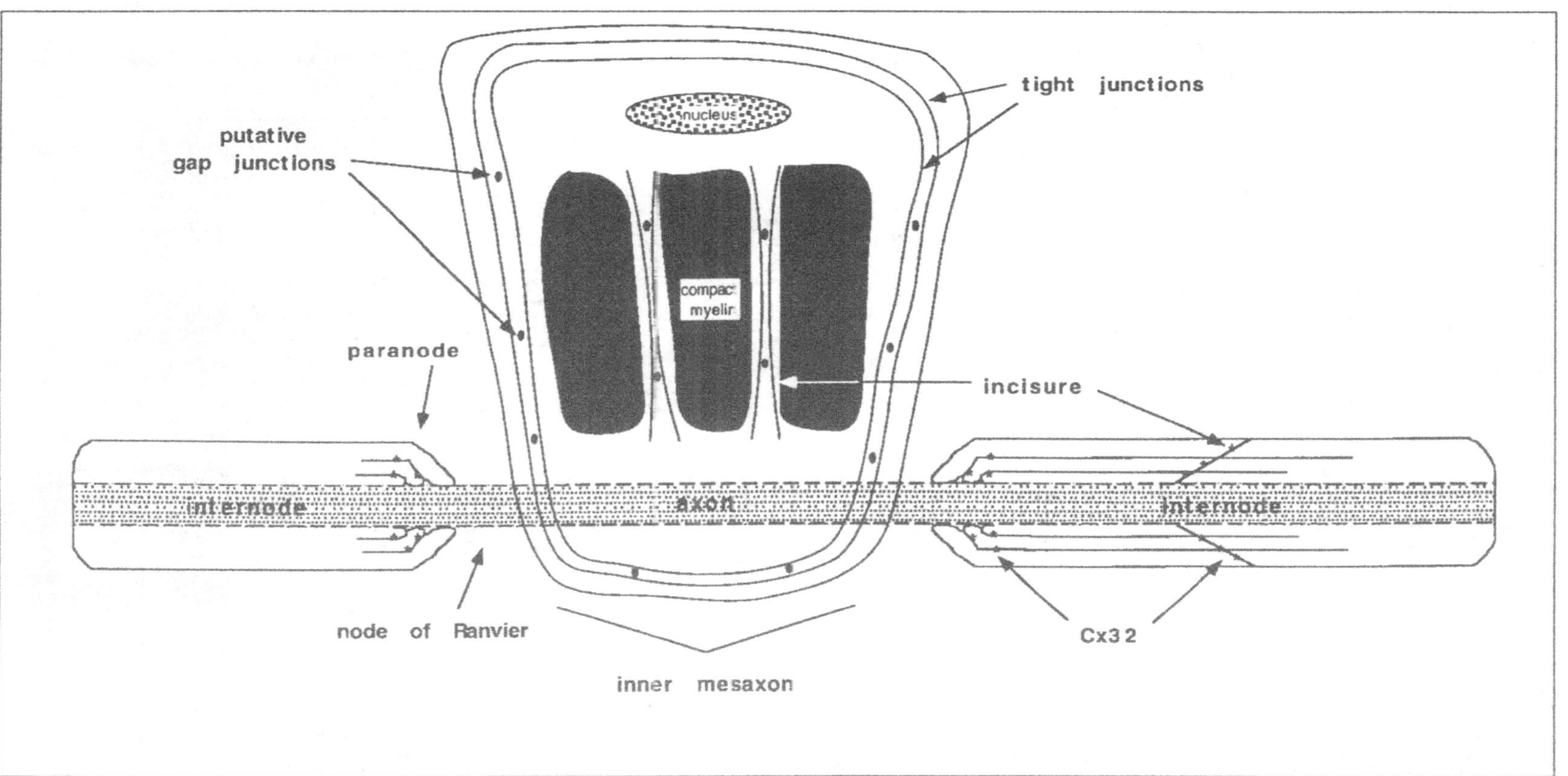

Fig. 13.5. Diagrammatic representation of Cx32 in the PNS myelin sheath. Shown are three myelinating Schwann cells, one unwrapped to show the hypothetical distribution of gap junctions between continuous rows of tight junctions. Asterisks mark the location of Cx32 at the paranodes and an incisure.

likely that Cx32 mutations alter the function of gap junctions in vivo. By adversely affecting the function of gap junctions, these mutations may also affect the structural integrity of incisures and paranodes, and ultimately the health of myelinated axons.

Cx32 is also expressed by oligodendrocytes,[80,82-84] which are the only other cell type that makes myelin. Cx32-immunoreactivity is found in oligodendrocyte cell bodies and processes, but does not appear to be found in compact myelin or incisures.[80,82] Cx32 mRNA and protein are expressed in parallel with other myelin-related genes during CNS development,[80] indicating that common transcriptional mechanisms regulate the expression of Cx32 and other myelin-related genes.

6. WHY DO CX32 MUTATIONS PRIMARILY AFFECT THE PNS?

Cx32 is expressed in many tissues,[82,85,86,87a] but the clinical manifestations of CMTX are generally limited to peripheral nerve.[20,21,32,42] A possible exception has been described in one CMTX patient who had white matter abnormalities in his brain.[87] In most cases, however, oligodendrocytes do not seem to be affected by Cx32 mutations, even though Schwann cells and oligodendrocytes both express Cx32. These data raise a number of important, interrelated questions. What is the function of Cx32, and how does it differ between Schwann cells and oligodendrocytes? Why are myelinating Schwann cells particularly susceptible to CMTX mutations? What mechanisms compensate for the loss of functional Cx32 in other tissues?

To explain the possible function of Cx32 in peripheral nerve, Bergoffen and colleagues[1] proposed a model in which connexons located between adjacent layers of noncompact myelin in the paranodes and incisures might form "reflexive" gap junctions that allow ions and other small molecules to diffuse radially through the myelin sheath, instead of taking the much longer circumferential pathway through the

Schwann cell cytoplasm (Fig. 13.5). If Cx32 mutations interrupt the function of these gap junctions, then this radial pathway would be eliminated, blocking the diffusion of small molecules across the myelin sheath. Both the nature of the crucial molecules that traverse these gap junctions and the relative importance of radial versus circumferential diffusion remain unknown.

The restriction of the CMTX phenotype to myelinating Schwann cells might be due to their failure to express another connexin that compensates for the deleterious effects of mutated Cx32. If this is the case, then a pure loss of function is more likely to be the cause of the CMTX phenotype than a dominant-negative interaction between mutant Cx32 and another connexin. Northern blot analysis of adult rat peripheral nerve did not show detectable Cx26 or Cx40 mRNA.[80] A low level of Cx43 mRNA was found, but probably originates from fibroblasts in peripheral nerve and not from Schwann cells.[80] As normal peripheral nerve also contains nonmyelinating Schwann cells, fibroblasts, and macrophages, it is difficult to determine the repertoire of connexins expressed by myelinating Schwann cells. A better model for this purpose may be forskolin-treated Schwann cells, which express moderate levels of most myelin-related genes, including Cx32.[88-91]

Gap junctions have also been described between nonmyelinating Schwann cells,[92] between "denervated" Schwann cells in nerves undergoing Wallerian degeneration,[93] and between cultured rat Schwann cells.[94,95] The connexin that forms these gap junctions is unlikely to be Cx32 for the following reasons: (1) The levels of Cx32 mRNA and protein fall after axotomy, in parallel with other myelin-related genes.[96,80] (2) Schwann cells cultured in the absence of forskolin express little Cx32.[80] (3) The electrophysiological properties of the gap junctions in Schwann cells cultured in the absence of forskolin differ from those of Cx32.[94] Gap junctional coupling between cultured Schwann cells is enhanced by the combination of forskolin and pituitary extract, and reduced by transforming growth factor-β or tumor necrosis factor-α.[95,96] The identity of the connexin expressed by cultured Schwann cells has not yet been determined.

7. ANIMAL MODELS AND CMTX

Perhaps the best way to address the issues raised above is to study CMTX mutations in vivo. As no naturally-occurring animal models for CMTX are known to exist, it is necessary to make Cx32 knockout and transgenic mice. If myelinating Schwann cells solely express Cx32, then the obliteration of this gene should result in the complete loss of functional gap junctions in peripheral nerve and the subsequent development of PNS manifestations. Nelles and colleagues[97] recently described the generation of a line of mice deficient in the expression of Cx32. Although no Cx32 mRNA or protein was detectable in liver or peripheral nerve, these animals had no peripheral nerve degeneration up to eight months following birth, suggesting that the expression or role of connexins may differ in the peripheral nerves of mice and humans. It is also possible that dominant-negative effects, rather than a simple loss of Cx32 function, may be necessary for the CMTX phenotype to be observed, or that other species differences, such as life span and nerve length, may protect mice from the neuropathologic effects of Cx32 mutations which are seen in humans.

Further information about the role of Cx32 in peripheral nerve may be gained by the generation of transgenic mice expressing CMTX mutations known to exhibit dominant-negative effects in vitro. If these mice develop the neuropathology of CMTX, then this would indicate that at least some CMTX mutations have dominant-negative effects in vivo. Such a transgenic model for CMTX would offer the opportunity for further elucidation of the pathogenesis and perhaps development of an effective treatment for this disorder.

ACKNOWLEDGMENTS

We are indebted to the CMTX families and their health care providers who are represented in this review, Dr. David Paul for providing antibodies and stimulating discussion, Drs. Angelica Hahn and Marvin Rozear for the nerve biopsies, and Shelly Whyatt for preparing the electron micrographs. This work was supported by grants from the Muscular Dystrophy Association and the National Institutes of Health (HG00233, NS08075, NS01565) to K.H.F. and S.S.S. and the Murray M. Stokely Award to S.S.S.

REFERENCES

1. Bergoffen J, Scherer SS, Wang S et al. Connexin mutations in X-linked Charcot-Marie-Tooth disease. Science 1993a; 262:2039-2042.
2. Charcot J-M, Marie P. Sur une forme particulière d'atrophie musculaire progressive souvent familial débutant par les pieds et les jambes et atteignant plus tard les mains. Rev Med 1886; 6:97-138.
3. Tooth HH. The Peroneal Type of Progressive Muscular Atrophy. London, H.K. Lewis and Co., Ltd., 1886.
4. Skre, H. Genetic and clinical aspects of Charcot Marie Tooth's disease. Clin Genet 1974; 6:98-118.
5. Matsunami N, Smith B, Ballard L, et al. Peripheral myelin protein-22 gene maps in the duplication in chromosome17p11.2 associated with Charcot-Marie-Tooth 1A. Nat Genet 1992; 1:176- 179.
6. Patel PI, Roa BB, Welcher AA et al. The gene for the peripheral myelin protein PMP-22 is a candidate for Charcot-Marie - Tooth disease type 1A. Nat Genet 1992; 1:159-165.
7. Timmerman V, Nelis E, Van Hul W et al. The peripheral myelin protein gene *PMP-22* is contained within the Charcot-Marie-Tooth disease type1 A duplication. Nat Genet 1:1992; 171-175.
8. Valentijn LJ, Bolhuis PA, Zorn I et al. The peripheral myelin gene *PMP-22/GAS-3* is duplicated in Charcot-Marie-Tooth disease type 1a. Nat Genet 1992; 1:166-170.
9. Valentijn LJ, Baas F, Wolterman RA. Iden-

tical point mutations of PMP-22 in *Trembler-J* mouse and Charcot-Marie-Tooth disease type 1A. Nat Genet 1992; 2:288-291.

10. Roa BB, Garcia CA, Suter U et al. Charcot-Marie-Tooth disease type 1A: Association with a spontaneous point mutation in the *PMP22* gene. N Engl J Med 1993; 329:96-101.

11. Hayasaka K, Himoro M, Sato W et al. Charcot-Marie-Tooth neuropathy type 1B is associated with mutations of the myelin *Po* gene. Nat Genet 1993; 5:31-34.

12. Hayasaka K, Takada G, Ionasescu VV. Mutation of the myelin *Po* gene in Charcot-Marie-Tooth neuropathy type 1B. Hum Mol Genet 1993; 2:1369-1372.

13. Hayasaka K, Ohnishi A, Takada G et al. Mutation of the myelin *Po* gene in Charcot-Marie-Tooth neuropathy type 1. Biochem Biophys Res Commun 1993; 194:1317-1322.

14. Himoro M, Yoshikawa H, Matsui T et al. New mutation of the myelin *Po* gene in a pedigree of Charcot-Marie-Tooth neuropathy 1. Biochem & Mol Biol Intl 1993; 31:169-173.

15. Kulkens T, Bolhuis PA, Wolterman RA et al. Deletion of the serine 34 codon from the major peripheral myelin protein *Po* gene in Charcot-Marie-Tooth disease type 1B. Nat Genet 1993; 5:35-39.

16. Herringham WP. Muscular atrophy of the peroneal type affecting many members of a family. Brain 1888; 11:230-236.

17. Ionasescu VV, Ionasescu R, Searby C. Screening of dominantly inherited Charcot-Marie-Tooth neuropathies. Muscle Nerve 1993; 16:1232-1238.

18. Ionasescu VV, Trofatter J, Haines JL. Heterogeneity in X-linked recessive Charcot-Marie-Tooth neuropathy. Am J Hum Genet 1991; 48:1075-1083.

19. Ionasescu VV, Trofatter J, Haines JL et al. X-linked recessive Charcot-Marie-Tooth neuropathy—clinical and genetic study. Muscle Nerve 1992; 15:368-373.

20. Rozear MP, Pericak-Vance MA, Fischbeck K et al. Hereditary motor and sensory neuropathy, X-linked: A half century follow-up. Neurology 1987; 37:1460-1465.

21. Hahn AF, Brown WF, Koopman WJ et al.

22. Nicholson GA, Nash J. Intermediate nerve conduction velocities define X-linked Charcot-Marie-Tooth neuropathy families. Neurology 1993; 43:2558-2564.

23. Kaku DA, Parry GJ, Malamut R et al. Nerve conduction studies in Charcot-Marie-Tooth polyneuropathy associated with a segmental duplication of chromosome 17. Neurology 1993; 43:1806-1808.

24. Harding AE, Thomas PK. The clinical features of hereditary motor and sensory neuropathy types I and II. Brain 1980; 103:259-280.

25. Allan W. Relation of hereditary pattern to clinical severity as illustrated by peroneal atrophy. Arch Intern Med 1939; 63:1123-1131.

26. Erwin WG. A pedigree of sex-linked recessive peroneal atrophy. J Hered 1944; 35:24-26.

27. Campeanu E, Morariu, M. Les relations entre genotype et phenotype dans la maladie de Charcot-Marie-Tooth. Rev Roum Neurol 1970; 7:47-56.

28. Fryns JP, van den Berghe, H. Sex-linked recessive inheritance in Charcot-Marie-Tooth disease with partial clinical manifestations in female carriers. Hum Genet 1980; 55:413-415.

29. Woratz G. Neurale Muskelatrophie mit dominantem X-chromosomalem Erbgang. Abh Dtsch Akad Wissensch Berl Nr. 2, 1964.

30. de Weerdt CJ. Charcot-Marie-Tooth disease with sex-linked inheritance, linkage studies and abnormal serum alkaline phosphatase levels. Eur Neurol 1978; 17:336-344.

31. Iselius L, Grimby L. A family with Charcot-Marie-Tooth's disease, showing a probable X-linked incompletely dominant inheritance. Hereditas 1982; 97:157-158.

32. Phillips LH, Kelly TE, Schnatterly P et al. Hereditary motor-sensory neuropathy (HMSN): Possible X-linked dominant inheritance. Neurology 1985; 35:498-502.

33. Gal A, Mucke J, Theile H, Wieacker PF et al. X-linked dominant Charcot-Marie-Tooth disease: Suggestion of linkage with a cloned

DNA sequence from the proximal Xq. Hum Genet 1985; 70:38-42.

34. Beckett J, Holden JJA, Simpson NE et al. Localization of X-linked dominant Charcot-Marie-Tooth disease (CMT2) to Xq13. J Neurogenet 1986; 3:225-231.

35. Fischbeck KH, ar-Rushdi N, Pericak-Vance M et al. X-linked neuropathy: Gene localization with DNA probes. Ann Neurol 1986; 20:527-532.

36. Kelly TE, Lunt P, Schnatterly P et al. Evidence that bulbospinal neuropathy (BSN-X) lies distal to DXYS1 and X-linked Charcot-Marie-Tooth disease (CMT-X) lies proximal to DXYS1 on Xq. Am J Hum Genet 1987; 41:A101.

37. Goonewardena P, Welinhinda J, Anvret M et al. A linkage study of the locus for X-linked Charcot-Marie-Tooth disease. Clin Genet 1988; 33:435-440.

38. Ionasescu VV, Burns TL, Searby C et al. X-linked dominant Charcot-Marie-Tooth neuropathy with 15 cases in a family: Genetic linkage study. Muscle Nerve 1988; 11:1154-1156.

39. Haites N, Fairweather N, Clark C et al. Linkage in a family with X-linked Charcot-Marie-Tooth disease. Clin Genet 1989; 35:399-403.

40. Fischbeck KH, Ritter A, Shi Y et al. X-linked recessive and X-linked dominant Charcot-Marie-Tooth disease. In: Lovelace, R.E., Shapiro, H.K. Charcot-Marie-Tooth Disorders: Pathophysiology, Molecular Genetics, and Therapy. New York: Wiley-Liss, 1990:335-341.

41. Mostacciuolo ML, Miller E, Fardin P et al. X-linked Charcot-Marie-Tooth disease—A linkage study in a large family by using 12 probes of the pericentromeric region. Hum Genet 1991; 87:23-27.

42. Ionasescu VV, Trofatter J, Haines JL et al. Mapping of the gene for X-linked dominant Charcot-Marie-Tooth neuropathy. Neurology 1992a; 42:903-908.

43. Bergoffen J, Trofatter J, Pericak-Vance MA et al. Linkage localization of X-linked Charcot-Marie-Tooth disease. Am J Hum Genet 1993b; 52:312-318.

44. Fain PR, Barker DF, Chance PF. Refined genetic mapping of X-linked Charcot-Marie-Tooth neuropathy. Am J Hum Genet 1994; 54:229-235.

45. Cochrane S, Bergoffen J, Fairweather ND et al. X-linked Charcot-Marie-Tooth disease (CMTX1)—A study of 15 families with 12 highly informative polymorphisms. J Med Genet 1994; 31:193-196.

46. Brown CJ, Sekiguchi T, Nishimoto T et al. Regional localization of *CCG1* gene which complements hamster cell cycle mutation BN462 to Xq11-Xq13. Somat Cell Molec Genet 1989; 15:93-96.

47. Noguchi M, Yi H, Rosenblatt HM et al. Interleukin-2 receptor gamma chain mutation results in X-linked severe combined immunodeficiency in humans. Cell 1993; 73:147-157.

48. Puck JM, Deschênes SM, Porter JC et al. The interleukin-2 receptor gamma chain maps to Xq13.1 and is mutated in X-linked severe combined immunodeficiency, SCIDX1. Hum Mol Genet 1993; 2: 1099-1104.

49. Corcos IA, Lafrenière RG, Begy CR et al. Refined localization of human connexin32 gene locus, GJB1, to Xq13.1. Genomics 1992; 13:479-480.

50. Raimondi E, Gaudi S, Moralli D et al. Assignment of the human connexin32 gene *(GJB1)* to band Xq13. Cytogenet Cell Genet 1992; 60:210-211.

51. Cherryson AK, Yeung L, Kennerson ML et al. Mutational studies in X-linked Charcot-Marie-Tooth disease (CMTX). Am J Hum Genet 1994; 55(suppl):1261.

52. Fairweather N, Bell C, Cochrane S et al. Mutations in the connexin32 gene in X-linked dominant Charcot-Marie-Tooth disease (CMTX1). Hum Mol Genet 1994; 3:29-34.

53. Ionasescu V, Searby C, Ionasescu R. Point mutations of the connexin32 (GJB1) gene in X-linked dominant Charcot-Marie-Tooth neuropathy. Hum Mol Genet 1994; 3: 355-358.

54. Orth U, Fairweather N, Exler MC et al. X-linked dominant Charcot-Marie-Tooth neuropathy: valine-38-methionine substitution of connexin32. Hum Mol Genet 1994; 3:1699-1700.

55. Tan CC, Ainsworth PJ, Hahn AF, MacLeod

PM. Novel mutations in the connexin32 gene associated with X-linked Charcot Marie Tooth disease. Human Mutation 1996; 7:167-171.

56. Bone LJ, Dahl N, Lensch MW et al. New connexin32 mutations associated with X-linked Charcot-Marie-Tooth disease. Neurology 1995; 45:1863-1866.

57. Kumar N, Gilula NB. Cloning and characterization of human and rat liver cDNAs coding for a gap junction protein. J Cell Biol 1986; 103:767-776.

58. Omitted in proofs.

59. Miller T, Dahl G., Werner R. Structure of a gap junction gene: rat connexin32. Biosci Reports 1988; 8:455-464.

60. Hennemann H, Kozjek G, Dahl E et al. Molecular cloning of mouse connexins26 and -32: similar genomic organization but distinct promoter sequences of two gap junction genes. Eur J Cell Biol 1992; 58:81-89.

61. Neuhaus IM, Dahl G, Werner R. Use of alternate promoters for tissue-specific expression of the gene coding for connexin32. Gene 1995., 158:257-262.

62. Revel J-P, Hoh JH, John SA et al. Aspects of gap junction structure and assembly. Semin Cell Biol 1992; 3:21-28.

63. Stauffer KA, Unwin N. Structure of gap junction channels. Semin Cell Biol 1992; 3:17-20.

64. Werner R, Levine E, Rabadan-Diehl C et al. Gating properties of connexin32 cell-cell channels and their mutants expressed in *Xenopus* oocytes. Proc R Soc Lond 1991; 243:5-11.

65. Rabadan-Diehl C, Dahl G, Werner R. A connexin32 mutation associated with Charcot-Marie-Tooth disease does not affect channel formation in oocytes. FEBS Lett 1994; 351:90-94.

66. Omori Y, Mesnil M, Yamasaki H. Connexin32 mutations from x-linked Charcot-Marie-Tooth disease patients: Functional defects and dominant negative effects. Molecular Biology of the Cell 1996; 7:907-916.

67. Bruzzone R, White TW, Scherer SS et al. Null mutations of connexin32 in patients with X-linked Charcot-Marie-Tooth disease. Neuron 1994; 13:1253-1260.

68. Nicholson B, Dermietzel R, Teplow D et al. Two homologous protein components of hepatic gap junctions. Nature 1987; 329:732-734.

69. Meda P, Pepper MS, Traub O et al. Differential expression of gap junction connexins in endocrine and exocrine glands. Endocrinol 1993; 133:2371-2378.

70. Barrio LC, Suchyna T, Bargiello T et al. Gap junctionss formed by connexins 26 and 32 alone and in combination are differently affected by applied voltage. Proc Natl Acad Sci USA 1991; 88:8410-8414.

71. Lemke G. Myelin and myelination. In: Hall, Z.W. Molecular Neurobiology. Sunderland, MA: Sinauer, 1992:281-312.

72. Lemke G. The molecular genetics of myelination—an update. Glia 1993; 7:263-271.

73. Hudson L. Molecular biology of myelin proteins in the central and peripheral nervous systems. Sem Neurosci 1990; 2: 487-496.

74. Chance PF, Pleasure D. Charcot-Marie-Tooth syndrome. Arch Neurol 1993; 50:1180-1183.

75. Suter U, Welcher AA, Snipes GJ. Progress in the molecular understanding of hereditary peripheral neuropathies reveals new insights into the biology of the peripheral nervous system. Trends Neurosci 1993; 16:50-56.

76. Peters A, Palay SL, Webster HD. The Fine Structure of the Nervous System. New York, Oxford University Press, PP.494; 1991.

77. Apostolski S, Sadiq SA, Hays A, Corbo M et al. Identification of Gal(b1-3)GalNAc—bearing glycoproteins at the nodes of Ranvier in peripheral nerve. J Neurosci Res 1994; 38:134-141.

78. Fannon AM, Sherman DL, Ilyina-Gragerova G et al. Novel E-cadherin mediated adhesion in peripheral nerve: Schwann cell architecture is stabilized by autotypic adherens junctions. J Cell Biol 1995; 129:189-202.

79. Trapp BD, Quarles RH. Presence of the myelin-associated glycoprotein correlates with alterations in the periodicity of peripheral myelin. J Cell Biol 1982; 9: 2877-882.

80. Scherer SS, Deschênes SM, Xu Y-T et al. Connexin32 is a myelin-related protein in

the PNS and CNS. J Neurosci 1995, 15:8281-8294.

81. Sandri C, van Buren JM, Akert K. Membrane morphology of the vertebrate nervous system: a study with freeze-etch technique. In: Progress in Brain Research, Vol. 46. 2nd rev. ed. New York, Elsevier Biomedical Press, 1982:201-265.

82. Dermietzel R, Traub O, Hwang TK et al. Differential expression of three gap junction proteins in developing and mature brain tissues. Proc Natl Acad Sci USA 1989; 86:10148-10152.

83. Micevych PE, Abelson L. Distribution of mRNAs coding for liver and heart gap junction proteins in the rat central nervous system. J Comp Neurol 1991; 305:96-118.

84. Robinson SR, Hampson ECGM, Munro MN et al. Unidirectional coupling of gap junctions between neuroglia. Science 1993; 262:1072-1074.

85. Dermietzel R, Hwang TK, Spray DC. The gap junction family: Structure, function and chemistry. Anat. Embryol. 1990; 182:517-528.

86. Bennett MVL, Barrio LC, Bargiello TA et al. Gap junctions: New tools, new answers, new questions. Neuron 1991; 6:305-320.

87. Paulson H, Deschênes S, Scherer S et al. CNS white matter abnormalities in a patient with X-linked Charcot-Marie-Tooth disease (CMTX). Am J Hum Genet 1994; 55(suppl):A89.

87a. Kumar NM, Gilula NB. Molecular biology and genetics of gap junction channels. Semin Cell Biol 1992; 3:3-16.

88. Sobue G, Pleasure D. Schwann cell galactocerebroside induced by derivatives of adenosine 3',5'-monophosphate. Science 1984; 224:72-74.

89. Lemke G, Chao M. Axons regulate Schwann cell expression of the major myelin and NGF receptor genes. Development 1988; 102:499-504.

90. Morgan L, Jessen KR, Mirsky R. The effects of cAMP on differentiation of cultured Schwann cells: progression from an early phenotype (04+) to a myelin phenotype (Po+, GFAP-, N-CAM-, NGF-receptor-) depends on growth inhibition. J Cell Biol 1991; 112:457-467.

91. Morgan L, Jessen KR, Mirsky R. Negative regulation of the Po gene in Schwann cells: suppression of Po mRNA and protein in cultured Schwann cells by FGF2 and TGFb1, TGFb2 and TGFb3. Development 1994; 120:1399-1409.

92. Konishi T. Dye coupling between mouse Schwann cells. Brain Res 1990; 508:85-92.

93. Tetzlaff W. Tight junction contact events and temporary gap junctions in the sciatic nerve fibres of the chicken during Wallerian degeneration and subsequent regeneration. J Neurocytol 1982; 11:839-858.

94. Chanson M, Chandross KJ, Rook MB et al. Gating characteristics of a steeply voltage dependent gap junction channel in rat Schwann-cells. J Gen Physiol 1993; 102:925-946.

95. Chandross KJ, Chanson M, Spray DC et al. Transforming growth factor-b1 and forskolin modulate gap junctional communication and cellular phenotype of cultured Schwann cells. J Neurosci 1995b; 15:262 273.

96. Chandross KJ, Kessler JA, Spray DC et al. Altered gap junction expression after peripheral nerve injury. Mol Cell Neurosci 1996; in press.

97. Nelles E, Jung D, Gabriel H-D et al. Characterization of connexin32 deficient mice generated by gene targeting. 1995 Gap Junction Conference; 1995, March 5-10; Ile des Embiez, France.

CHAPTER 14

REGULATION OF CONNEXIN EXPRESSION IN SCHWANN CELLS

Karen J. Chandross, David C. Spray and John A. Kessler

1. SUMMARY

Myelinating Schwann cells in adult peripheral nerve express the gap junction protein connexin32 (Cx32), which is localized to paranodal loops and Schmidt-Lantermann incisures.[1] Cx32 is thought to form reflexive contacts within individual Schwann cells providing direct pathways for intracellular ionic and metabolic exchange from the cell body to the innermost periaxonal cytoplasmic regions. However, nerve injury induces coordinated phenotypic and functional changes in the distal Schwann cells no longer in contact with the viable nerve, including the downregulation of Cx32 mRNA and protein[2,3] and the upregulation of connexin46 (Cx46).[2] The cascade of Schwann cell responses seen after the injury-induced decrease in Cx32, and the observation that cultured Schwann cells express Cx46 and are coupled by a connexin(s) distinct from Cx32,[4-6] raise the intriguing possibility that there are coordinated changes in Schwann cell proliferation and junctional coupling and that Cx46 may be involved in Schwann cell injury responses. Consistent with this hypothesis, dividing Schwann cells are preferentially dye coupled compared to nondividing cells,[6] directly demonstrating an association between proliferation and coupling. Moreover, the strength of junctional coupling among cultured Schwann cells is modulated by a number of inflammatory cytokines to which Schwann cells are exposed to in vivo and Cx46 mRNA and protein expression correlate with the degree of junctional coupling.[5,6]

2. INTRODUCTION

In adult peripheral nerve, Schwann cells are mitotically quiescent, ensheath or myelinate axons, and contribute towards normal nerve function.[7-9] Following nerve injury, axons distal to the site of injury are

Gap Junctions in the Nervous System, edited by David C. Spray and Rolf Dermietzel.
© 1996 R.G. Landes Company.

dissociated from viable cell bodies and undergo Wallerian degeneration, depriving the Schwann cells that formerly ensheathed them of axonal contact. In contrast to the limited plasticity of neurons, Schwann cells re-enter the cell cycle, rapidly proliferate, and dedifferentiate in response to injury.[10-12] Coordinated phenotypic changes that occur in the "distal" Schwann cells associated with degenerating axons[10,13] elicit functional changes that are thought to be instrumental for neuron survival and axon regeneration (Table 14.1, Fig. 14.1). For example, during Wallerian degeneration Schwann cells express proteins normally found in mature nonmyelinating cells or characteristic of developing Schwann cells; members of the nerve growth factor (NGF) family of proteins, low affinity nerve growth factor receptor (L-NGFR), glial fibrillary acidic protein (GFAP), and neural cell adhesion molecules (N-CAMs) are upregulated or re-expressed at relatively high levels.[14-19] Moreover, in previously myelinating Schwann cells, expression of the peripheral myelin proteins P_0, myelin basic protein, and peripheral myelin protein-22 rapidly

declines.[20-25] During the later stages of nerve regeneration, re-establishment of axonal contact elicits another increase in Schwann cell proliferation[26,27] and induces new basal lamina formation.[28,29] Overlying Schwann cells stop dividing and re-establish an intimate association with the axon. Moreover, these re-differentiated Schwann cells decrease expression of L-NGFR, GFAP, and N-CAM.[16,30] During Schwann cell redifferentiation, axons greater than 1 μm in diameter promote an increase in synthesis of myelin-specific proteins and remyelination.[31-35]

Injury-related Schwann cell phenotypic and functional changes are stimulated, in part, by the upregulation and local release of peptide factors during the inflammatory response. However, the specific signals involved in initiating Schwann cell responses to nerve injury and the mechanisms underlying the propagation of information from the site of injury to surrounding Schwann cells remain elusive. Moreover, the mechanisms modulating the temporal and regional differences in Schwann cell responses, such as the early and late

---

Table 14.1. Antigenic markers of Schwann cell phenotype

| Antigen | Differentiated Schwann Cells | | Dedifferentiated Schwann Cells | |
	Myelinating*	Nonmyelinating	Distal to Injury	Cultured (no neurons)
P_0	+	–	↓	↓
MBP	+	–	↓	↓
PMP-22	+	–	↓	↓
MAG	+	–	↓	↓
Cx32	+	+	↓	↓
Galactocerebroside (galC)	+	+	↓	↓
S100	+	+	+	+
Laminin	+	+	+	+
GFAP	–	+	↑	↑
L-NGFR	–	+	↑	↑
N-CAM	–	+	↑	↑
Proliferation	–	–	↑	↓

* Stimulation of cAMP generating pathways has been shown to produce similar phenotypic effects.
–, absent/low; ↓, decreased; +, present/high; ↑, increased

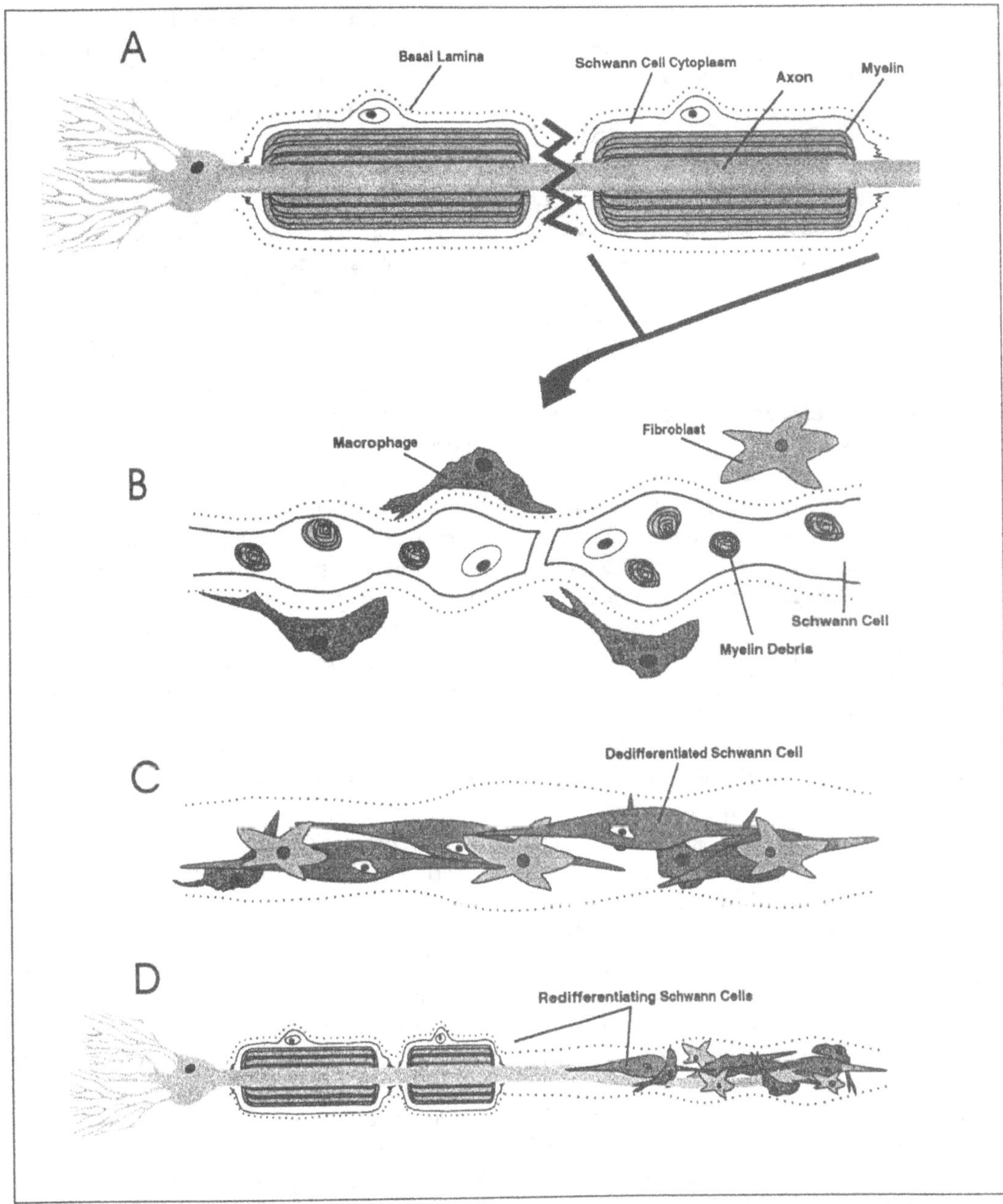

Fig. 14.1. Schwann Cell Responses to Nerve Injury. (A) Normal myelinated nerve fiber showing two Schwann cells and the intervening node of Ranvier, surrounded by Schwann cell basal lamina. (B) Within 3-4 days after injury, the distal region of the axon degenerates, the myelin sheath breaks down and forms ovoids within the Schwann cell cytoplasm, myelin gene synthesis is decreased, and Schwann cells begin to divide. Moreover, numerous macrophages and fibroblasts localize to the injured region and adhere to the Schwann cell basal lamina. (C) After 1 to 2 weeks, Schwann cells are rapidly dividing and form Büngner bands within their original basal lamina tubes. Moreover, Schwann cells increase expression of neurotrophins, the low affinity nerve growth factor receptor (L-NGFR), glial fibrillary acidic protein (GFAP), and neural cell adhesion molecule (N-CAM). Macrophages and fibroblasts have infiltrated the Schwann cell tubes. Most of the myelin debris has been cleared by macrophages, and to a lesser extent by Schwann cells. Moreover, many extratubal endoneurial macrophages now contain large amounts of myelin debris and lipid droplets. (D) By 3 weeks, Schwann cells make contact with the outgrowing axon and they begin to redifferentiate (e.g., decrease expression of L-NGFR, GFAP, and N-CAM and increase myelin gene expression). Moreover, in the more proximal region Schwann cells have remyelinated axons of appropriate diameter.

Schwann cell proliferative responses localized to the degenerating region or the subsequent withdrawal from the cell cycle after recontact with the regenerating axon, are not known. Our own studies, as well as the work of others,[5,6,36,37] suggest that some of the critical regulatory signals may be cytokines to which Schwann cells are exposed to during injury responses. We have examined the effects of inflammatory cytokines on cultured Schwann cell connexin expression, junctional coupling strength, proliferation, and antigenic phenotype. These studies demonstrate coordinated changes in Schwann cell phenotype[2,5,6] that parallel changes that occur after nerve injury. A specific hypothesis that has emerged is that regulation of gap junctional communication among Schwann cells helps to coordinate responses to nerve injury.

3. CONNEXINS AND GAP JUNCTION CHANNELS IN PERIPHERAL NERVE

Several cellular elements of adult peripheral nerve, such as fibroblasts and various blood vessel wall cells, including vascular smooth muscle cells and endothelial cells, express the gap junction protein, connexin43 (Cx43).[38,39] Moreover, Cx43 is transiently expressed in endoneurial fibroblasts following sciatic nerve crush injury.[2] Additionally, cultured mouse macrophages express Cx43[40,41] and this connexin is briefly expressed in peripheral nerve macrophages during the early stages of Wallerian degeneration (unpublished observations). It has been hypothesized that the function of Cx43 in macrophages may be in ATP-induced permeabilization[42,43] (however, see also ref. 44).

In adult rat peripheral nerve Cx32 is localized to myelinating Schwann cells at the noncompacted region of the myelin including the lamellar layers at the nodes of Ranvier and Schmidt-Lantermann incisures.[1,45] Mutations in the gene for Cx32 are responsible for the inherited human X-linked form of Charcot-Marie-Tooth disease, implicating this gap junction protein in normal adult nerve function.[1,46-48] The

resulting peripheral neuropathy is characterized by progressive nerve demyelination and impeded nerve conduction.

The presence of gap junctions across Schwann cell internodes in normal myelinated nerve has not been demonstrated, suggesting that the primary role of Cx32 is in intracellular metabolic and ionic exchange across the myelin layers.[1] The resulting gap junction channels would provide a direct transcytoplasmic route for ionic and metabolic exchange between the cell body to the innermost para-axonal cytoplasmic region. However, after sciatic nerve crush injury there is a rapid decrease in Cx32 mRNA and protein in the crush and distal segments, with the subsequent re-expression of Cx32 in these regions as regeneration proceeds.[2,3] Moreover, the changes in expression of Cx32 coincide with other phenotypic changes that Schwann cells undergo during Wallerian degeneration, including changes in L-NGFR[2] and myelin expression.[3]

Although gap junction expression between mitotically quiescent Schwann cells surrounding intact nerve has not been reported, ultrastructural studies show gap junction formation between Schwann cells forming the bands of Büngner during Wallerian degeneration in the chick sciatic nerve[49] and following experimental allergic neuritis.[50] These junctions become much less numerous with the onset of remyelination. Mouse Schwann cells are dye coupled when they are cocultured with dorsal root ganglia but are not coupled after onset of myelination.[51] Moreover, Schwann cells are coupled when cultured alone[4-6] or when expressing a nonmyelinating phenotype in coculture with dorsal root ganglia.[51] In acutely dissected rat sciatic nerve myelinating Schwann cells are uncoupled or are weakly coupled.[52] However, distinct changes in Schwann cell intercellular coupling are observed after injury; by 7 days after crush injury proliferating Schwann cells in the distal region are coupled[52] (unpublished observations). Moreover, after sciatic nerve crush injury Cx46 mRNA and protein are upregulated in the crush and distal regions.[2] These results suggest a possible role for Schwann cell gap junction channels, distinct from Cx32, in intercellular coupling and modulating the coordinated responses of these cells to nerve injury.

4. IN VITRO ANALYSIS OF SCHWANN CELL GAP JUNCTIONS

Electrophysiological, molecular, and immunocytochemical analyses demonstrate that both cultured Schwann cells and Schwann cells in the distal segment of crush injured peripheral nerve express a gap junction channel(s) distinct from Cx32.[2,4-6] Moreover, Cx46 mRNA and protein expression is rapidly induced in injured nerve,[2] where Schwann cells are dedifferentiating and are re-entering the cell cycle. In order to examine the connexin(s) expressed in Schwann cells and to establish whether a link exists between functional phenotype and expression of gap junction channels we developed an in vitro paradigm using Schwann cell culturing techniques first described by Brockes et al.[53]

Initially we looked for expression of gap junction channels and functional coupling among purified rat sciatic nerve Schwann cells. In vitro, Schwann cells assume a dedifferentiated phenotype (see Table 14.1; e.g., low levels of P_0, increased expression of both L-NGFR and GFAP).[30] Moreover, these cells are relatively mitotically quiescent and are weakly coupled by gap junction channels.[4,6] Schwann cell pairs exhibit two distinct gap junction channel sizes with single channel currents corresponding to a population with unitary conductance of about 44 pS and a second, less frequent, population of about 26 pS. However, phosphorylating treatments (e.g., forskolin), which have been shown to shift unitary conductance and modify voltage dependence kinetics connexin43 channels,[54] do not effect events of either size or change the electrophysiological properties of the Schwann cell junctional channels.[4-6] These results suggest either that Schwann cells express more than one functional gap junc-

tion channel, or that channel properties are modified by other regulatory agents, such as physiological changes in intracellular pH or Ca^{2+}.

Schwann cell junctional channels are highly voltage sensitive.[4] In response to either depolarizing or hyperpolarizing transjunctional voltage (V_j) steps, currents decay exponentially, with time constants ranging from <1-10 sec. The relationship between normalized steady state junctional conductance (G_{ss}) and V_j is well described by a Boltzmann relation with a steeper voltage sensitivity than seen for any other known mammalian gap junction channel $(V_0$, the V_j at which g_j is reduced by 50%, is about 14 mV; n, the equivalent gating change, is about 2.4). At high V_j, G_{ss} is virtually zero, a property unique among gap junction channels studied to date. Furthermore, the open probability is the primary determinant of macroscopic voltage dependence. These data demonstrate that cultured Schwann cells express a gap junction channel(s) that exhibits properties that are unique among connexins. For comparison, mammalian cells stably transfected with Cx32 express channels that have larger unitary conductances (about 120-150 pS), and that are only about half as voltage sensitive as Schwann cell junctional channel(s).[4,55,56] Moreover, Cx43 channels exhibit smaller unitary conductance values (e.g., 60-70 and 90-110 pS), but they are only weakly voltage sensitive.[54] Connexin45 channels have comparably low unitary conductances, although macroscopic voltage sensitivity is lower than in Schwann cells.[57]

Reverse transcription PCR sequence analyses demonstrate that Schwann cells express mRNA for a Cx(s) that is at least 99% identical to Cx46 in the amplified region[6] (Fig. 14.2; amplified region extends from the second to third transmembrane domains, including the most diverse region among the connexins, the cytoplasmic loop; primers were first described by Haefliger et al[58]). Moreover, in untreated cultures, only a Group II (or α) band is present, and none of the isolated clones show high sequence homology to the Group I (or β) connexins, including Cx32 (see ref. 59 for classification of connexins). Immunocytochemical and Western blot analyses of cultured Schwann cells confirm the presence of Cx46.[6] Moreover, expression of Cx46 protein correlates with the proliferative state of the cells; dividing Schwann cells express higher, whereas non dividing cells express lower levels of this protein (see below).[6] However, the functional properties exhibited by Schwann cell gap junction channels differ from those obtained following expression of Cx46 cDNA in *Xenopus* oocytes, where junctional channels are characterized by low voltage sensitivity.[60] One explanation for this anomalous behavior is that oocyte and Schwann cell Cx46 undergo different posttranslational modifications. Evidence for this reasoning is seen by a marked difference in mobility of lens versus oocyte Cx46 on western blots.[60,61] Moreover, unlike in lens,[61] alkaline phosphatase does not reduce the size of the Schwann cell Cx46 species.[6] However, Schwann cells may be coupled by several different connexins whose identity and function have yet to be discovered.

5. CYTOKINES AND SCHWANN CELL GAP JUNCTIONS

In order to identify signals that may modulate cellular responses, we examined the effect of cytokines released after nerve injury on the morphologic, biochemical and functional phenotype of Schwann cells in vitro (Table 14.2). Moreover, to begin to examine mechanisms regulating Schwann cell responses to nerve injury we examined the effects of inflammatory cytokines on the expression of gap junction channels. For example, the cytokine transforming growth factor (TGF)-β_1 is present in normal adult sciatic nerve Schwann cells.[62] Following nerve injury there is increased production of TGFβ_1 by Schwann cells,[63] injured dorsal root ganglion neurons,[64] fibroblasts,[65] and activated macrophages.[36] As axons regenerate, levels of TGFβ_1 decline to levels comparable to those of normal nerve.[62] Moreover, TGFβ_1 is present in cultured Schwann cells[62] and this cytokine has been shown to exert a variety of effects on these cells.[5,63,66-69] Treatment of Schwann cell cultures with TGFβ_1 significantly decreases both Schwann cell electrical and dye coupling and coordinately decreases Cx46 expression.[5,6] Moreover, this cytokine has a minimal effect on proliferation and it enhances Schwann cell differentiation (see Table 14.1); immunolabeling for both L-NGFR and GFAP are reduced.[5,66] Additionally, TGFβ_1 decreases the expression of P_0,[5,62,66,68] suggesting that this treatment induces differentiation towards the mature nonmyelinating phenotype.

After nerve injury plasma levels of the cytokine tumor necrosis factor (TNF)-α

Table 14.2. Phenotypic changes in Schwann cells in culture by inflammatory peptides

Phenotype	TFBβ_1	TNFα	F-BPE	NDFβ
L-NGFR	↓	NE	↓	NE
GFAP	↓	↓	↑↑	NE
P_0	NE/↓	↓	↑↑	↓
Proliferation	NE	↓	↑↑	↑
Process, SOMA	Multipolar FLAT	Bipolar ROUNDED	Multipolar ROUNDED	Bipolar SPINDLE
Dye/Junctional Coupling	↓	↓	↑↑	↑/NE*
Cx46	↓	↓	↑	↑
γ_j	NE	NE	NE	LARGER

↓, decreased as compared to untreated cultures; ↑, increased as compared to untreated cultures; NE, no/minor effect; * initially decreased coupling but values returned to control by 48 hr.

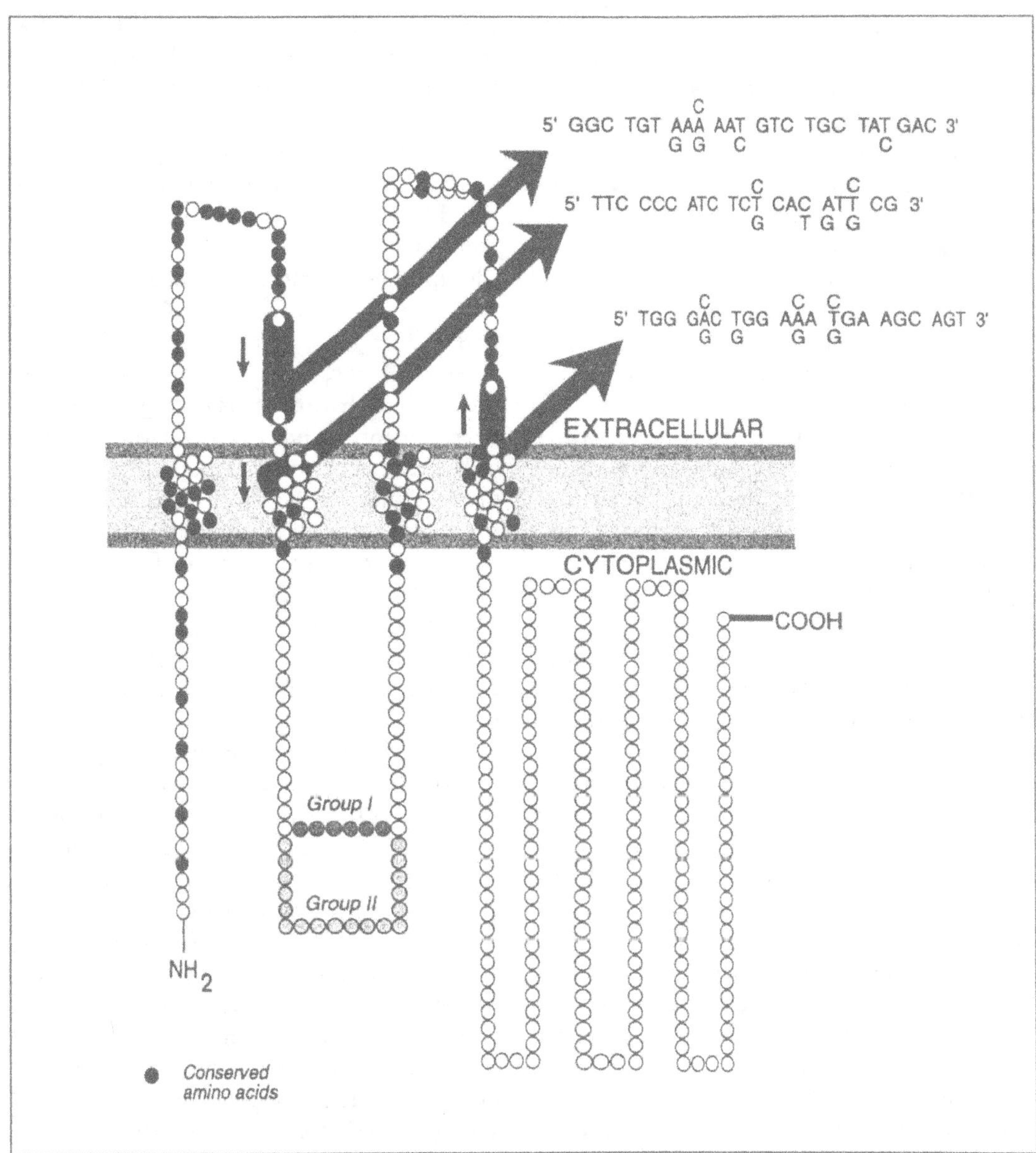

Fig. 14.2. Members of the connexin gene family have diverged into two distinct groups; Group I (or β) and Group II (or α). The Group I/β connexins include Cx26, Cx30, Cx30.3, Cx31, Cx31.1 and Cx32. The Group II/α connexins include Cx33, Cx37, Cx38, Cx40, Cx42, Cx43, Cx45, Cx45.6, Cx50 and Cx56. Sequence analysis reveals that Group I/β and Group II/α connexins differ in the length of their cytoplasmic loop. Consequently, primers to nucleotide sequences conserved among family members can be used to amplify and sequence a highly diverse region between the second and third transmembrane.

increase, peaking at 2 days and again between 11 days and 13 days, immediately before the decline of the two waves of Schwann cell proliferation.[70] TNFα is released by activated macrophages[71,72] and is found during Wallerian degeneration in the degenerating region of rat peripheral nerve.[37] Schwann cell proliferation during injury responses decreases temporally just after the peaks of TNFα, implicating this cytokine in cellular withdrawal from the mitotic cycle. In culture, TNFα significantly decreases Schwann cell junctional coupling, Cx46 protein expression, and proliferation without affecting cell viability.[6] Moreover, TNFα decreases both P_0 and GFAP immunoreactivity,[6] suggesting enhanced differentiation towards the non-myelinating phenotype (Tables 14.1 and 14.2).

Although it has been suggested that increased cyclic AMP levels in injured nerve are associated with cellular proliferation[73] and early re-establishment of function after injury,[74] the physiological signals responsible for the increased levels remain unclear. After injury elevated levels of cyclic AMP have been reported in Schwann cells in the distal region.[73,75] Moreover, in vitro experiments have shown that axonal membrane fragments are mitogenic to Schwann cells and that a rise in cyclic AMP occurs after stimulation with PC12 or dorsal root ganglion neuron-derived membranes.[76,77] However, there is still some controversy as to whether cyclic AMP levels are significantly elevated in the nerve after injury.[78,79] In combination with serum or growth factors, increases in this second messenger molecule induce Schwann cell proliferation in culture.[80,81] Treatment of Schwann cells with the mitogenic combination of forskolin, which increases intracellular levels of cyclic AMP, and bovine pituitary extract, which synergizes with forskolin (F-BPE)[80,81] coordinately increases junctional and dye coupling, Cx46 protein expression, and proliferation. Interestingly, although F-BPE enhances GFAP immunoreactivity, it reduces immunolabeling for L-NGFR and in-

creases expression of P_0, suggesting that this treatment promotes only partial differentiation of Schwann cells (Tables 14.1 and 14.2).

Several members of the glial growth factor (GGF) family of proteins, including neu differentiation factor (NDF)-β, are potent mitogens for rat Schwann cells.[82-85] The mitogenic effects are thought to be mediated by tyrosine phosphorylation of $P185^{neu/erbB2}$ (refs. 85,86) and possibly of other erbB receptors.[87] $P185^{neu/erbB2}$ is expressed in Schwann cells during Wallerian degeneration, and in cultured Schwann cells after exposure to forskolin.[86,87] Importantly, although TGFβ₁, TNFα, and F-BPE treatments have distinct effects on Schwann cell coupling strength, none of these treatments change the size or electrophysiologic characteristics of the junctional channels that are expressed between the cells.[5,6] By contrast, treatment with NDFβ alters the properties of the junctional channels; markedly larger unitary conductances and decreased voltage sensitivity are observed.[6] Initially, NDFβ reduces junctional coupling but has little effect on cell division or antigenic phenotype. However coupling increases concomitantly with increased expression of Cx46 protein, enhanced effects on proliferation and decreased P_0 immunoreactivity (Table 14.2).[6] The properties of the junctional channels change after NDFβ treatment with markedly larger unitary conductances (e.g., 60 pS and 100 pS) and decreased voltage sensitivity.

6. CONCLUSIONS

Inflammatory cytokines have direct effects on Schwann cell phenotype and function in vitro, suggesting that they may play an instrumental role in responses to nerve injury. Moreover, Schwann cell proliferation and junctional coupling are coordinately modulated. In vitro, factors that stimulate proliferation enhance junctional coupling and Cx46 expression, whereas factors that induce differentiation of basal state cells or inhibit proliferation decrease both junctional coupling and Cx46 expression. Moreover, after nerve injury, proliferating Schwann cells in the distal region are coupled and Cx46 mRNA and protein are upregulated. A specific hypothesis that has developed from these studies is that cytokine regulation of connexin expression and the strength of gap junctional communication among Schwann cells may play a critical role in coordinating Schwann cell responses to nerve injury and successful nerve regeneration, and indicate the involvement of gap junction proteins in addition to Cx32. As part of the initial response to injury, perhaps mediated by inflammatory cytokines, mitotically inactive Schwann cells decrease expression of Cx32 and may upregulate expression of Cx46. The enhanced degree of intercellular coupling would allow the transfer of metabolic and second messenger molecules between Schwann cells and may help to ensure more uniform and synchronous proliferative and dedifferentiating responses. After recontact with the regenerating axon, differentiating signals (e.g., cytokines) may simultaneously reduce coupling between cells and promote conversion to the differentiated, mitotically quiescent phenotype, characterized by re-expression of Cx32 in myelinating cells. Understanding the role of Schwann cells in nerve injury responses will therefore require a thorough understanding of the specific signals involved in regulating injury responses, the connexins expressed by these cells, and mechanisms regulating functional communication via gap junctions.

ACKNOWLEDGMENTS

Thanks to Dr. Ricky I. Cohen for his help in the preparation of this manuscript. This work was supported by NIH grants NS20013 (J.A.K), NS20778 (J.A.K), and NS34931 and NS07512 (D.C.S). K.J.C. was supported, in part, by NIH training grant T32 DK 07513. D.C.S. was also supported by a grant from the Muscular Dystrophy Association.

REFERENCES

1. Bergoffen J, Scherer SS, Wang S et al. Connexin mutations in X-linked Charcot-Marie-Tooth disease. Science (Wash) 1993; 262:2039-2042.
2. Chandross KC, Kessler JA, Cohen RI et al. Altered connexin expression after peripheral nerve injury. MCN 1996; 7:501-518.
3. Scherer SS, Deschénes SM, Xu Y-T et al. Connexin32 is a myelin-related protein in the PNS and CNS. J Neurosci 1995; 15:8281-8294.
4. Chanson M, Chandross KJ, Rook MB, Kessler JA et al. Gating characteristics of a steeply voltage-dependent gap junction channel in rat Schwann cells. J Gen Physiol 1993; 102:1-22.
5. Chandross KJ, Spray DC, Kessler JA. Transforming growth factor-β_1 and forskolin modulate gap junctional communication and cellular phenotype of cultured Schwann cells. J Neurosci 1995; 15:262-273.
6. Chandross KJ, Spray DC, Cohen RI et al. TNFα inhibits Schwann cell proliferation, connexin46 expression, and gap junctional communication. MCN 1996; 7:479-500.
7. Shrager P. Sodium channels in single demyelinated mammalian axons. Brain Res 1989; 483:149-154.
8. Kirkpatrick LL, Brady ST. Modulation of the axonal microtubule cytoskeleton by myelinating Schwann cells. J Neurosci 1994; 14:7440-7450.
9. Dugandzija-Novakovic S, Koszowski AG, Levinson SR, Shrager P. Clustering of Na$^+$ channels and node of Ranvier formation in remyelinating axons. J Neurosci 1995; 15:492-503.
10. Cajal SR. Degeneration and Regeneration of the Nervous System. In: May RM, ed. Oxford University Press, Oxford, UK. 1928; 769.
11. Pellegrino RG, Ritchie JM, Spencer JS. The role of Schwann cell division in the clearance of nodal axolemma following nerve section in the cat. J Physiol Lond 1982; 334:68-69.
12. Clemence A, Mirsky R, Jessen KR. Non-myelin-forming Schwann cells proliferate rapidly during Wallerian degeneration in the rat sciatic nerve. J Neurocytol 1989; 18:185-192.
13. Bunge RP. Tissue culture observations relevant to the study of axon-Schwann cell interactions during peripheral nerve develop-

ment and repair. J Exp Biol 1987; 132:21-34.

14. Rush RA. Immunohistochemical localization of endogenous nerve growth factor. Nature (Lond) 1984; 312:364-367.

15. Taniuchi M, Clark HB, Johnson EM Jr. Induction of nerve growth factor receptor in Schwann cells after axotomy. Proc Natl Acad Sci USA 1986; 83:4094-4098.

16. Taniuchi M, Clark HB, Schweitzer JB. Expression of nerve growth factor receptors by Schwann cells of axotomized peripheral nerves: ultrastructural location, suppression by axonal contact, and binding properties. J Neurosci 1988; 8:664-681.

17. Martini R, Schachner M. Immunoelectron microscopic localization of neural cell adhesion molecules (L1, N-CAM, and myelin-associated glycoprotein) in regenerating adult mouse sciatic nerve. J Cell Biol 1988; 106:1735-1746.

18. Jessen KR, Morgan L, Stewart HJS et al. Three markers of adult nonmyelin-forming Schwann cells, 217c (Ran-1), A5E3 and GFAP: development and regulation by neuron-Schwann cell interactions. Development (Camb) 1990; 109:91-103.

19. Martini R. Expression and functional roles of neural cell surface molecules and extracellular matrix components during development and regeneration of peripheral nerves. J Neurocytol 1994; 23:1-28.

20. Mirsky R, Winter J, Abney ER. Myelin-specific proteins and glycolipids in rat Schwann cells and oligodendrocytes in culture. J Cell Biol 1980; 84:483-494.

21. Poduslo JF. Regulation of myelin: biosynthesis of the major myelin glycoprotein by Schwann cells in the presence and absence of myelin assembly. J Neurochem 1984; 42:493-503.

22. Gupta SK, Poduslo JF, Mezei C. Temporal changes in P_0 and MBP gene expression after crush-injury of the adult peripheral nerve. Mol Brain Res 1988; 4:133-141.

23. Trapp BD, Hauer P, Lemke G. Axonal regulation of myelin protein mRNA levels in actively myelinating Schwann cells. J Neurosci 1988; 8:3515-3521.

24. Welcher AA, Suter U, De Leon M et al. A myelin protein is encoded by the homologue of a growth arrest-specific gene. Proc Natl Acad Sci USA 1991; 88:7195-7199.

25. Snipes GJ, Suter U, Welcher AA. Characterization of a novel peripheral nervous system myelin protein (PMP-22/SR13). J Cell Biol 1992; 117:225-238.

26. Wood PM, Bunge RP. Evidence that sensory axons are mitogenic for Schwann cells. Nature (Lond) 1975; 256:662-664.

27. Pellegrino RG, Spencer PS. Schwann cell mitosis in response to regenerating peripheral axons in vivo. Brain Res 1985; 341:16-25.

28. Armati-Gulson P. Schwann cells, basement lamina, and collagen in developing rat dorsal root ganglia in vitro. Dev Biol 1980; 77:213-217.

29. Bunge MB, Williams AK, Wood PM. Neuron-Schwann cell interaction in basal lamina formation. Dev Biol 1982; 92: 449-460.

30. Jessen KR, Mirsky R. Schwann cell precursors and their development. Glia 1991; 4:185-194.

31. Spencer PS, Politis MJ, Pellegrino RG et al. Control of Schwann cell behavior during nerve degeneration and regeneration. In: Gorio A, Millesi H, Mingrino S, eds. Posttraumatic Peripheral Nerve Regeneration: Experimental Basis and Clinical Applications. New York: Raven Press, 1981: 411-426

32. Politis MJ, Sternberger N, Ederle K. Studies on the control of myelinogenesis. IV. Neuronal induction of Schwann cell myelin-specific protein synthesis during nerve fiber regeneration. J Neurosci 1982; 2:1252-1266.

33. Ratner N, Elbein A, Bunge MB et al. Specific asparagine-linked oligosaccharides are not required for certain neuron-neuron and neuron-Schwann cell interactions. J Cell Biol 1986; 103:159-170.

34. LeBlanc AC, Poduslo JF, Mezei C. Gene expression in the presence or absence of myelin assembly. Mol Brain Res 1987; 2:57-67.

35. Mitchell LS, Griffiths IR, Morrison S et al. Expression of myelin protein gene transcripts by Schwann cells of regenerating nerve. J Neurosci Res 1990; 27:125-135.

36. Assoian RK, Fleurdelys BE, Stevenson HC et al. Expression and secretion of type β

transforming growth factor by activated human macrophages. Proc Natl Acad Sci USA 1987; 84:6020-6024.

37. Stoll G, Jung S, Jander et al. Tumor necrosis factor-α in immune-mediated demyelination and Wallerian degeneration of the rat peripheral nervous system. J Neuroimmunol 1993; 45:175-182.

38. Larson DM, Haudenschild CC, Beyer EC. Gap junction messenger RNA expression by vascular wall cells. Circ Res 1990; 66:1074-1080.

39. Moore LK, Beyer EC, Burt JM. Characterization of gap junction channels in A7r5 vascular smooth muscle cells. Am J Physiol 1991; 260:C975-C981.

40. Levy JA, Weiss RM, Dirksen ER, Rosen MR. Possible communication between murine macrophages oriented in linear chains in tissue culture. Exp Cell Res 1976; 103:375-385.

41. Ori M, Shiba Y, Kanno Y. Facilitation of cell coupling formation between mouse macrophages with an increase in the external calcium ion concentration. Cell Struct Funct 1980; 5:259-263.

42. Beyer EC, Steinberg TH. Evidence that the gap junction protein connexin43 is the ATP-induced pore of mouse macrophages. J Biol Chem 1991; 266:7971-7974.

43. Beyer EC, Steinberg TH. Connexins, gap junction proteins, and ATP-induced pores in macrophages. In: Hall JE, Zampighi GA, Davis RM, eds. Progress in Cell Research. Vol 3. 1993;71-74.

44. Alvez LA, Coutinho-Silva R, Persecchini PM et al. Are there functional gap junctions or junction hemichannels in macrophages? Blood, (in press).

45. Spray DC, Dermietzel R. X-linked dominant Charcot-Marie-Tooth disease and other potential gap-junction diseases of the nervous system. Trends Neurosci 1995; 18:256-262.

46. Chance PF, Pleasure D. Charcot-Marie Tooth syndrome. Arch Neurol 1993; 50:1180-1184.

47. Fairweather N, Cell C, Cochrane S et al.. Mutation in the connexin32 gene in X-linked dominant Charcot-Marie-Tooth disease (CMTX2). Human Mol Gen 1994; 3:29-34.

48. Deschénes SM, Bone LJ, Fishbeck KH, Scherer SS. Cx32 and X-linked Charcot Marie Tooth Disease. This volume.

49. Tetzlaff W. Tight junction contact and temporary gap junctions in the sciatic nerve fibers of the chicken during Wallerian degeneration and subsequent regeneration. J Neurocytol 1982; 11:839-852.

50. Bonnaud-Toulze EN, Raine CS. Remodeling during remyelination in the peripheral nervous system. Neuropath Applied Neurobiol 1980; 6:279-290.

51. Konishi T. Dye coupling between mouse Schwann cells. Brain Res 1990; 508:85-92.

52. Chandross KJ, Spray DC, Dermietzel R et al. Changes in gap junctional communication coordinate Schwann cell responses to injury. Soc Neurosci Res 1994; 20:1112.

53. Brockes JP, Fields KL, Raff MC. Studies on cultured rat Schwann cells. I. Establishment of purified populations from cultures of peripheral nerve. Brain Res 1979; 65: 105-118.

54. Moreno AP, Saez JC, Fishman GI. Human connexin43 gap junction channels. Regulation of unitary conductances by phosphorylation. Circ Res 1994; 74:1050-1057.

55. Moreno AP, Eghbali B, Spray DC. Connexin32 gap junction channels in stably transfected cells: unitary conductance. Biophys J 1991; 60:1254-1266.

56. Moreno AP, Eghbali B, Spray DC. Connexin32 gap junction channels in stably transfected cells: equilibrium and kinetic properties. Biophys J 1991; 60:1267-1277.

57. Moreno AP, Laing JG, Beyer EC. Properties of gap junction channels formed of connexin45 endogenously expressed in human hepatoma (SKHep1) cells. Am Phsyiol Soc 1995; 268:(Cell Physiol 37)C356-C365.

58. Haefliger JA, Bruzzone R, Jenkins NA. Four novel members of the connexin family of gap junction proteins. Molecular cloning, expression, and chromosome mapping. J Biol Chem 1992; 267:2057-2064.

59. Bennett MVL, Barrio LC, Bargiello TA, Spray DC et al. Gap junctions: new tools, new answers, new questions. Neuron 1991; 6:305-320.

60. Paul DL, Ebihara L, Takemota LJ et al. Connexin46, a novel lens gap junction

protein, induces voltage-gated currents in nonjunctional plasma membranes of *Xenopus* oocytes. J Cell Biol 1991; 115:1077-1089.

61. Jiang JX, Paul DL, Goodenough DA. Post-translational phosphorylation of lens fiber connexin46: a slow occurrence. Invest Opthalmol Vis Sci 1993; 34:3558-3565.

62. Scherer SS, Kamholz J, Jakowlew SB. Axons modulate the expression of transforming growth factor-betas in Schwann cells. Glia 1993; 8:265-276.

63. Ridley AJ, Davis JB, Stroobant P et al. Transforming growth factors-β1 and β2 are mitogens for rat Schwann cells. J Cell Biol 1989; 109:3419-3424.

64. Rogister B, Delree P, Leprince P et al. Transforming growth factor β as a neuro-noglial signal during peripheral nervous system response to injury. J Neurosci Res 1993; 34:32-43.

65. Anzano MA, Roberts AB, Sporn MB. Anchorage-independent growth of primary rat embryo cells is induced by platelet-derived growth factor and inhibited by type-beta transforming growth factor. J Cell Physiol 1986; 126:312-318.

66. Mews M, Meyer M. Modulation of Schwann cell phenotype by TGF-β_1: Inhibition of P_0 mRNA expression and downregulation of the low affinity NGF receptor. Glia 1993; 8:208-217.

67. Morgan L, Jessen KR, Mirsky R. Negative regulation of the P_0 gene in Schwann cells: suppression of P_0 mRNA and protein induction in cultured Schwann cells by FGF2 and TGF beta 1, TGF beta 2, and TGF beta 3. Development (Camb) 1994; 120: 1399-1409.

68. Peulve P, Laquerriere A, Paresy M et al. Establishment of adult rat Schwann cell cultures: effect of b-FGF, alpha-MSH, NGF, PDGF, and TGF-beta on cell cycle. Exptl Cell Res 1994; 214:543-550.

69. Einheber S, Hannocks M-J, Metz CN et al. Transforming growth factor-β1 regulates axon/Schwann cell interactions. J Cell Biol 1995; 129:443-458.

70. Wells MR, Racis SP, Jr, Vaidya U. Changes in plasma cytokines associated with peripheral nerve injury. J Neuroimmunol 1992; 39:261-268.

71. Matthews N. Timor-necrosis factor from the rabbit. V. Synthesis in vitro by mononuclear phagocytes from various tissues of normal and BCG-injected rabbits. Brit J Cancer 1981; 44:418-424.

72. Chensue SW, Remick DG, Shmyr-Forsch C et al. Immunocytochemical demonstration of cytoplasmic and membrane-associated tumor necrosis factor in murine macrophages. Am J Pathol 1988; 133:564-572.

73. Appenzeller O, Palmer G. The cyclic AMP (adenosine 3',5'-phosphate) content of sciatic nerve: changes after nerve crush. Brain Res 1972; 42:521-524.

74. Pichichero M, Beer B, Clody DE. Effects of dibutyryl cyclic AMP on restoration of function of damaged sciatic nerve in rats. Science (Wash) 1973; 182:724-725.

75. LeBlanc AC, Windebank AJ, Poduslo JF. P_0 gene expression in Schwann cells is modulated by an increase inn cAMP which is dependent on the presence of axons. Mol Brain Res 1992; 12:31-38.

76. Ratner N, Glaser L, Bunge RP. PC12 cells as a source of neurite-derived cell surface mitogen, which stimulates Schwann cell division. J Cell Biol 1984; 98:1150-1155.

77. Ratner N, Bunge RP, Glaser L. A neuronal heparin sulfate proteoglycan is required for dorsal root ganglion neuron stimulation of Schwann cell proliferation. J Cell Biol 1985; 101:744-754.

78. Meador-Woodruff JH, Lewis BL, DeVries GH. Cyclic AMP and calcium as potential mediators of stimulation of cultured Schwann cell proliferation by axolemma-enriched and myelin-enriched membrane fractions. Biochem Biophys Res Commun 1984; 122:373-380.

79. Poduslo JF, Walikonis RS, Domec MC et al. The second messenger, cyclic AMP, is not sufficient for myelin gene induction in the peripheral nervous system. J Neurochem 1995; 65:149-59

80. Raff MC, Hornby-Smith A, Brockes JP. Cyclic AMP as a mitogenic signal for cultured rat Schwann cells. Nature (Lond) 1978; 273:672-673.

81. Porter S, Clark MB, Glaser L et al. Schwann cells stimulated to proliferate in the absence

of neurons retain full functional capability. J Neurosci 1986; 6:3070-3078.

82. Lemke GE, Brockes JP. Identification and purification of glial growth factor. J Neurosci 1984; 4:75-83.

83. Brockes, J.P. Assay and isolation of glial growth factor from bovine pituitary. Methods Enzymol 1987; 147:217-225.

84. Goodearl AD, Davis JB, Mistry K et al. Purification of multiple forms of glial growth factor. J Biol Chem 1993; 268: 18095-18102.

85. Marchionni MA, Goodearl ADJ, Chen MS et al. Glial growth factors are alternatively spliced erbB2 ligands expressed in the nervous system. Nature (Lond) 1993; 362: 312-318.

86. Cohen JA, Yachnis AT, Arai M et al. Expression of the *neu* proto-oncogene by Schwann cells during peripheral nerve development and Wallerian degeneration. J Neurosci Res 1992; 31:622-634.

87. Levi ADO, Bunge RP, Lofgren JA et al. The influence of heregulins on human Schwann cell proliferation. J Neurosci 1995; 15:1329-1340.

CHAPTER 15

GAP JUNCTION EXPRESSION IN THE OLFACTORY SYSTEM

Fernando Miragall, Otto Traub
and Rolf Dermietzel

1. INTRODUCTION

The olfactory system represents a unique region in the nervous system in terms of plasticity and proliferative as well as regenerative capacities.[1] The reason for these exceptional features resides in the existence of two pools of stem cells which persist in adult animals. One pool is localized at the basal layer of the olfactory epithelium and cells of this pool are called basal cells. The basal cells of the olfactory epithelium continuously divide and differentiate into mature bipolar olfactory neurons,[1,2] the axons of which establish synaptic contacts with their target cells in the glomeruli of the olfactory bulb.[3,4]

A second pool of undifferentiated cells is localized in the subependymal or ventricular layer of the olfactory bulb.[5-7] These cells originate in the subependymal region of the lateral ventricles, and from there they migrate rostrally into the olfactory bulb where they form its central core.[5-7] Due to its migratory capacities, this central layer of the bulb has also been termed the rostral migratory stream.[6,8,9] The cells of the rostral migratory stream show considerable proliferative activity even in adulthood[5,6,10] and they have been described to be bipotential since they can differentiate into both glia and neurons.[7,11,12]

For all these reasons the olfactory system has become an attractive model to investigate plasticity, proliferation and regeneration of neural tissues. In addition, there is increasing evidence that gap junctions are involved in developmental as well as regenerative processes.[13-15] There are therefore good reasons to believe that gap junction-mediated communication is also implicated in the regulation of the proliferative activity of the olfactory system.

Gap Junctions in the Nervous System, edited by David C. Spray and Rolf Dermietzel.
© 1996 R.G. Landes Company.

In this article we review the localization of gap junctions and the expression of connexins (Cxs) in the developing and mature olfactory system including the olfactory epithelium, nerves and bulb as well as the bulbar extension of the subependymal layer. To investigate the localization of gap junctions the freeze-fracture technique has been used. The expression of connexins (Cx26, Cx32, Cx40 and Cx43) was determined by immunocytochemical methods at the light and electron microscopic levels. Specific attention has been paid to the expression of connexins in proliferating cells of the olfactory system by combining the bromodeoxyuridine (BrdU) method with the immunocytochemical detection of connexins.[14].

2. EXPRESSION OF GAP JUNCTIONS AND CONNEXINS IN THE OLFACTORY EPITHELIUM

The pseudostratified neuroepithelium of the nasal olfactory region is organized in three layers (Fig. 15.1). From apical to basal these are: (1) the layer of the supporting cells; (2) the central layer occupied by the cell bodies of the olfactory neurons and (3) the basal layer consisting of two types of cells, the dark and globose basal cells (Fig. 15.1). The basal cells are considered to be the progenitor cells of the olfactory neurons and they have been described to divide continuously.[2,16] Evaluation of the proliferative activity can be done by the administration of the thymidine analog BrdU. We found a preferential incorporation of BrdU into basal cells (Fig. 15.1). Some BrdU-positive nuclei were also localized in the apical layer of the olfactory epithelium (Fig. 15.1) corresponding to supporting cells, which have been shown to divide in situ within this layer and which constitute a separate cell lineage.[16,17]

Freeze fracture electron microscopy is a convenient technique to investigate biological membranes and their associated structures such as intercellular junctions.

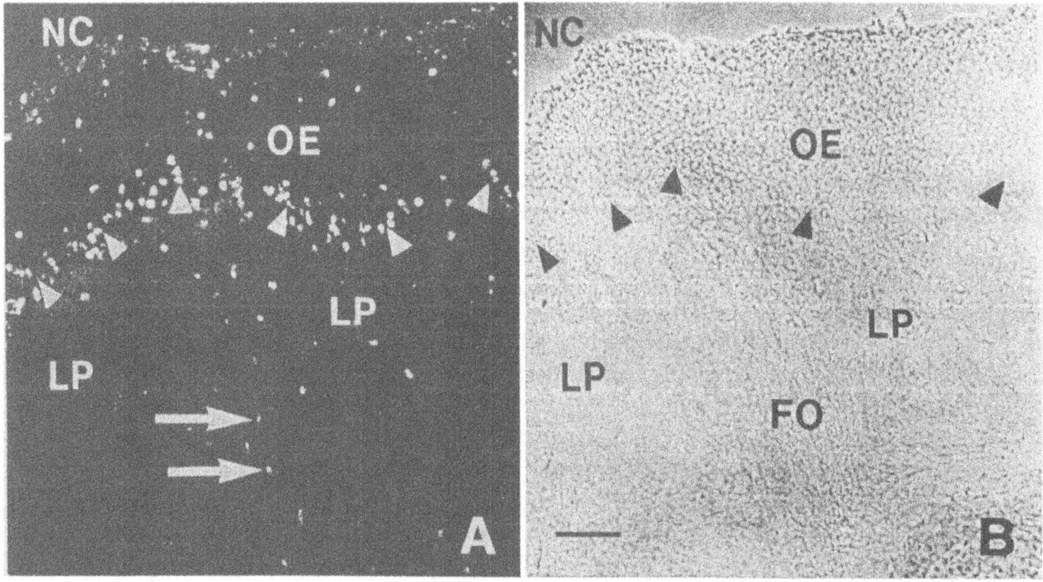

Fig. 15.1. (A) Olfactory epithelium (OE) of the P7 mouse after incorporation of BrdU. Proliferating cells after incorporating BrdU show a characteristic nuclear labeling (white arrowheads and arrows). Proliferating cells are mostly localized at the basal layer (arrowheads) of the olfactory epithelium. Some proliferative activity is also localized in other layers of the neuroepithelium and in the underlying lamina propria (LP) including the fila olfactoria (arrows). (B) is the phase contrast image corresponding to the fluorescence image (A). FO: Fila olfactoria. NC: Nasal cavity. Bar = 50 µm.

Freeze-fracture studies have shown that gap junctions are present in the olfactory epithelium of a variety of species, such as mouse,[18,19] newt,[20] rat[21] and chicken.[22] Some of these studies include the analysis of the olfactory epithelium during development.[18,22] These developmental studies reveal that gap junctions appear very early at the stage of olfactory placode, which corresponds to 3-day-embryos in chickens and to 9th day of gestation in mice.

In the olfactory epithelium of all species examined, gap junctions are localized between supporting cells.[18-21] Neither olfactory sensory neurons nor their precursors, the basal cells of the olfactory epithelium, possess gap junctions.[19] Figure 15.2 illustrates gap junctions in the olfactory epithelium of the mouse visualized by freeze fracture electron microscopy. Gap junctions are frequently localized in association with the apical tight junctional complex of the olfactory neuroepithelim (Fig. 15.2A,B). Due to their particular appearance at the branches of the tight junctional network they were named angular gap junctions.[18] In the developing olfactory epithelium of the chicken gap junctions in association with tight junctions are a transient phenomenon since they disappear from this particular location around day 7 after birth, which correlates with the period of formation of tight junctions.

Gap junctions are also present at the lateral plasma membranes of supporting cells (Fig. 15.2C,D). As suggested by Kerjaschki and Horandner[18] the strategic location of gap junctions between supporting cells is physiologically meanful and provides a means for the ready spread of electrical events among these cells. In contrast, olfactory sensory neurons completely lack gap junctions and must therefore be regarded as electrically uncoupled.

To investigate the expression and localization of connexins in the olfactory system, immunocytochemical methods at the light and electron microscopic level have been extensively used.[19,23-25] Studies on the expression of connexins in the olfactory

epithelium of the mouse have shown that Cx43 is the most prominent connexin expressed in this unique part of the nervous system (Figs. 15.3, 15.4 and 15.5). Consistent with the freeze fracture observations, Cx43 appears at very early stages of development, being detected at the stages of the olfactory placode and pit;[25] (Fig. 15.3A,B). Cx43 in the olfactory epithelium is consistently expressed throughout embryonic and postnatal development, as well as in adulthood (Figs. 15.3, 15.4 and 15.5). Also consistent with freeze fracture data is the observation that Cx43 localizes at the apical regions of the epithelium, in association with the apical junctional complex as demonstrated by double immunocytochemistry using polyclonal Cx43 antibodies and monoclonal antibodies against the tight junction associated protein ZO-1 (Fig. 15.4A-C). Additionally, Cx43 is localized in the connective tissue of the underlying lamina propria (Fig. 15.4D-F) including the bundles of the olfactory axons labeled with antibodies against the cell adhesion molecule L1 (Fig. 15.4E).

Interestingly, in the olfactory epithelium Cx43 expression exhibits strong regional differences from the stage of E18 onward (Fig. 15.3C,D). There are portions of the nasal olfactory region showing high levels of Cx43 expression (Figs. 15.3C, 15.4D, 15.5C,E). Cx43 immunolabeling is characterized by small punctate and/or fibrillar fluorescence, mostly corresponding to the gap junctional plaques observed by freeze-fracture. Other portions of the nasal olfactory region showed no immunoreactivity for Cx43 (Figs. 15.3C, 15.5A). Since the olfactory epithelium displays continuous neurogenesis, and the activity of this neurogenesis has been described to be heterogeneous (active and quiescent regions of neuronal cell proliferation)[16] differences in Cx43 expression were initially related to differences in the proliferating activity of neuronal precursor cells.[25] To test this hypothesis double immunocytochemistry was carried out using antibodies against Cx43 and either the embryonic form of N-CAM

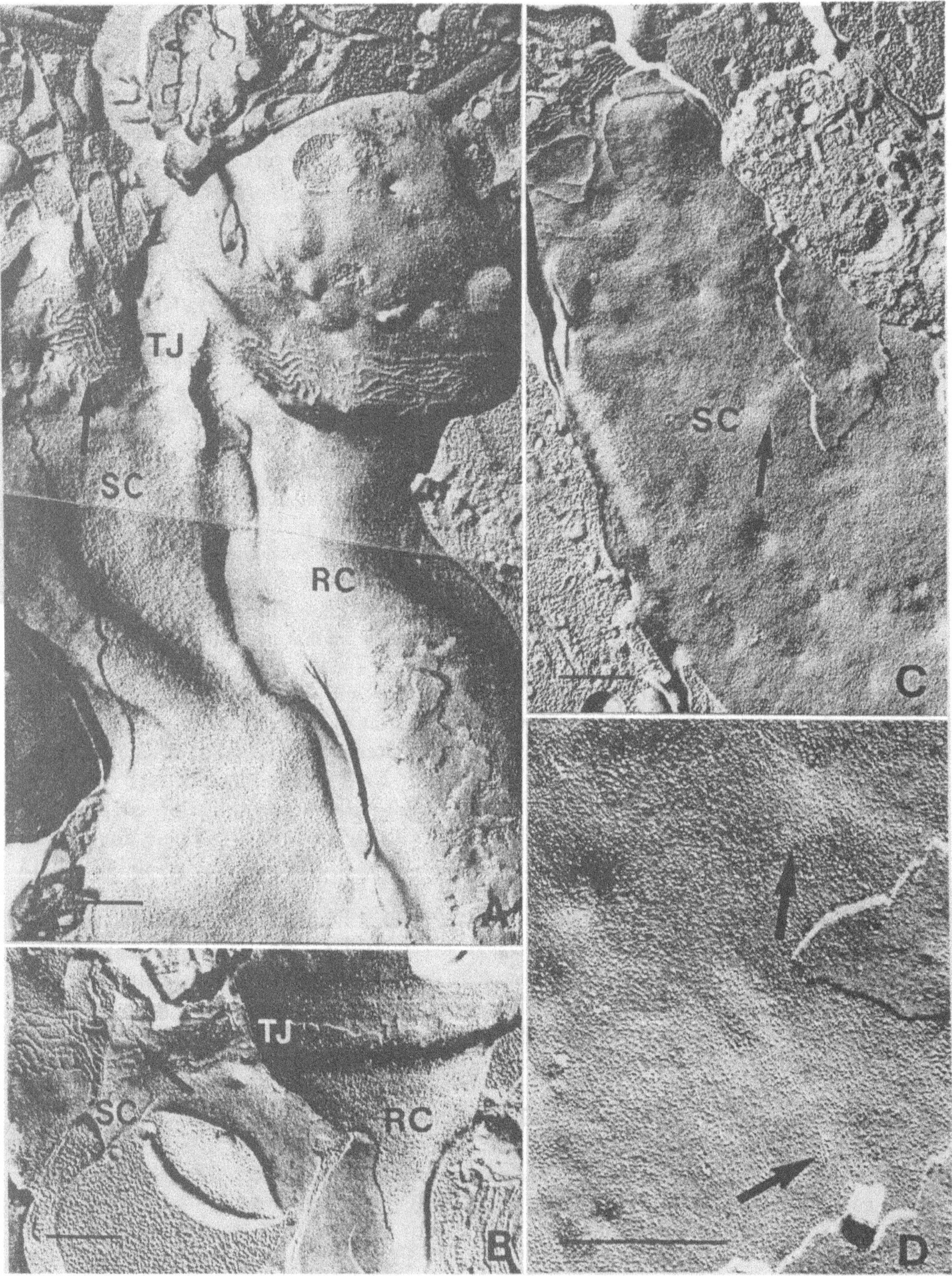

Fig. 15.2. *Freeze fracture microscopy of the olfactory epithelium of the mouse. (A) and (B) show the apical portions of the olfactory neuroepithelium. Small gap junctions (arrows) are localized within the tight junctional belt (TJ) on supporting cells (SC). Receptor cells (RC) do not exhibit gap junctions. (C, D) Gap junctions (arrows) are also localized at the lateral plasma membranes of the supporting cells (SC). D is a higher magnification of C. Bars = 0.25 µm.*

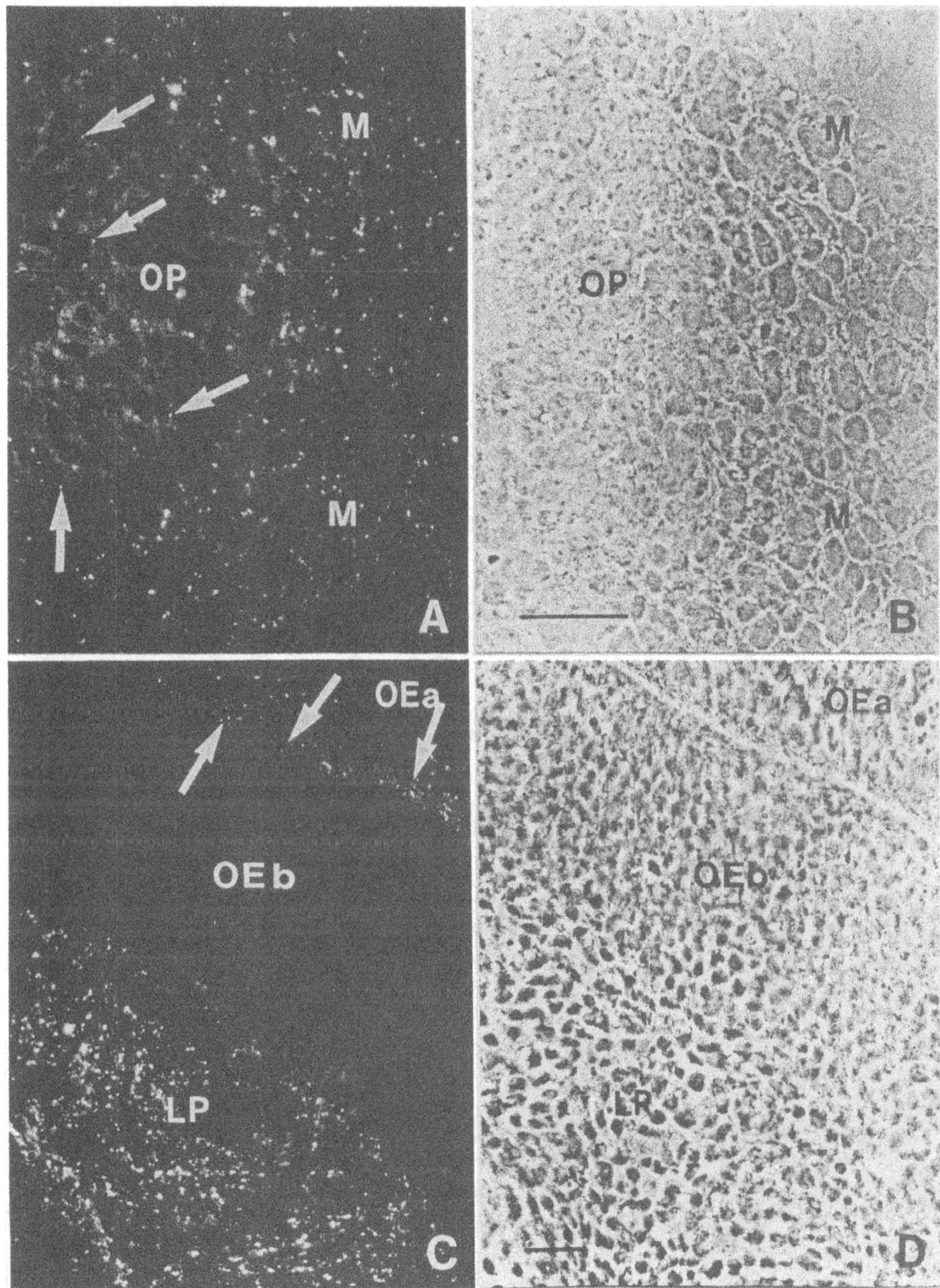

Fig. 15.3. Immunofluorescence of Cx43 in the olfactory pit at E11 (A, B) and olfactory epithelium at P0 (C, D). (A, B) Cx43 immunoreactivity is present both in the olfactory pit (OP) and in the underlying mesenchyme (M). (C, D) Regional differences in the expression of Cx43 can be observed in the olfactory epithelium (OE). Note that the upper region of the olfactory neuroepithelium (OEa) displays clear Cx43 immunoreactivity whereas in the lower portion of the neuroepithelium (OEb) no immunoreactivity for Cx43 is observed, although the corresponding lamina propria is strongly positive (LP). Bars = 25 μm.

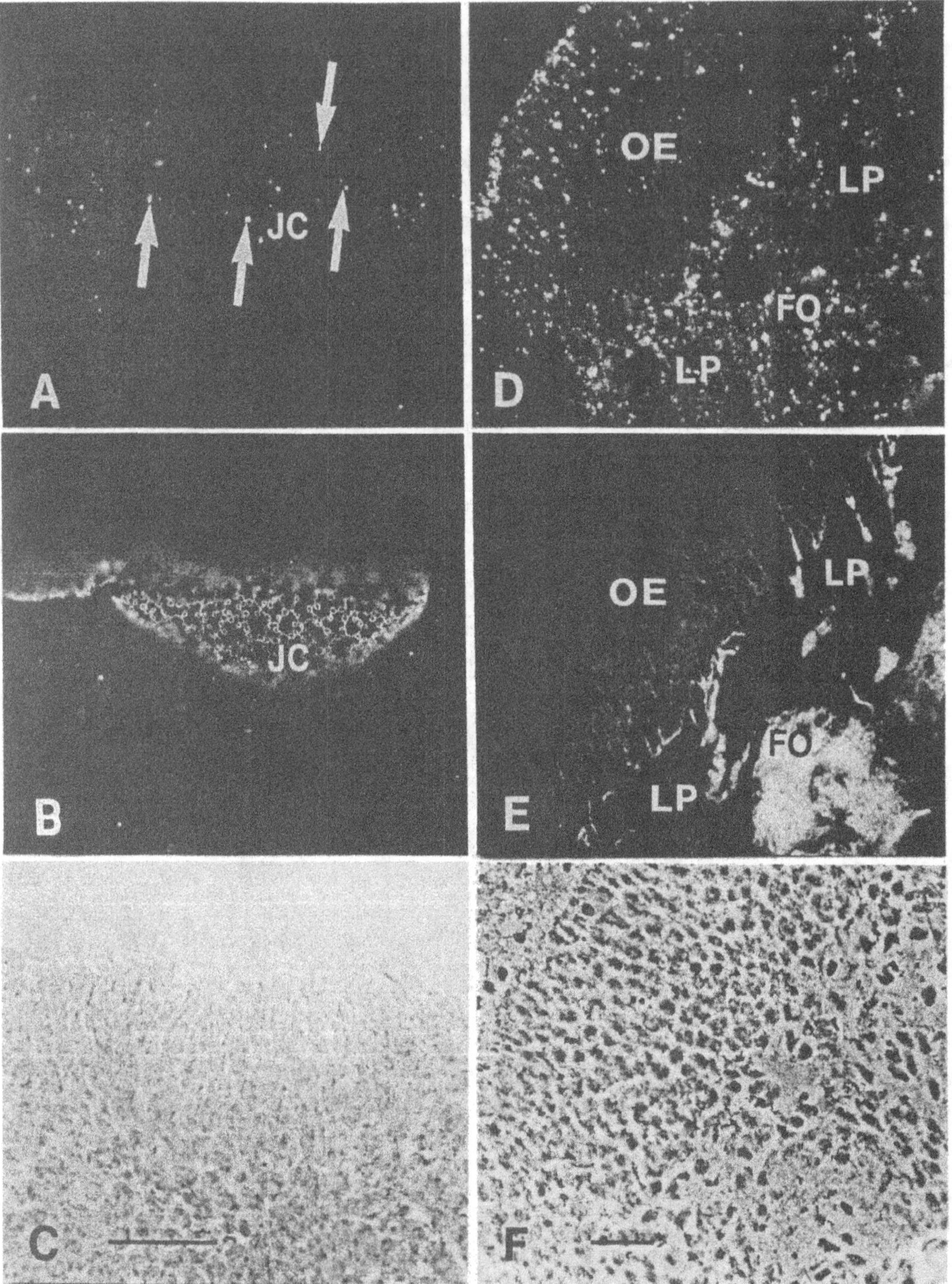

Fig. 15.4. Double immunofluorescence analysis of the olfactory epithelium and lamina propria with polyclonal antibodies against Cx43 (A, D) and monoclonal antibodies either against the tight junction associated protein ZO-1 (B) or the adhesion molecule LI (E).
(A, B) Cx43 (arrows) is localized in association with the junctional complex (JC). (D, E) Cx43 localizes not only in regions of the olfactory epithelium (OE) but also within the lamina propria (LP) including the fila olfactoria (FO). Bars = 25 μm.

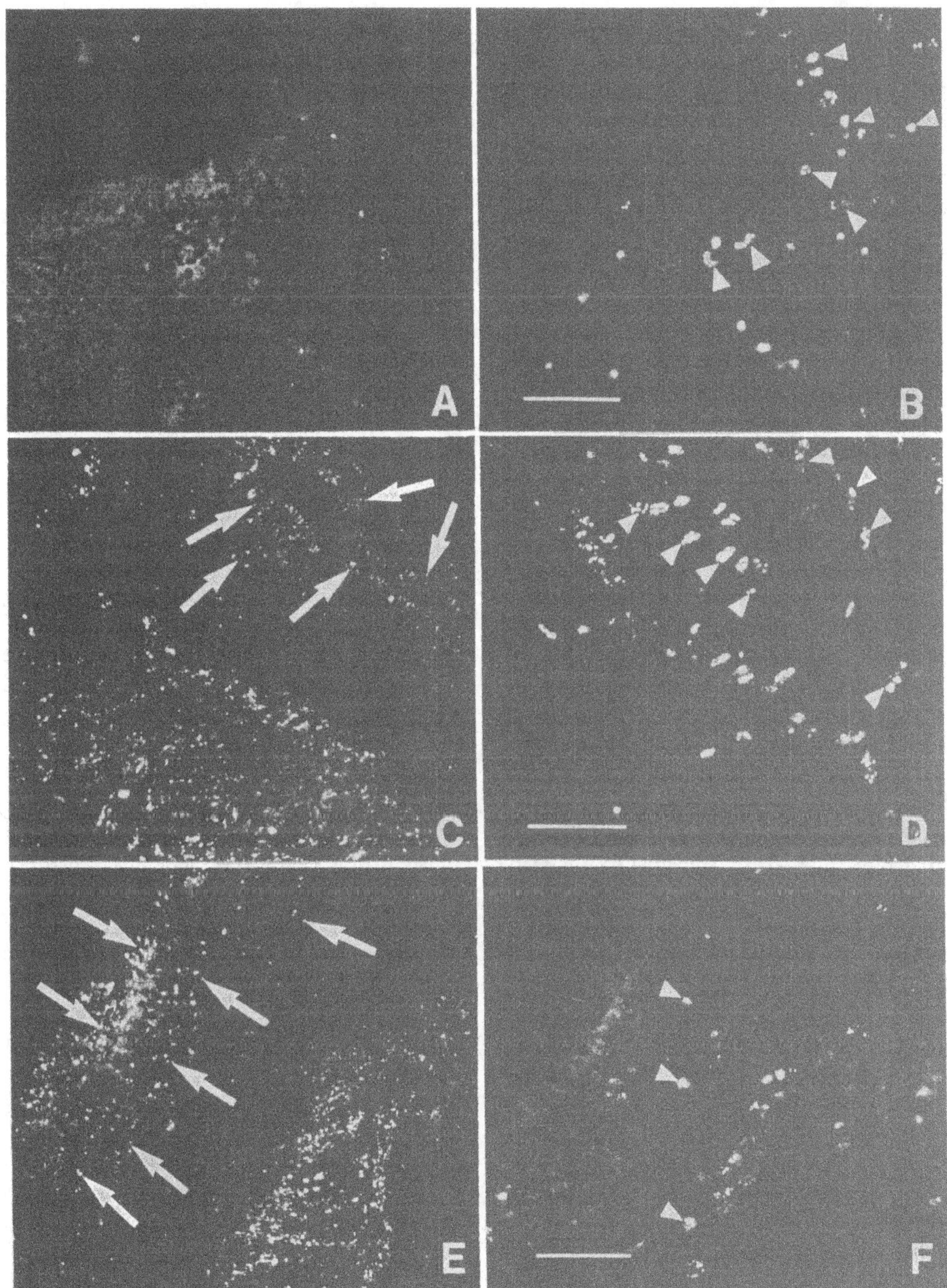

Fig. 15.5. Double immunofluorescence analysis of Cx43 (A, C, E) and BrdU (B, D, F) of the murine olfactory neuroepithelium at P0. (A, B) Portion of the neuroepithelium lacking Cx43 immunoreactivity (A) and showing strong proliferative activity (arrowheads). (C, D) Portion of the olfactory neuroepithelium showing prominent Cx43 immunoreactivity (arrows) and strong proliferative activity (arrowheads). (E, F) Portion of the olfactory epithelium with high levels of Cx43 expression (arrows) and reduced cell proliferation (arrowheads). Bars = 50 μm.

or BrdU, the latter in order to label proliferating neuronal precursor cells and juvenile neurons[25] (Fig. 15.5). No correlation was detected between levels of Cx43 expression and proliferating activity of neuronal precursor cells in the olfactory epithelium. The heterogeneous pattern of expression Cx43 in the olfactory epithelium resembled that described for the pattern of certain olfactory receptor molecules.[26] Moreover, since the appearance of this characteristic heterogeneous pattern correlates with the onset of olfactory function,[27] it is tempting to speculate that connexin43 expression may be related to olfactory receptor function. In fact, supporting cells have been reported to play a role in the neutralization and inactivation (clearance) of hydrophobic odorants subsequent to their binding to the receptor cells. This inactivation of odorants is carried out by an enzymatic complex partially produced by supporting cells and is a prerequisite to activation by further odorants.[28] Gap junctions between supporting cells may therefore help to coordinate their clearance efficacy.

Western blot analysis of the olfactory epithelium including the underlying connective tissue of the lamina propria and the olfactory nerves using polyclonal antibodies against Cx43 showed a triplet of bands around 43 kDa corresponding to three different phosphorylated isoforms of Cx43.[29] In the olfactory epithelium and nerves the high molecular isoforms are more prominent than in the bulb, suggesting that in the peripheral olfactory system these forms of Cx43 dominate.[25] Phosphorylation of Cx43 has been reported to occur within specific cellular compartments and the different isoforms may represent different functional states of connexin assembly.[29]

The expression of Cx26, Cx32 and Cx40 in the murine olfactory epithelium has also been investigated.[19,25] Although Cx26 and Cx32 were not detected,[25] weak but unambiguous immunoreactivity for Cx40 was present indicating that Cx40 is also expressed in this tissue. The expression of a second type of connexin in the olfactory epithelium is not unexpected since coexpression of at least two connexins has been reported in various types of mammal tissues.[30]

3. EXPRESSION OF GAP JUNCTIONS AND CONNEXINS IN THE OLFACTORY NERVES

At the basal layer of the olfactory epithelium and in the underlying lamina propria the olfactory axons fasciculate and form bundles of variable thickness. These bundles are called the fila olfactoria and they form the first cranial nerve. The olfactory axons of the fila olfactoria are covered by characteristic glial cells, the so-called olfactory ensheathing cells.[31] These particular glial cells show a typical phenotype that differs from that of astrocytes and Schwann cells.[31] The olfactory ensheathing cells do not form myelin and do not wrap around axons individually, but rather ensheath groups or bundles of many olfactory axons.[31] The olfactory ensheathing cells express glial fibrillary acidic protein[32] together with the adhesion molecule L1[33,34] and the tight junction associated protein ZO-1.[35] In this respect the olfactory ensheathing cells represent an intermediate phenotype between astrocytes and Schwann cells.

After leaving the cribiform plate the olfactory nerves enter the olfactory bulb where they establish synaptic contacts with their target cells (mainly the mitral and tufted cells of the bulb), in the form of special globose structures called olfactory glomeruli.[4] This first relay station of the olfactory pathway is characterized by a strong convergence, with a ratio of about 1000 primary olfactory axons to one target cell.[4]

Immunofluorescence analysis of Cx26 showed that this connexin localizes at the leptomeninges including the cells ensheathing the olfactory nerves in the PNS-CNS transitional zone of the olfactory bulb (Fig. 15.6A,B). Freeze-fracture analysis of the olfactory nerves reveals the presence of numerous gap junctions between ensheathing cells (Fig. 15.6E). Immuno-

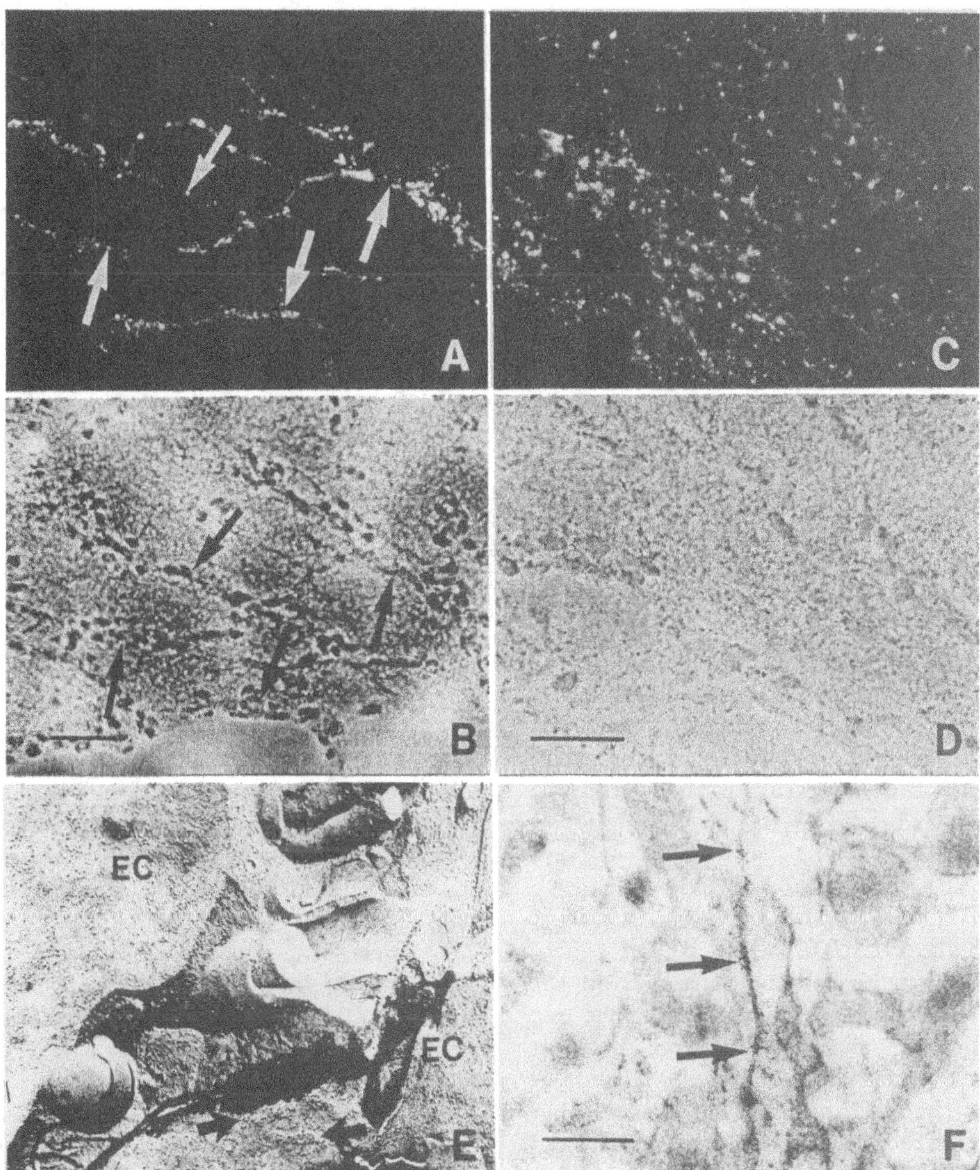

Fig. 15.6. *Expression of connexins and gap junctions in the olfactory nerves of the mouse. (A, B) Immunofluoresence analysis of connexin26 in the olfactory nerves at the transitional region of the bulb between peripheral and central olfactory system. In this region the meninges surround the olfactory nerves. Cx26 localizes at the meninges (arrows). (C, D) Cx43 immunofluoresce is observed within the olfactory nerves. (E) Freeze fracture electron micrograph showing a gap junction (short arrows) on the plasma membrane of an olfactory ensheathing cell (EC). (F) Immunogold electron microscopy of Cx43 in the olfactory nerves show that gap junctions at this location express Cx43 (arrows). B and D are the corresponding phase contrast pictures to fluorescence images A and C respectively. Bars in B and D = 25 μm. Bars in E and F = 0.2 μm.*

gold electron microscopy using antibodies against Cx43 (Fig. 15.6F), demonstrated that gap junctions within the olfactory nerves consists predominantly of Cx43 and that gap junctions within the olfactory nerves are localized between olfactory ensheathing cells (Fig. 15.6B).

Developmental studies have shown that Cx43 is detectable in the peripheral part of the olfactory nerves at E18 whereas at their central portions, i.e., the olfactory nerve layer of the bulb, Cx43 is detectable initially at P0.[25] Since in the olfactory nerves Cx43 is exclusively expressed by glial cells, this observation gives support to previous studies by Miragall et al[33,35] and Doucette[31] reporting differences between peripheral and central glia of the olfactory nerves. Expression of Cx43 in the olfactory nerves remains very strong throughout postnatal development and in adulthood (Fig. 15.6C,D).

In addition, some immunoreactivity in the olfactory nerves was observed with antibodies against Cx40 (data not shown). In contrast to the presence of these group II (α) connexins, Cx32 was not detected.

4. EXPRESSION OF GAP JUNCTIONS AND CONNEXINS IN THE OLFACTORY BULB

Previous morphological studies on the olfactory bulb have demonstrated that gap junctions are common features of this part of the telencephalon. Using the freeze-fracture technique Landis et al[36] showed that granular cells are coupled to mitral and tufted cells via gap junctions. Aggregates of granule cells have also been reported to be coupled via gap junctions.[24] The olfactory bulb is very rich in small bioactive substances such as glutamate, aspartate, glycine, taurine, alanine, dopamine, DL-homocysteic acid and γ–amino butyrate.[37] All these molecules are small enough to pass through gap junctional channels. It is therefore plausible that gap junctions in the olfactory bulb may mediate important biological processes such

as propagation of currents or regulation of neural plasticity.

Furthermore, studies by Reyher et al[24] and Miragall et al[25] have shown that gap junctions are also very abundant in other regions of the bulb including the olfactory nerve layer, the glomerular and periglomerular region, external plexiform layer and also the subependymal or periventricular layer. Gap junctions are localized on neurons,[24,36] glial cells[19,25] or their precursor cells of the subependymal layer.[25]

Consistent with the high frequency of gap junctions observed by the freeze-fracture technique, immunocytochemical investigations have corroborated the presence of gap junction proteins in the olfactory bulb.[19,23-25] Immunocytochemical analysis of the olfactory bulb using Cx43 antibodies showed that expression of Cx43 in the olfactory bulb is developmentally regulated[25] (Figs. 15.7, 15.8). Thus, during early embryonic and perinatal stages of development expression of Cx43 remained low (Fig. 15.7A,B). A gradual increase in the expression of this connexin occurs during postnatal development so that expression of Cx43 becomes high at late postnatal stages of development and adulthood[25] (Figs. 15.7C-F, 15.8). At mature stages Cx43 is present in nearly all layers of the olfactory bulb including the olfactory nerve layer (Fig. 15.7C), glomerular layer (Fig. 15.7C,E), granule cell layer (Fig. 15.8A-C), external and internal plexiform layers (Fig. 15.8D-F) and the bulbar extension of the subependymal layer (Fig. 15.9). As mentioned above, the subependymal layer represents the central core of the bulb and its relative thickness and proliferative activity decreases with the advent of maturity.[5,6] Some proliferative activity persists in the adult subependymal layer.[10] The subependymal layer shows strong immunoreactivity for Cx43[25] (Fig. 15.9). Quantification of the intensity of Cx43 immunoreaction by confocal laser microscopy and rates of cell proliferation, as measured by immunostaining of incorporated BrdU, showed

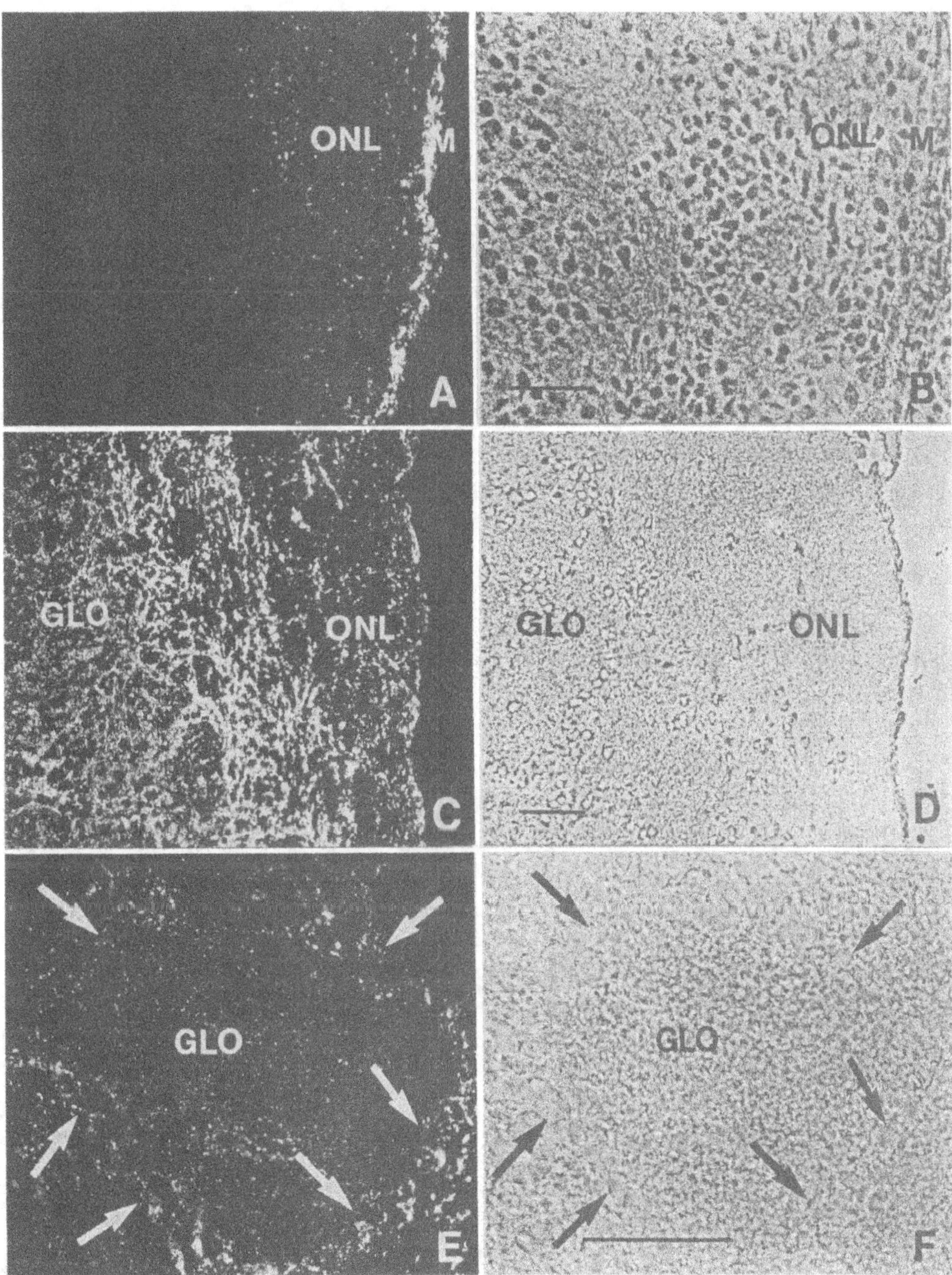

Fig. 15.7. Immunofluorescence analysis of the mouse olfactory bulb. (A, B) At P0 Cx43 is expressed at low levels in the peripheral layers of the bulb. (C, D) In the mature olfactory bulb Cx43 is strongly expressed at all layers of the bulb. ONL: olfactory nerve layer; GLO: olfactory glomeruli. (E, F) Detail of an olfactory glomerulus showing faint punctate immunolabeling both within the glomerulus (GLO) and at the periglomerular region (arrows). B, D and F are the corresponding phase contrast pictures to fluorescence images A, C and E respectively. Bars = 50 μm.

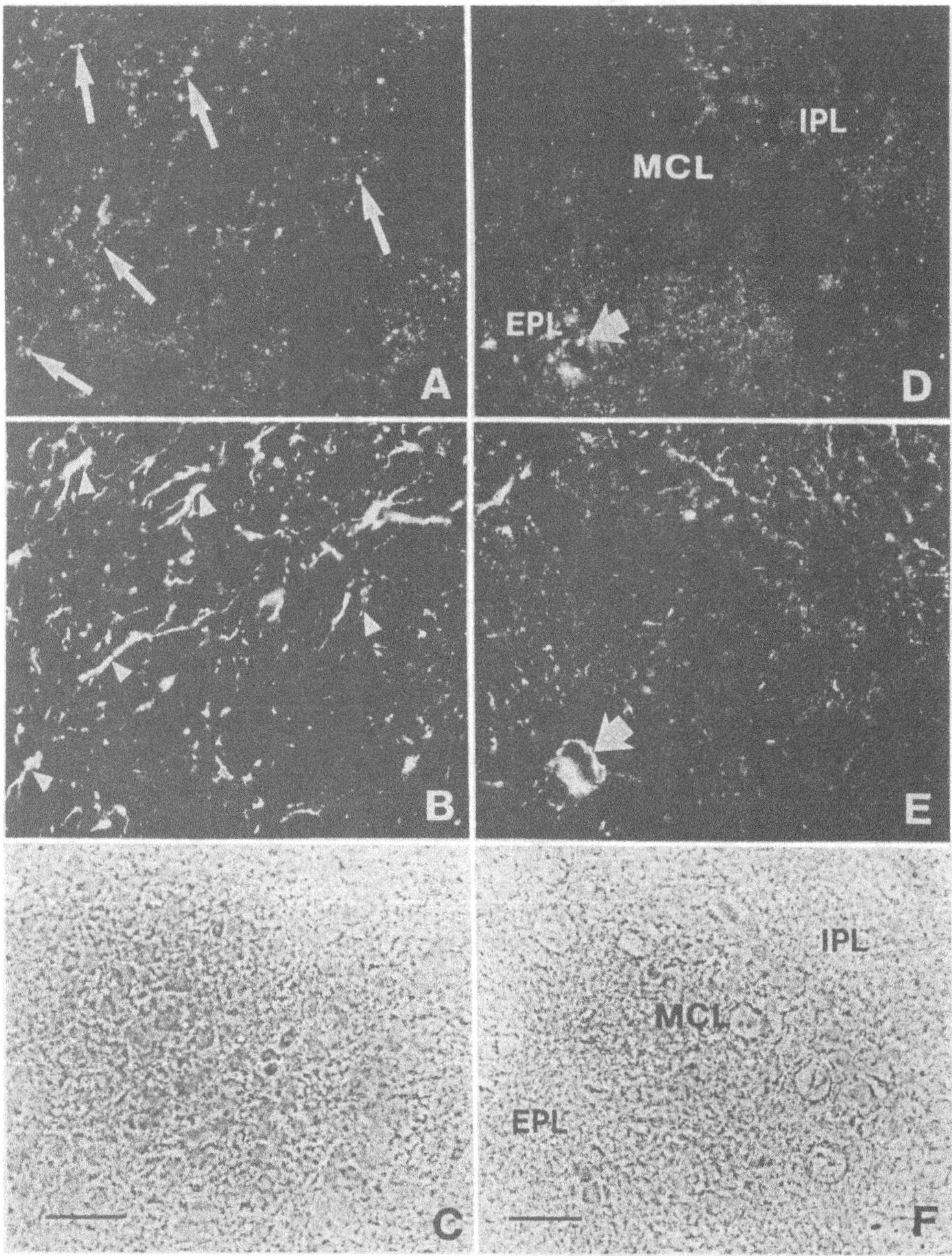

Fig. 15.8. Double immunofluorescence analysis of Cx43 (A, D) and GFAP (B, E) in the mature olfactory bulb. (A, B) Granule cell layer. Cx43 immunoreactivity (arrows) is frequently localized on GFAP-positive structures (arrowhead). (D, E) Mitral cell layer (MCL) and external (EPL) and internal plexiform (IPL) layers. Thick arrows indicate a blood vessel. C and F are the corresponding phase contrast micrographs to A, B and D, E respectively. Bars = 25 μm.

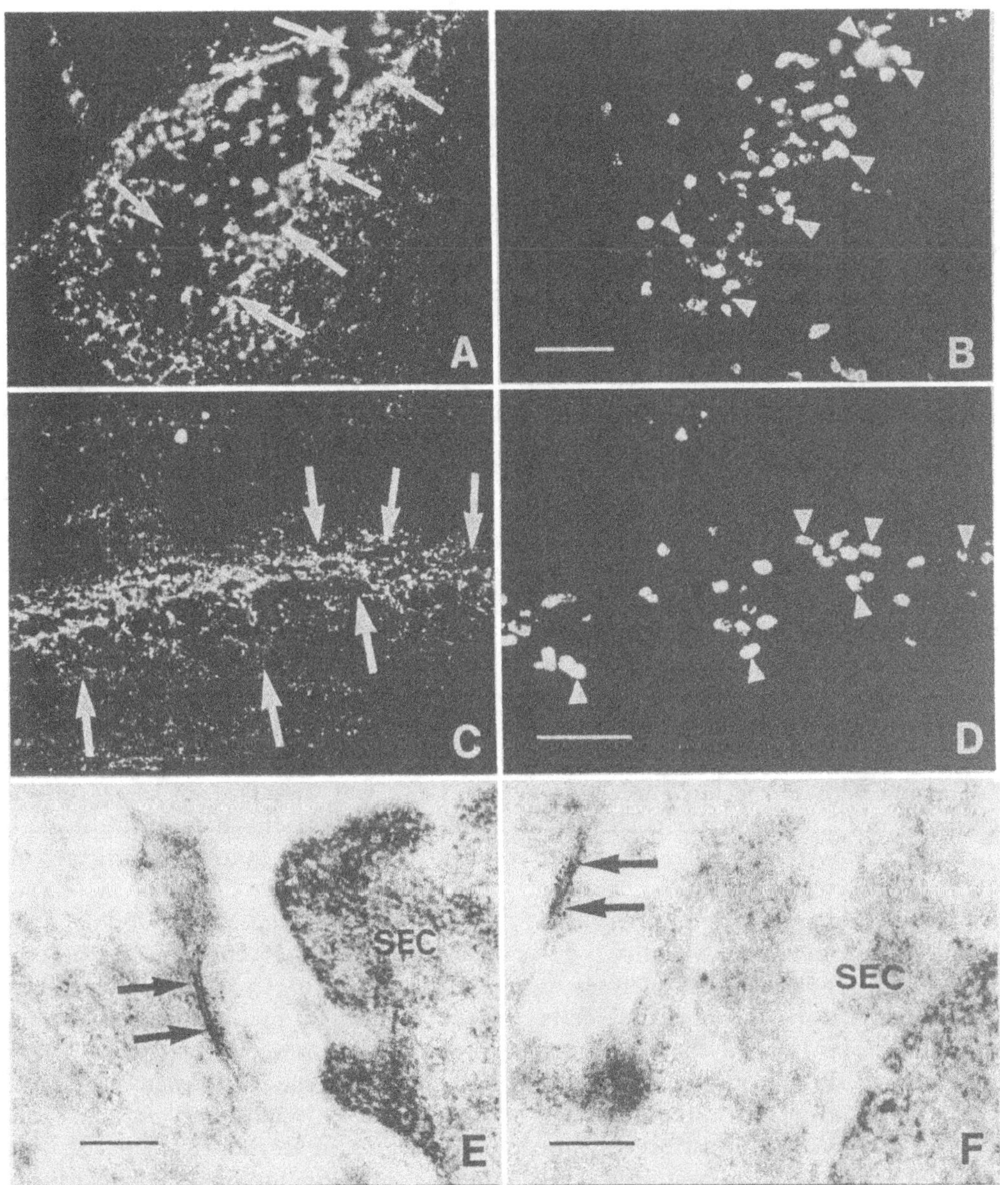

Fig. 15.9. (A, D) Double immunofluorescence analysis of Cx43 (A, C) and BrdU (B, D) in the subependymal layer. Groups of proliferating cells (arrowheads) mostly display lower Cx43 immunoreactivity (arrows) than groups of nonproliferating cells. (E, F) Immunogold electron microscopy of Cx43 in the subependymal layer. Note Cx43 immunolabeling associated to gap junctions (arrows) between subependymal cells (SC). Bars in B and D = 50 µm. Bars in E and F = 0.2 µm.

a clear inverse correlation during postnatal development between these two parameters in the subependymal layer and the rostral migratory stream (unpublished data). These observations strongly suggest that Cx43 expression and cell proliferation are related phenomena.

Western blot analysis of the olfactory bulb from the mouse using polyclonal antibodies against Cx43 also showed the typical triplet consisting of three bands around 40 kDa to 43 kDa corresponding to the three different phosphorylated isoforms of Cx43. In contrast to peripheral olfactory tissues, in the olfactory bulb a predominance of the low molecular weight isoform was prevalent.[25] Apparently, the olfactory system constitutes an organ in which Cx43 reveals local differences in its phosphorylation forms (olfactory epithelium and peripheral olfactory nerves versus olfactory bulb). Similar regional differences have been described in the heart by Kadle et al.[38] On the other hand Western blot analysis of embryonic (E18) olfactory bulb showed the high molecular weight isoforms to be stronger than the low molecular weight isoform. This indicates a shift of the isoforms of Cx43 during development, implicating differences in synthesis and/or assembly mechanisms of Cx43 between embryonic and adult olfactory bulbs.

Double immunofluorescence analyses of Cx43 and markers specific for neural cells showed that Cx43 immunolabeling is associated with GFAP positive structures (Fig. 15.8) indicating that in the olfactory bulb Cx43 is expressed by astrocytes and the GFAP-positive glial ensheathing cells of the olfactory nerve layer. This observation was confirmed by immunoelectron microscopy.[19,25] In some instances, Cx43 immunoreactivity was also observed in close association with neurons. However, since neurons are often tightly surrounded by thin lamellar astrocytic processes, it was not possible, even at the electron microscopical level, to definitely confirm whether Cx43 appeared on neuronal membranes or on thin glial lamellae surrounding neurons. Since previous freeze fracture studies of

Landis et al[36] have revealed gap junctions on plasma membranes of mitral cells, it was expected that these and probably other types of neurons express connexins. In situ hybridization allows clearer visualization of the histochemical signal since mRNA localizes within the cytoplasm instead of at small plaques in the plasma membrane. In situ hybridization analysis using specific oligonucleotides for Cx43-mRNA revealed clear cytoplasmatic labeling of Cx43-mRNA in the mitral, tufted and granule cells of the bulb (unpublished results) indicating that connexin43 is also expressed by neurons and that Cx43 gap junctional channels are likely to be involved in intercellular coupling among granule cells and between mitral and granule cells.

Cytoplasmic labeling was also observed in the mitral cells for Cx26 by immunocytochemistry (Fig. 15.10A). Cx26 was also observed in other regions of the olfactory bulb including the periglomerular region (Fig. 15.10B) and the bulbar projection of the subependymal layer (Fig. 15.10C,D) and the leptomeninges (Fig. 15.10E).

Cx32 was localized on some cells in several zones of the bulb including the glomerular external plexiform, mitral cell, internal plexiform and granule cell layers at late postnatal stages of development and adulthood[25] (Fig. 15.10F). Due to the time of appearance of Cx32, which correlates with the onset of myelination, and the specific location of Cx32-positive cells near myelin tracts, these cells were tentatively identified as oligodendrocytes.[25] In addition, studies of Reyher et al[24] have shown immunolabeling for Cx32 among granule cells in the olfactory bulb of several mammalian species including rat, gerbil and marmoset, which also indicates neuronal expression of Cx32 in the olfactory bulb. This finding for the olfactory bulb is coherent with other studies reporting that Cx32 is present in oligodendrocytes and some subsets of neurons in the mature brain.[39,40]

Cx40 is another member of the connexin family which has been recently described.[41,42] Cx40 localizes preferentially in

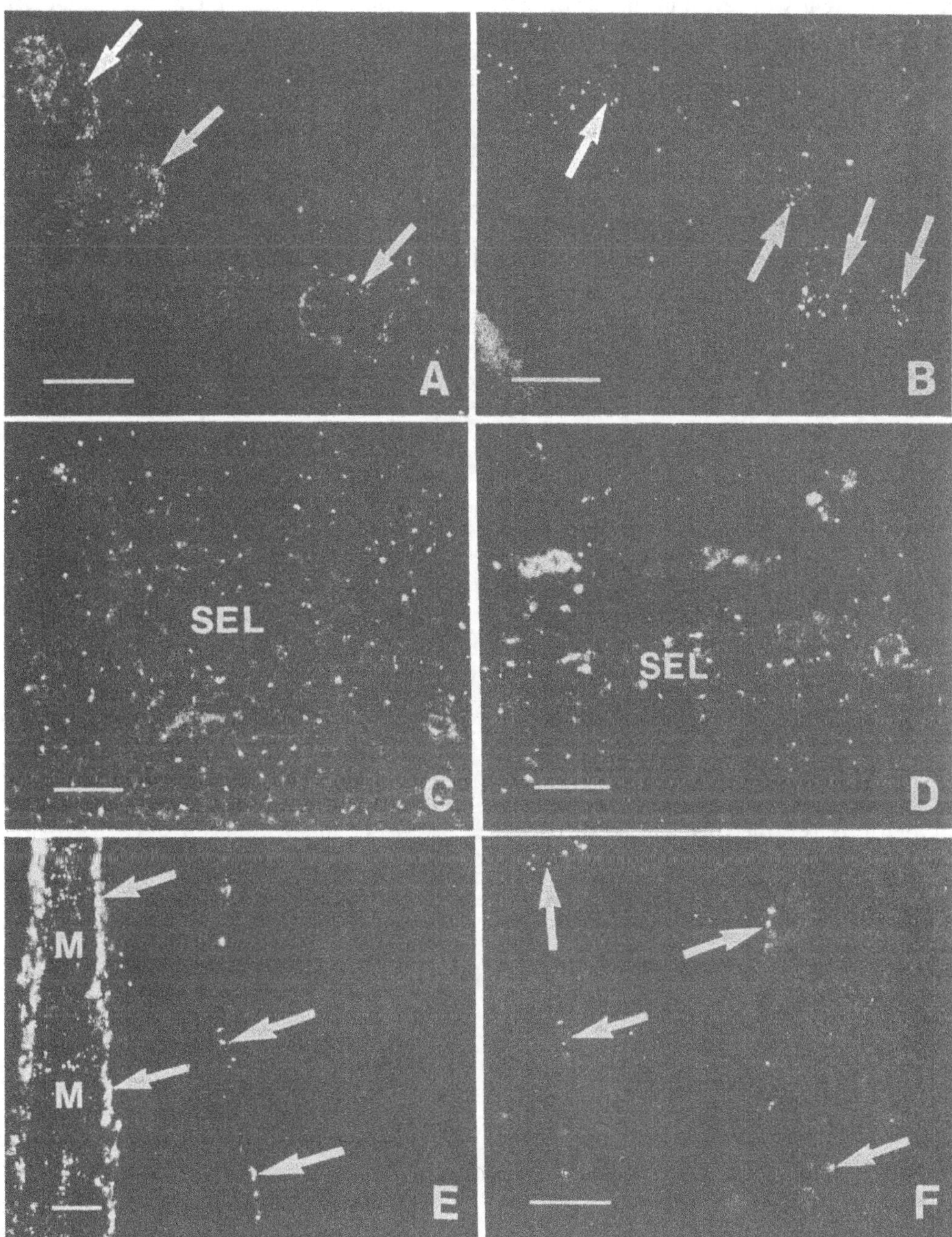

Fig. 15.10. Immunofluorescence analysis of Cx26 (A-E) and Cx32 (F) in the olfactory bulb. (A) Mitral cells depict faint Cx26 immunoreactivity within the cytoplasm (arrows). (B) Glomerular region. Cx26 localizes on some periglomerular cells (arrows). (C, D) Cx26 immunolabeling (arrows) in the subependymal layer at P0 (C) and adulthood (D). (E) Cx26 (arrows) is strongly expressed at the meninges (M) of the olfactory bulb. (F) Cx32 is expressed by some cells (arrows) which appear scattered within the olfactory bulb. These cells are identified as oligodendrocytes. Bars = 25 μm.

the lung and vascular endothelium, and is also present at lower levels in the brain.[41,42] This connexin was expressed in the bulbar meninges as well as in the olfactory nerve layer of the bulb. Cx40 immunoreactivity was also observed in the subependymal layer.

5. SUMMARY

Gap junctions and their molecular constituents, the connexins, are very abundant in the olfactory epithelium, nerves and bulb. With the exception of the primary olfactory neurons and their progenitor cells, all cell types of the olfactory system are coupled via gap junctions. As in other regions of the brain, connexins show cell type specific expression. Thus, Cx43 is mainly expressed by supporting cells, ensheathing cells and astrocytes, some subsets of neurons and undifferentiated cells of the subependymal layer, whereas Cx32 is mainly expressed by oligodendrocytes. Moreover, connexins in the olfactory system are also developmentally regulated, showing a pattern similar to that in other portions of the developing brain. In the mature olfactory bulb, however, the expression of Cx26, which is typical of embryonic brain, is retained by some cells, such as periglomerular and subependymal cells. This persistent expression of Cx26 might underly the uncommon embryonic and regenerative capacities of this system.

An inverse correlation between levels of expression of Cx43 and rates of cell proliferation could be demonstrated in the subependymal layer and its rostral migratory stream during postnatal development, suggesting a relationship between these two phenomena.

ACKNOWLEDGMENTS

The authors express their gratitude to Dr. M. Schachner for antibodies against L1, and to Karin Dassler and Anita Piringer for superb photographic work. The excellent technical assistance of Irmgard Hertting is also gratefully acknowledged. This work was supported by a grant of the DFG (Schwerpunkt Glia to RD).

REFERENCES

1. Monti Graziadei GA, Graziadei PPC. Neurogenesis and neuron regeneration in the olfactory system of mammals. II. Degeneration and reconstitution of the olfactory sensory neurons after axotomy. J Neurocytol 1979; 8:197-213.
2. Miragall F, Monti Graziadei GA. Experimental studies on the olfactory marker protein. II. Appearance of the olfactory marker protein during differentiation of the olfactory sensory neurons of mouse: An immunohistochemical and autoradiographic study. Brain Res 1982; 239:245-250.
3. Ramon y Cajal R. Histologie du Systeme Nerveux de Homme et des Vertebres. Maloines, 1911.
4. Graziadei PPC, Monti Graziadei GA. Principles of organization of the vertebrate olfactory glomerulus: An hypothesis. Neuroscience 1986; 19:1025-1035.
5. Hinds JW. Autoradiographic study of histogenesis in the mouse olfactory bulb. Time of origin of neurons and glia. J Comp Neurol 1968; 134:305-322.
6. Altman J. Autoradiographic and histological studies of postnatal neurogenesis. IV. Cell proliferation and migration in the anterior forebrain, with special reference to persisting neurogenesis in the olfactory bulb. J Comp Neurol 1969; 137:433-458.
7. Miragall F, Kadmon G, Faissner A et al. Retention of J1/tenascin and the polysialylated form of the neural cell adhesion molecule (N-CAM) in the adult olfactory bulb. J Neurocytol 1990; 19:899-914.
8. Frazier-Cierpal L, Brunjes PC. Early postnatal cellular proliferation and survival in the olfactory bulb and rostral migratory stream of normal and unilateral odor-deprived rats. J Comp Neurol 1989; 289: 481-492.
9. Miragall F, Dermietzel R. Immunocytochemical localization of cell adhesion molecules in the developing and mature olfactory system. Microsc Res Technique 1992; 23:157-172.
10. Morshead CM, van der Kooy D. Postmitotic death is the fate of constitutively proliferating cells in the subependymal layer of the adult mouse brain. J Neurosci 1992; 12:249-256.

11. Levison SW, Goldman JE. Both oligodendrocytes and astrocytes develop from progenitors in the subventricular zone of postnatal rat forebrain. Neuron 1993; 10: 201-212.

12. Morshead CM, Reynolds BA, Craig CG et al. Neural stem cells in the adult mammalian forebrain: A relatively quiescent subpopulation of subependymal cells. Neuron 1194; 13:1071-1082.

13. Warner AE, Guthrie SC, Gilula NB. Antibodies to gap junction protein selectively disrupt junctional communication in early amphibian embryo. Nature 1984; 311: 127-131.

14. Dermietzel R, Yancey SB, Traub O et al. Major loss of the 28-kDa protein of gap junction in proliferating hepatocytes. J Cell Biol 1987; 105:1925-1934.

15. Fraser SE, Green CR, Bode HR et al. Selective disruption of gap junctional communication interferes with a patterning process in Hydra. Science 198;237:49-55.

16. Graziadei PPC, Monti Graziadei GA. Neurogenesis and neuron regeneration in the olfactory system of mammals. I. Morphological aspects of differentiation and structural organization of the olfactory sensory neurons. J Neurocytol 1979; 8:1-18.

17. Caggiano M, Kauer JS, Hunter DD. Globose basal cells are neuronal precursors in the olfactory epithelium: a lineage analysis using a replication-incompetent retrovirus. Neuron 1994; 13:339-352.

18. Kerjaschki D, Horandner H. The development of mouse olfactory vesicles and their cell contacts: a freeze-etching study. J Ultrastruct Res 1976; 54: 420-444.

19. Miragall F, Kremer M, Dermietzel R. Intercellular junctions via gap junctions in the olfactory system In: Kurihara K, Suzuki N, Ogawa H, eds. Olfaction and Taste XI. Tokyo/Heidelberg: Springer Verlag, 1994: 32-35.

20. Usukura J, Yamada E. Observations on the cytolemma of the olfactory receptor cell in the newt 1. Freeze-replica analysis. Cell Tissue Res 1978; 188:83-98.

21. Menco B. Qualitative und quantitative freeze-fracture studies on olfactory and nasal respiratory epithelial surfaces of frog, ox, rat and dog. III. Tight junctions. Cell Tissue Res 1980; 211:361-373.

22. Mendoza AS, Miragall F, Breipohl W. Intercellular junctions during the development of the olfactory epithelium in the chick. A freeze-etching study. J Submicros Cytol 1980; 12:29-41.

23. Yamamoto T, Ochalski A, Hertzberg EL et al. On the organization of astrocytic gap junctions in rat brain as suggested by LM and EM immunohistichemistry of connexin43 expresssion. J Comp Neurol 1990; 302:853-883

24. Reyher CKH, Lobke J, Larsen WJ et al. Olfactory bulb granule cells aggregates: Morphological evidence for interperikaryal electronic coupling via gap junctions. J Neurosci 1991; 11:1485-1495.

25. Miragall F, Hwang TK, Traub O et al. Expression of connexins in the developing olfactory system of the mouse. J Comp Neurol 1992; 325:359-378.

26. Ressler KJ, Sullivan SL, Buck LB. A zonal organization of odorant receptor gene expression in the olfactory epithelium. Cell 1993; 73:597-609.

27. Greer CA, Stewart WB, Teicher MH et al. Functional development of the olfactory bulb and a unique glomerular complex in the neonatal rat. J Neurosci 1982; 2: 1744-1759.

28. Krieger J, Breer H. Der Geruchsinn: Die Grundlage der Duftwahrnehmung. Biologie in unsere Zeit 1994; 24,2:70-76.

29. Musil LS, Goodenough DA. Biochemical analysis of connexin43 intracellular transport, phosphorylation, and assembly into gap junctional plaques. J Cell Biol 1991; 115:1357-1374.

30. Dermietzel R, Hwang TK, Spray DC. The gap junction family: Structure, functions and chemistry. Anat Embryol 1990; 182: 517-528.

31. Doucette R. Glial influences on axonal growth in the primary olfactory system. Glia 1990; 3:433-449.

32. Barber PC, Lindsay RM. Schwann cells of the olfactory nerves contain glial fibrillary acidic protein and resemble astrocytes. Neuroscience 1982; 7:3077-3090.

33. Miragall F, Kadmon G, Husmann M et al.

Expression of cell adhesion molecules in the olfactory system of the adult mouse: Presence of the embryonic form of N-CAM. Dev Biol 1988; 129:516-531.

34. Miragall F, Kadmon G, Schachner M. Expression of L1 and N-CAM cell adhesion molecules during development of the mouse olfactory system. Dev Biol 1989; 135: 272-286.

35. Miragall F, Krause D, de Vries et al. Expression of the tight junction protein ZO-1 in the olfactory system: Presence of ZO-1 on olfactory sensory neurons and glial cells. J Comp Neurol 1994; 341:433-448.

36. Landis DMD, Reese TS, Raviola E. Differences in membrane structure between excitatory and inhibitory components of the reciprocal synapse in the olfactory bulb. J Comp Neurol 1974; 155:67-92.

37. Halasz N, Shepherd GM. Neurochemistry of the vertebrate olfactory bulb. Neuroscience 1983; 10:579-619.

38. Kadle R, Zhang JT, Nicholson BJ. Tissue-specific distribution of differentially phosphorylated forms of Cx43. Mol Cell Biol 1991; 11:363-369.

39. Dermietzel R, Traub O, Hwang TK et al. Differential expression of three gap junction proteins in developing and mature brain tissues. Proc Natl Acad Sci USA 1989; 86:10148-10152.

40. Dermietzel R, Spray DC. Gap junctions in the brain: where, what type, how many and why? Trends Neurosci 1993; 16:186-192.

41. Hennemann H, Suchyna T, Lichtenberg-Frate H et al. Molecular cloning and functional expression of mouse connexin40, a second gap junction gene preferentially expressed in lung. J Cell Biol 1992; 117: 1299-1310.

42. Bastide B, Neyses L, Ganten D et al. Gap junction protein connexin40 is preferentially expressed in vascular endothelium and conductive bundles of rat myocardium and increased under hypertensive conditions. Circ Res 1993; 73:1138-1149.

CHAPTER 16

TEMPORAL EXPRESSION OF GAP JUNCTIONS DURING NEURONAL ONTOGENY

Renato Rozental and David C. Spray

1. INTRODUCTION

Although Ramón y Cajal suggested that cellular contiguity was the basis of neural function, it was only by the mid-1950s that the detailed morphology of synapses were unveiled by electron microscopy. As a result, the term "synapse" (from Greek *synapsis* meaning "to clasp") has been widely used to designate specialized sites of transmission that can be either chemically or electrotonically mediated between cells. While discussion about the relative contribution of these forms of cellular interactions in the developing brain has been controversial, electrotonic coupling among neurons seems to diminish greatly at the time when chemical synaptic interactions are established. Electrotonic synapses have thus been suggested to provide the interactions necessary for neuronal pathfinding, chemical synaptogenesis and establishment of neuronal circuitry. Nevertheless, the mechanisms of progression between these types of synaptic interactions during development remains unclear. From the standpoint of function, although fast transmission is best achieved throughout gap junction channels, inhibitory modulation is best regulated chemically; thus, coexistence of mixed synapses could offer distinct advantageous performances during brain ontogeny. One recent example of coexistence of these modes of synaptic interaction is neurons cultured from second trimester human fetal brain, where the high level of coupling among neurons appears to compensate for the poor expression of chemical synaptic inputs[1] (Fig. 16.1). Since this high incidence of neuronal coupling is sustained during early postnatal (3-16 months) stages of human brain maturation, gap junctions may play a broader physiological role in the

Gap Junctions in the Nervous System, edited by David C. Spray and Rolf Dermietzel.
© 1996 R.G. Landes Company.

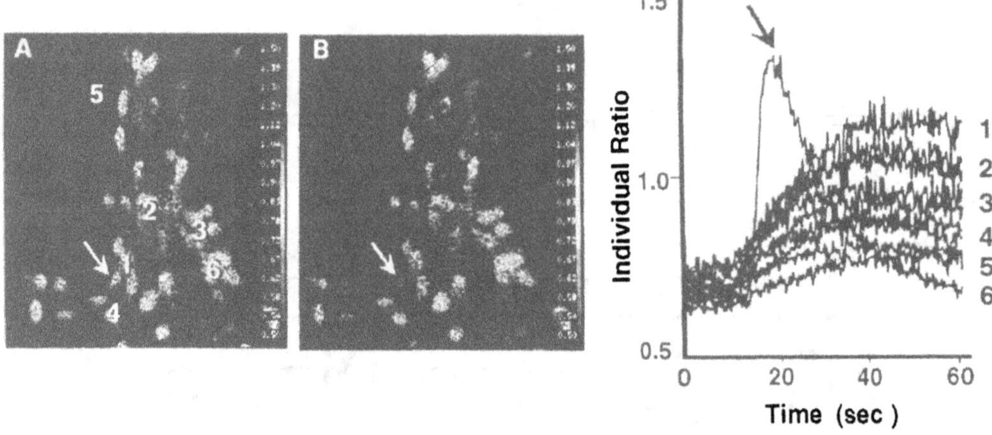

Fig. 16.1. Temporal-spatial dynamics of Ca²⁺ spread in human fetal brain neurons by the second trimester (see Institutional Approvals)[2,3,35,69]. Ratio images of [Ca²⁺]ᵢ were acquired from neurons loaded with 1.5 µM Indo-1 AM using Real Time dual emission confocal microscope (Nikon RCM-8000). Excitation at 351 nm was from an argon ion laser and the two emission bands from Ca²⁺-bound and Ca²⁺-free Indo-1 molecules were 405 and 485 nm, respectively. One cell was mechanically stimulated (arrow in A) and seven cells were simultaneously observed (B) and analyzed and their individual ratios were plotted over time (60 sec)(C). Notice how cells distant from the stimulated cell exhibit progressive increase in the [Ca²⁺]ᵢ levels (B,C). When these neurons are exposed to 2 mM heptanol, a well described uncoupling agent, the mechanical stimulation of a cell do not result in "signal" transfer to neighboring cells. For Ca²⁺ indicators, see Grynkiewicz et al.[77]

maturation of the CNS than is generally appreciated.[1-3]

Functional studies and those using immunocytochemical techniques have revealed that intercellular coupling and the distribution of gap junction proteins are much more common in the mature CNS than previously suggested by means of ultrastructural techniques.[4-11] However, the functional role of gap junctions in differentiated neurons under physiological and physiopathological conditions is not yet fully understood. Because gap junctions are overexpressed in brain tissues removed from patients with seizure disorders and may also mediate signaling between glia and neurons in the brain, it has been speculated that these junctional channels may synchronize the abnormal neuronal discharge within the epileptic foci.[12,13] However, rigorous examination of this and other issues of neuronal gap junction function have been delayed due to the difficulties that have been encountered in identifying

neuronal gap junction proteins and the lack of appropriate tissue culture models in which the properties of neuronal gap junctions could be examined. Indeed, the most basic property of a neuron, excitability, is not yet clearly defined with regard to developmental expression due to the lack of a suitable preparation in which changes could be followed over the developmental window of interest.

Knowledge of neuronal functional properties has expanded dramatically with the advent of immortalized precursor cells that can be manipulated in vitro towards graded stages of differentiation. We believe that enormous insight into neuronal gap junctions will be gained from developmental studies which follow the descendants of these identified precursors cells. In this chapter we discuss our recent results concerning the identity of the neuronal gap junctions expressed in developing cells in culture, the temporal interrelationship among gap junctions and voltage- and ligand-gated responses during neuronal ontogeny, and the influence of increased junctional communication in neuronal differentiation using immortalized neuroblasts and primary cultured neurons.

2. GAP JUNCTIONS IN DEVELOPMENT

A fascinating aspect of gap junctions that has attracted considerable interest is their potential role in coordinating tissue growth, through establishment of coupled communication compartments, and with cell uncoupling allowing differentiation.[7,14-19] Specific developmental functions in which gap junctions may participate include: (1) coordinating responses to inductive morphogens by relaying second messengers among the cells;[14,15] (2) modulating cellular proliferation;[16,20] and (3) facilitating pattern formation by helping to establish rostral/caudal and ventral/dorsal gradients of pH and second messengers.[14,21]

Although there had long been numerous correlations between gap junction expression and cell development,[21] a direct test for the hypothesis of a cause and effect relationship was first provided ten years ago by studies in which antibodies that blocked junctional communication were injected into early *Xenopus* embryonic cells.[22] These antibodies were found to lead to consistent developmental malformations, implying that normal development required functional gap junctions. Subsequent studies utilizing both anti-connexin antibodies[23] and antisense oligonucleotide constructs[24] have substantiated this view. Despite the elegance and impact of these studies on the gap junction field, concerns regarding their interpretation are raised by the finding that numerous connexins are expressed during early embryogenesis[25] and by the recent findings of grossly normal organogenesis of the brain and most other organs in mice in which connexin43 or connexin32 had been completely eliminated by homologous recombination.[26,27] Nevertheless, developmental malformations in the cardiovascular system, so severe that the animals die at birth, characterize the connexin43

knockout mouse,[26] axis malformations result from injection of connexin43 antibody into *Xenopus*[28] and the axis disorder atriovisceral heterotaxia has been associated with human mutations of the Cx43 gene.[29] Thus, although gross morphology of the brain is normal in connexin knockout animals, it remains to be determined whether connectivity and rate of neuronal maturation are normal. As is indicated below, we would predict such acceleration of neuronal maturation in the connexin43 knockout mouse based on our studies of immortalized rodent hippocampal cell lines.

3. GAP JUNCTIONS IN BRAIN DEVELOPMENT

At present, there is immunocytochemical evidence that at least six of the dozen known mammalian gap junction proteins are expressed in the brain (Cx26, Cx32, Cx37, Cx40, Cx43 and Cx45), either in embryonic or adult tissues, and Northern blotting of brain tissue has additionally indicated the presence of mRNA for Cx46 and Cx31, although cell types are not yet assigned and protein expression is not yet verified.

The connexins expressed by astrocytes (Cx43, possibly also Cx45 and Cx40), pinealocytes (Cx26), ependyma (Cx26, Cx43), leptomeninges (Cx26, Cx43) and oligodendrocytes (Cx32) have been determined in adult animals.[30-35] However, the connexins responsible for coupling in developing or adult *neurons* have not been identified unambiguously, although there have been indications for Cx26, Cx32, Cx33, Cx37, Cx40 and Cx43 in certain cell populations and under certain conditions.[3,7,18,33,35] Despite the lack of information regarding the identity of the neuronal connexin(s), numerous studies have demonstrated that the incidence of interneuronal coupling (and, by inference, the expression of connexins) dramatically decreases during the process of brain maturation.[5,7,18,19,36,69] In addition, the relative expression of certain connexin transcripts in fetal brain is higher than in adults, although other connexins (in particular, Cx32) show a developmentally regulated increase in abundance.[31] Such findings suggest that gap junction expression in the brain is diverse and cell-specific; moreover, the cellular specific distribution of gap junctions may coordinate different functions at distinct developmental stages.

4. TEMPORAL EXPRESSION OF GAP JUNCTIONS AND DEVELOPMENT OF VOLTAGE- AND LIGAND-GATED RESPONSES IN DIFFERENTIATING NEUROBLASTS

Since the pioneering nerve cell culture studies developed by Harrison,[37-39] numerous reports have extensively described the morphological and functional properties of primary cultured neurons and glial cells.[40] However, although cultured neurons retain their identities and properties, they are already committed in their differentiation. Characteristic features of mature neurons are their ability to express neurofilament proteins (NF), and to respond to voltages applied across their membranes and to neurotransmitters present in the extracellular medium. At the embryonic stage, however, neuronal precursor cells are morphologically undifferentiated, do not express characteristic cytoskeleton features and do not express excitable or chemically responsive membranes. Consequently, it has been difficult to follow cell programs controlling stage-specific gene expression, cell-cell interactions and chemical signaling between particular subsets of neuronal cells in primary culture and brain explants early during ontogeny. However, one approach that has been successfully employed in developmental neurobiology to circumvent these problems has been the production of immortal neuronal cell lines using a temperature-sensitive form of the SV40 large T antigen and an antibiotic resistant gene as a selectable marker.[41-43] These models offer an additional method of controlling gene expression, and thereby controlling cell proliferation and differentiation.

4.1. GENERATION OF NEURONAL CELL LINES

To investigate evolutionary properties of neurons early during ontogeny, we have developed developmental studies in a series of conditionally immortalized murine hippocampal progenitor cells with retrovirus constructs containing a temperature-sensitive form of the SV40 large T antigen and a neomycin-resistance gene (selectable marker) in collaboration with Drs. M. Mehler and J.A. Kessler (Albert Einstein College of Medicine) and Drs. E.M. Eves and M.S. Rosner (University of Chicago). In contrast to primary cultured neurons, these experimental systems are genetically homogeneous with self-renewing progenitor populations that can be induced to progressively undergo terminal differentiation when switched to the temperature not permissive for T-antigen expression (39°C). Using this approach, we showed that treatment of the immortalized hippocampal progenitor cells with different cytokines causes progressive neuronal differentiation, as defined by morphological criteria (neurite extension, presence of growth cones, nuclear maturation, increase in somatal diameter and nonneuronal cellular process outgrowth), successive expression of increasingly mature neurofilament proteins (Nestin, NF-66 and the high molecular weight phosphorylated neurofilament protein NFH-P), and generation of membrane excitability (tetrodotoxin-resistant or -sensitive)[7,18,43] (Fig. 16.2).

In addition, such strategy enabled us to evaluate the expression and identity of gap junction proteins during neuroblast proliferation (33°C) and commitment and differentiation (39°C) under different differentiating pharmacological treatments, as described below.

4.2. STUDIES ON IMMORTALIZED MOUSE HIPPOCAMPAL PROGENITOR CELLS (MK31 CELLS)

A series of conditionally immortalized neural stem and progenitor cell lines from murine embryonic hippocampus (MK31) were immortalized as described in our

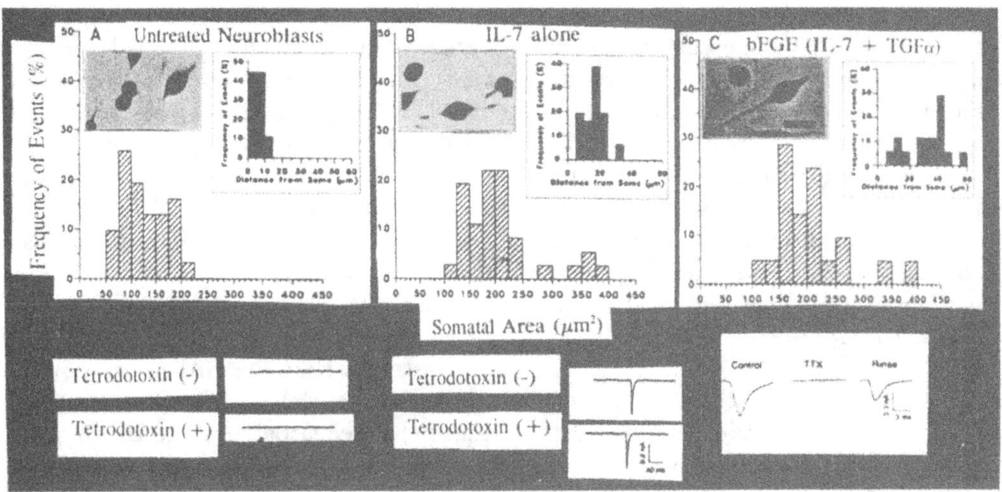

Fig. 16.2. Maturation of MK31 cells as assessed by morphometric, immunocytochemical and electrophysiological assays. A. Untreated neuroblasts. B. Neuroblasts treated with IL-7 alone. C. Neuroblasts treated with bFGF and then IL-7 + TGFα. Top Insets: immunostaining for NF-66 (A) and NFH-P (B,C); Measurements of neurite length. Bottom: Untreated neuroblasts initially are not excitable (A), but express inward currents that are TTX-resistant (TTX -) following IL-7 (B) and become TTX-sensitive (TTX+) (C) after combined treatment, leading to maximal differentiation.[7,43]

previous publications.[7,18,43] The general hypothesis that has guided our preliminary studies is that the same pattern of cytokines are essential for acquisition of progressive developmental stages in both nervous and hematopoietic systems, and that both utilize related signal transduction pathways.[43-45]

We showed that these progenitor cell lines are genetically homogeneous, with a self-renewing progenitor population and a manipulable experimental window that under the influence of cytokines (interleukins 5, 7, 9 and 11; bFGF and TGFα) may differentiate in vitro into neuron-like cells as evidenced by morphological, immunocytochemical and functional assays (Fig. 16.2). Among the interleukins evaluated, IL-7 was one of the most effective in inducing morphological changes characteristic of the neuronal phenotype.[7,43] In addition, IL-7 is present in vivo in adult and developing murine brain.[46] Two specific factors, bFGF and TGFα, were chosen for these studies on the basis of previous demonstrations of neuronal activity, expression in the developing brain and synergism for

obligatory expression of more mature neuronal phenotype.[43,47,48] For subsequent differentiation studies, therefore, we have evaluated neuroblasts treated with either IL-7 alone or in combination with TGFα and bFGF.[7,18]

Treatment of these cells at 39°C (temperature not permissive for T-antigen expression) with cytokines causes progressive neuronal differentiation as defined by morphological criteria successive expression of increasingly mature neurofilament proteins (nestin, NF-66 and NFHP), and the hallmark of the neuronal phenotype, the generation of membrane excitability (tetrodotoxin (TTX)-resistant inward currents that in time become TTX-sensitive). These results are illustrated in Figure 16.2.

We have evaluated mechanisms regulating the expression of intercellular coupling, development of membrane excitability, and neurotransmitter responsiveness during neuronal ontogeny using the immortalized cells described above. Together, our results indicate that certain cytokines may orchestrate the progressive expression of functional neuronal phenotype in vitro, in which the gradual disappearance of intercellular coupling (see Fig. 16.3) parallels the onset of voltage-dependent responses and both of which precede the expression of neurotransmitter chemosensitivity. Our findings represent the first direct interrelationship among gap junctions, voltage- and ligand-gated responses during neuronal ontogeny and substantiate conclusions derived from several studies of neuronal differentiation.[49-57]

Ligand-receptor interactions have been suggested to regulate brain neuronal differentiation.[53,58,59] For example, GABA and glutamate influence hippocampal pyramidal neuronal outgrowth during development as early as two days in culture.[58] In contrast, ACh-induced responses have been observed only later during neuronal ontogeny, i.e., in hippocampal neurons kept in culture from seven to twenty-five days.[60] It was suggested that the role of the GABA system in perinatal animals differs from that in adult CNS.[61] In fetal hippocampal neurons, for example, GABA has been shown to elicit excitatory effects,[62] thereby increasing $[Ca^{2+}]_i$ that could modulate neuronal growth and differentiation.[58] In agreement we these results, we have demonstrated that the GABA receptor was the earliest of the traditional receptor-operated channel superfamily to be functionally expressed in hippocampal neurons developing in vitro.[7] Interestingly, GABA (10 μM) has also been shown to induce increases in $[Ca^{2+}]_i$ (200-800 nM) in the immortalized GnRh secreting neuronal cell

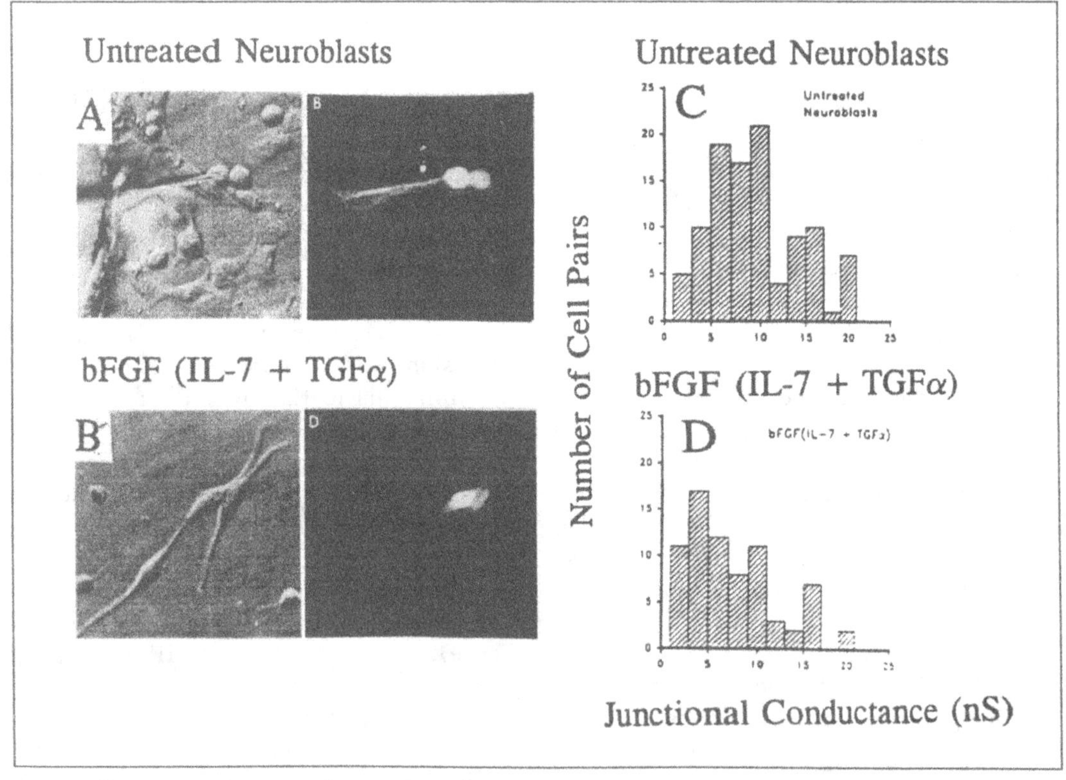

Fig. 16.3. Lucifer yellow dye-coupling (A,B) and magnitude of junctional conductance (g_j) (C,D) between MK31 cells decrease during cytokine-mediated differentiation. 92% of untreated neuroblasts are dye-coupled (A), with an average of g_j of 8 ± 0.6 nS (C). Following bFGF plus IL-7 and TGFα for 5 days, only 30% of cells are coupled (B), and g_j in the coupled pairs is lower (5 ± 1.5 nS) (D). These measurements indicate that cell uncoupling parallels cell differentiation.[7,18]

line at early times in culture.[63] In the same developmental window, glutamate application alone or in the presence of glycine did not induce $[Ca^{2+}]_i$ increases during the maturation of neuroblasts. Moreover, high resolution recordings from cells at these time points failed to reveal conductance changes in response to glutamate, glycine or ACh. Thus, our data provide direct evidence that electrical coupling is reduced and voltage-gated responses appear prior to the onset of chemosensitivity for some of the most important CNS neurotransmitters present in the adult mammalian brain.[7] Interestingly, GABA- and NMDA-mediated responses, but not acetylcholine, were also found in human fetal brain neurons by the second trimester.[1,2]

In additional experiments performed on these cells we found that action potentials were neither a specific requirement for the neuronal expression of more mature forms of neurofilament proteins (Nest-NF66-NFHP), nor were necessary for the onset of the progressive morphological changes that occur in vitro; action potentials, at least, are not the "endogenous signal" transmitted among neurons that govern the extent of neurite extension in vitro.[7] In parallel, the question arose whether intracellular steady-state Ca^{2+} levels were modulated in parallel with the morphological and functional changes observed during neuronal ontogeny. Although our measurements of basal intracellular Ca^{2+} levels (100 nM) from mature neurons are consistent with previous investigations, it has been shown that electrical activity increases calcium levels in neuronal growth cones.[64-67] Thus, our finding of a lack of differences in the average intracellular somatic steady-state Ca^{2+} levels during early stages of neuronal maturation suggests that homeostatic mechanisms capable of buffering neuritic Ca^{2+} changes are already established at early developmental stages, even prior to the onset of neuronal membrane excitability. In addition, our data suggest that neuronal uncoupling during ontogeny is not due to increased $[Ca^{2+}]_i$ levels.[7]

4.3. Gap Junction Expression During Maturation of Mouse Neuroblasts

Using molecular biological (Northern and Southern blots, RT-PCR), immunocytochemical (antibodies to Cx26, Cx32, Cx37, Cx40, Cx43 and Cx45) and electrophysiological techniques, we have shown that in our murine culture system, the progressive loss in gap junctions during differentiation were caused by decreased expression of connexin43 along with the progressive expression of the nonfunctional connexin33.[18,68,69] In addition to the progressive decline in strength of coupling between the cells (Fig. 16.3), the intrinsic junctional properties (i.e., voltage-dependence and single channel conductances) of cells treated with IL-7 alone differed from both untreated cells and those treated with a combination of growth factors (Fig. 16.4). Furthermore, because Cx33 is reported to form less functional channels when paired with Cx43 in oocytes, these findings raise the possibility that the progressive decrease in coupling observed during neuronal differentiation is due in part to the expression of a gap junction protein with a dominant negative function. Further, the presence of Cx33 may have an additional functional role in development by contributing to formation of compartments of clusters of cells during histogenesis.

A possible explanation for this change in junctional channel properties is that the expression of connexin-types exhibiting different physiological properties may change during ontogeny. (Although it might also be that properties of the individual connexin type would change as a result of posttranslational modifications, our studies of Cx43 properties in many systems under both phosphorylating and dephosphorylating conditions indicate that their plasticity is insufficient to account for this change). Connexin37 and connexin40 both show larger unitary conductance and more pronounced voltage-sensitivity than does connexin43 and could account for the channel population appearing after IL-7,

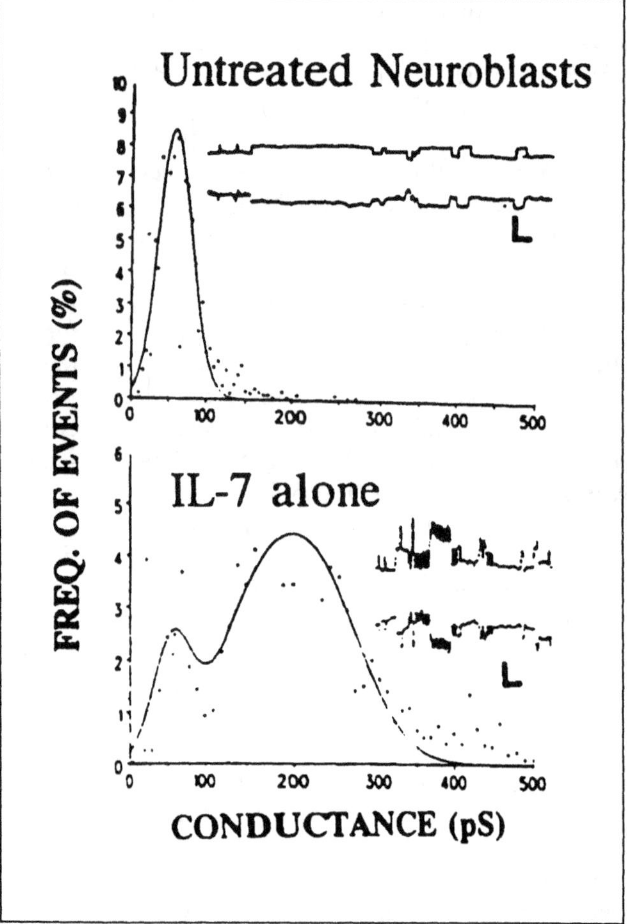

Fig. 16.4. Histograms and recordings of unitary conductance (γ_j) recorded between untreated and cytokine treated pairs of MK31 cells. (A) Naive neuroblasts show a single population of γ_j values of about 60 pS. (B) An additional unitary population of about 200 pS is expressed when neuroblasts are treated with IL-7. Insets: unitary I_j (driving forces: 30 mV). Vertical bar: 5 pA (top); 10 pA (bottom); horizontal bar: 2 sec.[7,18]

although we have not yet been able to detect mRNA for these connexins using either Northern blots or RT-PCR (we used "universal primers" that were designed to correspond to two conserved regions of connexin sequences).[18,69] Because Cx40 does not functionally pair with Cx43, its expression could offer the possibility for segregation of coupled compartments from one another.

4.4. STUDIES ON IMMORTALIZED RAT HIPPOCAMPAL CELL LINE EXPRESSING BCL-2, AN ANTIAPOPTOTIC GENE (HBCL-2 CELLS): A SIMPLEST MODEL

On Immortalized Rat Hippocampal Cell Line Expressing bcl-2, An Antiapoptotic Gene (Hbcl-2 Cells): A Simplest Model.

Our studies on MK31 cells have revealed two major difficulties involved in using the murine cells for detailed analysis of changes in cell properties during differentiation. First, these cells are quite susceptible to apoptotic cell death, thus limiting the extent of differentiation that can be achieved. Second, the expression of multiple connexins in these cells makes it more difficult to interpret changes in coupling strength during the process of maturation. As a result of our appreciation of these drawbacks, we have explored other models, one of which is described below.

Rat embryonic day 17 (E17) hippocampal cells were immortalized by retroviral transduction of a temperature-sensitive SV40 large T antigen[41] using constructs and techniques similar to those described above.[43] In order to reduce apoptosis,[70] human bcl-2 cDNA sequences were introduced into these cells via retroviral transduction in a pMV12 vector containing the hygromycin-resistance gene. Cells were maintained at 33°C in DMEM and 10% FBS. Among clones, we have concentrated on characterizing one termed H19-7-bcl-2-R10 (Hbcl-2) which exhibited the most pronounced responses to differentiating agents at 39°C.[19,69]

We have assessed the extent of maturation of the neuronal cells under three different culture conditions that resulted in the most dramatic phenotypic changes: (1) Permissive conditions for T-antigen expression (33°C in a humidified 5% CO_2 atmosphere), in DMEM containing 2 mM glutamine and 10% FBS (in addition, because cells were selected with both 200 µg/ml G418 and 100 µg/ml hygromycin, these antibiotics are included in all treatments to retard reversion); (2) Nonpermissive condition for T antigen expression (39°C) in N2 media (5 µg/ml insulin, 20 nM progesterone, 16 µg/ml putrescine, 300 nM selenium, 100 µg/ml transferrin, and 0.1% ovalbumin) and (3) at 39°C in DMEM containing 10 ng/ml bFGF and 0.3 µM retinoic acid.[43,71] These cells express the anti-aptoptotic gene bcl-2 and this expression allows their survival in culture for extended periods (>4 wks).[72] Hbcl-2 cells were kept in culture for only 5-6 days, a period during which neurons differentiate in response to bFGF and retinoic acid.[19,69,71] In serum-containing medium at 33°C, Hbcl-2 cells assumed a rounded morphology, expressed very weak immunopositivity for the low molecular weight neurofilament protein NF-66, failed to exhibit any immunoreactivity with anti-GFAP antibodies (astrocytic markers) and did not exhibit membrane excitability. Dye- and electrotonic coupling among these cells were observed in 100% of injections into clusters (Fig. 16.5) and intercellular coupling in cell pairs evaluated by the dual voltage-clamp method was quite strong (averaging about 20 nS). Connexin43 was also strongly expressed in these cells as evidenced by

using RT-PCR techniques and by immunostaining assays, which revealed a pattern of staining virtually surrounding the individual cells (Fig. 16.6). The absence of expression of a second connexin-type was evidenced by RT-PCR assays[19] and by recordings of unitary junctional currents which exhibited biophysical properties similar to those described for Cx43. Upon shifting the cells to a nonpermissive temperature (39°C), the cells differentiated, exhibiting striking neuronal-like morphologies (Fig. 16.5). In contrast to our findings in MK31 cells, RT-PCR assays did not detect expression of another connexin-type during Hbcl-2 cell differentiation.[69]

The degree of apparent differentiation depended on the treatment. At 39°C but in serum-free medium for 4-5 days, cells in clusters were less differentiated than isolated cells as evidenced by less elaborate neuritic processes, and with less intense neurofilament staining. Particularly interesting was the observation that, in contrast to clustered cells, isolated cells expressed more defined cellular processes and growth cones, more intense immunoreactivity to NF-66 antibodies and about 35% of these cells were excitable (evidenced by inward currents in voltage-clamp recordings). Notably, electrical excitability was never observed in recordings from cells in clusters. Cells in clusters exhibited a high incidence of moderately strong coupling with respect to diffusion of the dye Lucifer yellow (80%) and double-whole cell voltage-clamp measurements (about 8 nS). Cx43 was also expressed by cells treated with serum-free medium, as evaluated by immunostaining with anti-Cx43 antibodies and by RT-PCR assays. Although the immunoreactivity was most prominent at interfaces between cells, it was much less extensive than that observed in cells kept at 33°C (Fig. 16.6). Treatment of these cells with 10 ng/ml bFGF and 0.3 µM retinoic acid at 39°C led to more pronounced immunostaining for NF-66 and more frequent cellular excitability, the hallmark of neuronal maturation. In contrast to what was found for the other experimental groups, both "clustered" and

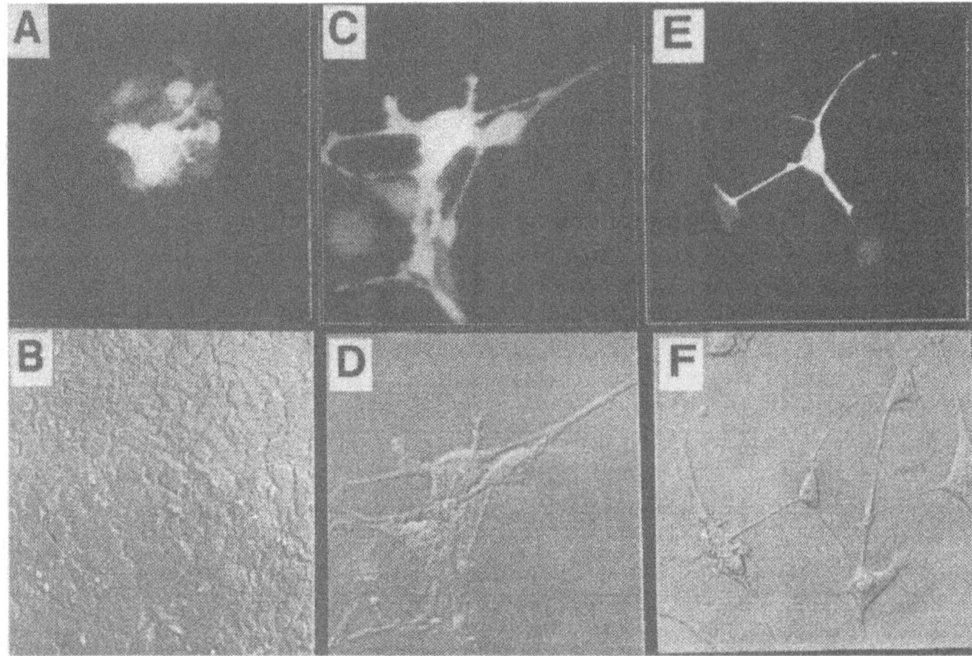

Fig. 16.5. Lucifer-yellow transfer between Hbcl-2 cells after 5 days of treatment. A,B. Cells at 33° are round and are strongly coupled. C,D. Cells at 39°C in serum-free media have begun to extend processes and are still coupled. E,F. Cells at 39°C but in serum-containing media + bFGF + retinoic acid exhibit typical neuronal phenotype and cells are weakly coupled. Top panels are correspondent fluorescence, bottom corresponding phase micrographs.[19,69]

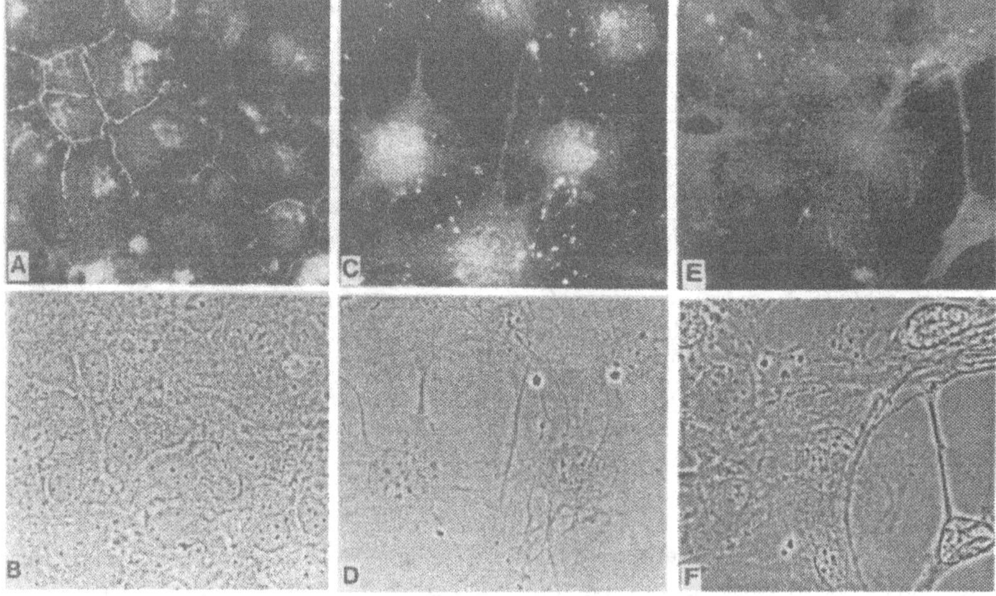

Fig. 16.6. Distribution of anti-Cx43 immunostaining antibodies in Hbcl-2 cells. A,B. Cells at 33°C. C,D. Cells at 39°C but in serum-free media. E,F. Cells in serum-containing media + retinoic acid + bFGF (39°C). Note the progressive decrease in the intensity of staining from A to E; at all stages, however, most of the fluorescence is observed near the cell appositions, where gap junction proteins are typically located.[19,69]

isolated cells were excitable, although the incidence was still higher in isolated cells (60% compared to 30%). Cx43 was again revealed in these cells using immunostaining, but it was confined to less abundant, smaller discrete plaques on the surface of the Hbcl-2 cells. The decrease in the magnitude of intercellular communication was further evidenced by the reduction of g_j to about 2 nS. In this experimental model, it was apparent from RT-PCR and from the unitary conductances of junctional channels that Cx43 downregulation was not compensated by the expression of another connexin-type.

4.5. EFFECTS OF CONSTITUTIVE CONNEXIN43 EXPRESSION ON HBCL-2 CELL DIFFERENTIATION

In order to test whether maturation of the Hbcl-2 cells required Cx43 downregulation, we transiently expressed human connexin43 in neuroblasts, using cotransfection with green fluorescent protein (GFP) as a tool to identify the transfected cells.[19,69,73] To express GFP, a KpnI-EcoR1 fragment from plasmid TU#65[73] was inserted into VR1012, a modified CMV expression vector (Vical, CA), resulting in pCMV-GFP. Human Cx43 was expressed using plasmid pGF1, which utilizes the Rous sarcoma virus terminal repeat; high levels of constitutive expression of exogenous hCx43 by transfectants has been previously described.[74] As a control for efficiency of cotransfection, the marker gene β-galactosidase was also coexpressed with GFP; the expression vector (β-geo) was prepared by subcloning a HindIII-XbaI fragment from pSAβgeo neomycin phosphotransferase-βgalactosidase fusion protein[75] into the pCMX vector, resulting in pCMVβgeo. Prior to transfection with these constructs, Hbcl-2 cells were maintained in proliferative state (33°C) in DMEM with 10% FBS. When cells reached 70-90% confluence, they were transfected by electroporation (250 V, 1000 μF) with a total of 20 μg DNA, containing 4 μg pCMV-GFP and 16 μg pGF1 or pCMVβgeo. Following electroporation, cells were seeded onto glass coverslips in 24 well dishes and maintained overnight at 33°C. Twenty-four hours later, cells were switched to differentiating conditions.

Cells expressing GFP were identified by standard epifluorescence, using either a Nikon Axiophot microscope equipped with a fluorescein filter set (Fig. 16.7), or by confocal microscopy, using a Nikon RCM-8000 real time confocal microscope. Transfected cells were switched to 39°C and treated with medium containing serum plus NGF and RA. In contrast to pCMVβgeo transfected cells that underwent the normal sequence of morphological changes and expressed membrane excitability after 3-5 days of treatment, the coupled pGF1-GFP transfected cells did not differentiate (i.e., were never excitable). Thus, increasing intercellular coupling by the constitutive expression of Cx43 in transfectants appears to delay or prevent the expression of the neuronal phenotype.

4.6. ACCELERATION OF MATURATION BY CELL UNCOUPLERS

Since constitutive coupling appeared to retard differentiation, we reasoned that cell uncoupling might accelerate this process of maturation. To test this hypothesis, the effects of uncoupling agents in cellular differentiation were tested in Hbcl-2 cells switched to 39°C kept in serum-free media from 9-12 hr. Because of the short duration of this assay, "control" cells remained primarily round in shape and were highly coupled as evaluated by dye-transfer. The interruption of the cell-cell transfer of Lucifer yellow observed after treatment with 3 mM heptanol coincided with onset of expression of membrane excitability by these cells; such events were not observed when the cells were treated with 3 mM hexanol, which was used to control for nonspecific alcohol-induced effects (Fig. 16.8). Another treatment that has been shown to uncouple Cx43-expressing cells is to culture them in the presence of antibodies prepared against amino-acid sequences corresponding to the presumed extracellular loop of Cx43 (Fig. 16.8).[28] To

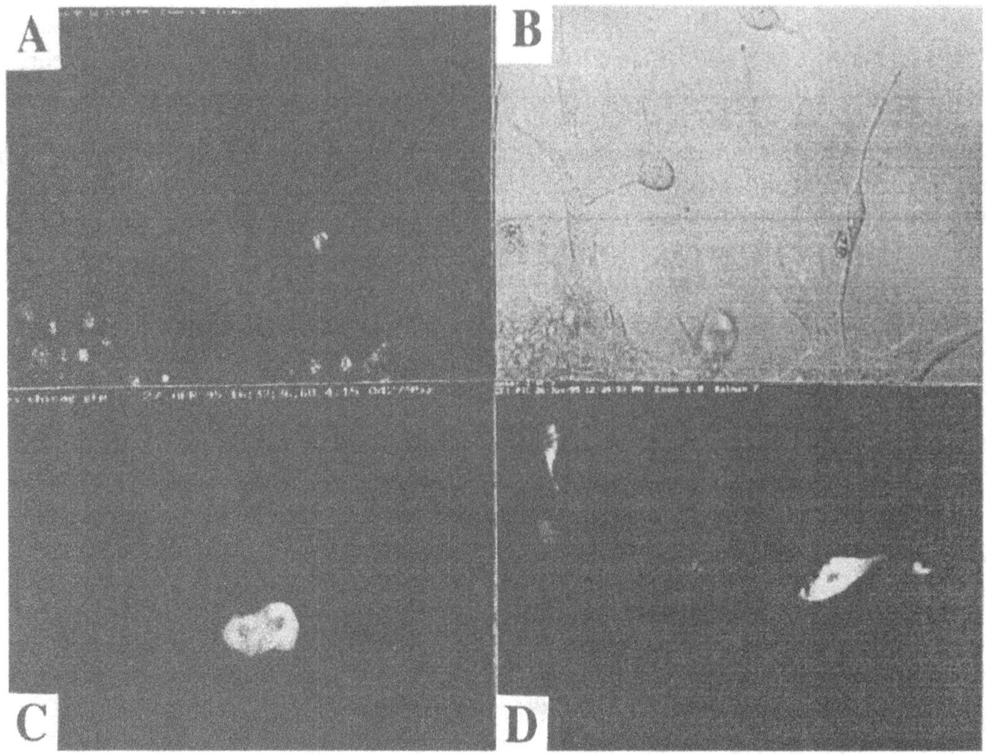

Fig. 16.7. Transfection of Hbcl-2 cells with connexin43 retards the acquisition of neuronal phenotype. Cells were cotransfected with GFP for identification by epifluorescence. A,B. Nontransfected cells. C. Coupled pGF1-GFP transfected Hbcl-2 cells remained round after differentiating (and were not excitable). D. Isolated pGF1-GFP transfected cells were morphologically differentiated.

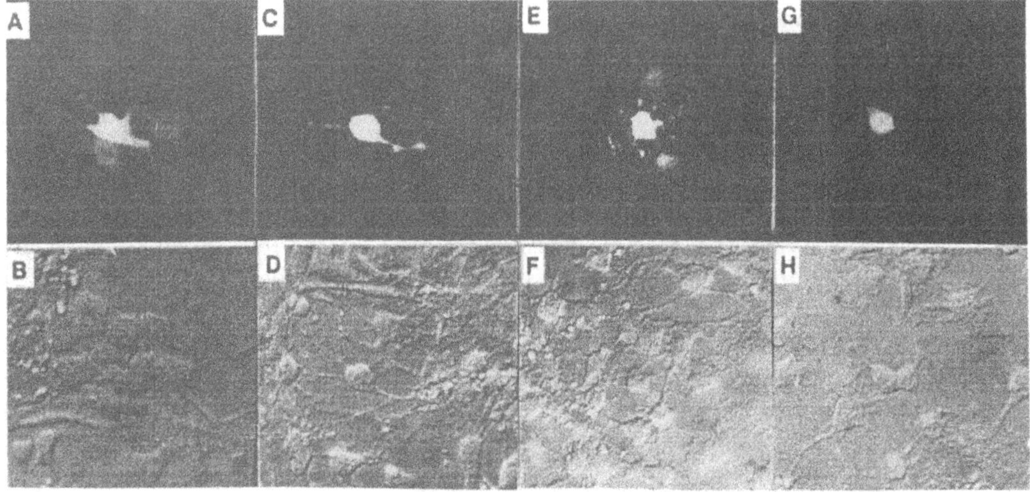

Fig. 16.8. Cell uncouplers accelerate Hbcl-2 cell maturation. Cells kept at 39°C in serum-free media were evaluated 9-12 hr after each treatment. A,B. Controls. C,D. 3 mM Heptanol. E,F. 3 mM Hexanol, not an uncoupling agent, was a control for alcohol effects. G,H. Cx-43 extracellular loop antibody uncouples.

test the efficacy of such reagents, we added extracellular loop antibodies (0.1-10 µg/ml). After 12-72 hr, coupling among treated cells was substantially reduced or blocked, and membrane excitability and neurite extension were more common and elaborated than in untreated controls. Together, these preliminary results provide further evidence for the hypothesis that uncoupling is a permissive or triggering step for neuronal differentiation.

5. CONCLUSIONS AND PROSPECTS

The studies presented herein represent the first evaluation of the temporal interrelationship among gap junctions and voltage- and ligand-gated responses early during neuronal ontogeny (see Table 16.1). Although previous studies were carried out with neuronal clonal cell lines, they were restricted to specific developmental stages and did not unveil a temporal correlation

Table 16.1. Time table of expression

Phenotypic Marker	Naive Neuroblasts	IL-7 singly	treated w/ bFGF then (IL7 + TGFα)	Primary Cultured Neurons
Appearance	rounded cells	neurite extension	simplex networks	complex network
Neurofilament protein	NF-66	NFH-P	NFH-P	NFH-P
Steady state $[Ca^{2+}]_i$	90 nM	95 nM	91 nM	85 nM
Dye-coupling	93%	80%	30%	9%
Junctional Conductance	7.6 ± 0.6 nS	6.6 ± 0.6 nS	5.4 ± 1.3 nS	1.6 ± 0.6 nS
Unitary Junctional Conductance	60 pS	60 pS, 200 pS	?	?
Connexin-type expressed (RT-PCR)	Cx43[‡]	Cx43[‡], Cx33 Cx37?, Cx40?	Cx43[‡], Cx33 Cx?	connexin-type?
Excitability	inexcitable	excitable (10%)	excitable (20%)	excitable (100%)
Tetrodotoxin sensitivity	inexcitable	TTX-resistant	TTX-sensitive	TTX-sensitive
Ligand-Gated Responses	(-)	(-/+)	(+)	(+++)

Summary of phenotypic changes that we have observed in transformed neuroblasts (MK31) following treatment compared to the murine neuronal phenotype expressed in primary culture.
Time line of events observed during progressive neuronal maturation in the presence of cytokines. Excitability was assessed on cells under voltage- or current-clamp conditions.[7,18,43] Excitable cells were either insensitive or sensitive to 1 µM tetrodotoxin (TTX). Distinct ligand-gated responses (LGR) were observed in neuroblasts treated with cytokines or in murine hippocampal primary cultured neurons (7-14 days in culture). For each experimental condition, the most mature neurofilament protein expressed is listed: Nestin, NF-66 (low molecular-weight neurofilament protein) and NFH-P (high molecular-weight neurofilament protein). During intermediate developmental stages (3-5 DIV), cells became responsive to GABA (100 µM); glutamate -induced responses were also observed (14 DIV).[7] The connexin-type expressed in these cultures have been identified using RT-PCR analysis and double-immunostaining[‡] for Cx-43 and NF-66.

among different functional properties. Understanding these processes, with their resultant effects on the expression of more mature neuronal phenotype and emergence of neuronal membrane excitability, will help provide a basis for understanding graded stages of neuronal differentiation as well as neuronal dysfunctions due to abnormal electrical activity.

In summary, our data indicate that neuronal progenitor cells are highly coupled by gap junctions (connexin43) during early ontogeny, preceding the functional expression of membrane electrical excitability, and the chemoresponsiveness to a variety of excitatory neurotransmitters. Moreover, as neuroblasts exhibit progressively neuronal phenotype coupling decreases. In addition, this temporal relationship seems to imply a functional connection since forced connexin43 expression on neuroblasts antagonize cellular differentiation. However, further studies using the same paradigms are still required to evaluate the behavior of stable transfected neuroblasts in clusters, a more representative cytoarchitecture of the developing brain.

HUMAN TISSUE ACQUISITION AND INSTITUTIONAL APPROVALS

The data regarding human fetal brain neurons presented in this chapter is part of an ongoing research program that has been approved by the Albert Einstein College of Medicine Committee on Clinical Investigations and the City of New York Health and Hospitals Corporation. Informed consent was obtained from all participants. Fetal tissues from abortuses of normal women were collected after elective pregnancy termination. The age of the abortuses was determined by multiple parameters including the date of the last menstrual period by history, uterine size by bimanual and abdominal examination, ultrasonography using predominantly the maximum biparietal diameter and, post-abortally, by measurement of fetal foot length. If a conflict arose between the various measures of gestational age, fetal foot length was accepted as the standard.[76]

ACKNOWLEDGMENTS

The authors would like to acknowledge Drs. F-.C Chiu, M. Meller, J. Kessler, E.M. Eves, M.S. Rosner and R. Dermietzel for their collaboration in some of the experiments illustrated in this chapter. We wish to thank Drs. E. Hertzberg (AECOM) and R. Dermietzel (Inst. Anat., Regensburg, Germany) for providing Cx43 antibodies for fixed tissue and for the functional assays (antibodies against the extracellular loop of Cx43), respectively. In addition, we wish to express our gratitude for the technical support provided by Ms. M. Urban, Ms. D.M. Vieira and Mr. H. Rubin. We also thank Nikon Inc. for their assistance with the Nikon RCM 8000 Real Time Confocal Laser Microscope. This work has been supported by grants from National Institutes of Health, Pew Charitable Trust and CNPq. This chapter is dedicated to Adriano and Amanda, our children.

REFERENCES

1. Chiu F-C, Rozental R, Bassallo C et al. Human fetal neurons in culture: Intercellular communication and voltage- and ligand-gated responses. J Neurosci Res 1994; 38:687-697.
2. Rozental R, Gebhard D, Padin C et al. Purification of cell populations from human fetal brain using flow cytometric techniques. Dev Brain Res 1995; 85:161-170.
3. Rozental R, Padin C, Spray DC et al. Purification of human fetal neurons from primary dissociated cultures. 1995: (submitted).
4. Kita H, Armstrong W. A biotin-containing compound N-(2-aminoethyl)biotinamide for intracellular labeling and neuronal tracing studies: comparison with biocytin. J Neurosci Methods 1991; 37:141-150.
5. Lo Turco JJ, Kriegstein AR. Clusters of coupled neuroblasts in embryonic neocortex. Science 1991; 252:563-566.
6. Peinado A, Yuste R, Katz LC. Extensive dye coupling between rat neocortical neurons during the period of circuit formation. Neuron 1993; 10:103-114.
7. Rozental R, Mehler, MF, Morales, M et al. Differentiation of hippocampal progenitor

cells in vitro: Temporal expression of intercellular coupling and voltage- and ligand-gated responses. Dev Biol 1995; 167: 350-362.

8. Shiosaka S, Yamamoto T, Hertzberg EL et al. Gap junction protein in rat hippocampus: correlative light and electron microscope immunohistochemical localization. J Comp Neurol 1989; 281:282-297.

9. Stewart WW. Functional connections between cells as revealed by dye-coupling with a highly fluorescent naphthalimide tracer. Cell 1978; 14:741-759.

10. Vaney DI. Many diverse types of retinal neurons show tracer coupling when injected with biocytin or neurobiotin. Neurosci Lett 1991; 125:187-190.

11. Yamamoto T, Shiosaka S, Whittaker ME et al. Gap junction protein in rat hippocampus: light microscope immunohistochemical localization. J Comp Neurol 1989; 281:269-281.

12. Naus CCG, Bechberger, JF, Paul DL. Gap junction gene expression in human seizure disorder. Exp Neurol 1991; 111:198-203.

13. Perez-Velazquez JL, Valiante TA, Carlen PL. Modulation of gap junctional mechanisms during calcium-free induced field burst activity: a possible role for electrotonic coupling in epileptogenesis. J Neurosci 1994; 14:4308-4317.

14. Caveney S. The role of gap junctions in development. Annu Rev Physiol 1985; 47:319-335.

15. Loewenstein WR. Junctional communication and the control of growth. Biochem Biophy Acta 1979; 560:1-65.

16. Loewenstein WR, Kanno Y. Intercellular communication and the control of tissue growth. Lack of communication between cancer cells. Nature 1966; 209:1248-1249.

17. Mehta PP, Bertram JS, Loewenstein WR. Growth inhibition of transformed cells correlates with their junctional communication with normal cells. Cell 1986; 44:187-196.

18. Morales M, Rozental R, Mehler MF et al. Changes in gap junction properties of an immortalized hippocampal cell line induced by differentiation. (1996, submitted).

19. Rozental R, Urban M, Fishman GI et al.

Gap junction downregulation is required for neuronal differentiation. Soc Neurosci Abst 1995; 21:2000.

20. Loewenstein WR, Kanno Y. Intercellular communication and tissue growth. I. Cancerous growth. J Cell Biol 1967; 33: 225-234.

21. Bennett MVL, Spray DC, Harris AL. Electrical coupling in development. Am Zool 1981; 21:413-427.

22. Warner AE, Guthrie SC, Gilula NB. Antibodies to gap junctional protein selectively disrupt junctional communication in early amphibian embryo. Nature 1984; 311: 127-131.

23. Lee S, Gilula NB, Warner AE. Gap junctional communication and compaction during preimplantation stages of mouse development. Cell 1987; 51:851-860.

24. Bevilacqua A, Loch-Caruso R, Erickson RP. Abnormal development and dye coupling produced by antisense RNA to gap junction protein in mouse preimplantation embryos. Proc Natl Acad Sci USA 1989; 86:5444-5558.

25. DeSousa PA, Valimarsson G, Nicholson BJ et al. Connexin trafficking and the control of gap junction asssembly in mouse preimplantation embryos. Development 1993; 117:1355-1367.

26. Reaume AG, DeSousa PA, Kulkarni S. Cardiac malformation in neonatal mice lacking connexin43. Science 1995; 267: 1831-1834.

27. Nelles E, Jung D, Gabriel HD. Characterization of connexin32 deficient mice generated by gene targeting. In: The Role of Connexin Diversity. Gap Junction International Conference, L'Ile des Embiez, France, 1995.

28. Davids M, Heydrich U, Hofer A et al. Microinjection of antibody to connexin43 into early *Xenopus* embryo results in specific defects during development. (Submitted).

29. Britz-Cunningham SH, Shah BSMM, Zuppan BSCW et al. Mutations of the connexin43 gap junction gene in patients with heart malformations and defects of laterality. New England J Medicine 1995; 332:1323-1329.

30. Batter DK, Corpina RA, Roy C et al. Heterogeneity gap junction expression in astro-

cytes cultured from different brain regions. Glia 1992; 6:213-221.

31. Dermietzel R, Traub O, Hwang TK. Differential expression of three gap junction proteins in developing and mature brain tissues. Proc Natl Acad Sci USA 1989; 86:10148-10152.

32. Dermietzel R, Hertzberg EL, Kessler JA et al. Gap junctions between cultured astrocytes: immunocytochemical, molecular, and electrophysiological analysis. J Neurosci 1991; 11:1421-1432.

33. Dermietzel R, Spray DC. Gap junctions in the brain: where, what type, how many and why? Trends Neurosci 1993; 16:186-192.

34. Spray DC, Moreno AP, Kessler JA. Characterization of gap junctions between cultured leptomeningeal cells. Brain Res 1991; 568:1-14.

35. Spray DC, Peinado A, Dermietzel R et al. Interactive Panel: Gap Junctions in the nervous system: What's new and what do they do. 28th Winter Conference on Brain Research Abst. 1995; 28:68.

36. Connors BW, Bernardo LS, Prince DA. Coupling between neurons of the developing rat neocortex. J Neurosci 1983; 3:773-782.

37. Harrison RG. Observations on the living developing nerve fiber. Anat Rec 1907; 1"116-118.

38. Harrison RG. The outgrowth of the nerve fiber as a mode of protoplasmic movement. J Exp Zool 1910; 9:787-846.

39. Harrison RG. The cultivation of tissues in extraneous media as a method of morphogenetic study. Anat Rec 1912; 6:181-193.

40. Banker G, Goslin K. Culturing Nerve Cells. MIT Press, MA, 1992.

41. Eves EM, Tucker MS, Roback JD et al. Immortal rat hippocampal cell lines exhibit neuronal and glial lineages and neurotrophin gene expression. Proc Natl Acad Sci 1992; 89: 4373-4377.

42. Frederiksen K, Jat PS, Valtz N. Immortalization of precursor cells from the mammalian CNS. Neuron 1988; 1:439-448.

43. Mehler MF, Rozental R, Dougherty M et al. Cytokine regulation of neuronal differentiation of hippocampal progenitor cells. Nature 1993; 362:62-65.

44. Jacks T, Fazeli A, Schmitt EA. Effects of an Rb mutation in the mouse. Nature 1992; 359:295-300.

45. Lee EY-H, Chang C-Y, Nanopin H et al. Mice deficient for Rb are noviable and show defects in neurogenesis and haematopoiesis. Nature 1992; 359:288-330.

46. Michaelson MD, Xu H, Mehler MF et al. Interleukin-7 is a neuronal growth factor. Soc Neurosci Abstr 1993; 19:1102.

47. Cattaneo E, McKay R. Proliferation and differentiation of neuronal stem cells regulated by nerve growth factor. Nature 1990; 347:762-765.

48. Lazar LM, Blum M. Regional distribution and developmental expression of epidermal growth factor and transforming growth factor-α mRNA in mouse brain by a quantitative nuclease protection assay. J Neurosci 1992; 12:1688-1697.

49. Baker RE, Corner MA, Habets AMMC. Effects of chronic supression of bioelectric activity on the development of sensory ganglion evoked responses in spinal cord explants. J Neurosci 1984; 4:1187-1192.

50. Bergey GK, Fitzgerald SC, Schrier B. et al. Neuronal maturation in mammalian cell cultures is dependent on spontaneous bioelectric activity. Brain Res 1981; 207:49-58.

51. Corner MA. Localization of the capacities for functional development in the neural plate of *Xenopus*. J Comp Neurol 1964; 123:243-255.

52. Corner MA, Crain SM. Patterns of spontaneous bioelectric activity during maturation in culture of fetal rodent medulla and spinal cord tissues. J Neurobiol 1972; 3:25-45.

53. Moody WJ, Simoncini L, Coombs JL et al. Development of ion channels in early embryos. J Neurobiol 1991; 22:674-684.

54. Spitzer NC. Ion channels in development. Ann Rev Neurosci 1979; 2:363-397.

55. Spitzer NC. Development of voltage-dependent and ligand-gated channels in excitable membranes. In: van Pelt J, Corner MA, Uylings HBM, Lopes da Silva FH, eds. Progress in Brain Research. Elsevier Science BV, 1994:169-179.

56. Spitzer NC, Lamborghini JE. The development of the action potential mechanism of amphibian neurons isolated in

cell culture. Proc Natl Acad Sci USA 1976; 73:1641-1645.

57. Xie H, Ziskind-Conhaim L. Blocking Ca^{2+}-dependent synaptic release delays motoneuron differentiation in the rat spinal cord. J Neurosci 1995; 15:5900-5911.

58. Mattson MP. Neurotransmitters in the regulation of neuronal cytoarchitecture. Brain Res Rev 1988; 13:179-212.

59. Rashid NA, Cambray-Deakin MA. N-Methyl-D-Aspartate effects on the growth, morphology and cytoskeleton of individual neurons in vitro. Dev Brain Res 1992; 67:301-308.

60. Pereira EFR, Reinhardt-Maelicke S, Schrattenholz A et al. Identification and functional characterization of a new agonist site on nicotinic acetylcholine receptors of cultured hippocampal neurons. J Pharmacol Exp Ther 1993; 265:1474-1491.

61. Laurie DJ, Wisden W, Seeburg PH. The distribution of thirteen $GABA_A$ receptor subunit mRNAs in the rat brain. III. Embryonic and postnatal development. J Neurosci 1992; 12:4151-4172.

62. Alger BE, Nicoll RA. Pharmacological evidence for two kinds of GABA receptor on rat hippocampal pyramidal cells studied in vitro. J Physiol (Lond) 1982; 328:125-141.

63. Yagodin S, Holtzclaw LA, Barker JL et al. GABA-A receptor mediated Cl-flux induces intracellular calcium increase in LHRH secreting neuronal cell line. Biophy Soc Abstr 1993; 64,A325.

64. Garyantes TK, Regehr WG. Electrical activity increases growth cone calcium but fails to inhibit neurite outgrowth from rat sympathetic neurons. J Neurosci 1992; 12:96-103.

65. Kater SB, Mattson MP, Cohan CS et al. Calcium regulation of the neuronal growth cone. Trends Neurosci 1988; 11:315-321.

66. Kater SB, Mills LR. Regulation of growth cone behavior by calcium. J Neurosci 1991; 11:891-899.

67. Mattson MP, Kater SB. Calcium regulation of neurite elongation and growth cone motility. J Neurosci 1987; 7:4034-4043.

68. Chang M, Dahl G, Werner R. A role for an inhibitory connexin is testis? Dev Biol 1996; 175:50-56.

69. Rozental R, Giaume C, Nedergaard M et al. Understanding the function of gap junction signalling in the CNS: New insights for brain development and dysfunction. 29th Winter Conference on Brain Res 1996: 29:55.

70. Davies AM. The Bcl-2 family of proteins, and the regulation of neuronal survival. Trends Neurosci 1995; 18:355-358.

71. McBurney MW, Reuhl KR, Ally AI et al. Differentiation and maturation of embryonal carcinoma-derived neurons in cell culture. J Neurosci 1988; 8:1063-1073.

72. Eves EM, Boise LH, Thompson CB et al. Bcl-xL inhibits apoptosis induced differentiation in an immortalized central nervous system cell line. J Neurochem, in press.

73. Chalfie M, Tu Y, Euskirchen G et al. Green fluorescent protein as a marker for gene expression. Science 1994; 263802-805.

74. Fishman G, Spray DC, Leinwand LA. Molecular characterization and functional expression of the human cardiac gap junction channel. J Cell Biol 1990; 111:589-598.

75. Friedrich G, Soriano P. Insertional mutagenesis by retroviruses and promoter traps in embryonic stem cells. Methods Enzymol 1993; 225:681-701.

76. Hern WM. Correlation of fetal age and measurements between 10 and 26 weeks of gestation. J Obstet Gynecol 1994; 63:26-32.

77. Grynkiewicz G, Poenie M, Tsien, RY. A new generation of Ca^{2+} indicators with greatly improved fluorescence properties. J Biol Chem 1985; 260:3440-3450.

DENDRITIC GAP JUNCTIONS IN DEVELOPING NEOCORTEX: A POSSIBLE ROUTE FOR WAVE-LIKE PROPAGATION OF NEURONAL ACTIVITY

Alejandro Peinado

1. INTRODUCTION

The notion of large scale synchronized neuronal activity advancing at slow speeds through neural tissue in the form of a traveling wave does not resonate with current concepts of how activity spreads through the nervous system during its normal mode of operation. Such a notion, however, is not new and has often been discussed in the context of various disorders of the nervous system. More recently, slow waves of synchronous activity have been described in the developing retina in vitro[1,2] and in a "sleeping" lateral geniculate nucleus (LGN) slice preparation,[3] all of which suggests that large-scale synchronization of neurons occurring in the form of traveling waves may play important roles in normal, as well as abnormal, brain function. How ubiquitous traveling waves actually are in the nervous system is not known, but the more widespread use of assorted multisite electrophysiological recording and imaging techniques will make it possible in the future to uncover other examples where they may exist.

A particularly dramatic, and somewhat eerie, illustration of a transient visual disorder associated with wave-like activity can be found in a classic paper by Karl Lashley[4] published in 1941. In it Lashley described

Gap Junctions in the Nervous System, edited by David C. Spray and Rolf Dermietzel.
© 1996 R.G. Landes Company.

his own personal experience of visual disturbances associated with migraine: a visual scotoma confined to one half of the visual field and symmetric for both eyes, starting as a disturbance of vision in the vicinity of the macula and spreading toward the temporal field as a crescent-shaped area, which he inferred reflected the perceptual correlate of a physiological process advancing over the primary visual cortex at a rate of about 3 mm per minute. At the leading edge of the enlarging visual scotoma was a band of scintillating white or colored lines of various orientations forming arrays of parallel lines, open angles and polygons.

While Lashley's description is commonly cited in review articles on the phenomenon known as spreading depression (SD) in animal models,[5,6] and the similarities between the two phenomena are striking indeed, as first pointed out by Milner,[7] it is also worth emphasizing the point made by Lashley himself that the enlarging scotoma and the advancing edge of scintillating lines can sometimes occur independently of each other in different regions of the visual field even during a single episode of migraine. If spreading depression underlies the phenomenon described by Lashley, this raises the interesting notion, not usually addressed in studies of SD, that it should be possible to dissociate the depression of neural activity typical of SD from the intense synchronous neuronal firing that occurs at the leading edge of the advancing depression. If this is so, the phenomenon of SD may be related to other forms of traveling synchronous activity (Table 17.1), which raises the interesting question of whether the mechanism by which SD propagates also applies to other forms of

Table 17.1. Comparison of slow wave-type events in several CNS models

Wave Type (Refs.)	Age/Species	Speed Range	Recording Method (Resolution)	Pharmacology
Retinal waves (1,2)	Early Postnatal/ Ferret	50-140 μm/sec	Extracellular 61-electrode array (Single-unit)	Blocked by TTX: Not blocked in 0 Ca^{2+}
LGNd waves (3)	Adult/Ferret	0.5-1.0 mm/sec	Extracellular 8-electrode array (Multi-unit)	Blocked by ACPD, NE, ACh
Hippocampal Spreading Depression (8,14)	Adult/Rat	20-150 μm/sec	Extracellular 2-electrode recording (Filed pot)	Not blocked by TTX or EAA agonists
4-AP-&GABA-dependent Neocortical waves (28,29)	Adult/Human, Guinea Pig	4.5-11.8 mm/sec	Extracellular 3-electrode recording (Field pot)	Blocked by TTX and BMI; Requires 4-AP; Not blocked by AP-5 or CNQX
Low-Ca^{2+} Seizure-like event waves (hippoc.) (30)	Adult/Rat	1.7 mm/sec	Extracellular 2-electrode recording (Field pot.)	Requires $\leq$0.2 mM Ca^{2+}
TEA-dependent Neocortical waves (15)	Early Postnatal/Rat	0.1-3 mm/sec	Multi-neuron Imaging with Fura-2 (Single cell)	Blocked by TTX; Not blocked by AP-5, CNQX.

traveling waves. Herreras et al[8] have suggested that in SD, synchronization of neuronal firing at the leading edge might be mediated by gap junctions, implying that the progressive recruitment of neurons that takes place at the advancing edge may require gap junctions in order to propagate from its site of initiation.

2. GAP JUNCTIONS IN DEVELOPING NEOCORTEX

It has been known for over 15 years, mostly as a result of dye-coupling experiments using intracellular injections of Lucifer yellow, that neuronal gap junctions are present in neocortex, particularly during early postnatal development.[9,10] More recently, Peinado et al[11] have shown, using the intracellular gap junction tracer Neurobiotin rather than Lucifer yellow, that there is a much higher incidence of neuronal coupling in developing neocortex than was once recognized on the basis of the Lucifer yellow experiments (see Fig. 17.1a). The incidence of dye coupling as well as the distribution of dye-coupled cells in the Neurobiotin experiments support the idea that coupling links neurons into large networks, extending over large areas of cortex rather than forming small, local arrays of coupled neurons. It is this aspect of neuronal coupling that makes gap junctions relevant to understanding traveling waves, since a neuronal syncytium could provide the connectivity necessary for activity to propagate as a slow wave over long distances.

Also supported by evidence obtained with different methods is that most neuronal gap junctions are on dendrites and somata rather than on axons. Figure 17.1 (b and c) shows two examples, at the light-microscopic level, of putative gap junction contacts between the injected neuron's dendrite and the soma or dendrite of neurons coupled to them. This feature of neuronal coupling is relevant to a possible involvement of gap junctions in the mechanism of spread of traveling waves insofar as it implies a restriction of the spatial sphere of influence of any particular neuron strictly to other neurons situated nearby, within reach of each other's dendrites.

Most gap junctions between neocortical neurons probably also consist of a small number of functional channels. This, in

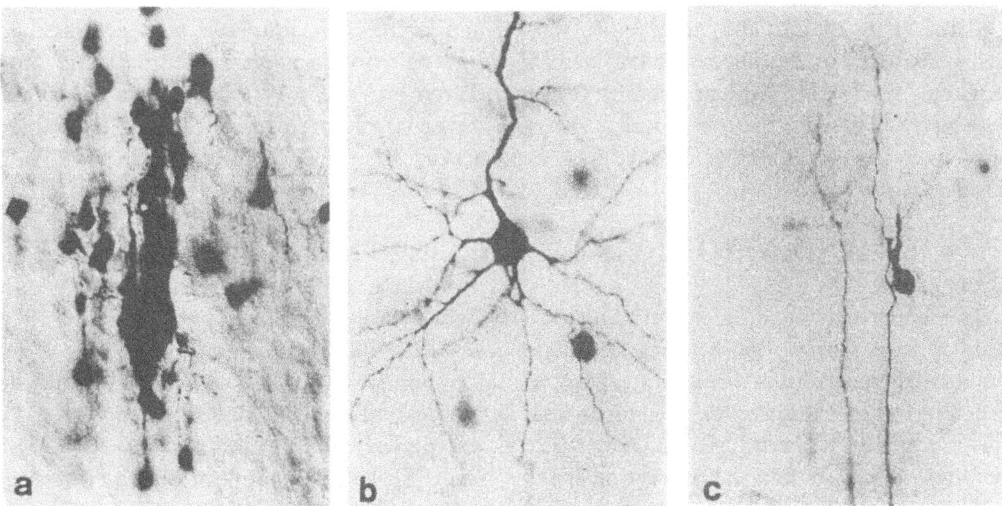

Fig. 17.1. Dendritic coupling is extensive in early postnatal neocortex. (a) Before PND 6 it is common for injections of Neurobiotin into a single neuron to result in labeling of numerous cells, forming a vertical arrangement, near the injected cell. (b, c) At later ages, from PND 14 onwards, the number of coupled cells that are labeled following an injection is drastically reduced. At these ages it becomes easier to see that the somata of coupled cells is often in direct contact with a basal (b) or an apical (c) dendrite of the injected neuron.

turn, is bound to further restrict the influence of any particular neuron on a coupled neighbor, making it unlikely that, for example, a single action potential occurring in only one neuron would depolarize a coupled neighbor enough to elicit an action potential (except in cases where the neighbor is very close to threshold). Instead, it seems more likely that for a neuron in the network to reach threshold, the contribution of many coupled neighbors is required. This in turn removes the "instantaneous" quality from this type of gap junctional communication since both the time constant and the space constant of the membrane, insofar as they determine how fast these contributions spread electrotonically within a neuron, would be expected to gain importance in the way such signaling propagates from neuron to neuron. This is in contrast to the situation in many invertebrate nervous systems, where the distinction between axons and dendrites is often blurred by the fact that both types of processes (often indistinguishable even morphologically) are fully capable of sustaining action potential activity. In such cases it is not unusual to have action potentials transmitted from one neuron to another via gap junctions.

Thus, the extent and nature of neuronal coupling in developing neocortex is consistent with the idea that activity can propagate through the neuronal syncytium, over large regions of cortex, and at relatively slow speed.

3. WAVES IN DEVELOPING NEOCORTEX

Fluorescence imaging techniques are ideal for monitoring the activity of large numbers of neurons in brain slices, often with single-cell resolution, and thus can provide a different perspective on wave-like phenomena than can be derived from single- or even multi-electrode recordings. The results described here were obtained by imaging slices of developing rat neocortex stained with the membrane permeant acetoxymethyl ester (AM) form of the calcium indicator dye Fura-2. Staining

was done as previously described.[12,13] Slices were then placed submerged in a temperature-controlled perfusion chamber on the stage of an Axioskop FS (Zeiss) upright microscope fitted with 20X and 40X water-immersion objectives and equipped with epifluorescence optics. Imaging was done with a 512 x 1024 pixel, frame transfer, cooled CCD device (Princeton Instruments). Sequences of images were acquired using a 12-bit, 1 Mpixel/sec digitizing mode and transferred directly to memory using IP Lab software (Signal Analytics) running on a Macintosh Quadra 840AV computer. Pixel binning, whereby neighboring pixels are combined into superpixels prior to readout from the CCD chip, was used in order to increase the number of frames that could be held in RAM during a single acquisition sequence, and to improve signal/noise.

For some experiments, two patch clamp amplifiers (Axopatch 200A; Axon Instruments) were also used to simultaneously monitor whole cell currents in pairs of neurons. Neurons were patched under a 40X objective (Zeiss Achroplan water-immersion objective; 1 mm working distance) using two Huxley-Wall type micromanipulators (Sutter Instrument Co.) to guide the micropipettes onto selected target neurons separated by small distances. Patch pipettes (5-10 MΩ) were filled with internal solution containing (in mM): CsCl, 70; CsF, 70; Hepes, 10; EGTA, 10; pH 7.2. Current traces were digitized and stored in a Macintosh II computer equipped with an ITC-16 data acquisition interface (Instrutech Corp.) using Synapse electrophysiology software (Synergistic Research Systems).

Spontaneous waves in developing neocortical slices were induced by changing the perfusion solution from standard artificial CSF (aCSF; composition in mM: NaCl, 125; KCl, 5; KH$_2$PO$_4$, 1.25; MgCl$_2$, 1.3; NaHCO$_3$, 26; Dextrose, 10; CaCl$_2$, 1.4; pH 7.2) to aCSF containing 10 mM tetraethylammonium (TEA) chloride. Under these conditions and without any further external manipulation, slices cut in the coronal plane exhibit waves of activity sweeping along the horizontal axis of the slice and occurring several times a minute. At any particular recording site (imaged area is approximately 370 μm x 370 μm when using a 20X objective) it is possible to record waves traveling in either direction (Fig. 17.2), which indicates that these events originate in multiple sites within a slice. For any particular wave event the calcium-induced changes in Fura-2 fluorescence last approximately 5-10 seconds and individual waves advance at speeds of 0.1-3 mm per second (Fig. 17.3).

Whole-cell voltage-clamp recordings made under identical conditions reveal that the wave-like activity recorded with Fura-2 is associated with large currents. Dual whole-cell recordings reveal that these currents occur simultaneously in neighboring neurons. Interestingly, when the gap junction blocking agent halothane is added to the aCSF at concentrations that abolish Neurobiotin dye-coupling, the large synchronous inward currents are abolished and the activity recorded in pairs of neurons becomes highly uncorrelated (Fig. 17.4), an effect that is consistent with gap junctions playing a role in the propagation of synchronous activity.

4. MECHANISMS INVOLVED IN WAVE PROPAGATION

To what extent the waves described here share mechanisms in common with other wave phenomena such as spreading depression, waves of synchronous bursts in developing retina, spindle oscillations in LGN slices, and the march of Jacksonian epileptic fits, remains to be established.

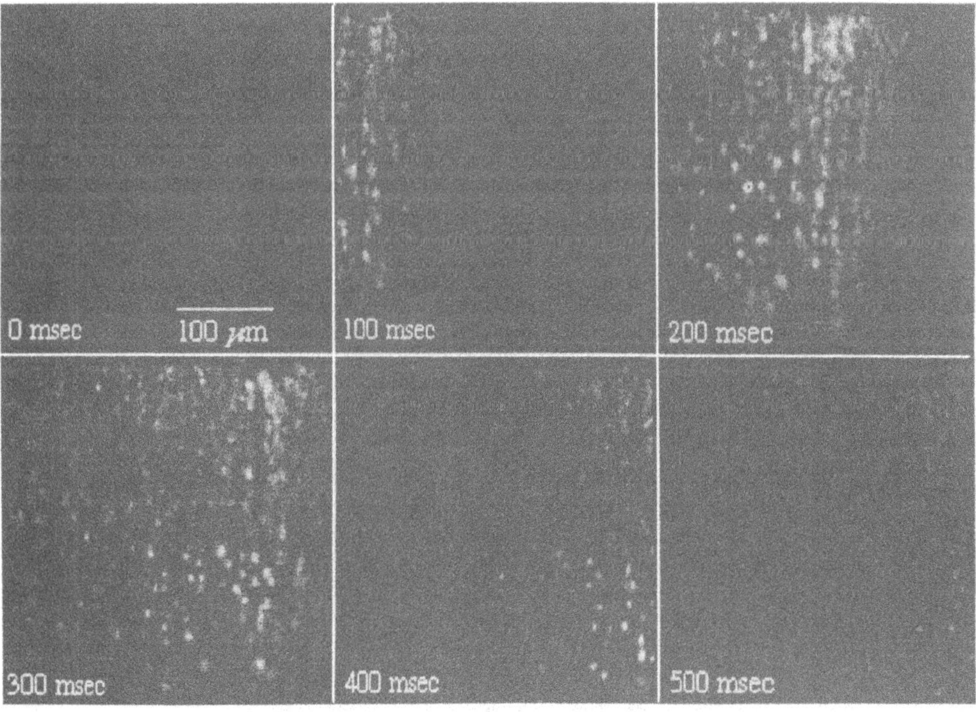

Fig. 17.2. A wave of activity sweeping horizontally, from left to right, along a developing neocortical slice. Image-processed sequence of a wave in PND 4 neocortex captured using fluorescence optics on a Fura-2 AM-stained brain slice perfused with TEA-containing aCSF. Each frame in the sequence shown represents the arithmetic subtraction of two consecutive images (Image$_n$ – Image$_{n+1}$) in the original raw sequence, where each frame was a record of fluorescence integrated over a period of 100 msec. White regions represent those where cells had the largest negative change in fluorescence (and by implication, the largest increase in [Ca^{2+}]$_i$) between two consecutive frames.

Some differences are already apparent, as for example in the speed of propagation. TEA-induced waves in neocortex travel at speeds comparable to that of the LGN spindle oscillations (0.5-1 mm/sec),[3] and somewhat faster than the retinal waves (80-140 µm/sec)[2] or spreading depression in hippocampus (80-125 µm/sec).[14] There are also pharmacological differences, as in the sensitivity to the sodium channel blocker tetrodotoxin (TTX): Spontaneous TEA-induced waves in neocortex cease in the presence of TTX[15] whereas spreading depression—at least the *depression* aspect of it—is resistant to this action potential blocker.[16]

Perhaps the most notable difference between spreading depression and other wave phenomena is, of course, that in these there is no obvious depression of activity, at least not to the extent that there is in SD, where an almost complete loss of ion homeostasis in neurons precludes any further excitability for periods of up to several minutes.

Despite these and other differences between the examples of traveling waves thus far known in brain, there is also much that unifies these diverse processes. Indeed, notwithstanding the precipitous loss of excitability characteristic of SD there is a rather striking similarity between all

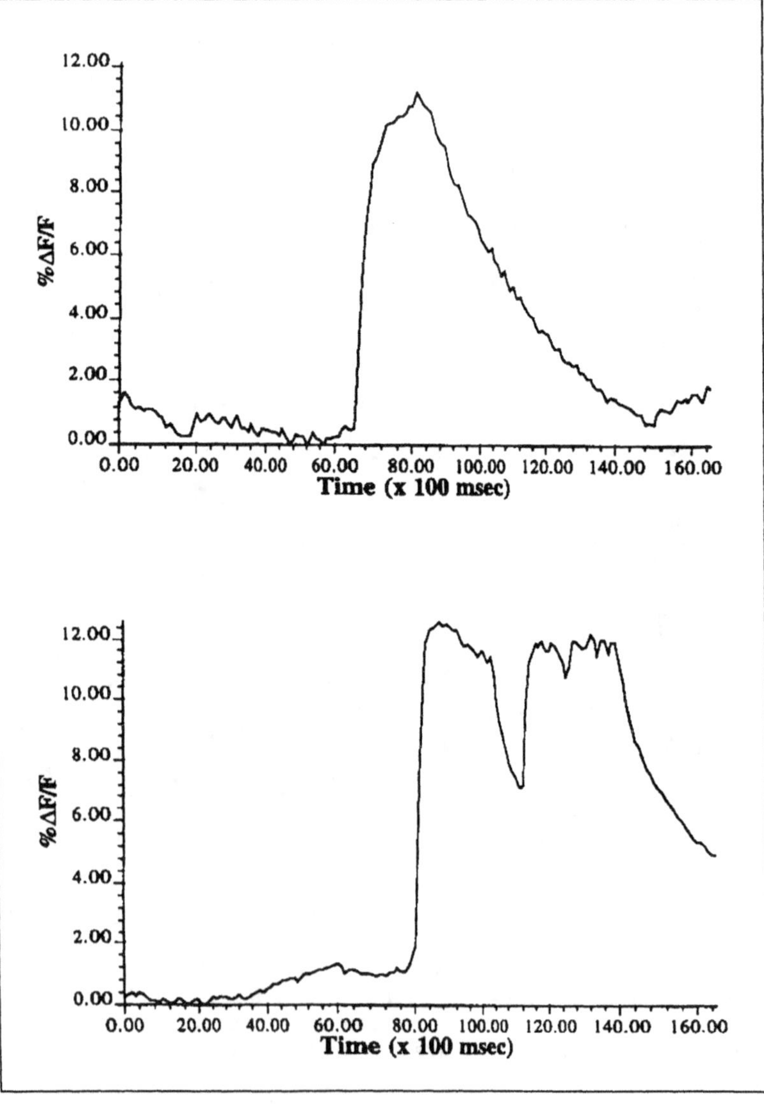

Fig. 17.3. Changes in Fura-2 fluorescence (obtained at 380nm excitation) during a wave take several seconds to return to baseline. A second wave, however, can propagate through the same region before the effects of the first one have recovered completely, suggesting that, unlike with spreading depression, there is very little if any refractory period. The top graph shows the change in fluorescence in a small region of cortex (approx. 50 x 50 µm) resulting from a single wave passing through. In the bottom graph one wave (peak at approx. 8.2 seconds) is followed by a second wave (peak at approx. 11.8 seconds) that invades the region before the fluorescence change from the first has recovered to baseline. The two waves traveled in opposite directions in this example.

wave-type phenomena in that all involve a short burst of activity lasting a few seconds that travels through the tissue and which involves the participation of large numbers of neurons. In the case of SD this corresponds to the prodromal population spike burst which typically precedes the depressed activity phase of this phenomenon. One wonders whether SD may not be a special case among traveling waves, one in which, perhaps due to the particularly stressful conditions usually required to elicit it (i.e., mechanical damage, low pO₂, very high KCl levels), events that

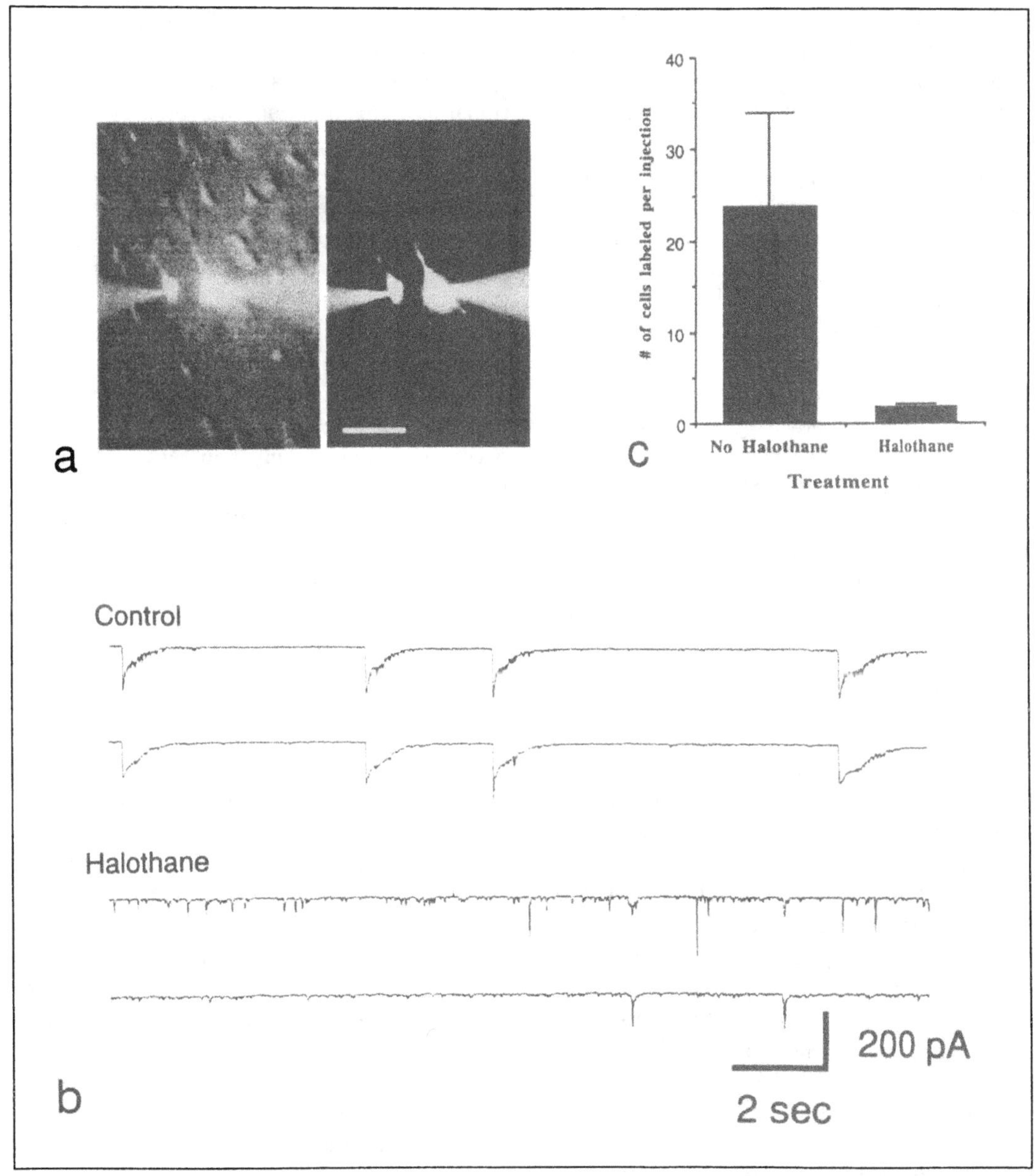

Fig. 17.4. Synchronous activity is abolished in the presence of the gap junction blocker Halothane. (a) Pairs of closely situated neurons (here shown dye-filled for the purpose of visualization) were whole-cell voltage-clamped in a PND 7 neocortical slice perfused with TEA-containing aCSF. (b) Currents were recorded before and after addition of Halothane to the aCSF. Note that while much activity remains after addition of Halothane, it is no longer synchronous in both cells. The concentration of Halothane was the same as that required to block dye coupling in Neurobiotin injection experiments, as shown in (c). Scale bar in (a), 50 μm.

would normally generate a propagating population burst suddenly degenerate into an inability of neurons and/or glia to maintain normal ion homeostasis. Unlike SD, some of the other wave phenomena occur spontaneously and under more or less normal physiological conditions.

In the case of the TEA-induced waves in developing neocortex described here it remains to be determined whether the K⁺ channel blocker is actually *inducing* the traveling wave or whether it is in fact only enhancing our ability to detect, using the calcium imaging approach, a wave event that can also occur in the absence of TEA. TEA could act by either increasing the frequency of such waves or enhancing calcium influx during these events, both of which would make them more easily detectable with calcium-sensitive flourescence imaging. It is clear, however, that in the absence of TEA (or other agents that block K⁺ channels) waves detected with Fura-2 imaging in slices are at best extremely infrequent (data not shown). On the other hand, it is possible that in the absence of TEA a different type of burst activity,[17] with less calcium influx associated with it, may occur and go undetected. Alternatively, since there is no shortage of mechanisms for modulating potassium channel activity in neurons in vivo (e.g., ACh, NE and glutamate via mGluRs decrease K⁺ currents in neocortical neurons), it is possible that the effects seen with TEA may occur under more normal physiological conditions in vivo, where the ACh and NE innervation of cortex is fully functional.

In these experiments, potassium channel blockers were employed as a way of enhancing electrotonic interactions via gap junctions. Although it is impractical to measure this enhancement directly in the cortical slice preparation, such an effect has been demonstrated at an identified electrical synapse in Aplysia.[18] And K⁺ channel blockers are occasionally used in mammalian brain slice preparations to enhance electrical coupling in neurons.[19,20] Clearly however, as with halothane, the actions of TEA are not re-stricted to their effect on gap junctional communication.

TEA is bound to exert its effect on at least three mechanisms, all of which theoretically could be involved in the initiation and/or propagation of waves in grey matter: synaptic transmission, gap junctional communication, and activity-dependent increases in extracellular potassium. Although most agents that modulate synaptic transmission can also affect gap junctional communication, the effects on the latter are often neglected in many studies in which such communication could be important for synchronization of activity, such as during seizures and spreading depression. Such lapses are not surprising given the difficulty of demonstrating extensive neuronal coupling in most mammalian CNS preparations. Recent findings, however, ought to encourage a revisiting of some previous interpretations since it is becoming increasingly apparent that neuronal gap junctions are more prevalent than previously thought, and this may even be true in mature brain.[21,22]

Also increasingly apparent, as more knowledge is gathered on the modulatory actions of various agents on gap junctional communication, is the idea that neuronal coupling is likely to be a transient phenomenon, highly regulated by activity, and also subject to the influence of many experimental manipulations which are sometimes assumed to affect exclusively chemical synaptic mechanisms. Thus, in astrocytes for example, which are more highly coupled than neurons and hence more amenable to such studies, it has been shown that high K⁺ as well as glutamate are capable of upregulating gap junctional communication with a timecourse of several minutes.[23] More recently, an activity-dependent increase in electrotonic coupling between the Mauthner neuron and its afferents has been found to be mediated by NMDA receptors.[24] Other modulatory neurotransmitters have also been shown to affect gap junctional communication in neuronal as well as nonneuronal cells (cf. ref. 26) And recent Neurobiotin experiments have implicated dopamine and acetylcholine in the modulation of neuronal coupling in developing neocortex.[25]

5. CONCLUDING THOUGHTS

Chemical synaptic transmission can undoubtedly contribute to neuronal excitability during wave propagation, and may indeed turn out to be exclusively responsible for some types of wave-like phenomena. The same can be said of transient elevations in extracellular K^+ which no doubt are present to greater or lesser extents during periods of high neuronal activity. We must keep in mind, however, that chemical synaptic transmission is often not necessary for propagation of wave-like activity (Table 17.1) and that the role of elevations in $[K^+]o$ remains a matter of correlation due to the absence of means to effectively manipulate this variable experimentally.

Together, the three factors namely chemical synapses, gap junctions, and extracellular effects, in varying proportions, must underlie not only slow wave phenomena but fast-spreading seizure activity as well. To what extent agents known to affect epileptiform activity (i.e., excitatory amino acid receptor antagonists, potassium channel blockers, high $[K^+]o$) are doing so through effects on gap junctional communication,[27] remains a difficult question due to the complex interrelationship between these three factors. The role gap junctions might play in the initiation and/or propagation of various forms of synchronous activity is, therefore, still very much in need of investigation.

The notion that gap junctional communication as a common mechanism could bring unity to a set of seemingly disparate phenomena as are the different types of waves, is an exciting prospect; all the more if waves turn out to have a role in the normal development and physiology of the brain as is suggested by the observations in retina and LGN. One hopes that a combination of imaging techniques capable of sampling large numbers of cells and electrophysiological techniques with fine-grained temporal resolution, along with careful pharmacological manipulations, will yield new insights on the relative contribution of chemical synapses, gap junctions, and activity-related changes in $[K^+]o$ to the propagation of wave-like activity in the CNS.

REFERENCES

1. Meister M, Wong ROL, Baylor DA, Shatz CJ. Synchronous bursts of action potentials in ganglion cells of the developing retina. Science 1991; 252:939-943.
2. Wong ROL, Meister M, Shatz CJ. Transient period of correlated bursting activity during development of the mammalian retina. Neuron 1993; 11:923-938.
3. Kim U, Bal T, McCormick DA. Spindle waves are propagating synchronized oscillations in the ferret LGNd in vitro. J Neurophysiol 1995; 74:1301-23.
4. Lashley KS. Patterns of cerebral integration indicated by the scotomas of migraine. Arch Neurol Psychiatr 1941; 46:331-339.
5. Leao AAP. Spreading depression of activity in the cerebral cortex. J Neurophysiol 1944; 7:359-390.
6. Laurintzen M. Cortical spreading depression as a putative migraine mechanism. Trends Neurosci 1987; 10:8-12.
7. Milner PM. A note on the possible correspondence between the scotoma of migraine and spreading depression of Leao. Electroencephalogr Clin Neurophysiol 1958; 10:705.
8. Herreras O, Largo C, Ibarz JM et al. Role of neuronal synchronizing mechanisms in the propagation of spreading depression in the in vivo hippocampus. J Neurosci 1994; 14:7087-7098.
9. Gutnick MJ, Prince DA. Dye coupling and possible electrotonic coupling in the guinea pig neocortical slice. Science 1981; 211:67-70.
10. Connors BW, Bernardo LS, Prince DA. Coupling between neurons of the developing rat neocortex. J Neurosci 1983; 4:1324-1330.
11. Peinado A, Yuste R, Katz LC. Extensive dye coupling between rat neocortical neurons during the period of circuit formation.

Neuron 1993; 10:103-114.

12. Yuste R, Katz LC. Control of postsynaptic Ca2+ influx in developing neocortex by excitatory and inhibitory neurotransmitters. Neuron 1991; 6:333-344.

13. Yuste R, Peinado A, Katz LC. Neuronal domains in developing neocortex. Science 1992; 257:665-668.

14. Herreras O, Somjen GG. Propagation of spreading depression among dendrites and somata of the same cell population. Brain Res 1993; 610:276-282.

15. Peinado A. Waves of synchronous neuronal activity in neonatal rat cortical slices. Soc Neurosci Abs 1995.

16. Sugaya E, Takato M, Noda Y. Neuronal and glial activity during spreading depression in cerebral cortex of cat. J Neurophysiol 1975; 38:822-841.

17. Jones RSG, Heinnemann U. Spontaneous and evoked NMDA-receptor mediated potentials in the entohrinal cortex of the neonate rat in vitro. In: Excitatory Amino Acids and Neuronal Plasticity; Y. Ben-Ari (ed.), 1990. New York: Plenum Press.

18. Rayport SG, Kandel ER. Epileptogenic agents enhance transmission at an identified weak electrical synapse in *Aplysia*. Science 1981; 213:462-464.

19. Christie MJ, Williams JT, North RA. Electrical coupling synchronizes subthreshold activity in locus coeruleus neurons in vitro from neonatal rats. J Neurosci 1989; 9:3584-3589.

20. Michelson HB, Wong RKS. Synchronization of inhibitory neurons in the guinea-pig hippocampus in vitro. J Physiol 1994; 477:35-45.

21. Parnavelas JG, Nadarajah B. Gap junctions are prevalent in the developing cerebral cortex. Soc Neurosci Abs 1994; 20:1672.

22. Parnavelas JG, Nadarajah B. Gap junctions may provide an important means of communication between cells in the cerebral cortex. Soc Neurosci Abs 1995; 21:1507.

23. Enkvist MOK, McCarthy KD. Astroglial gap junction communication is increased by treatment with either glutamate or high K$^+$ concentration. J Neurochem 1994; 62: 489-495.

24. Pereda A, Faber DS. NMDA-mediated activity-dependent short-term plasticity of electrotonic coupling. Soc Neurosci Abs 1995; 21:2006.

25. Peinado A, Yuste R, Katz LC. Gap junctional communication and the development of local circuits in neocortex. Cereb Cortex 1993; 3:488-498.

26. Roerig B, Klausa G, Sutor B. Modulatory neurotransmitters reduce dye-coupling between developing lamina II/III pyramidal neurons in rat neocortex. Soc Neurosci Abs 1995; 21:1509.

27. Perez-Velazquez JL, Valiante TA, Carlen PL. Modulation of gap junctional mechanisms during calcium-free induced field burst activity: A possible role for electrotonic coupling in epileptogenesis. J Neurosci 1994; 14:4308-4317.

28. Aram JA, Michelson HB, Wong RKS. Synchronized gabaergic IPSPs recorded in the neocortex after blockade of synaptic transmission mediated by excitatory amino acids. J Neurophysiol 1991; 65:1034-1041.

29. Avoli M, Mattia D, Siniscalchi A, Perreault P, Tomaiuolo F. Pharmacology and electrophysiology of a synchronous gaba-mediated potential in the human neocortex. Neuroscience 1994; 62:655-666.

30. Konnerth A, Heinnemann U, Yaari Y. Nonsynaptic epileptogenesis in the mammalian hippocampus in vitro. I. Development of seizurelike activity in low extracellular calcium. J Neurophysiol 1986; 56:409-423.

ELECTRIC COUPLING IN EPILEPTOGENESIS

Peter L. Carlen, Jose L. Perez-Velazquez, Taufik A. Valiante,
Shokrollah S. Jahromi and Berj L. Bardakjian

1. INTRODUCTION

One of the hallmarks of epileptiform activity is neural synchrony. There are several putative mechanisms for creating neural synchrony in a neural network including the chemical synaptic actions of decreased inhibition or increased excitation, extracellular ionic and volume shifts, and changes in electric coupling. Electric coupling includes ephaptic electrical field effects and direct interneuronal electrotonic (cable-like) coupling via gap junctions. Electric field coupling is governed by cell morphology, propagation velocity of depolarization waves, the rate of change to the transmembrane voltage, and extracellular resistivity. Such an effect can be approximately represented as a capacitative pathway via the extracellular fields. Gap junctional coupling can be represented by low resistive pathways through the adjacent connexins. Excellent reviews of electrical interactions between neurons have been published.[1-3]

The EEG of normal awake brain function is thought to be governed by chaotic dynamics,[4] whereas seizures are characterized by periodicity and fewer degrees of freedom.[5] The nature of the electrical activity patterns in the brain suggests that the brain behaves electrically like a population of coupled nonlinear oscillators. The most prominent feature of interaction between coupled oscillators is a change in frequency that can either be in the form of frequency pulling (e.g., chaos, quasiperiodicity, waxing and waning) or frequency entrainment (e.g., synchrony), depending on the differences in intrinsic oscillator parameters, and the manner in which the oscillators are coupled. Specifically, the interaction among coupled nonlinear oscillators is governed by two main factors: (1) intrinsic oscillator properties and (2) coupling mechanisms.[6] Two mechanisms for electrical coupling between excitable cells have been proposed: (1) electrotonic coupling via low resistance pathways between cells such that

Gap Junctions in the Nervous System, edited by David C. Spray and Rolf Dermietzel.
© 1996 R.G. Landes Company.

local circuit currents can readily spread from one cell to the next[7] and (2) field coupling such that excitation is transferred between excitable cells by the electric field that develops between two or more nearby cells.[8]

This chapter will concentrate on the role of electric coupling in causing neural synchrony and epilepsy in isolated neural systems, emphasizing recent data. Arising from both electrotonic and electric field coupling, we wish to stress the following points: (1) gap junctions, when open, can play a large role in neural synchronization; (2) gap junctional conductance can be rapidly modified; (3) closely aggregated neurons, even when gap junctional channels are closed, can be a substrate for capacitive coupling through neuronally generated extracellular fields; (4) a slower propagation velocity of the surface depolarization wave along the axons and the dendritic trees increases the extracellular potentials responsible for electric field coupling; (5) a faster rate of rise of the transmembrane voltage also increases the field coupling; (6) a significant difference between the resting transmembrane potentials in adjacent sites can prevent frequency synchronization between them. Some of these concepts have been developed in excitable cells other than neurons (i.e., cardiac and smooth muscle).

2. HISTORICAL DEVELOPMENT

As described by Eccles,[9] histologists studying the nervous system in the nineteenth century postulated that the central nervous system was an interlacing net-like structure called a reticulum. In the first half of the twentieth century, the concept of the synapse evolved. Then there was much debate as to whether central synaptic transmission was electrical or chemical, especially since peripheral chemical transmission had been well established by Loewi.[10] Intraneuronal electrophysiological recordings revolutionized our understanding of central synaptic transmission, particularly with the work of Eccles,[9] who demonstrated the importance of chemical transmission in the mammalian CNS. The

role of electrotonic transmission made a comeback in the 1970s with the work of Bennett[1] and others who showed that in the CNS electrotonic synapses characterized by gap junctions could be found in several CNS regions. Dye coupling was established in the cerebral cortex,[11] and in hippocampal pyramidal cells and dentate granule neurons.[12,13] Although the presence of gap junctional electrotonic coupling was well established in the early 1980s, its functional importance was questioned. MacVicar and Dudek[12] were the first to demonstrate, with dual intracellular recordings in central mammalian (hippocampal CA3) neurons, direct intercellular passage of current between the recorded neurons. These dual recordings are difficult, and when achieved, direct intercellular current transfer is usually difficult to demonstrate. However, in 47 of >400 simultaneous impalements, there was evidence of electrotonic coupling. Some of these recordings were from the same neuron because of a high coupling ratio, but other recordings were clearly from two different neurons since there was a low coupling ratio between the recorded neurons, and injections of HRP through both electrodes showed two stained cells. Knowles et al[14] could not find electrotonic coupling potentials from 101 pairs of CA1 pyramidal neurons, nor could they find short latency depolarizations or fast prepotentials (FFPs) in more than 75 neurons in response to antidromic activation. MacVicar and Dudek[13] showed direct electrotonic coupling in only 1 of 19 simultaneously recorded hippocampal dentate granule neuronal pairs and FPPs in 18% of recorded cells. FPPs were classically considered to represent dendritic spike activity,[15,16] but MacVicar and Dudek[12] showed in weakly coupled CA3 neurons that action potentials in one cell were usually associated with FPPs in the other coupled cell.

FPPs were also suggested to be related to electrotonic coupling in superficial neocortical neurons,[11] although in CA1 pyramidal cells, Andrew et al[17] showed FPPs in some cells without dye coupling. However fluorescent dye may not traverse all gap junctions for such reasons as gap junctions being present but closed, the dye molecule being too large for the gap junction in question, or the junction being far from the injection site. There is morphological evidence of gap junctions in interneurons in the hippocampal CA1 region[18] and in the polymorph layer of the dentate gyrus.[19] Spikelets have been observed in hippocampal type II hilar interneurons in normal perfusate.[20] Electrotonic coupling of interneurons could be critical for epileptogenesis. Finally there is some morphological and electrophysiological evidence for dual electrical-chemical synapses in the vertebrate CNS,[21] although little work has been done in this regard recently.

3. CONNEXINS

Gap junctions are formed by proteins called connexins, six of which aggregate together to form a connexon, which forms a hemichannel in the cell membrane as discussed elsewhere in this book. The connexon joins with another connexon of an adjacent cell to create a gap junction with the channel running through the center of the two connexons. There are more than a dozen types of connexins identified to date using molecular biological techniques. Their distribution in the CNS at the cellular level is presently being worked out. In the CNS the best characterized gap junctional protein is connexin43 which is presently thought to be mainly localized to glia. Not only could gap junctions between neurons play an important role in epileptogenesis, but one must not ignore the role of the more numerous glia which are well known to be coupled to each other by gap junctions.[22] Recently there is evidence for direct neuronal-glial communication,[23,24] even though dye coupling was not observed between neurons and glia.[23] How glia are specifically involved in epileptogenesis is unclear. Naus et al[25] showed that in peritumoral tissue removed at operation, connexin mRNA expression was increased in those cases who had sei-

zures originating from the tumor focus, and not in those cases without seizures.

4. GAP JUNCTIONS AND EPILEPTOGENESIS

Given that dye coupling reflects intercellular electrical communication via gap junctions, what is/are the role(s) of gap junctions in epileptogenesis? Intuitively it seems obvious that direct intercellular electrotonic interaction between CNS neurons should facilitate neural synchrony and thence epileptogenesis. Electrotonic junctions are considerably rarer than chemical synapses in the mammalian CNS. But as pointed out by Dermietzel and Spray,[26] the nonjunctional conductance of the entire neuron may be similar to the conductance of a single gap-junction channel which can range between 50-150 pS (picosiemens). Modeling demonstrates that the degree of electrotonic coupling can either increase or decrease the frequency of a coupled oscillator depending on its membrane potential wave form, the state of the neuron to which it is coupled, and the strength of the coupling.[27] Rayport and Kandel[28] showed in *Aplysia* that epileptogenic agents (pentylenetetrazole, strychnine, tetraethylammonium) enhanced transmission at a weak electrical synapse. Baimbridge et al[29] showed that epileptiform bursting from depolarizing currents in CA1 neurons was seen only in those neurons which demonstrated dye coupling.

In electronic models of oscillator-to-oscillator communication[30] it was demonstrated that entrainment of oscillators having different intrinsic frequencies can be achieved by resistive and/or resistive-capacitive coupling pathways. Using resistive coupling pathways, the entrained frequency was always less than or equal to the highest intrinsic frequency of the oscillators, whereas for resistive-capacitive coupling pathways, it could be higher than the highest intrinsic frequency. In fact we have shown by simultaneous whole cell and extracellular recordings in the CA1 region, that when field bursting was blocked by intracellular acidosis with propionate (see below), in some neurons the spontaneous firing frequency did indeed increase (Fig. 18.3A, Perez-Velazquez et al[31]). This suggests that resistive-capacitive coupling pathways may be present in the brain since entrained frequencies higher than the intrinsic frequencies can be observed. This also further reinforces the concept that an aggregate of gap junctions can be represented by a resistive-capacitive pathway.

5. ELECTRIC COUPLING DURING SEIZURES

The role of electric coupling and gap junctions during seizure activity is unclear. Taylor and Dudek (1982) demonstrated antidromically-evoked epileptiform activity in hippocampal slices bathed in manganese (2.3 mM) and lowered calcium (0.5 mM), a perfusate which blocked orthodromically-evoked chemical synaptic transmission. Simultaneous intracellular and extracellular recordings showed synchrony of neuronal action potentials with the extracellularly recorded population spikes. Measurement of the transmembrane potential indicated that the transient extracellular fields contributed to the observed synchrony by an ephaptic interaction. They also suggested that electrotonic coupling and changes in extracellular ion concentration could have also played a role in the epileptiform synchrony. Traub and Wong[33] modeled epileptiform activity in a small neuronal network. When electrotonic junctions were added to this model, epileptiform bursts tended to be blocked because of shunting of excitatory synapses. However, as mentioned above, electrotonic coupling can have different effects depending on the conditions of the coupled neurons.[27] Later modeling showed that electrical field effects contribute to the shape of the epileptiform field potential.[34] Computer simulations showed that the synchronized action potentials seen in this model of epilepsy could be explained by extracellular currents given excitable neurons and sufficiently high resistivity of the extracellular medium.[35] This supports a role for ephaptic transmission in epilepto-

genesis. In computer models used to investigate the extracellular potentials of excitable cells having a depolarization wave propagating on the cell's surface,[36] it was demonstrated that greater extracellular potentials (and hence greater field coupling between cells) can be achieved if the following features were present: (1) slow propagation velocities especially near sites of coupling; (2) fast rates of depolarization of the transmembrane voltage; and (3) low extracellular conductivites. Such a mechanism is relevant in neurons where surface depolarization waves travel along axons and dendritic trees.

Reducing the size of the extracellular space is the most obvious way to increase the extracellular resistance, which would enhance neuronal electrical interactions. This occurs particularly in hyposmolar states such as water intoxication and hypoxia-ischemia. Hyposmolar states are well known to be associated with clinical seizures and are used to create or augment in vitro models of epilepsy.[37-39] With synaptic transmission blocked and seizures established in hippocampal CA1 neurons using zero calcium perfusate, decreasing the osmolality, which decreases the extracellular volume, enhanced epileptiform bursting.[37] Conversely, increasing the extracellular volume with membrane-impermeant solutes, reduced or blocked epileptiform discharges in the zero-calcium seizure model,[37] and in an in vitro model of high potassium-induced seizures.[40] How changes in extracellular volume effect gap junctional function is not known. It is possible that with cell swelling, increased peri-gap junctional apposition could lead to increased capacitive coupling between nearby neurons. Alternatively, changes in extracellular volume could have a direct effect on extracellular potentials and consequently on electric field coupling.

6. INTRINSIC PROPERTIES AND ENTRAINMENT

Oscillatory behavior has been described in neurons of several brain regions; e.g., thalamus,[41] hippocampus,[42,43] entorhinal cortex,[44] nucleus accumbens.[45] The frequency, amplitude and waveshape of the intrinsic oscillations have been known to affect the interaction between coupled nonlinear oscillators. In addition, a difference in the intrinsic level between coupled oscillators can: (1) ultimately determine whether or not they achieve entrainment; (2) be one of the factors governing their phase lag characteristics; and (3) can produce a wide range of frequencies depending on coupling strength. These effects can be seen in computer models of coupled oscillators having the same or different intrinsic frequencies.[6] The effects of an intrinsic resting level gradient on the phase lags between oscillators are of particular interest since both the amount of phase lag and the order of oscillator lead can be affected. In computer models of coupled nonlinear oscillators, it was demonstrated[6] that: (1) an intrinsic resting level gradient is sufficient to produce many features of interaction between coupled oscillators even in the absence of an intrinsic frequency gradient; (2) the greater the intrinsic resting level differences, the stronger the coupling needed for entrainment; i.e., if there is a significant difference in the intrinsic resting membrane potentials between two groups of cells, then coupling between these groups can be blocked; and (3) if both an intrinsic frequency and intrinsic resting level gradient are present then both of them contribute to the observed features of interaction between coupled oscillators.

Application of extracellular anodic currents to the hippocampal CA1 region for 50-75 msec was shown to be effective in suppressing neuronal bursts in two in vitro hippocampal models of epilepsy in the CA1 area; evoked bursts in the presence of penicillin,[46] and spontaneous bursts with perfusion of high potassium.[47] As well, application of much briefer extracellular current pulses caused a large decrease in the size of epileptiform spikes.[48] This suggests that regional differences in the intrinsic resting membrane potentials were caused by the current stimuli which prevented frequency entrainment and neural synchrony.

7. MODULATION OF GAP JUNCTIONS AND SEIZURES

We have recently demonstrated a putative role for electrotonic coupling via gap junctions in the modulation of calcium-free induced field burst activity in the hippocampal CA1 region.[31] Using a fast flow perfusion system at 35°C, it was possible to switch perfusates and observe physiological effects within minutes. The calcium-free induced epileptiform activity was rapidly abolished by maneuvers to decrease intracellular pH; 25 mM propionate perfusate, bubbling with 90% CO_2, or by washout of ammonium chloride; and was enhanced by increasing intracellular pH by wash-in of 10 mM ammonium chloride, 60 mM bicarbonate, or trimethylamine. These effects, which occurred within 1-2 min, are thought to be mediated by changes in intracellular pH; acidosis blocking and alkalosis enhancing gap junctional conductance. Other agents, halothane (2%) and octanol (0.2 mM), thought to block gap junctional communication by non-pH dependent mechanisms, also rapidly diminished the calcium-free induced epileptiform activity, usually with no decrease in individual cell spiking frequency, suggesting uncoupling of neurons. Hence it is possible that mammalian CNS gap junctional communication can be rapidly modified by biochemical means. It was also found that dye coupling measured by Lucifer yellow was greatly enhanced by calcium-free medium and was greatly diminished by an acidic medium.

In clinical epilepsy, hyperventilation is used to bring out seizures. Hyperventilation blows off CO_2 causing a systemic alkalosis, which could secondarily enhance gap junctional conductance and neural synchrony by an intracellular alkalosis. Conversely, the ketogenic diet which has recently regained favor in the treatment of intractable seizures, might work via an intracellular acidosis and the uncoupling of gap junctions.

Second messenger systems are known to modulate gap junctional conductance, as are some neurotransmitters. Calcium ions can diminish electrotonic coupling,[49] as can arachidonic acid,[50] diacylglycerol,[51] phorbol esters[52] and acidosis.[53] Dopamine has been shown to decrease penicillin-induced epileptiform activity in hippocampal pyramidal neurons[54] and to uncouple nucleus accumbens neurons.[55] Dopamine uncouples retinal neurons[56] through a cyclic AMP kinase.[57] The relationship of these modulatory effects on gap junctions and epileptogenesis is not known. Spreading depression is thought to occur with or following seizures. Recently, Herreras et al[58] concluded that in the in vivo hippocampus, the initial evolution of spreading depression (SD), evoked by microdialysis or focal microinjection of high potassium solution, was associated with changes in neuronal excitability which preceded the regenerating depolarization of SD by several seconds. They proposed that opening of normally closed electronic junctions between neurons could best explain the long-distance synchronization between neurons and the current flow that occurs in the leading edge of a propagating wave of SD.

8. SPIKELETS: POSSIBLE CAPACITIVE COUPLING

The characteristics of spikelets suggested that they could be a result of capacitive coupling; a relatively unrecognized concept for CNS transmission. We investigated spikelets using the calcium-free model of seizures.[59] The presence of spikelets was highly correlated with the presence of dye coupling (e.g., Fig. 18.1) and their pattern of appearance was identical to that of spikes in recorded neurons. Our most recent data in calcium-free perfusate show that 25 CA1 neurons without spikelets were not dye coupled, whereas of 13 neurons which showed spikelets, 9 were dye coupled. The spikelet amplitude did not vary with changes in membrane potential. Spikelets were a few mV in amplitude and briefer (a few msec) than intracellularly measured spikes or extracellular population spikes. They were usually biphasic in shape with a rapid depolarization followed by a slower hyperpolarization. Numerical integration of spikelets yielded waveforms indistinguishable

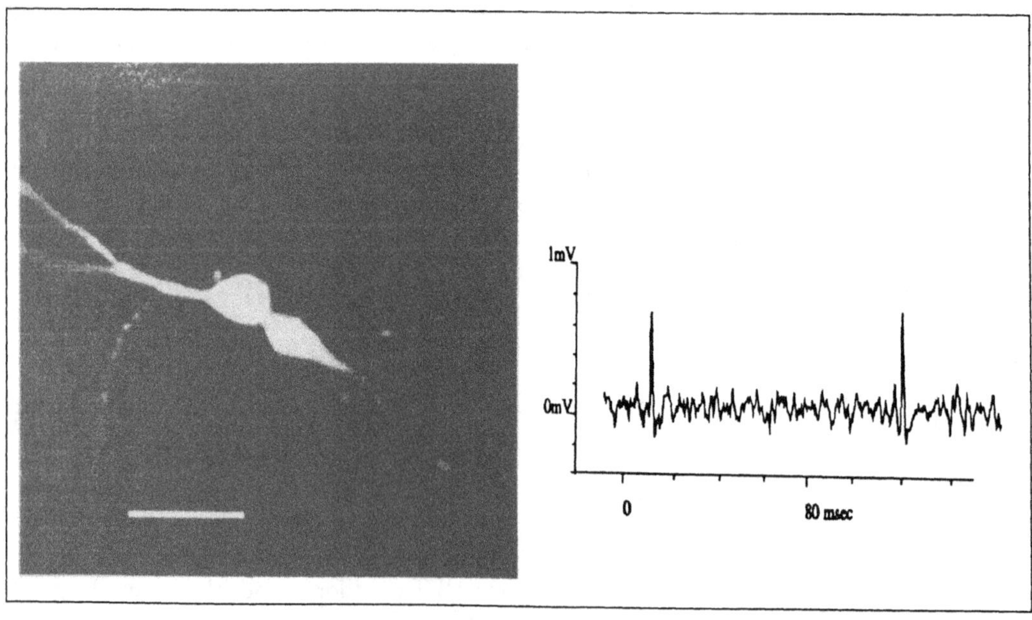

Fig. 18.1. Two cells dye-coupled in the CA1 area of the rat hippocampus. Traces show single spikelets recorded in one of these neurons during perfusion with calcium-free solution. Scale bar is 25 micrometers.

from postsynaptic action potentials; conversely, differentiation of action potentials resulted in waveforms which looked like spikelets (Fig. 18.2). Broadening of the action potential by the potassium channel blocker, tetraethylammonium chloride, also caused a widening of the spikelet, suggesting that spikelets could be due to both resistive and capacitive components from an action potential in an adjacent neuron.

Intracellular acidification decreased spikelet frequency and alkalinization increased spikelet frequency.

9. DIFFERENTIATION OF A PREJUNCTIONAL ACTION POTENTIAL

Between two closely opposed cells that are coupled through gap junctions, two mechanisms exist for direct electrotonic

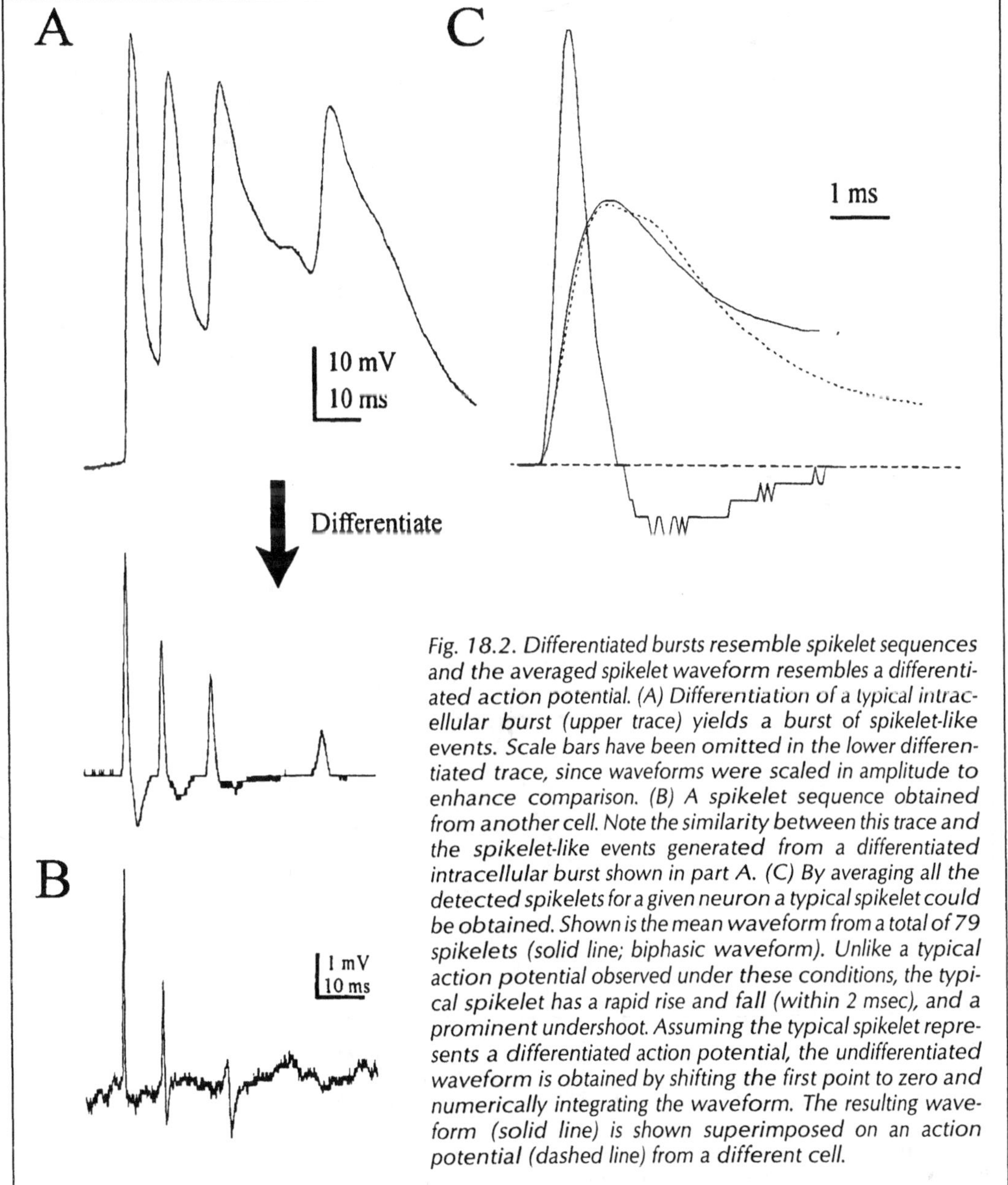

Fig. 18.2. Differentiated bursts resemble spikelet sequences and the averaged spikelet waveform resembles a differentiated action potential. (A) Differentiation of a typical intracellular burst (upper trace) yields a burst of spikelet-like events. Scale bars have been omitted in the lower differentiated trace, since waveforms were scaled in amplitude to enhance comparison. (B) A spikelet sequence obtained from another cell. Note the similarity between this trace and the spikelet-like events generated from a differentiated intracellular burst shown in part A. (C) By averaging all the detected spikelets for a given neuron a typical spikelet could be obtained. Shown is the mean waveform from a total of 79 spikelets (solid line; biphasic waveform). Unlike a typical action potential observed under these conditions, the typical spikelet has a rapid rise and fall (within 2 msec), and a prominent undershoot. Assuming the typical spikelet represents a differentiated action potential, the undifferentiated waveform is obtained by shifting the first point to zero and numerically integrating the waveform. The resulting waveform (solid line) is shown superimposed on an action potential (dashed line) from a different cell.

coupling. One pathway involves the low resistance electrical connection between the cells via the connexons themselves. which is primarily resistive. The second pathway is primarily capacitive and in theory can lead to a transfer of excitation from a cell actively generating an action potential to a neighboring quiescent cell, in the absence of low-resistance pathways.[60,61] The two pathways together functionally behave as a low pass filter. Adjacent neuronal source activity can induce a membrane current in a receiving cell that is essentially the first time derivative of the activity in the prejunctional cell;[62] i.e., a high-pass filter with a relatively short time constant. This appears to be an attractive explanation for the waveform of the spikelets observed in these experiments. An equivalent circuit representation of our experimental preparation during presumed coupling (Fig. 18.3) was adapted from the equivalent circuit presented for coupled myocytes.[62] With reasonable parameter values this model can generate spikelet-like waveforms given a current source that generates a spike like event in the active neuron. As can be appreciated from the circuit diagram, current flow between the two cells is either through the junctional resistance R_j or the junctional capacitance C_j. Since current flow through a capacitor is dependent on the rate of change of voltage (dV/dt), then capacitive effects only occur while voltage changes are occurring.

Modeling of the electrical interaction between closely opposed myocytes has suggested that electrical interactions can occur in the absence of low-resistance pathways.[60-62] In these models, prejunctional activity alters the extracellular potential in the junctional cleft, causing a change in the transmembrane potential of the postjunctional membrane (i.e., field coupling), independent of the rest of the cellular membrane. Since the amount of

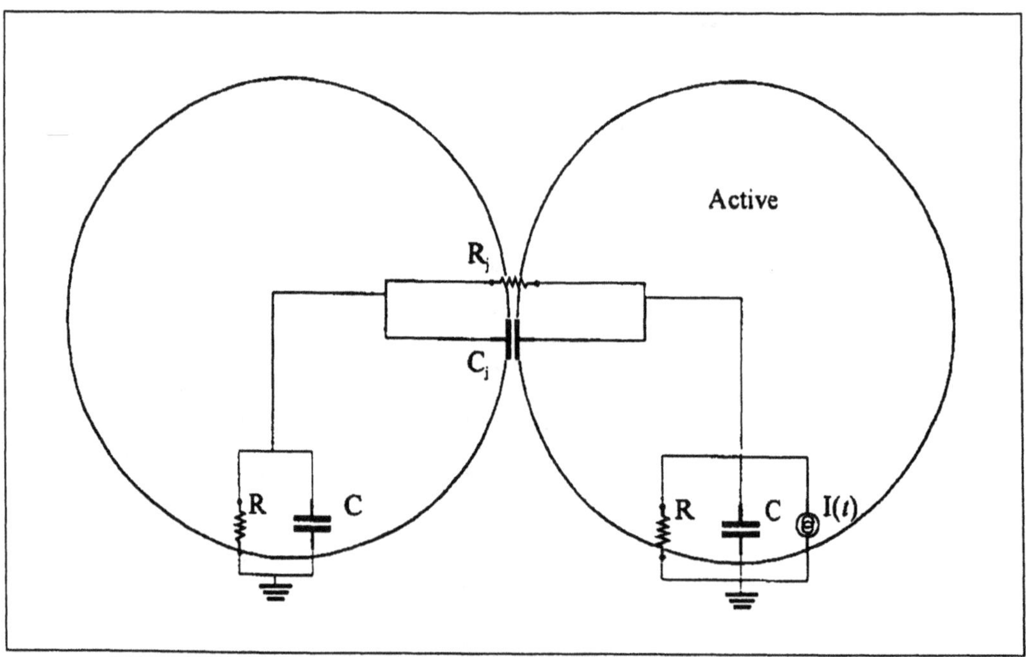

Fig. 18.3. Capacitive effects arising from activity in a neighboring neuron can explain the resemblance of spikelets to differentiated action potentials. Capacitive model adapted from Hogues et al (1992) for cardiac myocytes. The junctional capacitance C represents the field effects, and is assumed to be distinct from the membrane capacitance (C_m). R_m represents the membrane resistance of the neuron, R_j is a resistance representing the gap junctional resistive pathway for current flow. Two pathways for current flow exist between the two cells, one through R_j and the second through C_j.

extracellular polarization is dependent on the rate of change of the membrane potential of the prejunctional cell, the net result of this type of interaction is that an action potential in one cell can induce a differentiated current pulse in a neighboring cell.[62] Any factor that can increase the rate of rise of the source transmembrane voltage or decrease the propagation velocity, or increase the extracellular resistivity, will enhance this type of interaction through electric fields.

In conclusion, both functional and structural factors contribute to electric coupling in epileptogenesis. Functional factors arise from extracellular potentials caused by propagating surface depolarization and differences in intrinsic properties of neighboring neuronal regions. Structural factors arise from cellular morphological changes related to shrinkage of extracellular space between adjacent cells, and aggregates of gap junctions which include both resistive (connexons) and capacitive (from nearby neurons) pathways for local circuit currents. Anticonvulsants traditionally work through modulation of chemical synapses and intrinsic membrane ionic currents. Hence intractable drug-resistant epilepsy may be more related to pathological electric coupling.

ACKNOWLEDGMENTS

This work was supported by the MRC and NSERC.

REFERENCES

1. Bennett MVL. Electrical transmission: a functional analysis and comparison to chemical transmission. In: Kandel ER, ed. Handbook of Physiology, Section 1: The Nervous System, Vol. 1, Cellular Biology of Neurons, Part 1. Bethesda, Maryland: American Physiology Society, 1977: 357-412.
2. Korn H, Faber DS. Elictrical interactions between vertebrate neurons:field effects and electrotonic coupling. In: Schmitt FO, Worden FG, eds. The Neurosciences: Fourth Study Program. Cambridge: MIT Press, 1979:333-358.
3. Dudek FE, Snow RW, Taylor CP. Role of electrical interactions in synchronization of epileptiform bursts. In: Delgado-Escueta AV, Ward AA Jr, Woodbury DM, Porter RJ, eds. Advances in Neurology, Vol 44. New York: Raven Press, 1986:593-617.
4. Jansen BH. "Is it?" and "so what"-a critical view of EEG chaos. In: Duke DW, Pritchard WS, eds. Measuring Chaos in the Human Brain. Singapore, New Jersey, London, Hong Kong: World Scientific, 1991:83-96.
5. Babloyantz A, Destexhe A. Low dimensional chaos in an instance of epilepsy. Proc Natl Acad Sci USA 1986; 83:3513-3517.
6. Bardakjian BL, El-Sharkawy TY, Diamant NE. Interaction of coupled nonlinear oscillators having different intrinsic resting levels. J Theor Biol 1984; 106:9-23.
7. Bortoff A. Propagation of electrical activity in gastrointestinal smooth muscle: the case for propagation by local circuit current flow. J Gastrointest Motil 1991; 3:57-63.
8. Sperelakis N. Electrical field model: an alternative mechanism for cell-to-cell propagation in cardiac muscle and smooth muscle. J Gastrointest Motil 1991; 3:76-84.
9. Eccles JC. The Physiology of Synapses. New York: Springer-Verlag, 1964:1-316.
10. Loewi O. Problems connected with the principle of humoral transmission of nervous impulses. Proc Roy Soc B 1933; 118: 299-316.
11. Gutnick MJ, Prince DA. Dye coupling and possible electrotonic coupling in the guinea pig neocortical slice. Science 1981; 211: 67-70.
12. MacVicar BA, Dudek FD. Electrotonic coupling between pyramidal cells: a direct demonstration in rat hippocampal slices. Science 1981; 213:782-785.
13. MacVicar BA, Dudek FD. Electrotonic coupling between granule cells of rat dentate gyrus: physiological and anatomical evidence. J Neurophysiol 1982; 47:579-592.
14. Knowles WD, Funch PG, Schwartzkroin PA. Electrotonic and dye coupling in hippocampal CA1 pyramidal cells in vitro. Neuroscience 1982; 7:1713-1722.
15. Spencer WA, Kandel ER. Electrophysiology of hippocampal neurons. IV. Fast prepotentials. J Neurophysiology 1961; 24:272-285.

16. Turner RW, Meyers DER, Barker JL. Fast pre-potential generation in rat hippocampal CA1 neurons. Neuroscience 1993; 53:949-959.

17. Andrew RD, Taylor CP, Snow RW et al. Coupling in rat hippocampal slices: dye transfer between CA1 pyramidal cells. Brain Res. 1982; 8:211-222.

18. Katsumaru H, Kosaka T, Heizmann C et al. Gap junctions on GABAergic neurons containing the calcium-binding protein parvalbumin in the rat hippocampus (CA1 region). Exp Brain Res 1988; 7:363-370.

19. Kosaka T, Neuronal gap junctions in the polymorph layer of the rat dentate gyrus. Brain Res 1984; 47-351.

20. Michelson HB, Wong RKS. Synchronization of inhibitory neurones in the guinea-pig hippocampus in vitro. J Physiol 1994; 4(77):35-45.

21. Shapovalov AI. Interneuronal synapses with electrical, dual and chemical mode of transmission in vertebrates. Neuroscience 1980; 5:1113-1124.

22. Nagy JL, Yamamoto T, Sawchuk MA. et al. Quantitative immunohistochemical and biochemical correlates of connexin43 localization in rat brain. Glia 1992; 51-9.

23. Murphy TH, Blatter LA, Wier WG et al. Rapid communication between neurons and astrocytes in primary cortical cultures. J Neurosci 1993; 13(6):2672-2679.

24. Charles AC. Glia-neuron intercellular calcium signaling. Dev Neurosci 1994; 16:196-206.

25. Naus CCG, Bechberger JF, Paul DL. Gap junction gene expression in human seizure disorder. Exp Neurol 1994; 111:198-203.

26. Dermietzel R, Spray DC. Gap junctions in the brain: where, what type, how many and why? Trends Neurosci 1993; 16:186-192.

27. Kepler TB, Marder E, Abbott LF. The effect of electrical coupling on the frequency of model neuronal oscillators. Science 1990; 248:83-85.

28. Rayport SG, Kandel ER. Epileptogenic agents enhance transmission at an identified weak electrical synapse in aplysia. Science 1981; 213:462-464.

29. Baimbridge KG, McLennan PMJ, Church J. Bursting response to currents-evoked depolarization in rat CA1 pyramidal neurons is correlated with lucifer yellow dye coupling but not with the presence of calbindin-D_{28k}. Synapse 1994; 7:269-277.

30. Bardakjian BL, Diamant NE. A mapped clock oscillator model for transmembrane electrical rhythmic activity in excitable cells. J Theor Biol 1994; 166:225-235.

31. Perez-Velazquez JL, Valiante TA, Carlen PL. Modulation of gap junctional mechanisms during calcium-free induced field burst activity: a possible role of electrotonic coupling in epileptogenesis. J Neuroscience 1994; 14:4308-4317.

32. Taylor CP, Dudek FE. Synchronous neural afterdischarges in rat hippocampal slices without active chemical synapses. Science 1982; 218:810-812.

33. Traub RD, Wong RKS. Synaptic mechanisms underlying interictal spike initiation in a hippocampal network. Neurology 1983; 33:257-266.

34. Traub RD, Dudek FE, Taylor CP et al. Simulation of hippocampal afterdischarges synchronized by electrical interactions. Neuroscience 1985; 14:1033-1038.

35. Traub RD, Dudek FE, Snow RW et al. Computer simulations indicate that electrical field effects contribute to the shape of the epileptiform field potential. Neuroscience 1985; 15:947-958.

36. Bardakjian BL, Vigmond EJ. Effects of the propagation velocity of a surface depolarization wave on the extracellular potential of an excitable cell. IEEE Trans Biomed Eng 1994; 41:432-439.

37. Dudek FE, Obenaus A, Tasker JG. Osmolality-induced changes in extracellular volume alter epileptiform bursts independent of chemical synapses in the rat: importance of nonsynaptic mechanisms in hippocampal epileptogenesis. Neuroscience 1990; 120:267-270.

38. Andrew RD. Seizure and acute osmotic change: clinical and neurophysiological aspects. J Neurol 1991; 101:7-18.

39. Roper SN, Obenaus A, Dudek FE. Osmolality and nonsynaptic epileptiform bursts in rat CA1 and dentate gyrus. Ann Neurol 1992; 31:81-85.

40. Traynelis SF, Dingledine R. Role of extra-

cellular space in hyperosmotic suppression of potassium-induced electrographic seizures. J Neurophysiol 1989; 61:927-938.

41. Deschenes M, Paradis M, Roy JP et al. Electrophysiology of neurons of lateral thalamic nuclei in cat: resting properties and burst discharges. J Neurophysiol 1984; 51(6):1196-1218.

42. Soltesz I, Bourassa J, Deschenes M. The behaviour of mossy cells of the rat dentate gyrus during theta oscillations in vivo. Neurosci 1993; 57(3):555-564.

43. MacVicar BA, Tse FW. Local neuronal circuitry underlying cholinergicrhythmical slow activity in CA3 area of rat hippocampal slices. J Physiol (Lond) 1989; 417: 197-212.

44. Klink R, Alonso A. Ionic mechanisms for the subthreshold oscillation and differential electroresponsiveness of medial entorhinal cortex layer II neurons. J Neurophysiol 1193; 70:144-157.

45. Leung LS, Yim CY. Rhythmic delta-frequency activities in the nucleus accumbens of anesthetized and freely moving rats. Can J Physiol Pharmacol 1993; 71(5-6):311-320.

46. Kayyali H, Durand D. Effects of applied current on epileptiform bursts in vitro. Exp Neurol 1991; 113:249-254.

47. Nagakawa M, Durand D. Suppression of spontaneous epileptiform activity with applied currents. Brain Res 1991; 567: 241-247.

48. Durand D. Electrical stimulation can inhibit sybchronized neuronal activity. Brain Res 1986; 382:139-144.

49. Rao G, Barnes CA, McNaughton BL. Occlusion of hippocampal electrical junctions by intracellular calcium injection. Brain Res 1987; 418(1-2):267-270.

50. Miyachi E, KatoC, Nakaki I. Arachidonic acid blocks gap junctions between retinal horizontal cells. Neuro Report 1994; 5(4): 485-488.

51. Bastide B, Herve JC, Deleze J. The uncoupling effect of diacylglycerol and gap junctional communication of mammalian heart cells is independent of protein Kinase C Exp Cell Res 1994; 214:519-527.

52. Lampe, P.D. (1994). Analyzing phorbol ester effects on gap junctional communication: A dramatic inhibition assembly. J Cell Biol 127(6),part 2:1895-1905.

53. Spray DC, Harris AL, Bennett MVL. Gap junctional conductance is a simple and sensitive function of intracellular pH. Science 1981; 211:712-715.

54. Suppes T, Kriegstein AR, Prince DA. The influence of dopamine on epileptiform burst activity in hippocampal pyramidal neurons. Brain Res 1985; 326:273-280.

55. ODonnell P, Grace AA. Dopaminergic modulation of dye coupling between neurons in the core and shell regions of the nucleus accumbens. J Neurosci 1993; 13:3456-3471.

56. McMahon DG. Modulation of electrical synaptic transmission in zebrafish retinal horizontal cells. J Neurosci 1994; 14(3): 1722-1734.

57. Lasater EM. Retinal horizontal cell gap junctional conductance is modulated by dopamine through a cAMP-dependent protein kinase. Proc Natl Acad Sci 1987; 84: 7319-7323.

58. Herreras O, Largo C, Ibarz JM, Somjen GG et al. Role of neuronal synchronizing mechanisms in the propagation of spreading depression in the in vivo hippocampus. J Neurosci 1994; 14(11):7087-7098.

59. Valiante TA, Perez Velazquez JL, Jahromi SS et al. Coupling potentials in CA1 neurons during calcium-free induced field burst activity. J Neurosci 1995; (in press).

60. Sperelakis N, Mann JE Jr. Evaluation of electrical field changes in the cleft between excitable cells. J Theor Biol 1977; 64:71-96.

61. Mann JE Jr, Sperelakis N. Further development of a model for electrical transmission between myocardial cells not connected by low-resistance pathways. J Electrocardiol 1979; 12:23-33.

62. Hogues H, Leon JL, Roberge FA. A model study of electric field interactions between cardiac myocytes. IEEE Trans Biomed Eng 1992; 39:1232-1242.

63. Sperelakis N, Rubio R. Ultrastructural changes produced by hypertonicity in cat cardiac muscle. J Molec Cell Cardiol 1971; 3:139-156.

SPREADING DEPRESSION— A GAP JUNCTION MEDIATED EVENT?

Maiken Nedergaard and Steven Goldman

1. INTRODUCTION

Neural activity is typically detected and expressed in terms of electrical activity. Yet, several recent studies suggest that an electrically silent signaling pathway, diffusion of second messengers across gap junctions, contributes to local communication in the brain. This chapter evaluates current knowledge concerning gap junction-mediated signaling in the brain.

1.1. WHAT DEFINES A GAP JUNCTION?

Plasma membrane-linked channels provide important pathways for the passage of electrolytes between a cell and its environment. Gap junctions are a subset of membrane channels, characterized by two unique features: First, they connect the interior of a cell, not with its surroundings, but to the interiors of its neighbors. Second, their large internal diameter (approximately 15 Å) allows not only ions, but also intracellular messengers, such as cAMP, IP_3, and small peptides, to diffuse freely across (Fig. 19.1). As such, gap junctions can mediate the synchronous behavior of coupled cells within a tissue, especially in response to external stimuli. Gap junctions are composed of proteins, connexins, which are encoded by a highly conserved multigene family. Many cells express several different connexins, although one connexin protein will often predominate.[1] Six connexins aggregate to form a hemichannel or connexon.[2] Presently, 13 connexin genes have been cloned and characterized from rodents, and homologues have been identified in humans, chicks and frogs. The connexins differ in their gating characteristics, and in their responses to voltage, pH, and phosphorylation.[3]

Gap Junctions in the Nervous System, edited by David C. Spray and Rolf Dermietzel.
© 1996 R.G. Landes Company.

1.2. Where in the Brain are Gap Junctions Located?

Gap junction coupling among neurons

Using dual intracellular recordings, MacVicar and Dudek[4] demonstrated that electrotonic coupling existed among over 10% of hippocampal CA3 neurons. Functional coupling has also been identified in the hippocampal CA1 region, as well as in the dentate gyrus.[5,6] Such gap junction coupling offers a low resistance pathway which might participate in neuronal synchrony. This may be of particular potential importance in the highly epileptogenic hippocampal fields. Even when closed, the apposed membranes of gap junctions can carry and discharge a capacitance current, which can promote neuronal synchrony during seizure-like events (chapter 18). Indeed, it was recently shown that dye coupling between CA1 pyramidal cells was increased 2-fold in hyperexcitable hippocampal slices, compared to controls.[6] However, the connexin phenotype responsible for intrahippocampal electrotonic coupling remain unclear: Although Cx32-immunoreactivity has been demonstrated among neurons of several brain regions, including the cerebral cortex and striatum[1] it is not expressed in the hippocampus. Thus, it remains to be determined which connexin is responsible for the electrical coupling of CA1 and CA2 pyramidal neurons.

Coupling among astrocytes

Immunocytochemical, electrophysiological and ultrastructural studies have suggested that in the adult brain, astrocytes are so abundantly interconnected by gap junctions that they may be regarded as a functional syntycium. Yet, until recently, the functional significance of interastrocytic gap junctions remained obscure. It had been suggested that the junctions participated in astrocytic buffering of

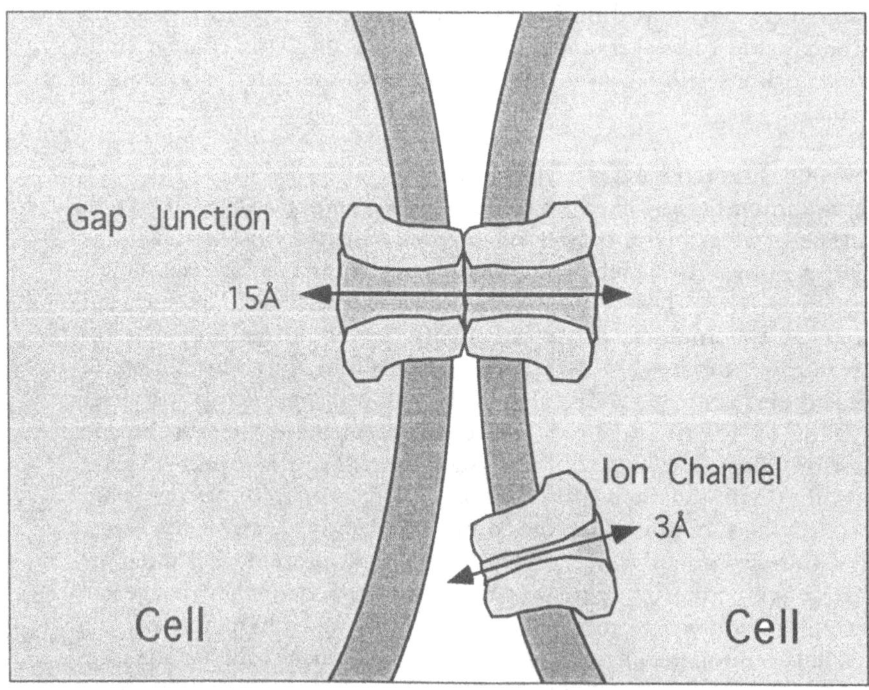

Fig. 19.1. Diagram of a gap junction. This diagram contrast the inner diameter of a gap junction (inner diameter 15 Å) with that of an ionic channel (inner diameter <5 Å). Gap junctions interconnect neighboring cells, while ionic channels connect a cell with its surrounding interstitial space.

K+ in regions of neuronal activity, but evidence for such spatial buffering of K+ is still lacking.[1] Although astrocytic gap junctions are assembled primarily of Cx43, Cx40 immunoreactivity has also been identified.[7]

2. GAP JUNCTION MEDIATED SIGNALING IN CULTURED BRAIN CELLS

Astrocytes are not electrically excitable, and have traditionally been regarded as passive support cells, their functions being restricted to interstitial ion homeostasis, maintenance of the blood-brain barrier, and neurotrophin-production.[8] This view of astrocytic function began to change in 1990, with the demonstration that glutamate can trigger calcium oscillations among cultured glia, with frequencies of roughly 0.1 Hz, which often resulted in propagating intercellular calcium waves.[9] Astrocytic calcium waves propagated with a velocity of 10-20 µm/sec, and were generated by the autocatalytic release of calcium from IP$_3$-sensitive intracellular stores.[10]

An intact gap junctional network appears to be required for cell-to-cell calcium signaling, in that junction inhibitors, such as octanol and halothane, were found to be effective blockers of calcium wave formation.[11] In this regard, transfection with a cDNA for Cx43 enabled a communication deficient glioma cell line, C6, to propagate robust intercellular calcium waves.[12] This observation, that gap junctions were required for the propagation of calcium signals among astrocytes, was the first demonstration of a functional role of glial gap junctions. Astrocytic calcium waves have since been evoked by either electrical or mechanical stimulation in cultured astrocytes, as well as by intense neuronal firing in cultured slices.[8,13,14]

2.1. LOCAL SIGNALING LOOP AMONG ASTROCYTES AND NEURONS

The notion that astrocytes might participate actively in neuronal transmission led to studies of calcium signaling in mixed neuronal and glial cultures. In these cultures, it was apparent that astrocytic calcium waves could trigger spike-like increases in neuronal cytosolic calcium.[14] Astrocytic calcium waves were evoked by focal electrical stimulation, at a distance of 50-300 µm from cocultured neurons. During the local stimulation, no calcium responses were observed in either astrocytes or neurons more than 30 µm from the stimulation electrode. However, the local increase in astrocytic calcium evoked a propagating astrocytic calcium wave, and when this wave approached neurons in the vicinity, their calcium levels increased 2- to 4-fold within seconds (Fig. 19.2). Thus, a signaling loop exists between cultured neurons and astrocytes: Astrocytes can modulate the calcium level, and thereby the firing pattern, of neurons in their midst. In turn, neurons can trigger astrocytic calcium waves by releasing glutamate. Accordingly, gap junction blockers are effective inhibitors of astrocytic calcium signaling to cocultured neurons, but glutamate receptor antagonists have been similarly reported to dampen though not to eliminate this signaling pathway.[15] Thus, several mechanisms might independently or synergistically act to mediate signaling between astrocytes and neurons.[8]

3. IS SPREADING DEPRESSION AN EXAMPLE OF GAP JUNCTION-MEDIATED SIGNALING WITHIN THE BRAIN?

The above results demonstrate that in vitro, astrocytes can propagate intercellular calcium signals, which can involve and include neurons. However, it remains unknown whether this signaling mechanism is operative in vivo, or whether it is an unique artifact of the culture environment.[16] Recently, we studied the role of gap junctions in the initiation and propagation of a possible in vivo manifestation of intracellular calcium waves, spreading depression (SD). SD shares a number of striking similarities with calcium waves in cultures (Table 19.1), suggesting that gap-junction-mediated diffusion might be responsible for SD in vivo. We found that SD is indeed reversibly

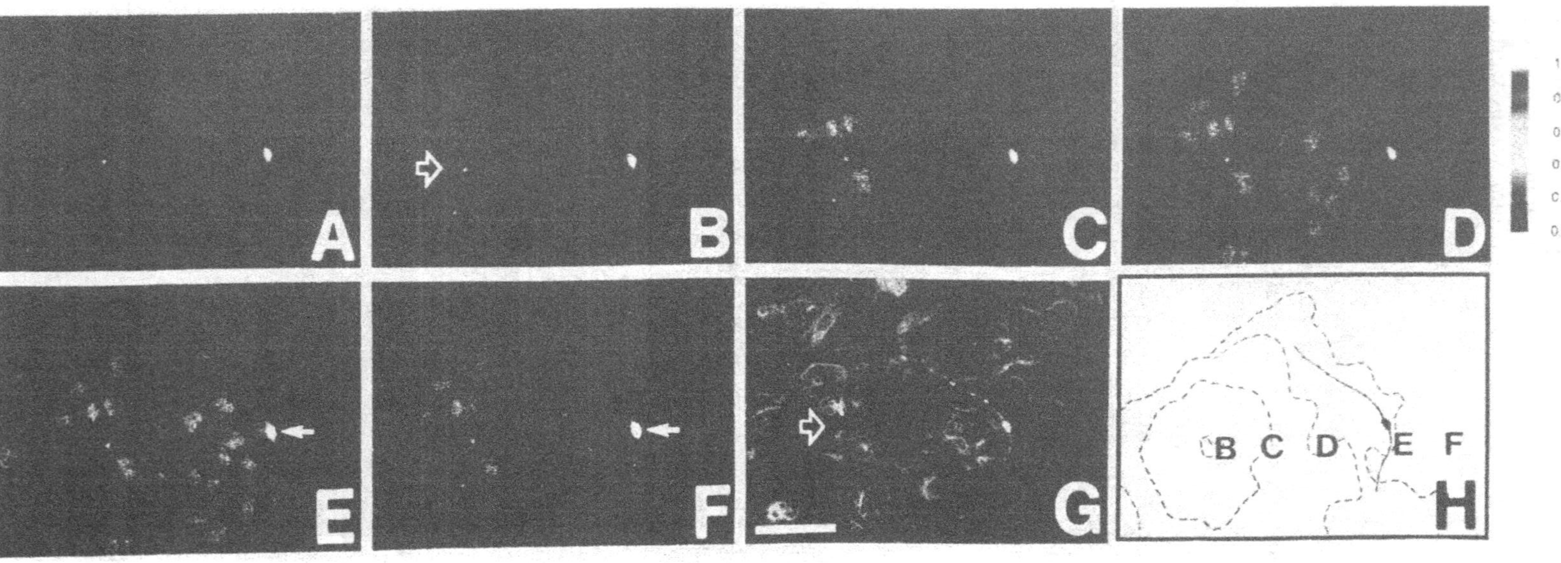

Fig. 19.2. Waves of astrocytic calcium elevation propagate and trigger elevations in neuronal cytosolic calcium. (A) A 18 days in vitro culture loaded with fluo-3. (B through F) Electric field stimulation delivered at the open arrow (15 μA, 0.5 sec) triggers a local elevation of astrocytic Ca²⁺. Images captured 1, 4, 7, 10, and 26 sec after electrical field stimulation (small arrow identify the neuron as cytosolic calcium rises). Elevations in neuronal calcium levels were first noted when the neuron was approached by the astrocytic calcium wave. (G) Detection of MAP-2 (texasred) and GFAP (fluorescein) immunoreactivity in the same field. (H) Line drawing illustrating propagation of the astrocytic calcium wave, from (A) through (F). The MAP-2 positive neuron in the field is outlined. Scale bar, 100 μm. Reprinted with permission from Science, 263:1768-1771, 1994.

inhibited by several gap junction blockers, consisting with the notion that astrocytic gap junctions carry the signal for SD. As such, we postulated that the neuronal depolarization of SD might be a sequela of local neuronal activation by spreading glial calcium waves.[17]

3.1. WHAT IS SD?

Spreading depression, classically described as the SD of Leao is a generalized response of vertebrate gray matter to a variety of noxious influences.[18] It constitute a slowly moving wave of tissue depolarization in the intact brain. SD is experimentally evoked by applying KCl or the excitatory neurotransmitter glutamate to exposed cortical tissue, or by electrical stimulation (Fig. 19.3). SD is characterized by a reversible cessation of neuronal activity which propagates slowly (20-80 μm/sec), and is accompanied by a loss of membrane potential and transmembrane ionic gradi-

ents.[19] The propagation of SD can in neocortex be followed by inserting ion or potential-sensitive electrodes in some distance from the focus of initiation (Fig. 19.3). Velocity of the SD wave can be calculated by dividing the distance between two electrodes with the lagtime between arrival of the wave front. Both neurons and astrocytes participate in SD, whose slow propagation suggests its mediation by the diffusion of soluble factors. Indeed, several studies have suggested that neuronal depolarization, triggered by the diffusion of glutamate and/or potassium into the interstitial space mediates the propagation of SD.[20,21] Astrocytes, in contrast, have not been implicated to have any role in SD.[22]

3.2. GAP JUNCTION INHIBITORS BLOCK SD

SD and astrocytic calcium waves share a number of fundamental characteristics, such as their initiating stimuli, velocities

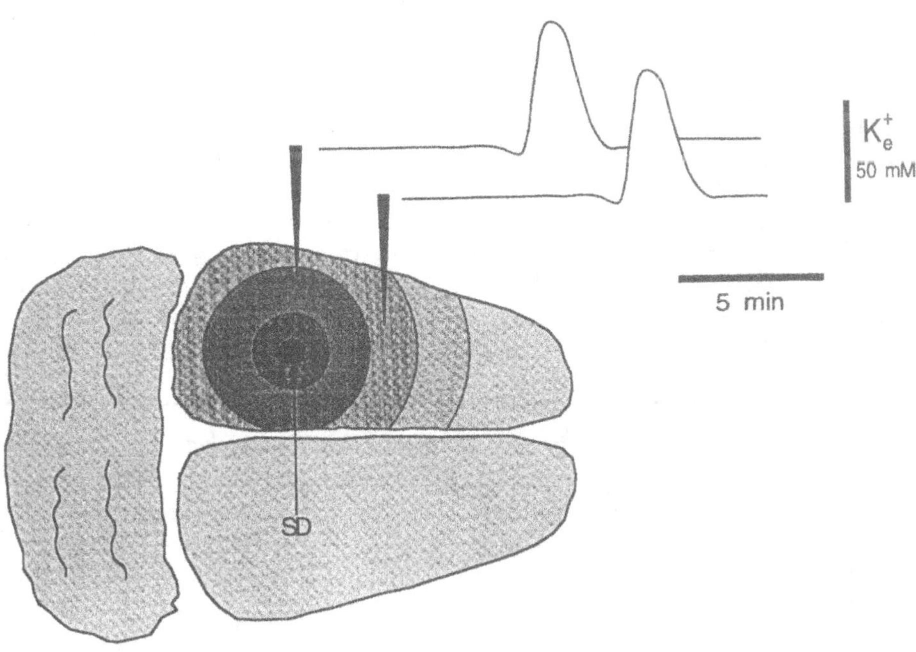

Fig. 19.3. SD in the adult rat forebrain. A wave of spreading depression was evoked in the right hemisphere (black circle). Propagation of the wave can be followed by potassium sensitive electrodes inserted in various distance from the point of origin. Velocity of the SD wave can be calculated by dividing the distance between two electrodes with the lagtime between arrival of the wave.

of propagation, and measured responses to a variety of pharmacological challenges (Table 19.1). Since gap junctions are necessary for the propagation of calcium waves, we next asked whether they are required for the propagation of SD. To this end, we examined SD in the isolated chick retina, a preparation in which the rate and extent of propagation are analogous to that of the cerebral cortex.[23]

In control preparations, SD propagated with a velocity of 1.1-1.6 mm/min, and could be elicited repeatedly. Treatment with a variety of gap junction inhibitors blocked SD (Table 19.2). The effects were all reversible, in that normal waves of SD could be evoked in each treated preparation after their removal (Fig. 19.4).

However, the gap junction blockers that we used also decrease current through glutamate-gated channels.[24-26] To exclude the possibility that these blockers inhibited SD by decreasing glutamate-associated calcium currents, we measured glutamate-associated calcium changes. To this end, we loaded isolated retinal cells with the calcium indicator dye fluo-3, and followed their calcium levels by confocal microscopy. We found that none of the gap junction blockers we employed, including octanol, halothane, heptanol, and 18α-glycyrrhetinic acid (18-αGA), affected either the level of glutamate-induced calcium increase, or the percentage of glutamate-responding cells, in concentrations at which they each reversibly block SD.[17] Thus, SD was inhibited by these agents despite their lack of influence on glutamatergic transmission.

3.3. Gap Junction Blockers Decrease the Diffusion of Lucifer Yellow in a Dose Dependent Manner

To test the efficiency of octanol, halothane, heptanol, and 18-AGA as gap junction blockers in the intact retina, we used a modification of the scrape loading/dye transfer method.[27,28] Using this technique, we found that retinal cells were extensively coupled: Lucifer yellow (LY), a small molecule (447 Da) that moves freely through gap junctions, diffused an average of 99 ± 14 μm (n = 28), 15 min after loading. In contrast, over the same period, the high MW conjugate rhodamine dextran (MW 10,000), which does not cross gap junctions, remained limited to the loaded cells. Octanol, halothane and heptanol each reduced the diffusion of LY in a dose dependent manner (Fig. 19.5); in each case, the minimal concentration of inhibitor that blocked SD yielded at least 50% reduction in gap junctional permeability. In this regard, the generation of SD is blocked by 2% halothane in cerebral cortex. This concentration of halothane is not sufficient to block depolarization-induced glutamate release from cortical neurons.[29] Thus, several lines of evidence suggest that gap junctions are required for the generation and propagation of SD.

Table 19.1 Comparison of Ca²⁺ waves in vitro and spreading depression

	In Vitro Calcium Waves	Spreading Depression
Initiation stimuli	K+, electrical	K+, glutamate, electrical
Velocity (μm/s)	15-50 μm/s	15-50 μm/s
Max. radius	<500 μm	>2-5 cm
Pattern of migration	radial	radial
Calcium homeostatsis	disrupted	disrupted
Refractory period	yes	yes
Participation of astrocytes	yes	yes
Participation of neurons	yes	yes
Gap junction blockers	inhibition	inhibition
Glutamate receptor antagonists	no effect	inhibition

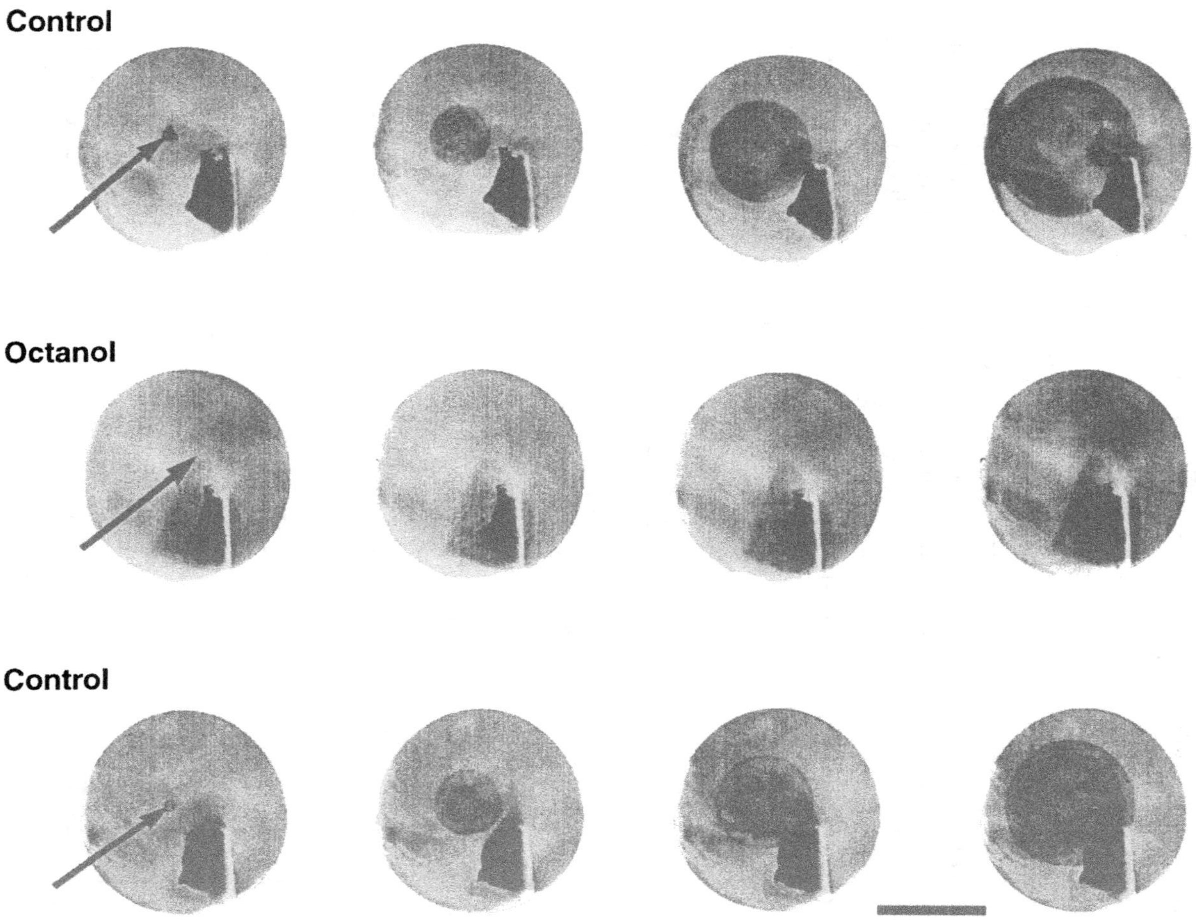

Fig. 19.4. Spreading depression in the retina is inhibited by gap junction blockade. Upper row, a mechanically induced wave of SD in the isolated retina propagated at a velocity of 20 μm/sec. Middle row, 1.1 mM octanol, inhibitor of gap junctions, completely blocked SD. Lower row, after removal of octanol, a new wave of spreading depression was elicited in the same preparation. Individual pictures were photographed at 20 sec, 80 sec, 120 sec, and 160 sec after elicitation of SD. Arrows indicate the point where SD was initiated. Scale bar, 3 mm. Reprinted with permission from J. Neurobiol 28:433-444, 1995.

3.4. Low pH is an Effective Inhibitor of SD

Elevated cellular acidity effectively inhibits gap junction function.[30] We found that at pH 6.7, SD generation was reversibly inhibited, concurrent with a reduction of over 70% in LY diffusion (Fig. 19.5). Notably, cellular acidity may influence calcium current through the NMDA receptor, so that glutamate-linked mechanisms might yet contribute to this phenomenon.[31,32] However, we could not assess effects of low pH upon glutamate-induced calcium increases, since applicable fluorescence indicators lose their calcium sensitivity at low pH.[33]

3.5. SD, Unlike Glial Calcium Waves, is Blocked by Glutamate-Receptor Antagonists

The NMDA receptor antagonists kynurenic acid, MK-801, AP-5 all block SD. Each does so in both the cerebral cortex as well as in the isolated chicken eye.[17,34,35] We found that none of these agents influenced gap junction permeability, but all blocked NMDA-channel linked calcium entry.[35] In contrast, neither the non-NMDA glutamate receptor antagonist NBQX, nor the sodium channel blocker tetrodotoxin, influenced the generation or velocity of SD. In contrast, calcium removal from the bathing medium

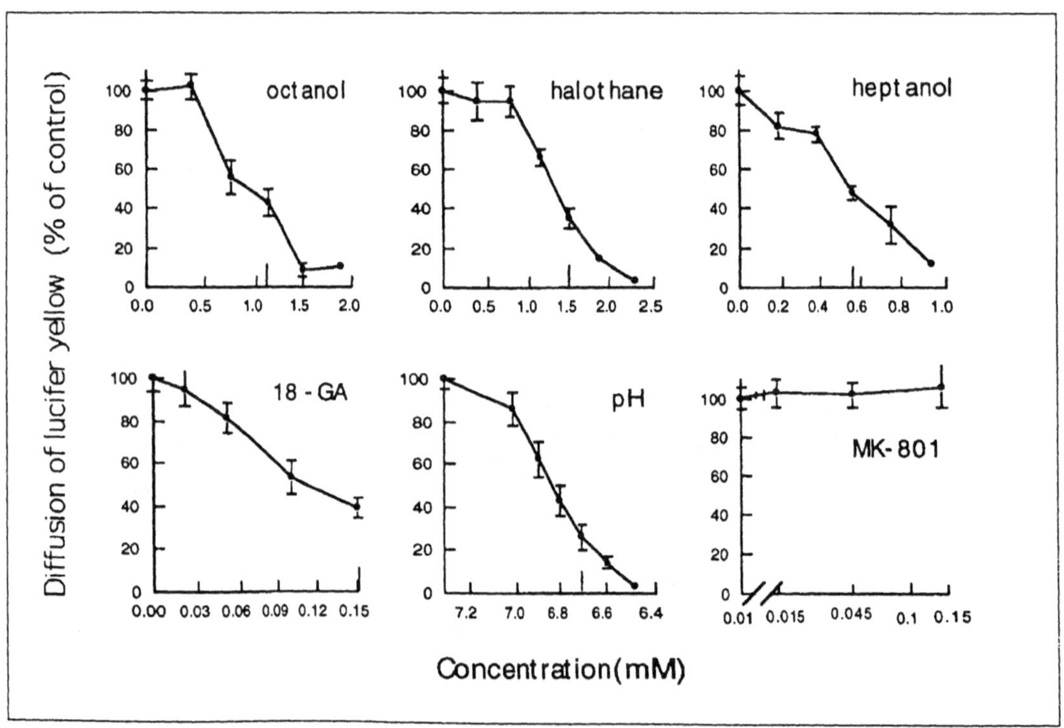

Fig. 19.5. Relationship between the diffusion of LY and the concentration inhibitors. The effects of increasing concentrations of octanol, halothane, heptanol, GA, pH, and MK-801 upon diffusion of lucifer yellow in the isolated retina. The arrows indicate the minimal concentration of inhibitor needed to reversibly block the propagation of spreading depression. The diffusion of Lucifer yellow was reduced by 50-70% at these concentrations. MK-801 did not block the diffusion of Lucifer yellow in retina and did thus block SD by a gap junction independent mechanism. Increasing concentrations of a selection of glutamate receptor antagonists, kynurenic acid (0-6 mM), AP-5 (0-3 mM), and NBQX (0-0.3 mM) did not affect intercellular diffusion of Lucifer yellow in retina (results not shown). Mean ± SEM. Reprinted with permission from J Neurobiol 28:433-444, 1995.

prevented SD generation (Table 19.2). Thus, calcium influx associated with NMDA receptor activation appears necessary for the generation and propagation of SD, even though glutamategic synaptic transmission per se is not.

3.6. DOES SD REPRESENT THE IN VIVO CORRELATE OF ASTROCYTIC CALCIUM WAVES?

Despite the similarities listed in Table 19.1, SD differs fundamentally from astrocytic calcium waves in that SD re-

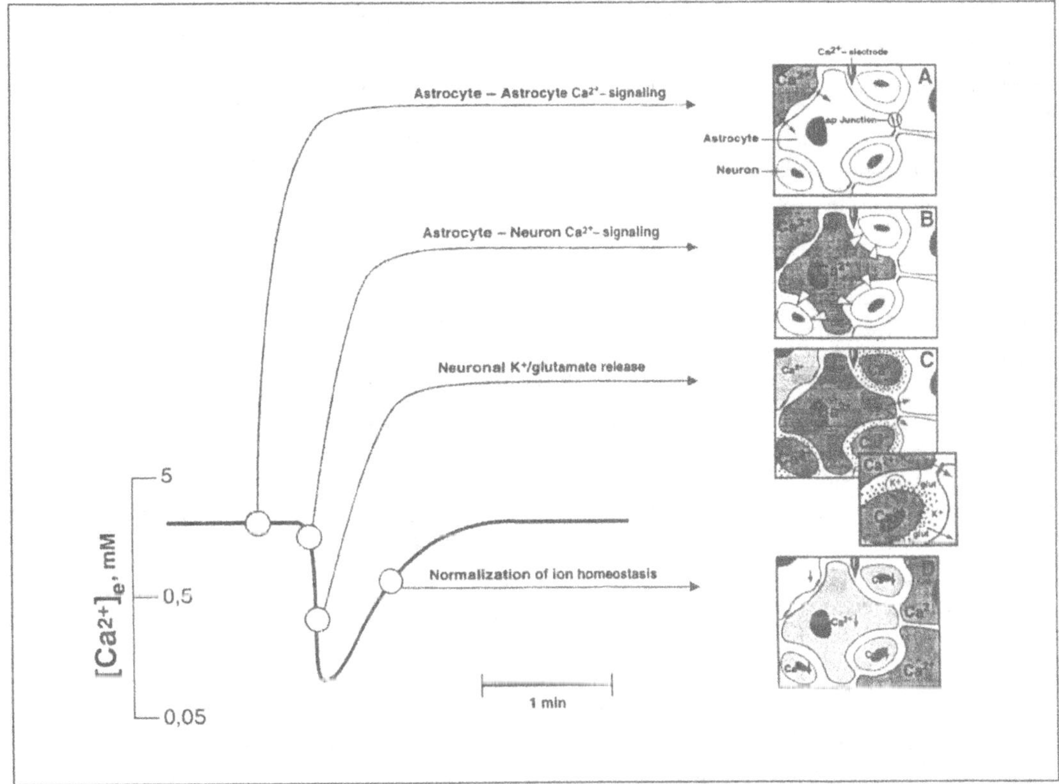

Fig. 19.6. A gap junction mediated model of spreading depression. In this illustration, a recording of extracellular calcium changes during SD is linked to diagrams which illustrate concurrent cellular events. The trace on the left represent the typical pattern of interstitial calcium concentration changes that occurs during propagation of spreading depression as recorded by a calcium sensitive electrode. The model illustrated in panels A to D includes two main components: gap junction-mediated calcium signaling, and a secondary amplification step of the intracellular calcium increases evoked by K⁺ and glutamate released during the SD wave. (A) Gap junction mediated calcium signaling from astrocyte to astrocyte constitutes the advancing edge of the SD wave (|). (B) Activated astrocytes then trigger a calcium increase in surrounding neurons (Δ). The means by which astrocytes signal to neurons is unclear but may occur through chimeric gap junctions composed of Cx43 (astrocytes) and Cx32 (neurons).[14] Alternatively, a calcium-mediated reversal of the astrocytic Na⁺/glutamate transporter might lead to astrocytic release of gluatamate, which might in turn stimulate surrounding neurons.[15] (C) Neuronal activation in turn may trigger the further release of both K⁺ and glutamate, which then lead to further calcium increases among the already activated astrocytes and neurons. This amplification of calcium increases by K⁺ and/or glutamate may be necessary for the continuous propagation of a wave of SD, since astrocytic calcium levels must reach at least 700 nM to allow propagation of the calcium wave from astrocyte to astrocyte (Nedergaard, unpublished observations). In addition, increased levels of glutamate and K⁺ might facilitate the downstream propagation of SD, since gap junctional coupling is increased by cellular depolarization.[38] (D) A recovery period evolves in the wave trough; it is characterized by normalization of both intracellular and interstitial calcium levels, while the wave front is invading new tissue. Reprinted with permission from J Neurobiol 28:433-444, 1995.

Table 19.2. **Inhibition of spreading depression**

| | A. SD (pre-test control) | | B. SD (experimental) | | C. SD (post-test control) | | |
	velocity ($\mu m/s$)	diameter (mm)	velocity ($\mu m/s$)	diameter (mm)	velocity ($\mu m/s$)	diameter (mm)	n
Kynurenic acid (3mM)	23±2	>10 mm	2±3*	0.3±0.1*	20±2	>10 mm	6
AP-5 (1mM)	27±5	>10 mm	2±2*	0.3±0.1*	28±3	8.9±1.3	6
MK-801 (15 µM)	23±3	>10 mm	2±3*	0.3±0.1*	25±3	>10 mm	6
NBQX (50 µM)	23±5	>10 mm	25±3	>10 mm	23±5	>10 mm	5
Calcium-free solution	22±5	>10 mm	2±2*	0.3±0.2*	22±3	>10 mm	8
Tetrodotoxin (1 µM)	20±3	>10 mm	22±2	>10 mm	20±2	>10 mm	8
Nifedipine (10 µM)	20±2	>10 mm	22±5	>10 mm	22±2	9.6±1.2	7
Octanol (1.1 mM)	25±8	>10 mm	3±3*	0.5±0.5*	25±7	9.4±1.1	7
Halothane (1.5 mM)	20±2	>10 mm	5±3*	0.2±0.1*	18±5	9.4±1.6	6
Heptanol (0.56 mM)	20±3	>10 mm	2±2*	0.2±0.1*	22±2	>10 mm	8
18A-GA	23±3	>10 mm	2±2*	0.3±0.5*	20±7	>10 mm	9
pH 6.7	23±2	>10 mm	3±2*	0.4±0.2*	22±2	>10 mm	7

Spreading depression was first evoked in control media (A. control), then in the presence of either channel- or gap junction blockers (B. experimental), then again in the control solution (C. control). Two-way analysis of variance was used to compare mean values, while Dunnett's multiple range test was used to determine which group changed characteristics. Significance was established at p<0.01 (*). Means ± SD.

quires glutamate-induced neuronal depolarization; furthermore, its distance of propagation is several order of magnitudes longer than that of calcium waves in vitro: While an astrocytic calcium wave rarely propagates more than 300 µm, a single wave of SD readily migrates several centimeters, without detectable decline in amplitude. It is possible that SD and astrocytic calcium waves are interdependent in that glutamatergic activation might constitute an amplification step required for the long-distance propagation of SD. Figure 19.6 illustrates our proposed model for SD and emphasize our postulate that a gap junction-mediated calcium waves constitute the advancing front of SD: In this illustration, a recording of extracellular calcium changes during SD is linked to diagrams which illustrate concurrent cellular events. These calcium waves trigger the secondary depolarization of surrounding neurons, which in turn release glutamate and K$^+$. The accumulation of glutamate and K$^+$ in the extracellular space further enhances the elevation of cytosolic calcium levels, a possible feedback required for the continued and nondecremental propagation of SD. With-

out this amplification step, calcium increments will gradually decrease as a function of the distance from the point of origin of the wave. Consequently, calcium signaling across gap junctions will diminish until a point where further wave propagation is abolished. Thus, while astrocytes are the primary instigators of the SD wave front, its continued propagation requires secondary glutamatergic stimulation within the advancing field. In this model, gap junction mediated signaling of calcium underly both the phenomena of SD and of astrocytic calcium waves. Secondary glutamatergic amplification might occur only in the three-dimensional structure of brain tissue, and not in monolayer cultures, at least in part due to the infinite volume of the bathing medium.

Two additional lines of work support our proposed model of SD: Using two electrodes inserted into the in vivo rat cortex, Sugaya et al[36] demonstrated that while astrocytic depolarization temporally coincides closely with the onset of SD, neuronal depolarization frequently is delayed by 10-30 sec. In addition, MacVicar[37] has recently demonstrated that astrocytic calcium increments detected by the fluorescence calcium indicator, calcium orange, precede neuronal depolarization during the propagation of SD in acute hippocampal slices from rat.

4. PROSPECTS FOR THE FUTURE

Studies of the last few years have shown that astrocytes express a variety of neurotransmitter receptors, the activation of which can initiate both intercellular calcium waves, and single cell oscillations in free calcium. In culture, these astrocytic calcium waves are transmitted to neighboring neurons. This neuro-glial transmission of the calcium wave provides a means by which astrocytes can directly affect neuronal firing patterns. As a result, it is now a plausible proposition that astrocytes might participate directly and actively in information processing within the CNS. Furthermore, the need for gap junctions in mediating both calcium waves in vitro and SD in vivo suggest a critical role of these intercellular junctions in both glio-glio and neuro-glial communication. With regards to the latter, gap functional signaling may represent one of several necessary conditions required for neuronal-glial signaling in vivo, glutamatergic transmission has already been identified as a concurrent process that is necessary, but not sufficient prerequisite for SD. Itself a model of intracellular electrotonic signaling in vivo. A major challenge for future work in this area will lie in understanding the communative interactions between astrocytes and neurons in intact brain tissue, particular in the neocortex. To estimate the relative significance of glial-neuronal communication in brain functioning, we will need to evaluate neuronal firing patterns before and after astrocytic stimulation. In this regard, recent studies suggesting that astrocytes might instigate SD are important, in that they indicate that astrocytes can activate neurons within a time-frame of seconds. However, SD is a phenomenon intimately linked to pathophysiological conditions, such as ischemia or trauma. As a result, both the spatial and temporal patterns of

interactions between neurons and glia in SD might differ substantially from those operative in normal brain. Thus, the major impetus for the coming years will be to better understand neuro-glial interactions during normal physiology in situ.

REFERENCES

1. Dermietzel R, Spray D. Gap junctions in the brain: Where? What type? How many and Why? Trends Neurosci 1993; 16(5):186-192.
2. Makowski L, Caspar D et al. Gap junction structure. II. Analysis of the X-ray diffraction data. J Cell Biol 1977; 77:629-645.
3. Spray D, White R et al. Gating of gap junctional conductance. J Biophys 1984; 45:219-230.
4. MacVicar B, Dudek F. Electrotonic coupling between pyramidal cells: a direct demonstration in rat hippocampal slices. Science 1981; 213:782-785.
5. MacVicar B, Dudek F. Electrotonic coupling between granula cells of rat dentate gyrus: physiological and anatomical evidence. J Neurophysiol 1982; 47:579-592.
6. Perez-Velazquez J, Valiante T et al. Modulation of gap junctional mechanisms during calcium-free induced field burst activity: a possible role for electrotonic coupling in epileptogenesis. J Neuroscien 1994; 14: 4308-4317.
7. Zheng X, Friedman L et al. Cx40 mRNA in astrocytes and neurons of rat brain. Soc of Neurosci Abst 1995; 21:231.5.
8. Smith SJ. Neuromodulatory astrocytes. Current Biology 1994; 4:807-810.
9. Cornell-Bell AH, Finkbeiner SM et al. Glutamate induces calcium waves in cultured astrocytes: long range glial signaling. Science 1990; 247:470-474.
10. Sanderson M, Charles A et al. Mechanisms and function of intercellular calcium signaling. Mol Cell Endocrin 1994; 98:173-187.
11. Finkbeiner S. Calcium waves in astrocytes—filling in the gaps. Neuron 1992; 8: 1101-1108.
12. Charles AC, Naus CCG et al. Intercellular calcium signaling via gap junctions in glioma cells. J Cell Biol 1992; 118(1): 195-201.
13. Charles A, Merrill J et al. Intercellular signaling in glial cell: calcium waves and os-

cillations in response to mechanical stimulation and glutamate. Neuron 1991; 6:983-992.

14. Nedergaard M. Direct signaling from astrocytes to neurons in cultures of mammalian brain cells. Science 1994; 263:1768-1771.

15. Parpura V, Basarsky TA et al. Glutamate-mediated astrocyte-neuron signaling. Nature 1994; 369:744-747.

16. Travis J. Glia: The brain's other cells. Science 1994; 266:970-972.

17. Nedergaard M, Cooper A et al. Gap junctions are required for the propagation of spreading depression. J Neurobiol 1995; 28:433-444.

18. Hansen A J. Effects of anoxia on ion distribution in the brain. Physiol Rev 1985; 65:101-148.

19. Nicholson C, Kraig RP. The behavior of extracellular ions during spreading depression. Amsterdam: Elsevier, 1981.

20. Grafstein B. Mechanism of spreading cortical depression. J Neurophys 1956; 19: 154-171.

21. Van Harreveld A, Fifkova E. Glutamate release from the retina during spreading depression. J Neurobiol 1970; 2:13-29.

22. Leibowitz D. The glial spike theory. I. On an active role of neuroglia in spreading depression and migraine. Proc R Soc Lond 1992; 250:287-295.

23. Martins-Ferreira H. Propagation of spreading depression in isolated retina. München-Wien-Baltimore, Urban & Schwarzen, 1993.

24. Johnston M F, Simon SA et al. Interaction of anaesthetics with electrical synapses. Nature 1980; 286:498-500.

25. McLarnon J G, Wong JHP et al. The actions of intermediate and lon-chain n-alkanols on unitary NMDA currents in hippocampal neurons. Can J Physiol Pharmacol 1991; 69:1422-1427.

26. Terrar D A, Victory JGG. Isoflurane depresses membrane currents associated with contraction in myocytes isolated from guinea-pig ventricle. Anesthesiology 1988; 69:742-749.

27. El-Fouly M, Trosko J et al. Scrape-loading and dye transfer. Exp Cell Res 1987; 168:422-430.

28. Giaume C, Marin P et al. Adrenergic regulation of intercellular communications between cultured striatal astrocytes from the mouse. Proc Natl Acad Sci 1991; 88: 5577-5581.

29. Saito R, Graf R et al. Anesthesia effects potassium evoked spreading depression in cats. J of Cerebral Blood Flow Metab 1993; 13(1): S86.

30. Spray DC, Bennett MVL. Physiology and pharmacology of gap junctions. Ann Rev Physiol 1985; 47:281-303.

31. Griffard R, Monyer H et al. Acidosis reduces NMDA receptor activation, glutamate neurotoxicity, and oxygen-glucose deprivation neuronal injury in cortical cultures. Brain Res 1990; 506:339-342.

32. Vyklicky L, Vlachova V et al. The effect of external pH changes on responses to excitatory amino acids in mouse hippocampal neurons. J Physiol 1990; 430:497.

33. Nedergaard M. Spreading depression as a contributor to ischemic brain damage. In: Siesjo BK, Wieloch T, eds. Advances in Neurology, Vol 71; Cellular and molcular mechanisms of ischemic brain damage. Lippincott-Raven Publisher, 1996; 75-85.

34. Lauritzen M, Hansen AJ. The effect of glutamate receptor blockade on anoxic depolarization and cortical spreading depression. J of Cereb Blood Flow Metab 1992; 12:223-229.

35. Sheardown M. The triggering of spreading depression in the chicken retina: a pharmacological study. Brain Res 1993; 607:189-194.

36. Sugaya E, Takato M et al. Neuronal and glial activity during spreading depression in cerebral cortex of cat. J Neurophysiol 1974; 38:822-841.

37. MacVicar B. New Insights about glia in neurological diseases. Winter Conference on Brain Research, 1996.

38. Enqvist M, McCarthy K. Astroglial gap junction communication is increased by treatment with either glutamate or high K+ concentration. J Neurochem 1994; 62: 489-495.

GPSR Compliance
The European Union's (EU) General Product Safety Regulation (GPSR) is a set
of rules that requires consumer products to be safe and our obligations to
ensure this.

If you have any concerns about our products, you can contact us on

ProductSafety@springernature.com

In case Publisher is established outside the EU, the EU authorized
representative is:

Springer Nature Customer Service Center GmbH
Europaplatz 3
69115 Heidelberg, Germany